Fire Officer's HANDBOOK of Tactics

Third Edition

Fire Officer's HANDBOOK of Tactics

Third Edition

JOHN NORMAN

Fire Engineering®

PennWell®

Disclaimer

The recommendations, advice, descriptions, and methods in this book are presented solely for educational purposes. The author and publisher assume no liability whatsoever for any loss or damage that results from the use of any of the material in this book. Use of the material in this book is solely at the risk of the user.

Copyright© 2005
PennWell Corporation
1421 South Sheridan Road
Tulsa, Oklahoma 74112-6600 USA
800.752.9764
+1.918.831.9421
sales@pennwell.com
www.pennwellbooks.com
www.pennwell.com

Director: Mary McGee
Managing Editor: Jerry Naylis
Production / Operations Manager: Traci Huntsman
Production Manager: Robin Remaley
Assistant Editor: Amethyst Hensley
Book Designer: Clark Bell

Library of Congress Cataloging-in-Publication Data Available on Request

Norman, John
Fire Officer's Handbook of Tactics, Third Edition by John Norman
p. cm.

ISBN 1-59370-061-X

Printed in the United States of America

1 2 3 4 5   09 08 07 06 05

To Dennis, Mikey, Ray, Pete and Butch

Acknowledgments

For the 404 members of the Fire Department, City of New York, who have made the supreme sacrifice in the line of duty since I was privileged enough to join their ranks in 1979.

For all my friends. So many of you are no longer with us, like Ex-Chief Butch Borfitz Sr., Deputy Chief Raymond Downey Sr., Chief of Department Peter Ganci and Lt. Dennis Mojica, guys I worked with and for in the good times. As the boys from Sheffield Avenue would put it, "You done good."

For my family: my wife Jeanne, who acted as my typist, editor, and general coauthor; my three sons, John, Patrick, and Conor, who were a constant source of inspiration; and my parents, who shaped my life so well, Ex-Chief John W. "Butch" Norman Jr., who got me started in this business in the first place, and my mother, Lillian.

Finally, for all of the members of all of my units. None of you have ever let me down. Without some of you, I wouldn't be here to write this now. I love you all. Stay safe.

Contents

PART I: GENERAL FIREFIGHTING TACTICS

PART II: SPECIFIC FIRE SITUATIONS

Foreword

This edition is written in the post 9/11 era of the New York City Fire Department, a department that for nearly 140 years as a career organization has prided itself on being an aggressive interior firefighting organization. The loss of 343 members of the department in the murderous attacks on the World Trade Center (WTC) has not changed that. What has changed is the recognition that this nation's firefighters are an integral part of our Homeland Security efforts, and the reality that training on a variety of new facets of our job is essential. Thus, I have added a chapter on Fire Department Roles in Terrorism and Homeland Security, gleaned from the many lessons we have learned over many years, as well as reflecting new threats and techniques. At the same time though, we must not lose sight of the fact that an average of 100 firefighters die each and every year in the United States from the same causes that have been killing firefighters for far more than 140 years.

Modern firefighting is a continually evolving science. New technologies are constantly being applied to the fire service, both from within and without. Some, like new chemical compounds or new locking devices, can increase our operating problems. Others, such as laptop computers and hydraulic forcible entry tools, can help us keep pace with these changes. The strategies of modern firefighting, however, haven't changed. The basics—protect life, confine the fire, and extinguish—are constants.

What has changed are the tactics. Strategy differs from tactics. Tactics are the actual hands-on operations that must be performed in the right time and place. Take, for example, one part of the overall strategy—protecting life. This can take several forms, from physically carrying a victim to safety, to placing a hoseline between the victim and the fire, to venting to draw fire away from the victim. It can be as simple as using a public address system to calm panicky occupants. All of these measures have been used to save lives. The challenge for the fire forces is to recognize the appropriate means and to properly, expeditiously employ them.

Technology is one of the primary forces that shape our tactics. For example, using a trench cut to stop a rapidly spreading cockloft fire wouldn't have been possible 35 years ago, before the development and proliferation of power saws. Today, it is a readily available option for the initial attack effort in certain structures. The same is true of positive-pressure ventilation, forcible entry using the new methods, water supply, and all else. The firefighter has a great responsibility to keep abreast of new developments in the field, for his life and the lives of others depend on his ability to recognize hazards and to take appropriate actions against them.

Tactics routinely used in the past for certain fires have either been replaced or relegated to positions of lower importance due to advances in equipment and our understanding of fire behavior and building construction. The use of self-contained breathing apparatus (SCBA) and power saws has reduced the usefulness of cellar pipes, while the occurrence of several catastrophes will (hopefully) prevent headlong attacks on well-involved truss-roof buildings. (When I first wrote that sentence in 1991, I thought that it might be true. Watching a videotape of an incident in 1995, however, proved to me that much of America's fire service still doesn't know enough about these and other deadly buildings.) Failure to adapt to these changes in tactics will lead to unnecessary loss of both life and property.

A number of excellent texts on firefighting have been published over the years, distilled from years of experience and building on the lessons that their authors had learned. This book is an attempt to improve still further on those texts, reflecting on what experience has taught us of those new ideas from 10, 20, or 30 years ago. In addition, it will introduce firefighters to new problems that they are only now encountering.

This book isn't intended to be a training manual to guide firefighters through physical tasks such as operating power saws and pumpers. Excellent training programs are available from various sources, the contents of which couldn't possibly be condensed into even two or three volumes of this size.

Instead, this book offers basic goals for handling specific types of emergencies and practical tips on how to achieve those goals. By necessity, many of the topics covered here aren't new, nor are some of the solutions offered. There isn't anything new, after all, that can be added to a fire department's basic goal: to protect life and property. To try to attribute each concept or guideline to a specific author or fire department is, if not impossible, certainly improper, since most departments encountering a similar problem will work out a solution that fits within their guidelines of acceptable practice.

What is offered here is a guide for the firefighter and the fire officer who, having learned the basic mechanics of the trade, are now looking for specific methods for handling specific situations. Although many texts will specify an action, such as "cut an 8-ft by 8-ft vent hole over the fire," few give any guidance on the practical problems involved.

I hope that, through my description of these problems and my suggestions of possible solutions, you will select the best alternatives when you encounter these kinds of situations. That is the only real value of such a text—to provide guidance to firefighters faced with a problem.

This book has two parts. **Part I, General Firefighting Tactics,** should be the basis of any department's operations. It is broken down into the following three main areas:

- general activities, goals, tasks, and size-up processes

- attack functions, methods, and duties (engine company operations)

- support functions, methods, and duties (ladder company operations)

Part I should establish the ground rules for most structural firefighting, and you should apply the concepts as part of your standard operating procedures. You shouldn't attempt to solve every problem with the same preprogrammed solution. The concepts discussed in Part I offer a variety of alternatives that could be adapted to the situation at hand, normally in succession—if Plan A fails, go to Plan B, and so on.

Part II, Specific Fire Situations, deals with fires and emergencies in the most common structures and occupancies. The constant repetition of these situations shows that we repeatedly face the same problems. In many cases, due to the vast experience (trial and error) that we've had with these most common fires, tactics have been developed specifically for particular situations. Other areas covered here are problems that are initially hidden (such as channel-rail construction) but that must be checked quickly to prevent a serious loss.

As I mentioned earlier, while some of these topics or their solutions aren't new, they are the basics that must be learned, relearned, and used as a foundation. A topic or an action is only old to someone who has seen, learned, and used it before. Until then, it remains an unknown that must be evaluated for the additional bits of information that it can yield. This edition also includes a new chapter on fires in garden apartments and townhouses, a growing problem in much of our ever-evolving country.

I must acknowledge the great influence that my service in the City of New York Fire Department (FDNY) has had on this text. Many of the topics covered, and their handling, are the result of my training by FDNY. In fact, one comment that has been made about this work is that it resembles a hardcover version of *Ladders 3*, FDNY's original tactical training manual for ladder companies operating at multiple-dwelling fires. I am honored to be considered in the same category as Battalion Chief John O'Regan (ret.), who wrote *Ladders 3*. Still, I feel that this text offers a unique perspective due to my own background and experiences. The most basic distinction involves staffing levels in FDNY and in other departments nationwide. In FDNY, a working fire in a private dwelling will have two six-person ladder companies at the scene to handle ladder company tasks. It used to be that, when I arrived at a house fire as a chief of my volunteer fire department on Long Island, I was lucky if I had four members to do all of the ladder company tasks that would by done by 12 members across the New York City border, only two blocks away. This required that all duplication of effort be minimized and that firm priorities be established to guide members to solve the most pressing problems first. That is one of my objectives in writing, and now revising, this text—to establish a way to evaluate problems, as well as to develop solutions that fit the situations that you face, with or without a large army of firefighters by your side.

Finally, this edition corrects a long-standing error, found in both the first and second editions, where the classification of the types of construction for ordinary and heavy timber was transposed. This error was pointed out by many alert readers, and was addressed by the publishers in the second edition by a correction page added to the book. Maybe after this third try I'll finally get it right! Stay safe.

PART 01 General Firefighting Tactics

01 General Principles of Firefighting

Modern firefighting techniques are founded on principles that predate almost all of their active practitioners. One of the most basic principles is that the firefighter's responsibility is to protect life and property. Whether paid or volunteer, there isn't a firefighter in the free world who is forced to join this profession. All do so by their own free will. When they take the oath of office, they commit themselves to uphold this ideal. It becomes their sworn duty. A further duty of all firefighters is to gain knowledge to perform new tasks. They must try to learn the reasons behind many of the functions that they perform. It may be sufficient for beginning firefighters to learn the manual skills of their trade and to operate under close supervision, relying on others for direction. A time soon comes, however, when all firefighters are expected to act independently.

The decisions that a firefighter makes will affect the outcome of a fire to a greater or lesser degree, whether that firefighter is initially assigned to an isolated position on the roof, to crawl along behind the nozzle, or to perform a role of leadership—the ultimate challenge. In some cases, lives hang in the balance—those of civilians, fellow firefighters, and their own. Firefighters who make decisions must be fully aware of the consequences of their actions. Outward actions must be based on sound information and must come out of a proper decision-making process. This process should be founded on sound understanding and an appreciation of a number of firefighting principles.

Consider these principles to be rules that you may break only under the most unusual circumstances rather than mere guidelines to disregard as you see fit. Some of these concepts may seem repetitious. In fact, they contain slight variations that need little explanation to the veteran. For those less experienced, the slight change in the conditions as described may serve to cover those special cases you'll encounter sooner or later. With this thought in mind, let's take a look at the ins and outs of fighting fires in the present-day setting.

The most basic principle of firefighting is that human life takes precedence over all other concerns. This rule may sound obvious, yet at times it can be overlooked (not intentionally, of course!). Still, on occasion, certain actions taken during the course of extinguishment can endanger occupants. Other times, actions that are overlooked can have the same effect. Many of these variables will be discussed in the chapters on engine and ladder company operations. Still more will be covered in the chapters dealing with fires in specific occupancies. For the present, a discussion of various principles that support the emphasis placed on human life will point out some of the generalities that apply under almost all circumstances.

When sufficient manpower isn't available to effect both rescue and extinguishment at the same time, rescue must be given priority.

Assume that a single engine company is first to arrive at a working fire on the first floor of a two-story private house. The unit is manned (undermanned) by a driver, an officer, and one firefighter. On arrival, an adult female is visible at the second-floor window. She is shrouded in smoke, screaming hysterically to the firefighters to save her baby. Before the members have even gotten off the apparatus, she and the child slump back into the room, apparently overcome by the heavy smoke that is pouring out of the window over their heads. What actions should this unit take while fighting this fire? Several options are possible, but with the limited manpower, simultaneous fire suppression and rescue isn't an option that will work.

In this instance, all three members will be required to perform the rescues. A portable ladder on the pumper provides the fastest, safest, and most direct access to the victims. In addition, unless fire is blowing out directly below the window, this is the only access that doesn't require hoseline support. Removing an adult victim via portable ladder is a very difficult task, requiring at least one person on the ladder and one in the room with the victim. In addition, you must remove the child. These three firefighters will have their hands full, but it can be done. If the victims survive, the firefighters will have been successful even if the house burns to the ground. In fact, rescue is the only practical alternative. You must at all times learn to put human life above property losses.

Attempting to approach the victims by way of the interior means passing the fire, which means stretching hoselines. If conditions are serious, it may not even be possible to get past the fire and up the stairs at all. Under the best conditions, one member must remain on the hoseline to protect the escape route while the others each remove a victim. This is a much slower and more dangerous approach. Due to limited manpower, it faces a much greater chance of failure. Similarly, trying a ladder rescue and simultaneously attacking the fire almost ensures that, with the three-member crew, neither operation will be successful, since there aren't enough firefighters to perform all of the needed tasks. But it is still an attempt at rescue. The last possible option would be to commit all of the members to a rapid knockdown of the fire, forcing the victims to await rescue by later-arriving personnel. This solution is unsatisfactory, since the victims will almost certainly perish before anyone can reach them. After 4 min without oxygen, the victims are likely to suffer brain damage. Any longer than that, death is nearly certain.

It is important to carefully consider the scenario described previously: this is a one-family house, with two known victims at a specific location and there are sufficient personnel (three) to perform this very limited task. That is what makes this scenario work. If the building is larger and contains more victims (say 8 or 10) or if the victims' locations are not immediately known, requiring an involved search, or are spread out, then the most important priority is to try to get the hoseline operating between the fire and the victims. Put the fire out if you can, but if that is not possible with the personnel at hand, at least protect the means of egress so that people can escape and rescuers can enter. This still places rescue as the highest priority; it just uses a different method to conduct it. The FDNY has an old saying that has been proven time and time again. Enginemen love it, and laddermen (truckies) hate it, but it's true: more lives are saved by properly positioned hoselines than by any other means! That is the guiding principle behind the next concept.

When you don't have sufficient manpower to perform all of the needed tasks, first perform those that protect the greatest number of human lives.

Although most firefighters would have a poor opinion of anyone who might suggest that they play God, the fact is that firefighters must occasionally make some hard choices between life and death. Sometimes conditions are such that people are going to die no matter what actions you take (Fig. 1–1). Although this is surely a stressful, disheartening situation, firefighters must realize that further lives may hang in the balance, requiring action based on rational decisions.

I recall responding to a multiple-alarm fire at which two 2½-story wood-frame houses were fully involved on arrival. Fire was already extending to two additional wood-frame houses on the left side of the fire, as well as to a six-story, non-fireproof apartment building on the right. All five buildings were fully occupied and, coupled with the lateness of the hour, presented high life hazards. The first engine and ladder to arrive, both manned by an officer and five firefighters, faced a difficult decision—where to operate first. The two frame exposures were definitely a great hazard, but so was the apartment house. The decision was made to concentrate on protecting that building, since it was home to 45 or 50 families. The two frame exposures received attention next, with a single outside line used in a holding action. Although the four families in these two buildings were severely threatened, the possible loss of more than 150 or 160 people in the apartment building far outweighed their possible loss. The occupants of the two original fire buildings received the lowest priority, since the buildings were solid flame from cellar to ridgepole. The word *savable* should be used in discussions of life hazards. A person in a room that is filled with fire isn't savable.

Fig. 1–1 When you don't have sufficient manpower to perform all of the needed tasks, first perform those that protect the greatest number of human lives first.

CONCEPT 03 Remove those in greatest danger first.

When you encounter more than one victim and don't have the resources to remove all of them simultaneously, you will have to establish some priorities. Generally, those people in the immediate vicinity of the fire are in the greatest danger and should be removed first. The next highest priority are those directly above the fire. They, too, must receive immediate attention. After these individuals, the removal may well skip a floor or more and shift to the top floor, since this is where smoke and heat will accumulate most rapidly.

A graphic example of this occurred at a fire on Manhattan's West Side, involving the apartment of a famous actor. The fire began in the actor's apartment and had flashed over one room prior to discovery. Occupants of the apartment called the FDNY while still evacuating the apartment. The next phone call reporting the fire on the 22nd floor came in moments later from the occupants of the adjacent apartment who were alerted by the commotion outside their door. The third phone call, reporting heavy smoke, came 2 min later from the occupants of an apartment on the 48th (top) floor! People below the fire are usually the last priority. You may have to vary this sequence somewhat as conditions warrant. For instance, a person on the fire floor, remote from the fire, may not be in as much danger as someone on a floor above. Also, someone who is emotionally agitated and threatening to jump may have to be removed immediately, even if he or she is in no apparent danger.

When sufficient personnel are available to perform both functions, they must carry out a coordinated fire attack.

An aggressive, coordinated fire attack may reduce or eliminate the life hazard. There are many ways to reduce the life hazard, including removing all of the victims, venting to draw fire away from the victims, and confining the fire to an isolated area. Quite often, however, the best way is to put out the fire. A coordinated attack uses the best of all of these methods to provide the highest level of life safety to the threatened occupants. It is a well-established practice in many departments to vent the roof immediately over the stairway to prevent mushrooming in apartment buildings. Simultaneously, a hoseline is stretched to the interior apartment door to prevent extension up the stairway. The apartment door is kept closed, confining the fire until all of the occupants are off the stairway. Then the hoseline is advanced for extinguishment. These simple acts, timed to occur in a smooth sequence, go a long way toward protecting life. Remember, the occupants visible at the window awaiting rescue aren't always the most seriously threatened. There may be many more inside who are overcome and can't get to a window to signal for help (Fig. 1–2).

Fig. 1–2 At fires where more than one person requires assistance, remove those in greatest danger first—generally those nearest the fire, those above the fire, and anyone in panic.

CONCEPT 05 When there is no threat to occupants, the lives of firefighters shouldn't be unduly endangered.

When there are savable lives at stake, firefighters must take extraordinary risks to save those lives. Acts such as dangling on ½-in. rope over the side of a building or pushing past a fire without fully darkening it down are taken under extreme circumstances. When no alternative exists, such acts are often performed. They don't always succeed, but if a firefighter is injured, it is usually recognized that the risk was justified by the circumstances. Yet, firefighters accustomed to performing this way sometimes behave in a similar fashion when there isn't any danger to occupants or when the occupants can no longer be saved. This must be recognized and avoided. As a young firefighter, I confess to having enjoyed the challenge of fires in vacant buildings. I regarded them as occasions where I could sharpen my skills and test myself without civilians being endangered. It was something like a trip to an amusement park, where I could experience all of the thrill and excitement without any of the distractions posed by concern for the occupants.

This attitude was extremely common in the fire departments in which I served. Then a string of tragedies occurred that started changing the firefighters' thinking. Probably none of them individually would have succeeded in effecting this change, but the combined weight of their loss awakened a number of the members. The loss of six firefighters to a roof collapse in a store where all of the occupants had been removed; the death of a lieutenant; the crippling of two firefighters in a vacant building, followed rapidly by the death of a chief and severe injury to other firefighters at yet another vacant building; the narrow escape of two firefighters when a collapsing wall of an unoccupied building sheared the bucket off of their platform, carrying them to the ground—all of these incidents served to change the attitude of our members toward vacant buildings.

Now firefighters, at least in the New York area, display an attitude of caution when operating at vacants. They no longer rush headlong into aggressive interior attack. More often than not, they assume a defensive mode, using an outside stream in conjunction with a careful survey of the stability of the structure. The officers in command must exercise tight control over their subordinates to ensure that they don't unnecessarily expose themselves to dangerous conditions. Otherwise, the lessons that these firefighters paid for with their lives will have been wasted. The real shame is that the lesson has only been learned locally, for it is still common in some areas for casualties to occur in buildings that are in such poor condition that they were barely standing prior to the fire and shouldn't have been entered in the first place (Fig. 1–3).

Fig. 1–3 Don't endanger firefighters' lives where the risks outweighs the benefits.

Sequence of Actions to be Taken

Locate, confine, extinguish

When a fire unit arrives at the scene of an incident, a number of items go through the minds of the firefighters—their size-up. Often they begin taking action to control the situation almost simultaneously. The actions that they take, regardless of the type of incident that they encounter, will almost always proceed in the following sequence: locate, confine, and extinguish. Firefighters may begin to take the first action, to locate the fire, before they complete their size-up. Actually, it constitutes part of the size-up, but it is a physical task that personnel must undertake. Some or all of the first members at the scene may have to be committed to locating the exact site of the fire even before the engine is committed to the block. Many times, fire units have arrived at the reported address only to find that the caller is looking at the fire through a rear window. Before committing yourself to operations, make sure that you know precisely where you are going. Equip the scouting members with portable radios. This will help ensure rapid hoseline placement at the correct location. These scouts can often find the best route to stretch the hoseline as they move toward the scene. It is much easier to climb back down the wrong stairway without a hose than with one.

The next action is usually to confine the fire. This involves limiting the fire's spread. In some texts, you will see a reference to *protecting exposures* and then a separate one to *confining the fire*. This is unnecessarily redundant. Firefighters must be constantly aware of these six directions for fire to spread: left, right, ahead, back, up, and down. Fire traveling in any direction isn't desirable, and it must be stopped. The idea behind protecting exposures before confining the fire to the fire building is based on the notion that it is better to lose one building than two. In its simplest form, that idea is fine, but situations arise where confining the fire within the fire building will be of far greater importance. Take the case of fire in a ground-floor classroom of a three-story, 150 x 300-ft ft brick and wood-joist school. Heavy fire is venting out the window, exposing a nearby one story, 30 x 50-ft warehouse. It is 3:00 AM, and both buildings are unoccupied. The first priority in this case should be to confine the fire in the fire building, since the total potential loss there outweighs that of the warehouse. The spread of fire within a building is usually faster than the rate at which fire spreads to an exposed building. However, when a large portion of a building or area is already heavily involved and is threatening another structure, it is reasonable to protect the exposed building first, since it might not be possible to prevent full involvement of the original fire building. For further discussion on this topic, see chapters 3 and 4.

It is ironic to some that the heart of firefighting, getting in and putting the fire out, is the last priority in our sequence of actions. Nevertheless, it is crucial that everyone involved understand that this sequence is how it must be. It is very frustrating to arrive first at a working fire and find out that you must take a defensive position, only to have a later-arriving unit go ahead and take *your* fire. Still, if all of the members thoroughly understand their firefighting tactics, they will appreciate the importance and significance of their position.

Once, at a fire in a frame row house, I was the officer of the second engine to arrive. We observed heavy smoke and a heat condition in the basement. The first engine had stretched a line to an outside basement entrance from which they could see that, in addition to the basement, a cellar located fully underground, below the basement, was also showing heavy fire. The situation was complicated by an open wooden staircase leading from the basement to the top (third) floor. Also, the house was in the middle of 18 attached homes, and a common cockloft spanned the entire row. Our choice was clear. We couldn't follow the first engine in and slug it out with the cellar fire. Instead, our line was needed on the first floor to protect the open interior stairway as well as to prevent extension up the partitions and in an enclosed light and airshaft. This was a hot location, being above two floors of fire. My crew didn't enjoy being stuck up above like that without getting a real share of the action. Still, they fought a vital battle to keep the fire from extending upward. If fire had reached the common cockloft, its rapid horizontal spread would have ensured heavy damage to at least several other buildings. The role of extinguishment was left to later-arriving units, while priority was placed on confining the fire. It succeeded.

The last rule for firefighting is a brief one: let circumstances dictate procedures. This statement simply means that it is impractical to specify actions in advance of every possible scenario. A good standard operating procedure (SOP) should handle the majority of incidents in which a unit is involved. When faced with a nonstandard situation, however, you must be flexible enough to adapt and develop a proper solution. Don't try to force an SOP to handle a situation for which it wasn't designed. Doing so creates problems. This is what leads companies to stretch 1¾-in. line when first due at a fully involved lumberyard because "that's what we always do." If an action is needed, go ahead and do it. When in doubt, err on the side of caution.

02 | Size-up

Size-up is a term often mentioned in firefighting. What exactly is size-up? Who does it and why? When does it begin? What do we use it for? Size-up can be described as an evaluation of problems and conditions that affect the outcome of a fire. All firefighters have performed size-up whether they were consciously thinking about it or not. The newest firefighter's perception of what is happening might not be as accurate as that of a veteran, but he is observing matters that affect him nonetheless. Just as perceptions of a situation change with experience, so do the number and degree of factors that must be considered as a given member's responsibilities increase. The officer in command of a fire situation has many more considerations than the junior member of an engine company. Regardless of the extent to which either is involved, both should perform a size-up. The information thus gained can determine where a firefighter operates, what he does, and when and how he does it. Quite literally, a firefighter's size-up can determine whether he lives or dies. That isn't a decision you would want to leave until the chief arrives.

Proper size-up begins from the moment the alarm is received, and it continues until the emergency is under control. Size-up should be thought of as an information-gathering process. As such, it should include information gathered during preplanning activities. Still, size-up isn't limited to this. Although such a survey can give much-needed physical information—such as the size and characteristics of a given building—there are many variables that can't be determined until the time of the alarm. Consider the following scenario, and list the items that you can determine for your size-up.

At 2:30 AM, your firefighters are alerted to the following dispatch message: "Attention, Engines 1 and 3, Ladder 2, Chief 1, and Rescue 1. Respond to a report of a store fire across from number 1510 Main Street." The first thing that should come to mind is the hour of the day. If you are sound asleep when the alarm sounds, chances are good that most other people are also. This translates into delayed alarms, a situation further evidenced by the phrase "across the street from." These two clues point to a working fire, since it has had time to gain headway before being noticed by neighbors. The next consideration is the location. Experienced firefighters make it a point to learn their district. They can generally pinpoint the local target hazards and at least describe the type of block to which they're responding. After several years in the same area, a firefighter can often tell you the specific building involved, using the address as a guide. In most cities, building addresses run consistently along the streets. For example, all odd-numbered addresses will be on the east side of streets that run north–south and on the north side of streets that run east–west.

There is no definite rule about which way street numbers are assigned, however, so it is advisable for firefighters to learn their own areas. With this information, a firefighter can now tell on which side of the street to find a given building. If your city uses a uniform block-numbering system (numbers 100 to 199 between First and Second Street; numbers 200 to 299 between Second and Third Street, etc.), you can easily pinpoint the block. Knowing that number 1510 would be closer to 15th Street than 16th Street further narrows down the choices. Such information in advance can help firefighters arrive at the nearest cross street. It allows them to pick the nearest hydrant and sometimes to get a glimpse of another side of the building on the way in. Just because the caller reported a store fire doesn't mean that there isn't any life hazard at 2:30 AM. Mixed commercial-residential buildings can be deadly. Often a fire originating in a store makes great headway while trapping sleeping occupants above.

The less obvious life hazard at this hour of the day is to the firefighter. One of the rules of safety in potential collapse situations is that, if a building (brick and wood joist or wood frame of standard construction; not lightweight) has been exposed to heavy fire for 20 min or more, it may be too dangerous to enter. For a fire to be discovered by passersby, it must often be of large proportions. The only fact that isn't known is how long the fire has been at that stage. Firefighters should be particularly aware of this and be sure to report any additional indications of potential collapse.

Other information received in the dispatch message concerned the fire department assignment—Engines 1 and 3, Ladder 2, Chief 1, and Rescue 1. Know the units that you should expect to receive for an alarm. Is this the standard response for a building fire in your area, or is something unusual occurring? Did the dispatcher add the rescue company to the response based on reports of people trapped or other unusual circumstances? Are Engine 2 and Ladder 1 normally assigned but not available at present? That could mean that the first unit operates without support for longer than they're accustomed. What are the special factors that affect the crews? Holidays are more significant in volunteer departments than they are in career departments, and this, too, can have an effect. This run could be occurring at 2:30 AM on January 1, when many of the members might still be out on the town. All such information should be registering in each member's head, and they should be starting the size-up process even while they're still getting dressed.

Although they provide a foundation, the alarm message, preplan information, and knowledge of the response area won't tell you everything you need to know to handle a working structural fire. What, then, are the vital elements needed to perform a size-up? A traditional listing of factors affecting size-up almost always includes the famous 13-point outline. This time-tested procedure has served many a fire chief well. It covers the majority of fireground considerations, and there is even a convenient acronym to help you remember all 13 of them— COAL WAS WEALTH. The traditional 13-point size-up includes the following:

C   Construction
O   Occupancy
A   Apparatus and manpower
L   Life hazard

W   Water supply
A   Auxiliary appliances
S   Street conditions

W   Weather
E   Exposures
A   Area
L   Location and extent of fire
T   Time
H   Height

At this point, many readers are surely asking, "What about hazardous materials (hazmat)?" I propose a slight modification to modernize the acronym. Combine area and height under the third *A* and let *H* stand for hazmats. After all, area and height are closely related.

Now, let's examine the importance of each element of the size-up in a more hierarchical order.

Life Hazard

Life hazard comes in the following two forms: them and us.

As mentioned in chapter 1, life hazard must be *the* deciding factor in determining tactics and procedures. Examine what constitutes a life hazard. The human body doesn't stand up well to direct-flame exposure, high heat, and toxic gases. In addition, having part of a structure collapse on top of you or being struck by projectiles from rupturing containers isn't exactly conducive to living a long, full life. You have to evaluate all of these threats to the occupants and, if possible, you must take steps to reduce or eliminate them. Of course, not all fires pose a severe life hazard, which is actually the result of several other size-up considerations. The time of day, occupancy, location, and extent of fire all combine to endanger victims.

The effect of time of day on the occupants is variable and related to the occupancy. People are likely to be asleep at 2 AM if the occupancy is any type of dwelling, and they may require removal if they are on the same level as or higher than a serious fire. With few exceptions, a change in the time of day, occupancy, or location and extent of fire will also change the life hazard. This change is due to human behavioral patterns. If people are aware of a fire, and if they are physically and mentally able to do so, they will flee before the fire engulfs them. The exceptions apply to cases where the fire threatens to outrun the occupants. The mere presence of a large number of people doesn't produce a severe life hazard. If, however, those people cannot flee a fire because of physical or mental impairments, poor exit facilities, or risk of rapid extension, then a life-threatening situation exists.

Firefighters have several methods of dealing with high life hazards, each of which might be used alone or in combination. These will be discussed in detail in the chapters on engine and ladder operations. The best method, however, is to reduce the life hazard long before the incident occurs. This can be done by imposing occupancy load restrictions; by improving exit facilities; by specifying fire doors and partitions; and finally, by the best method of all—installing a complete wet-pipe automatic sprinkler system. Such steps in fire prevention are usually outside the realm of the line firefighter. Each firefighter, however, has a responsibility to recognize hazardous conditions and to bring them to a safe conclusion by whatever means are necessary, even if that entails contacting an outside division or agency.

Finally, the life hazard comes in the following two forms: them and us—civilians and firefighters. Every fire has a potential life hazard once the fire department arrives: us. We bring our own life hazard with us, and even a vacant shell of an abandoned warehouse now has the potential for killing people once we commit fire personnel inside. A careful evaluation of the civilian life hazard should be made before deciding to put firefighters' lives at risk. A key component of this evaluation is the building's occupancy.

Occupancy

When responding to an alarm, firefighters often ask, "What kind of building is it? Is it a factory? An apartment?" What they actually want to know is not how the building is constructed but what is inside. As indicated previously, the occupancy has a considerable bearing on the extent of the life hazard, and this is dependent on the time of day. The life hazards associated with schools, for example, vary greatly with the time of day. Other occupancies, such as hospitals and apartment houses, pose a high potential life hazard around the clock. Still others, such as storage warehouses, have a uniformly low potential for life hazard regardless of time. It is only in response to a high civilian life hazard that we should be undertaking aggressive tactics.

Analysis of the annual national firefighter death and injury statistics reveals an absolutely shocking and disgraceful pattern. As vividly illustrated by the bar graph (shown in Fig. 2–1), we are killing firefighters at a devastating rate in buildings that for the most part we have no business being in.

As the chart shows, we lose almost four firefighters a year in residential structures for every 100,000 fires we fight in them. While every firefighter's line of duty death is a tragedy and is something I mourn deeply, at least in residential buildings, the firefighters were usually risking their lives in an effort to reach civilians trapped by fire. Nearly 80% of all civilian fire deaths occur in residential buildings, more than 3,000 people per year, thus justifying aggressive tactics in these buildings. Surely that is a national problem. Conversely, very few civilians die in fires in commercial buildings. In fact, more civilians die each year in car fires than die in store fires.

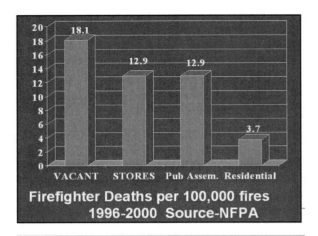

Fig. 2–1 Firefighters die needlessly in buildings where there is little or no civilian life hazard.

The most damning piece of information, though, is reflected in the rate that firefighters are killed in non-residential structures, especially vacant buildings, where we lose more than 18 firefighters for every 100,000 fires, more than four times the rate of residential buildings! By definition, a vacant building has no life hazard until we get there, so what the hell are we doing killing firefighters in them in such terrible numbers? Officers arriving at fires in these and other deadly occupancies (see chapter 15) need to impress the severe danger these occupancies pose on all their members. I strongly suggest a very frank instruction be given immediately on arrival. Grab them by their shoulders if necessary, huddle them up and tell them "Look, there is no civilian life hazard here, be careful! I want you all coming home with me in one piece."

The occupancy can also have a great effect on other aspects of your strategy. Both factories and warehouses typically have large, open floor areas, which may translate into fires that are beyond the scope of handline control. Such large spaces should also alert firefighters to the possibility of truss construction, a collapse hazard. Retail areas are usually smaller, but the relative constriction of these spaces may make it more difficult to advance handlines. Expect all of these types of occupancies to contain large amounts of materials, producing a heavy fire load.

The type of occupancy can also indicate the presence of hazardous materials. Expect hazmats in garden supply houses, paint stores, and extermination businesses. Some locations, however, aren't as obvious, and you may have to rely on information recorded during prefire inspections. When properly used, a computer-aided dispatching system (CADS) has proved to be the most effective tool by far for recalling this information. CADS surpasses the human memory and the three-ring binder in its ability to store information about dangerous conditions. Although firefighters usually know something about the most obvious and deadly hazards, hundreds of hazards may escape their initial attention. For these, a CADS-generated reminder could avert what might become a fatal encounter. The information can be recalled automatically whenever an alarm is transmitted from a building address that is on file. This serves to reduce the *crying wolf syndrome* that ensues when it is necessary to check manually each address to which you respond. CADS only displays hazard information when there is some data you should have.

Of course, any computer-generated information is only as good as the information that was entered. This means that field units must have some quick and simple method of reporting the conditions that they encounter. It also means that the information must be kept up to date. For this reason, any occupancy that is subject to CADS hazard display should be inspected annually (at least). Also, the message can be transmitted to field units by one of several ways. The dispatcher can read it to the units as part of the alert message, or you can flash the message on a video display terminal, teletype it to each unit, or have the message repeated over radio channels about 1 min after the initial alarm. The best possible choices are to teletype the alert message (which gives the field unit a hard copy to refer to) and to rebroadcast the message over the radio after the units are on the road. These methods serve to present the information to anyone who may have overlooked it as well as to verify the listing after the firefighters have had a moment to consider it (Figs. 2–2 and 2–3).

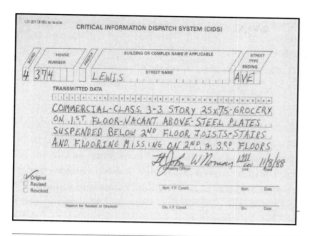

Fig. 2–2 Fire officers can input lifesaving information into a CADS that will save firefighters' lives for years to come.

```
CLASS 3-  1ST ALARM  — PUBLIC FACILITY
E236/E227 E281/E234 E282 L118/L111 L176
BC48/BC38 RS02

BOX 1612 - (=3-1612-4) 170 BUFFALO AVE

ST.MARKS AVE                    PROSPECT PL
H=HOSPITAL—WELLS FARGO MANUAL
HOSPITAL — ST. MARYS —8 STY 300X100 CL 1
FIRST ENG & LAD REPORT TO SECURITY CP
VIA EMERG RM ENT ON PROSPECT PL FOR USE
OF HOSP HT TO COMMUNICATE WITH ENGINEER
INCIDENT# 23

12/04/89    012336
```

Fig. 2–3 CADS entries should report information that cannot be known simply by looking at the building, such as special operational procedures.

Time

Time has many impacts on our firefighting operations. In reference to fire operations, time comes in many forms; time of day is one we've discussed already in relation to the life hazard. One can say that the time of day, coupled with the occupancy and the location and extent of the fire actually produces the life hazard. Remember the school at 2 PM on a school day vs the same building at 2 AM. Time of day can also impact us in other ways, for instance, response during rush hour traffic and late night alarms usually mean delayed alarms.

Time of year also impacts us directly. A church fire at 10 min after midnight usually signals a low life hazard, but at 10 min after midnight on December 25, you can have a serious problem. Certain times of the year can affect manpower, especially in volunteer or paid-on-call departments. Hunting and fishing seasons in many regions is a bad time to have a fire. Incident commanders (ICs) need to be able to factor such variables into their strategy. Always plan for the worst; that way you won't be surprised when it happens.

One night I responded as the chief of my volunteer department to a working fire on the top floor of a large vacant wood-frame boarding house. I was first to arrive and found a moderate volume of fire (it looked like two or three rooms worth) with possible extension to the cockloft. Ordinarily I would not be very concerned about this fire. Two lines ought to be able to make quick work of it, although I did order our tower ladder to be prepared for master stream operations on arrival. The time of year factor had a very large impact that night though, prompting me to call for additional help, beyond what I would have normally used. The fire occurred about an hour after our department's Christmas party! While the party was over, the bars were still open. I could not be certain that our normal late-night response would be up to its usual abilities. I immediately requested two engines and two ladders as mutual aid from neighboring departments. As it turned out, our people did a great job, and the fire was knocked down before any of the mutual aid units arrived. If you are the guy standing in the street with the white hat, you can't bet on that though. As one saying goes, "Hope is not an effective strategy." You have to plan for the bad things that can happen.

Another impact of time of year is the effect of the holiday season on the fire loading and life hazard in most commercial occupancies. Beginning around October 1st and running through the New Year, stores are packed to the rafters with both stock and customers. Throw in some highly flammable decorations like evergreen wreaths and some worn lighting, and the recipe for tragedy is complete.

Perhaps the biggest impact time has on firefighting is the elapsed time since the fire has begun. One of the key elements of our size-up is to estimate how long the fire has been burning and evaluate the structure for strength in terms of fire resistance, then we can estimate whether or not it is safe to operate in or near the structure. Elapsed burn time is a key indicator of the likelihood of structural collapse. The greater the elapsed time, the greater the danger. This part of the size-up is very much an art and not an exact science. It relies heavily on the experience (which is a very costly commodity), and close observation and reporting by all members on the fireground. This is due to the many variables that can be involved: the type of construction, fire loading, length of burn before the alarm is transmitted, alterations to the original structural design—either through routine structural alterations or deterioration due to exposure to weather or industrial processes, or from a catastrophic event such as when a jet airliner slams into a building taking out 40% of the structural columns.

The twin towers of the WTC had been designed with a *fire resistance rating* of 3 hrs, for key structural elements like columns. One tower stood up for less than 1 hr after impact and ignition, the other for 1 hr and 45 min. In both cases, the impact and fire loading played tremendous roles in hastening the collapses, but the fact is the Incident Commander (IC) had no reliable way of predicting how long either building would last under that type of fire exposure. The fire resistance ratings are not directly related to elapsed burn time and structural stability, especially in light of the increased use of plastics and their resulting impact on the time-temperature curve that was used to develop the fire resistance ratings. These ratings were developed many years ago, before plastics became such a large part of the fire load of modern buildings.

Structural collapse is the most difficult of the firemen traps to predict. One rule of thumb for predicting structural collapse is the 20-min rule. The 20-min rule is not a hard and fast rule at all. Rather it is a guideline that may be useful but requires a good bit of caution and common sense in applying it to actual fires. Knowing the class of construction is an important component of this evaluation.

The most common structures that we operate in, houses and other residential occupancies, as well as many smaller commercial buildings, are most often either wood frame (Class 5) or brick and wood joist (ordinary-Class 3). These buildings were built with relatively large floor and roof joists, typically 2x8 in. and larger. Both behave similarly under fire conditions, since the floor and roof supports are made of the same material in each—wooden joists. The interior furnishings and finish are nearly identical in most residential buildings as well, so we can predict a reliable pattern of fire behavior in either of these type buildings. The 20-min rule applies to these types of construction, and only to these two types of construction. If the fire is not under control in 20 min in one of these buildings, you should probably begin withdrawing your personnel and shift to exterior operations. Of course this is only a guide, and even as such, it must be applied carefully and with some modifiers.

A crucial element to understand is that the time frame does not start with your arrival at the scene (or even with the receipt of the alarm) but rather, when the fire has reached flashover and begun to attack the structural elements. How do you know how long the building has been at the post-flashover stage? Assuming you didn't witness the fire begin, you have to estimate this.

One method is to examine the building for fire venting through window and door openings. On most private homes, this is a fairly simple exercise, due to the number of windows on all sides of the building. A fire that has not vented from a single window has not been at the flashover stage for very long, generally under a minute or two, even with thermopane glass in the windows. (This assumes the fire began in a room and not in a void space, such as within a wall or ceiling space.) A fire that has vented out one or two windows, but has not vented from any other windows, is usually confined to one room and has been at flashover for anywhere from 1 to 5 min. If there are other non-vented windows on each side of the ones that are showing fire, that is a good sign. Go ahead and begin your attack. The more windows that fire is showing out, the longer the burn time. Fire that is venting out windows on two floors typically indicates one of the following two things: prolonged burning—greater than 10 min—or use of an accelerant to spread fire rapidly to both floors. Both are dangerous situations that should cause the attack forces to proceed cautiously (Figs. 2–4 and 2–5).

Fig. 2–4 Initial size-up indicates the home on the left has not been burning for very long, since fire has not vented from any openings. Beware of fires in cellars however.

Fig 2–5 The building above has been burning for an extended time, and should only be entered with great care, due to the potential for collapse. (Note: This 7AM fire had reports of trapped occupants, and the structure was entered with care after the fire was knocked down.)

Look for signs of advanced fire, such as fire burning *through* a wooden wall (not merely on the surface). This indicates the fire has been burning for a considerable time prior to your arrival and that collapse may be imminent, since the wooden walls support the floor and roof joists. Accelerants that cause the fire to spread extremely rapidly are another cause for major concern, since a rapid intense fire destroys wood more quickly than a slow smoldering fire. Liquid accelerants may also penetrate flooring and stairs they are poured on, resulting in fire attacking the joists or stair stringers from both top and bottom simultaneously. Many firefighters have been injured trying to climb staircases that collapse beneath them at arson fires. Also, try to be aware of the history of the building. A building which has had several fires in the past has suffered increasing damage and has only a limited time left before collapse. The same may be true of a building in which extensive renovations have been done, which have weakened the original structure.

If there is no outward indication of any of these situations, a standard interior attack is in order. If it does not succeed in a timely fashion, the IC must evaluate why. If the first attack crew exits the building after expending a 30-min air cylinder and their officer reports the main body of fire knocked down, you as the IC have to make a decision. Do you continue interior operations or shift to exterior? If *your* evaluation of conditions as the IC indicates that the attack crews report is correct (the main body of fire is knocked down), then continue with the operation.

On the other hand, if your overall perception is that there is still a considerable body of fire present that the interior crews have not put out yet, you have the responsibility to order them out. This is the most important decision that you will make. Only the IC can make it, based on reports he is getting from the various sectors. You must be able to trust the observations of your subordinates to accurately report conditions in their area, but you are likely to be the only one with *the big picture*. I recall operating with my engine company, knocking out a lot of fire on the first floor, when we were ordered out of the building. We were doing a great job, making a lot of headway, and reported that back to the IC. "Just a few more minutes, Chief, we got it." The chief sounded rather agitated when he repeated his order to "Get the hell out here. Now!" Imagine our surprise when we exited the building to find the top three floors blazing merrily away. *When you're ordered to retreat, follow the orders.*

At the height of a serious fire, it is very easy to lose track of elapsed time as so many activities demand our attention. We must have a system in place to help us keep track of time. In the past, many experienced chiefs used the air bottle method to help them keep track of time. A full *30-min* air cylinder usually only lasts 15–20 min under heavy firefighting use. A chief might continue with an aggressive attack until the first due units started to come out of the building with their air supplies depleted. If the fire was still not showing signs of being under control or at least knocked down, the ringing mask alarm bells acted as another type of alarm, sending the chief a message that he ought to think about pulling his people back out of harm's way.

That system works very well when properly applied, with a competent and experienced IC evaluating the overall control of the fire, in the types of buildings I described earlier. The *air bottle rule*, or the *20-min rule*, is only valid in *standard* wood frame or brick and wood joist buildings. It does not apply for example to Class 1 buildings of reinforced concrete construction. These buildings can withstand several hours of fire exposure and many teams of firefighters, working in relays expending several changes of air cylinders may be needed to complete final extinguishment in this type building. The same is true of Class 4 heavy-timber construction, although in practice, due to the combustible nature of these buildings, if you haven't put the fire out in one of these buildings after one or two teams have exhausted their air supply, something is very wrong and you are likely to be driven from the building very shortly by rapidly expanding fire.

The real problem with the air bottle rule or the 20-min rule is with lightweight construction. It is far too long to operate in any type of lightweight wood construction building! Buildings that were built using plywood I-beams, 2x4 wooden gusset plate trusses, or the composite steel and wood trusses, and most Class 2 buildings with metal *C-joists*, or bar joists, have all been proven to collapse with as little as 5 min of fire exposure! A comprehensive listing of all buildings in your area that are built with these techniques is critical to firefighter survival, along with a clear department policy that states that a fire that has reached the flashover stage in one of these structures will only be fought from defensive positions until the fire has been knocked down and ventilation and lighting conditions permit a careful examination of the structural elements that are trying to fall down on top of us or out from under us. This information should be recorded in a CADS message. The same rule should apply to the heavier wooden Bowstring trusses (which I have seen fall down in as little as 8 min after the fire reached flashover) and even seemingly heavy steel trusses. Unprotected steel can be expected to fail in as little as 5 min in a serious fire.

The air bottle rule has changed substantially in recent years as many departments have shifted from *30-min* cylinders to *45-min* cylinders. The *old-time* chief must adjust his or her timetable. Leaving the troops inside for the extra 6 or 8 min these cylinders allow may leave them inside just a little too long. This can put you dangerously beyond a safe margin if the IC were relying on that as the only way of tracking elapsed time and not evaluating the other warning signs of potential trouble.

To avoid the *tunnel vision* that sometimes sets in at chaotic incidents and refocus the IC's attention on elapsed time, some departments utilize a *time mark* system, where the dispatcher keeps track of the elapsed time and prompts the IC via radio. In some cases, this is done with a simple statement like, "Battalion 1, you are now 10 min into the incident." In New York City, the IC is required to transmit periodic reports on the incident, beginning 5 min after his or her arrival. After an additional 10 min, another *progress report* is due, continuing every 10 min for the first hour of the incident. The dispatcher will request the report if it is not given in time. Additional reports are required whenever a major change in incident status occurs, from fire escalation to fire knockdown or control or if a change in the mode of attack occurs from offensive to defensive or back. These reports serve the following purposes:

- they allow superior officers to monitor conditions via radio while in their offices or while responding
- units that may respond on additional alarms are also alerted to potential problems and actions that may be required

Above all, though, it forces the IC to recognize that time is passing and requires him to evaluate and verbalize what type of progress is or isn't being made in the firefight.

When deciding whether to commit troops to an interior attack in the first place or whether to pull them out of an ongoing effort, the primary mission of the fire service must come clearly to mind. To protect *life* and property. It is always *life* first, including firefighters' lives. Property can always be rebuilt, your people can't. Be constantly aware of what is happening to the building. Look for signs of structural weakness and/or alterations. Remember the time frames that you have to make a difference; they are set by the structural engineer that designed the building and not by your desire to save the structure. Remember that one of the primary indicators of an impending collapse is the elapsed time from flashover. Time may or may not be on our side. We need to constantly monitor what it means to us at each incident. Bring all your people home safely!

Construction

As pointed out previously, a critical size-up factor that should be included in any CADS hazard display program is the building's construction. Construction has many implications, the first of which is that the degree of compartmentation of a building can promote or thwart the spread of fire. Large, open floor spaces provide an opportunity for fire to spread readily throughout the exposed area.

The second implication is the degree to which the building itself contributes to the fire load. Older buildings are predominantly of wood construction, meaning that such a building in and of itself constitutes a heavy fire load. Newer buildings of metal construction add little or nothing to the fire load. The major exception to this statement, however, is the metal deck roof fire. This occurs when a fire exposes the underside of a metal roof deck on top of which hot, poured tar has been applied. The asphalt is used for water resistance and to help glue down the insulation that is applied above the tar. Fire exposing the bottom of the deck heats the tar, which then pours down through holes in the deck, raining liquid fireballs on people and objects below. (See chapters 9 and 15 for further discussions of the problems posed by metal deck roofs.)

The third concern related to a building's construction is the number of hidden voids within which fire can travel. Voids are responsible for the destruction by fire of more buildings than any other construction-related factor. Cocklofts, pipe chases, and channel-rail voids all provide concealed highways (often of wood construction) for the travel of fire. Planners in the 19th century designed mill construction buildings of wood, but they took special care not to incorporate voids. This meant that any fire originating in such buildings could be detected readily and be attacked with hose streams without a great need for extensive overhauling.

The fourth concern is probably the most important to firefighters: the ability of a building to resist collapse when threatened by fire. Buildings are composites of materials joined together in such a way as to resist the pull of gravity. Fire works with gravity to attack these materials, break their hold on each other, and cause them to fall. Certain materials are intrinsically cohesive and resist fire well. Poured-concrete buildings, for example, act as one-piece structures. Generally, any failure that occurs in them will be of another material, such as a hung ceiling. Buildings of steel columns and girders are well connected but not impervious to the effects of heat, as described under Class 2 construction.

Fire that attacks such buildings doesn't usually separate the connections; instead, the steel itself weakens and sags. Most other types of buildings, however, are joined by methods that are much more susceptible to fire. Nails, screws, gusset plates, glue, mortar, and sometimes gravity alone are all that keep many building elements in their upright position. All of these elements are readily attacked by fire and firefighting methods.

Finally, all materials have some point of overload. A building can be almost to that point and still stand. Once it has crossed that point, however, collapse is probable or imminent. The added weight of water used in firefighting or the weight of the firefighters themselves can be the straw that breaks the camel's back (Fig. 2–6).

A brief description of each of the major classes of construction is in order here, ranked by its resistance to collapse. Since there are many different building codes nationwide, it is necessary to establish a common reference. For this text, buildings are grouped in five classes, as found in the National Fire Protection Agency (NFPA) *Standard 220 Types of Building Construction, 1999 edition*, as follows.

Fig. 2–6 Many master streams operating into a building add tons of water per minute to the load that the structure must support—this, at a time when fire is weakening the structure.

Class 1: fire resistive

The walls, partitions, columns, floors, and roofs of these buildings are noncombustible. The various construction elements are designed to withstand the effects of fire for a *limited* time *and* prevent its spread. These are what some codes call *fireproof buildings*. As time has taught, however, an oven is also fireproof, but it can still have a great deal of fire inside of it. Buildings of poured or pre-cast concrete and steel-frame buildings with an applied fireproofing meet these general requirements. This type of construction is often used for high-rise buildings and, although the building may not fall down, the products of combustion often endanger both occupants and firefighters. Note also that for steel-framed construction, this is only true for as long as the *fireproofing* is able to protect the steel. Missing, inadequate, or misapplied fireproofing leaves the steel vulnerable to fire.

These structures are generally very sound in a fire, although a number of fires in recent years such as the three high-rise fires of September 11, 2001 and the fire at One Meridian Plaza in Philadelphia have shown that they can no longer be thought of as *fireproof* or impervious to the effects of fire (Fig. 2–7). Much of this is due to the shift toward lightweight construction in the post–World War II generation of high rises. As described in chapter 16, this weakness has long been recognized by the fire service, but we couldn't get anyone to listen to us until 9/11. In his 1977 text *High Rise Fire and Life Safety* (PennWell,) former FDNY Commissioner and Chief of Department John T. O'Hagan wrote, "Builders have therefore created a lightweight type of high rise construction with less fire resistance than the previous generation." The first and second editions of this book, published long before 9/11 quoted Assistant Chief Michael Cronin of the FDNY regarding the effects of an airplane striking a modern high-rise. He was off in only one aspect: the pieces ended up in Brooklyn, not Queens. A further discussion of Class 1 construction, especially its impact on high-rise buildings is found in chapter 16.

Fig. 2–7 Fire-resistive buildings can contain a lot of fire. They can be as deadly as other types under the right circumstances.

Class 2: noncombustible

The walls, partitions, columns, floors, and roofs in these buildings are also noncombustible, but they provide less fire resistance. In addition, they don't withstand the effects of fire, nor do they prevent its spread. *Noncombustible* is a term that is very misleading. It refers to the fuel contributed by the structural components, not its resistance to the spread of fire.

Noncombustible buildings are generally built with exposed metal floor and roof systems and metal or masonry walls (Fig. 2–8). Any sizable fire in the contents of the building will rapidly destroy the structural integrity of the unprotected steel. These structures are the least stable in terms of collapse when exposed to fire. Hundreds of thousands of stores, warehouses, and factories have been built with this type of metal deck on exposed steel-bar joists.

Fig. 2–8 Noncombustible buildings do not add fuel to the fire load but collapse at a rapid rate when exposed to contents fires.

Class 3: ordinary construction

Ordinary construction consists of masonry or other noncombustible walls with a 2 hr fire-resistance rating. Its floors, roofs, and interior partitions are of wood. The wood used in these buildings is smaller than heavy timber. It is therefore easier to ignite and offers less resistance to burn-through or collapse. This is how the majority of brick buildings are made. Ordinary construction can be found in anything from residential to commercial to manufacturing properties (Fig. 2–9).

Fig. 2–9 Buildings of brick and wood joist or ordinary construction generally do not fail early in a fire.

Class 4: heavy timber

The exterior walls of these buildings are of masonry or some other noncombustible material with at least a 2 hr fire-resistance rating. The interior columns, beams, and girders are of heavy timber (minimum 8x8), and the floors and roofs are of heavy planks (3x6 minimum). These buildings present an extremely heavy fire load, yet they are excellent buildings in terms of fire resistance. The key to these buildings is their mass. The sheer bulk of the material makes them difficult to ignite and, once ignited, the timbers stand up to flame better than exposed steel. In addition, the lack of any hidden voids makes firefighting less complex than in ordinary buildings. Once ignited, however, they require large quantities of water to cool the large surfaces exposed to flame. If the fire isn't checked early, the large, open combustible areas may quickly drive personnel out, producing a long, hot fire beyond the capabilities of manual firefighting (Fig. 2–10).

Fig. 2–10 Heavy-timber buildings have a heavy fire load built in. They produce tremendous quantities of heat and flame once fire gains control.

Class 5: wood frame

These structures have walls, floors, and roofs that are made wholly or in part of wood or some other combustible material. Obviously, a building of wood is more prone to extension than one of noncombustible material, but it might surprise some to learn that the wood-frame building poses less of a collapse hazard than does the noncombustible one.

A further discussion of specific problems related to construction can be found in the chapters dealing with specific occupancies. Typically, certain occupancies are built using certain construction styles. High-rise offices, for example, are built using Class 1 construction. Although there is some overlap, and it is impossible to specify an exact type for each occupancy, sufficient examples will be found to cover most of the common types of construction (Fig. 2–11).

Fig. 2–11 Wood-frame construction offers fire a myriad of combustible voids for fire to hide in.

Height and Area

The height and area of a given building are two obvious concerns during size-up, since they indicate the maximum potential fire area. Sometimes, however, these considerations aren't so obvious from the street. In some areas, it is very common to have structures built on a grade, with entrances located on two parallel streets. Units entering from the higher street might think that they are entering a six-story building, when in fact they may find themselves nine or more stories above the street at the rear. In such a case, they could be above the reach of aerial devices.

The height of the building may also let firefighters know what to expect as far as construction and auxiliary appliances are concerned. Many building codes specify that buildings over a certain height must be built of Class 1 construction. A similar requirement may exist regarding a required standpipe or sprinkler system. Firefighters arriving at a fire in a three-story building would know in this instance that they would have to stretch their handline from their apparatus since they wouldn't encounter a standpipe.

The frontage of a building is also important to firefighters, since an attack begins from the front. (We have to enter somewhere!) Additionally, this space provides opportunities for horizontal ventilation. For the most part, the frontage of a building also indicates the width and depth of the structure, but that isn't always the case. In some cases, one building will wrap around another; other times, a building will abut an adjoining structure but widen out toward the rear into an *L* or a *T* shape. What appears to be a 25-ft-wide structure from the street can turn out to be 100 ft deep from front door to rear wall. This could mean running short of hoseline or choosing a diameter that is too small. It could mean that you won't be able to advance into the face of a fire that turns out to be larger than anticipated. The way to combat this problem is to get early reports from the roof level. Party walls are usually visible from the roof, and they can tip off firefighters to irregular shapes and unexpected sizes (Figs. 2–12 and 2–13).

Unfortunately, something not visible from the roof creates another unusual circumstance: the presence of interconnecting openings between what appear to be two separate and distinct buildings. These openings effectively combine the structures into one large potential fire area. In the event of a serious fire, it is vital to check all of the adjoining areas for extension. Such openings between structures can be as large as a freight door or as small as a crack between the beams. Sufficient personnel with the proper tools are needed to open up and examine all of the adjoining areas if a heavy body of fire is present (Fig. 2–14).

Location and Extent of Fire

Unlike all of the previous factors, the location and extent of the fire cannot be determined until units arrive at the scene. The original reporting parties might be able to point you in the right direction, but occasionally their information isn't clear. The first reports of smoke may be remote from the source of the fire, particularly in buildings with central air conditioning.

For serious fires, the physical location of the fire can have a great influence on the tactics used in its control. Generally, the lower the fire is in a building, the more serious the hazard, since more of the building is exposed to the vertical, often rapid spread. There are several locations that create special firefighting problems, including the top floors of most ordinary brick and wood-joist buildings, as well as frame structures. The vast majority of these buildings have some type of void space above the top floor, an attic or cockloft that acts as insulation against heat and cold. In some cases, storage areas are located

Fig. 2–12 To be effective, size-up must include reports from roof level. From the street, the garage with the large 4 on the door appears to be a small building, only 20 ft wide.

Fig. 2–13 Seen from above, the building widens out dramatically, and goes back almost 300 ft, a different fire problem than what was thought to be a small garage.

Fig. 2–14 This 3x4 ft hole in the cinderblock wall will be hidden above a hung ceiling when the building is complete, making it difficult to locate during a fire.

there as well. For this reason, treat a fire on the top floor of such buildings with special care. Any fire that enters the void space may burn off the entire roof if the resources aren't present to put a quick halt to it (Fig. 2–15).

A second problematic location is below grade. Whether in a cellar, subway tunnel, or below deck on a ship, fires below grade are more complex than those above. There are two main reasons for this, the first being the lack of opportunities for horizontal ventilation. Stairways are thus turned into chimneys to vent the products of combustion. The second reason is that the entire operation, from entry through attack and even relief (if necessary), must be performed in this atmosphere, limiting the members' working time. A fire on the third floor usually allows members to stage at the level between the second and third floors. Here, masks may be donned immediately prior to entry into the area. Additionally, when a firefighter leaves the fire area, he only has to return here to be in clear air again.

Fig. 2–15 Hidden voids such as cocklofts, which extend over large areas are severe threats to a building's survival. Only a fast coordinated effort kept this cockloft fire from destroying an entire row of stores.

A below-grade fire forces a firefighter to wear self-contained breathing apparatus (SCBA) on entry and to keep it on until he has climbed back out into fresh air. Additional personnel are required to keep a handline advancing. Relief must be coordinated to make sure that fresh troops arrive at the nozzle in time to let the working crew get out without running out of air below grade. Fresh troops shouldn't be committed too early, since this results in a loss of useful work time.

A third trouble location is anyplace that a fire is beyond the reach of ladders for exterior ventilation and access. Such fires can be in high-rises or in low-rise, windowless buildings. The tactics used here are similar to below-grade fires. The conditions aren't usually quite as severe as cellar fires, however, since you can approach from below and may be able to ventilate from the interior in some cases. Still, the problems they present are severe.

At times, determining the extent of a fire can be as complicated as any other factor. Many smoldering fires give the appearance of working fires because of the volumes of smoke that they produce. A few useful hints on evaluating the smoke itself can sometimes indicate how serious a fire is, and they might indicate the location as well as what is burning. Items to evaluate quickly are the location of smoke and flame, as well as the color and movement of the smoke. The location of the smoke is often the first visible indication of a working fire, but it isn't always an accurate indicator of location. Remembering the normal smoke diffusion patterns—up and out—can point you in the right direction, but use your own judgment. Central air conditioning, as well as objects moving within shafts, can move smoke in ways that it doesn't usually go. For instance, an elevator dropping down a smoky shaft will push smoke down ahead of it, as will a train traveling in a tunnel. This may result in reports of smoke in areas that are remote from the actual fire.

Fortunately, these aren't common scenarios, and they can be recognized as soon as the contributing shaft or central air unit has been discovered. Normally, any floor with smoke on it will be at or above the origin of the fire. Don't bypass a floor with smoke without investigating the source, even if the original call reported the fire on another floor. If you encounter a floor that is full of hot smoke and find no fire visible, there is a good chance that you are right above the fire. This is a classic indicator of a cellar fire. If all of the floors above are pushing heavy smoke and there is no visible fire, expect it to be in the cellar. This suspicion can sometimes be confirmed by a quick look at the chimney, a shaft especially designed to vent smoke up and out of the cellar. Normally, heating units burn fairly cleanly. If heavy, dirty smoke is pushing out of the chimney, especially during non-heating seasons, look at the cellar right away.

The color of the smoke can often tell you what is burning. Black smoke suggests the presence of petroleum-based products. Large volumes of black smoke at the roof often signal involvement of the roofing materials. Lighter volumes of smoke at the same level could be the result of a defective oil burner. Light to moderate quantities of black smoke found in basements often indicate an oil burner malfunction. This should signal firefighters to bring a Class B extinguisher with them. Historically, black smoke from a residential building often meant the presence of an accelerant, but this premise is no longer reliable due to the proliferation of plastic furnishings, appliances, and piping.

Common Class A materials produce gray to light brown smoke when sufficient oxygen is present. Reduced amounts of oxygen produce large amounts of dark gray or yellow gray smoke. This is an indication of the potential for backdraft, especially if the smoke is issuing under pressure and is being drawn back into the building.

The movement of the smoke can tell other things about the intensity of the fire. Heavy, rolling clouds, violently twisting skyward, indicate extremely hot smoke from an intense fire within the building. This is frequently followed by the fire igniting through the openings from which the smoke is issuing. Firefighters should use extreme caution when entering to avoid being caught in a flashover. Wispy smoke, usually light in color, indicates a fire in the incipient stage. Quick use of an extinguisher should solve the problem before it becomes a danger. Smoke settling or hanging in low spots is cold smoke, and it is found at fires in sprinkled areas or where a fire is partially or fully extinguished. The cooling effect reduces the smoke's buoyancy; thus, it doesn't rise readily. Partially extinguished fires give off large amounts of carbon monoxide, so make sure that personnel use their SCBA.

Of course, the easiest way to know the extent and location of the fire is when it's blowing out of every opening in the structure. That tends to simplify this point in your size-up. A good firefighter can pick up all of the little signs that aren't as obvious, however, and develop a plan of action to suit the conditions.

Exposures

The protection of exposures is the firefighter's next priority following the life hazard. Although it is often possible in advance to determine likely exposure hazards, you must verify the extent to which they are threatened when determining the location and extent of the fire. The examination should include all six sides of the fire area—front and rear, left and right, top and bottom. On large structures, it is impractical to expect the IC to examine all of these areas personally. All members must be able to identify clearly the area in which they are operating and must be able to communicate accurately the conditions that they encounter. A common system of identifying areas within and around the fire building is needed. One system that has had much success consists of numbering the sides of the building from 1 through 4, beginning at the front and proceeding clockwise around the perimeter. Exposure 2 is the building to the left of the fire building (Fig. 2–16). Exposure 3 is the building to the rear. Exposure 4 is the building to the right of the fire building.

In rows of attached buildings, the same system can be used in conjunction with letters to describe buildings that are further away from the original fire building. In this way, you can avoid much confusion when people remote from the command post (CP) attempt to describe their location. In many areas, a similar system uses letters to do nearly the same thing within the original fire building. In this case, the four sides of the building are described as *sides*. *Side A* is the front of the building, *side B* is the left side of the building, *side C* the rear, and *side D* is the right side. It is critical that all personnel on the fireground are clear about the system in use and where they are located in relation to this system. You do not want to report you are trapped in side B if you meant that you were trapped in the building to the left of the original fire building (Fig. 2–17).

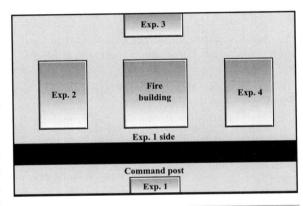

Fig. 2–16 A common system of reference is essential to effective communication on the fireground. Know the system used in your area.

Fig. 2–17 Attached buildings such as garden apartments and townhouses require a modification to the exposure identification system. These are complex fires that require everyone to have a common frame of reference for communications to succeed.

Once you have identified a potential threat to the exposures, you must take some actions to prevent extension. Often the best way to prevent extension is to put out the fire. Occasionally, though, conditions don't readily permit this approach. You may have to allow extremely large fires or fires involving pressurized flammable liquids or gases to burn out on their own. In some cases, it may be possible to move the exposure hazard. Truck and rail cars are both possible candidates for this. Don't forget that it might be possible to move a burning vehicle that is exposing a structure.

When I was a firefighter assigned to Ladder 103 one hectic summer evening, we pulled up to find a fully involved 30-yd dumpster exposing a one-story paper-box factory. The nearest engine company was coming from a considerable distance away. The windows in the factory were starting to crack, and we felt we were about to have extension to the piles of cardboard cartons inside the factory, when our chauffeur pulled the tow chain out of the apparatus, slung the dumpster to the rigs tow eyes, and drove down the street to a vacant lot where the nearest hydrant was located. It was a sight to see that blazing mass barreling down the street. But it was highly effective; the threat to the exposure was completely gone. When these options aren't practical, you must take a defensive stance. Remember the precepts about selecting the most important exposure, not necessarily the most severely exposed. (See chapters 3 and 4 for further discussions of exposure protection and methods.) At times, the condition that dictates taking a defensive mode isn't related to the building or the exposure but is rather a limitation on the following size-up points.

Apparatus and Manpower

All of the previous items relate primarily to the enemy: the fire. After you determine all that you can about those items, you must plan your strategy and tactics based on the available resources. This is where you must begin to develop solutions. How many engines are responding? What can you expect them to accomplish? Are they prepared to place the needed types of streams into operation quickly? How many and what types of aerial devices are responding? Are the units staffed with full crews, ready to be assigned their tasks, or will they be shorthanded, requiring additional companies just to get the basics accomplished? Further, what level of experience do the firefighters possess? Most chief officers, even if they won't admit it, feel a little more confident when they know that they have their first-string players in the lineup. Whether due to a lack of experience or inadequate training or staff changes, certain groups of individuals just don't perform as well as others. The fire doesn't know that, of course, but the officer in command should, and he should take that into account when orchestrating the incident.

Water Supply

Closely tied to apparatus and manpower is water supply. This is usually the primary function of engine companies, but the water supply entails much more than just having sufficient numbers of pumpers available. The water supply must be thought of as a system, meaning that there must be a source of water of sufficient volume to suppress the number of British thermal units (BTUs) being given off by the fire. Another element of the water supply system is the transport system. A typical water supply system consists of a hydrant supplying a feeder line to a pumper, which then supplies several handlines. Many other variations are possible, however, and it takes an alert officer to recognize them all.

Evaluating the water supply normally involves a degree of planning. In areas with uniformly good water supply and minimal fire hazards, the preplan may be as simple as just learning the hydrant locations, since all of the hydrants can supply the low flow anticipated. In other areas, however, planning for the water supply may involve performing flow tests around selected hazard areas or at least learning the location of large mains. Where residual flows aren't sufficient for the anticipated hazard, examine alternative supplies, including drafting sites, well points, relay operations, and tanker shuttles. Don't overlook such sources as private fire pumps taking suction from supplies separate from the municipal system.

How much water is required for a given fire area, and how much will protect the exposure? There are several methods and formulas to determine this. Stated most simply, the water flow depends on the fire load and the area involved. The formulas are based on scientific principles involving the amount of heat that one pound of fuel releases when it burns vs. how much heat water absorbs when its temperature is heated to vaporization. Most residential rooms have a fire load of 5 lbs of fuel per sq ft. Libraries, for example, have a loading of 25–30 lbs of combustible per sq ft. Each pound of ordinary combustibles gives off between 7,000 to 10,000 BTUs totally consumed. Each gallon of water absorbs about 9,275 BTUs when heated from 70° to turning to steam. Since fuel isn't consumed instantly, but burns over a period of time, we can spread the absorption of BTUs over time.

Fire flows are typically determined in gallons per minute (gpm). In theory, 1 gal of water will provide sufficient cooling to quench approximately 5 lbs of fuel at the average rate of surface burning. The fireground isn't a suitable location for attempting to use these rather unwieldy formulas. What is needed on the fireground are some general guidelines suited for the majority of cases. Tests performed by the NFPA, Factory Mutual, and others have determined that flows of 10 gpm for each 100 sq ft of fire area are sufficient to control fires in areas of light fire loads. Such areas include residential occupancies, classrooms, and offices. For ordinary hazard areas, such as most commercial occupancies, a flow of 20–30 gpm per 100 sq ft is recommended. For high fire loads, a flow of from 30 to 50 gpm per 100 sq ft is required.

Using these flows, it is possible to estimate the likely water demand and thus the apparatus and manpower requirements. For example, an engine company responding to a fully involved one-story private house fire that measures 20x30 ft would find that a flow of 60 gpm is sufficient for extinguishment. Similarly, assuming full involvement, a fire in a typical 20x75 ft store would require between 300 and 450 gpm. In still another example, a fire in a tire warehouse measuring 100x150 ft would require a flow of at least 7,500 gpm. How the flows are applied will vary widely. In the first example, the low flow will be discharged easily through a single 1½-in. or 1¾-in. handline. The second example would require either two 1¾-in. or 2-in. lines operating together or a single 2½-in. well-supplied line with either a 1¼-in. solid tip or a comparable gpm fog tip. The two smaller lines offer some advantages in flexibility, whereas the single large line offers benefits in added reach and power. Moreover, the single large line possibly requires one less firefighter. The third example requires a combination of large handlines and master streams.

In actual practice, however, and for a number of reasons, the theoretical flows outlined previously aren't the same as what is applied. Regarding the first example, the 60 gpm flow is no longer a common setting on handline nozzles where efficiency has been emphasized. Modern 1 ½-in. nozzles flow up to 125 gpm, while 1¾-in. nozzles flow 150–200 gpm. The nozzle team will find it advantageous to make use of this reserve capacity, for although 60 gpm will extinguish the fire, an interior advance at this rate will be slower and more punishing. If used properly, the added initial flow rate won't seriously affect the total flow required, since after the initial knockdown, a lower flow rate can be used intermittently to continue the extinguishment. This speedy knockdown is advantageous in reducing damage to structural elements as well as in aiding any search for victims.

Regarding the second example, a fully involved store fire usually means blowing out the front windows or venting through the roof. This means that a lot of fire will be showing and, in many cases, an officer may again opt for the fast knockdown, which often means using a pre-connected deck gun or other fast-acting master stream. This can be particularly attractive at store fires, since the fire is at ground level and there are often large windows to provide access for the stream.

In the case of the tire warehouse, the flow cited is rarely applied for reasons quite different from those given in the first two examples. For one thing, it would be rather unusual to find such an occupancy unprotected by automatic sprinklers, which are noted for limiting the size of fires. It would also be unusual for a fire to involve this large a building before fire units arrive. Catch it small and you'll use less water. For example, if only a quarter of the building were involved, a quick attack by several units applying 1,800–2,000 gpm would produce a good stop and nullify the need for the remainder of the flow. The last reason for not applying the theoretical fire flow is that it just isn't possible or practical. A 7,500-gpm demand is very large for all but the best water-delivery systems. This is why insurance companies and many building codes require automatic sprinklers in such high-hazard occupancies. They recognize that manual firefighting at this stage is a dubious effort.

A number of years ago, I was asked to give a deposition in a civil suit against a mid-western municipality that was being sued by the owners of a paint factory that was destroyed by a severe fire. The owners claimed that the city had failed to provide sufficient water to fight the fire in their plant, because only a single 8-in. main served the area. I testified that based on the volume of fire present in the building when the fire department arrived that there was simply no way to have saved the building, that a fire flow of at least 4,500 gpm would have been required, and that the building (built of Class 2, with cinder block walls and metal deck roof) would not stand up long enough for the fire department to have mounted an attack, even if the 4,500 gpm was available. It takes a lot of time and manpower to set up for that kind of fire flow. We won the case.

A discussion of alternatives is in order here. What happens when it isn't possible or practical to deliver the required fire flow? This could be the case when extraordinary relays are required that aren't likely to be in place before the building has been totally destroyed. In any event, the IC must make a decision as to how best to use his resources. He may dedicate all or part of the water supply to protecting the exposures, his highest priority. Or, he can use his water for public relations purposes, lobbing streams at the fire even though he knows that they are inadequate to extinguish it. He does this because there is a fire, and the public thinks that the department must do *something* even when the firefighters know that it will make no difference at all. This is supposed to soothe the public into saying that, "They really tried, but they couldn't save the old mill." In fact, the old mill was condemned because it lacked both an adequate water-delivery system and automatic sprinklers.

Consequently, there are two immediate ways to reduce the likelihood of such a loss. One is to survey your area in advance, determine the required minimum fire flows, and develop adequate water-delivery systems to apply it. The second much preferred method is to ensure that sprinklers are installed. Often, grading schedules developed by the Insurance Services Offices (ISO), the old Board of Fire Underwriters, have already pinpointed shortcomings for you. Contact your local ISO office for further advice. In the case of the paint factory described previously, the city had ordered a sprinkler system to be installed. The owner ignored the order, had his going-out-of-business fire after an Environmental Protection Agency (EPA) and Federal Bureau of Investigation (FBI) investigation of the business, and then tried to sue the city. Chutzpa!

Auxiliary Appliance

The presence and serviceability of auxiliary systems is an item that deserves high priority in any size-up. As mentioned in the previous section, if automatic sprinklers aren't provided, fires in large structures may exceed the capabilities of manual firefighting. In addition to sprinkler systems, a number of other auxiliary appliances exist, including standpipe systems and foam systems for bulk oil storage plants. Local appliances include halon, CO_2, and dry-chemical systems for special hazards such as restaurant grease ducts and spray paint booths.

Barring any immediate life or exposure hazard that must take priority, first-arriving units must determine whether auxiliary appliances are present. If a system is present, what is its operating status? Is it working automatically? If not, can it be operated manually or bypassed through a fire department connection (FDC)? You must often find the answers to these questions inside the building, and you should relay them to the IC as soon as possible. An out-of-service auxiliary appliance is often justification for an extra alarm, since the fire will have a great advantage while personnel struggle to do manually what the appliance failed to do. For example, a fire on the sixth floor of a building where the standpipe is inoperable will require all of the available members just to stretch the first handline up the staircase and into service (Fig. 2–18).

Fig. 2–18 The 3,000-gpm pump that supplies this test header is a crucial part of this building's fire protection system. First arriving units must quickly determine the status of such systems. Inoperable appliances may justify calling additional resources.

Weather Conditions

Extreme weather conditions usually have an adverse effect on firefighting efforts. It is often necessary to request additional assistance at incidents that would be relatively routine during moderate weather. High temperatures and humidity fatigue firefighters rapidly. Warm weather also brings many more people out into the street, especially in urban areas. Temperatures below freezing result in slower operations. As ice accumulates, members must take greater care in their movements to avoid falls and injuries. In addition, severe cold also causes mechanical failures as hydrants freeze and apparatus behave sluggishly or malfunction entirely. In one spectacular 10-alarm fire in Brooklyn in 1985, frozen basket piping on a telescoping platform allowed fire to extend to an eight-story factory. This occurred even though the tower ladder responded from heated quarters.

High winds can also have a catastrophic effect on fire attack, whipping fire into furnace-like proportions. Such winds can drive attack crews right off the fire floor. They can whip fire right past defensive measures such as hoselines and trench cuts (Fig. 2–19).

Fig. 2–19 Any extreme weather condition should prompt a request for additional resources.

Street Conditions

Street conditions can severely hamper a fire attack. Double-parked cars can slow or prevent maneuvering and hamper apparatus placement. Construction trenches for water, sewer, or gas mains halfway down the block may divide the apparatus from the fire building. In such a case, it is usually best to let the aerial apparatus onto the block, since it can reach over the trench while the engine company hand stretches a line down the sidewalk from the intersection. Other unusual street conditions, such as snow and ice, can slow response times, and deep snows can bury hydrants and block streets altogether. Again, it may be necessary for the apparatus to remain out on the cleared main artery while equipment is hand-carried into the snow-clogged side streets. All such operations are time-consuming and may require additional assistance. This may also require much more hose length than the normal *pre-connect*. Make sure you have enough hose to reach to the most remote part of the building. Special street conditions like trenches, road construction, or even deep snow may require a special arrangement of a hose bed to allow rapid stretching of hoselines of unusual length. Be alert for such *temporary* situations, and have a plan and the necessary hosebed arrangements to deal with them (Fig. 2–20).

Fig. 2–20 Be aware of conditions throughout your response area. Changes to SOPs may be required by conditions such as water main construction.

Hazmats

The presence of hazmats can be one of the most important factors in any size-up. By their very nature, they pose potential problems for firefighters, ranging from health hazards to accelerated fire extension. Hazmats are yet another factor that can be anticipated and entered into a CADS retrieval system. The presence of certain hazmats could mean little difference to firefighting units at, say, a flammable liquids fire. Other types of hazmats, notably poisons and explosives, could halt firefighting operations altogether. Information as to what specific hazards are present must be available to the IC if rational decisions are to be made (Fig. 2–21).

As you have seen, many, many variables come into play during size-up. At a given incident, any one of these items might be only another matter of detail; however, any one of them might also present justification for calling up additional resources when it presents a particular problem. Any item in the size-up can spell a change in the tactics used to control an incident. The importance of relaying all pieces of information to the IC cannot be overemphasized. You can only make the proper command decisions when armed with accurate intelligence from the field.

Fig. 2–21 The presence of a serious health hazard, indicated by the number 3 in the left hand (blue) box of this NFPA Diamond should prompt a very cautious approach to incidents in this building. All property owners should be encouraged to mark their hazardous properties.

03 Engine Company Operations

At the heart of modern firefighting is the ability to go aggressively after a serious fire and extinguish it. Up until the 19th century, firefighting was often limited to protecting exposures and removing goods and fuel in the fire's path. Through the 19th and into the early part of the 20th century, firefighters were able to project powerful streams onto a fire, thus controlling and eventually extinguishing it. Still, several types of fires, notably in cellars and large open-area structures, presented difficulties for these outside streams, which often couldn't reach the seat of the fire. What resulted were long-duration multiple alarms, lasting until either the unreachable areas burned themselves out or collapse occurred, exposing the fire to outside streams.

The use of breathing apparatus and protective clothing, as well as improvements in ventilation techniques and entry methods, have helped to improve our ability to get inside and put out a fire before it destroys the structure and kills the remaining occupants. Thus, interior firefighting has come into its own. This task falls mainly on the engine company, since it is their job to push the fight into enemy territory. While saving lives is the primary concern of all firefighters, the primary function of engine companies is to get sufficient water on the fire area to extinguish the fire. Saving lives may involve rescue via ladders or ventilation to draw fire away from the victims; still, in most cases, the threats to victims are best removed if we rapidly extinguish the fire. In fact, more lives are saved by properly placed streams than by any other means, since that is how we extinguish most fires. As Chief Fred Gallagher used to say when he was the captain of Rescue 2, "Put the fire out and everything else gets better."

In addition to extinguishing fires, hose streams have several other uses, such as during controlled burns. They are also used to protect exposures, to flush areas of flammable liquids and gases, and to absorb toxic fumes from dangerous chemical leaks. Yet all of these are secondary to the main purpose of an engine company, which is to put enough wet stuff on the red stuff to make it go out. At times, this simple objective is more easily said than done. Fire finds numerous ways of compounding the problem—by hiding in void spaces, for example, or attacking with such fury as to overwhelm an inefficient attack. Worse yet, it can threaten civilians, thereby detracting from our efforts to confront it directly. When faced with the situations encountered at structural fires, an engine company must operate within a set of guidelines that focus attention on the primary mission. These guidelines form the basis for the tactics used to extinguish the fire. Before discussing these guidelines, however, it is necessary to review some of the basic points of fire behavior and relate them to the tactics that will be developed.

Fire Behavior and Methods of Attack

Not all fires are alike, and methods of fire attack differ depending on the situation. Fires progress through the following three stages: the incipient stage, the free-burning stage, and the smoldering stage. Each stage is characterized by differences in room temperature and atmospheric composition, and each will respond differently when attacked by the various methods, namely the *direct method*, the *indirect method*, and the *combination method*. Certain types of fires may best be extinguished using one, two, or all three methods of fire attack, while others succumb best to only one particular method. A knowledge of which method best suits a given situation allows a firefighter to perform his duties in the safest and most effective manner.

The incipient stage

In this stage, the fire is still small. The heat and smoke conditions are light, and the fire is confined to its original area. Because of these conditions, attack personnel can approach quite close to the seat of the fire. Most often, the full flow of a handline isn't required (Fig. 3–1). In fact, a hand extinguisher will often halt further extension and prevent the development of flashover conditions. In any case, firefighters should approach the area with caution, using full protective clothing, since the room may suddenly flash over from hidden fire or the ignition of such items as aerosol spray cans. Crews should stretch a 1½- or 1¾-in. handline and, if necessary, charge it as a precaution. The fire will require more than one extinguisher to ensure complete extinguishment, particularly if it is in stuffed furniture. Charging the line provides a continuous supply even though only 5–10 gal may be required, and you can crack the nozzle as needed if the full flow of water isn't necessary. Apply the stream directly to the burning material—the direct method of attack—using either a straight stream or a very narrow fog pattern.

Fig. 3–1 The incipient stage fire should be fought with the direct attack. Use of a water extinguisher can often control such incidents, but a hoseline should be available as a safety measure.

The free-burning stage

In the second stage, the fire has greatly increased in intensity. The rooms are either approaching flashover or have passed it. The ceiling temperatures rise rapidly with the accumulation of hot gases. In a well-involved room, ceiling temperatures of more than 1,300°F and large volumes of smoke are common. The fire is receiving adequate supplies of oxygen to sustain open flaming. This oxygen may be coming from inside or outside the structure. This is the stage in which you will find most working fires (Fig. 3–2). Arriving as the first nozzle team at a free-burning fire is often a heart-pounding affair of swirling smoke and licking flames, plus lots of noise, excitement, and confusion. Now it's time to go to work! Still, before the actual attack begins, the nozzle team has a few tasks to accomplish. They must attempt to locate the occupants, account for them, and obtain information from them. The occupants may be able to tell you about other missing persons, the best way to the fire area, what is burning, and other significant information.

Fig. 3–2 The free-burning stage is a very dangerous period for firefighters. Rapid heat buildup, flashover and rapidly advancing flame fronts all threaten persons inside the structure.

On the way to the fire area, survey the layout of the structure for alternate means of escape, occupants still inside, and any fire that is remote from the main body (it could flare up and cut you off). Examine the layout of the floor below the fire quickly, particularly in multistory buildings such as apartment houses, hotels, hospitals, and some offices. You may be responding to a fire in room 515 on the fifth floor only to find when you get there that the hallway is filled with hot, black smoke. You could be 10 ft from the door to room 515, or you could be 150 ft away. Go down to the fourth floor and locate room 415 to orient yourself to its location from the stairway. A good way is to start from the stairs and count how many doors you pass on the fire side of the hall before you reach the room in question. Doors are among the few useful features that you can distinguish by touch in the dark. Even so, be careful not to pass one by.

The nozzle man must ensure that he arrives at the door to the fire area with sufficient hose to cover the entire area. Generally, one 50-ft length of hose is sufficient for most homes and apartments. Flake out spare hose where it can still be advanced as needed without kinking or causing some other sort of problem. Generally flake out the spare hose on the floor below the fire and on the stairway leading to the fire area, if the fire is above the ground floor. For ground floor fires, usually locate the spare hose outside the structure. The nozzle team should *stretch dry to a safe area* or get as near to the fire as possible before calling to have the line charged. The weight of the water in the hose makes it difficult to pull it over and around objects. For example, one length of charged 1¾-in. hose weighs about 80 lbs, whereas a dry length weighs only 40 lbs. An important distinction to be made is what constitutes a safe area. Usually, in frame or ordinary-construction buildings, this means the nozzle team can advance only to the entrance to the fire floor before opening the stream. For the ground floor of most average-size buildings, this means the front door. For multi-story buildings, it would be the stairway hall or landing. For below-grade fires, the safe area is often on the ground floor at the top of an enclosed stairway. The nozzle team shouldn't enter the immediate fire area without water except to save a life, and they should never start down a stairway into a working fire unless they have water. Coming too close could cause firefighters to abandon the line in the event of sudden extension. It could also result in burn injuries if the area flashes suddenly or if backdraft occurs (Fig. 3–3).

Fig. 3–3 Notice the ignition of combustible materials and gases ahead of the main body of fire. This phenomenon, known as rollover, is an indicator of dangerously rising temperatures at ceiling level, in spite of the fact that the thermometer at floor level only registers 60°!

The nozzle team should arrive at the fire area ready to go, with bunker gear buttoned and buckled, gloves on, and hose flaked out. Members should position themselves on the same side of the line and get down low, on their hands and knees or even on their bellies. There can be a difference of more than 200°F in the zone between 2 and 3 ft off the floor. While waiting for water, the firefighters should stay to the side of the doorway, out of the path of any venting fire, and they must use this time to full advantage. Quite often during this stage of the fire, there is a clear area several inches high below the smoke. The nozzle man and officer should take a few seconds to look in and get the layout of the area. Look for any overcome victims. If there aren't any, look for the glow of the fire, which sometimes reflects along the floor. Also look for obstructions such as furniture you'll have to move around. Look for fall hazards, open stairs or burned-through areas, and signs of fire below you, like flames coming up through the floor, especially around doorways and along walls. This may be your last moment of decent visibility for a while. Once the attack commences, the smoke and steam will bank right down to the floor until properly vented. It is far easier to move toward the fire if you have an idea where it is in advance rather than working strictly by touch.

As the line is charged, the nozzle man must ensure that the nozzle is slightly cracked to bleed off air from the line. When he receives water, the nozzle man should bleed further to ensure that a steady supply of water is available and that the nozzle is set at the correct pattern, either in a straight stream or a very tight fog. After a final check with

the backup man and an okay from the officer, the team can advance. When you begin to enter the fire area, try to stay off your knees to avoid painful burns. *Duckwalk* by squatting on your haunches. This is a very unstable position, and requires practice and teamwork to advance a flowing handline in this manner, but it greatly cuts the rate of burns to the knees. In this free-burning stage, fire can extend rapidly. The nozzle man must be alert to begin applying water, but he should also avoid opening the line on smoke. A good rule of thumb is to open up only when you see flame. An exception to this might be when extreme heat and no visible fire is preventing your advance, such as when attempting to descend a cellar stairway. If the visibility is zero, you will have to rely on your sense of hearing and the feel of the atmosphere to tell whether the line is having any effect. Listen for any violent reaction or hissing to indicate that you are hitting the fire area. In addition, when you hit the fire, you're likely to feel a blast of heat as steam and smoke are pushed back toward you. This should subside if you raise the angle of your stream. Use caution in this situation, however, since there's nothing like being able to see the enemy to know what it's doing. Position an observer at the entrance to the fire area to warn you of any danger arising behind you (Fig. 3–4).

Fig. 3–4 The combination attack uses the reach of the hose stream to cool the combustible gases at ceiling level ahead of the nozzle team's advance. (Note: This is a nozzle handling drill, not a real fire.)

The method of attack most often used in the free-burning stage is the combination method. This initially consists of sweeping the ceiling with the straight stream in a side-to-side or a clockwise circular motion. This ceiling-level attack is only done for 5–10 sec, just long enough to cool the hot gases that have built up overhead and to prevent them from igniting. After this short burst on the ceiling, lower the angle of the stream to sweep all of the burning materials in the room. After a few moments, the fire should be darkened. Then you can shut down the line and give the room a chance to lift. The remaining heat, steam, and smoke will escape and be replaced by fresh air as natural convection currents resume their flow.

The method described previously works well for the single residential room and contents fire, which is typically only about 10 × 12 ft, where the approach right up to the door is possible. Most times the fire can be extinguished right from the doorway, without even entering the room. Larger rooms or multiple rooms on fire are a different story however and require a continuing attack as the line is advanced to allow the stream to strike all of the involved surfaces. Begin the attack the same way as for single rooms, at the ceiling, for 5–10 sec, then lower the angle of the stream to sweep the contents of the room until the fire darkens down, usually only a matter of another few seconds if you're flowing enough water.

Before shutting down the nozzle though, use the straight stream setting to sweep the stream into the next involved room, and then use the straight stream to sweep the floor from side to side along the nozzle team's path of advance. *Sweeping the floor* with the straight stream cools any hot embers and physically washes away much debris such as broken glass, discarded hypodermic needles, and other detritus that you'd rather not have close contact with. In an advanced multi-room fire, the nozzle man should set up a steady pattern with the stream, sweeping the ceiling from side to side in a Z pattern, pushing the fire farther and farther away from the team. Then the nozzle man should lower the stream to sweep the room, repeating the Z motion, then drop down to sweep the floor, advance the line a few feet, and go back to the ceiling, starting the pattern all over again. This continues until the area is knocked down or until the nozzle man requires relief.

As an alternate to the *Z* pattern, the nozzle can be rotated *clockwise* from ceiling to right wall, to the floor, to the left wall, and back to the ceiling, continuing in this fashion as the line advances. When advancing, be careful not to bypass any side rooms that can trap the nozzle team if fire blows out of them after the nozzle team has passed. Fog streams in particular can draw substantial amounts of heat and smoke from behind the nozzle if a room that has not been completely extinguished is accidentally bypassed. If you are using a fog stream and find that you seem to be getting a lot of heat from behind you, halt your advance, switch to straight stream and try to hold your position while a member of the team investigates the situation behind you. Sometimes simply switching to straight stream relieves the situation as the draft is reduced.

The combination attack is highly suited for the second, free-burning stage, with its high ceiling temperatures and large volumes of flame, yet sufficient oxygen and relatively low floor-area temperatures. In striking the ceiling initially, the combination method cools the hot, flammable gases overhead, preventing ignition and subsequent flashover. Then, by dropping the stream, you cool the burning materials, causing them to cease production of any further gases and flame. Remember that flame is simply the burning of the gases released by a fuel. Until you cool the fuel sufficiently, it will continue to give off more flammable gases. In the well-involved room fire, the contents of the room, not the ceiling, are the main sources of fuel. The wall paneling, furniture, and carpeting are all producing more heat and flammable gases. By first bouncing the stream off the ceiling, you not only cool the area overhead, you also allow excess water to disperse on the main sources of fuel. This method has proved very successful and has the following advantages:

1. By using a straight stream or a very narrow fog to hit the ceiling first, you cool the fire gases nearest the line without creating a lot of steam, which could obscure your vision or force withdrawal. This process also stops the travel of fire and pushes it back toward its area of origin without causing a buildup of higher pressures in the fire area, which can occur if a wider fog pattern is used. By then dropping the angle of the stream, you cool the burning materials without having to be right on top of them.

    To begin a direct attack on a free-burning fire would endanger the nozzle team. If the stream were initially directed toward the burning materials, the fire would be driven up and along the ceiling over the hose stream, coming down either on top of or behind the nozzle team or igniting other objects in its path. The combination method avoids this rollover effect.

2. By shutting down the line when no further fire is visible, you won't disturb the thermal balance too severely within the fire area. Quite often, the ceiling temperatures in a fire room will reach 1,200–1,500°F, while those at the floor are in the range of 125–150°F, a survivable temperature. This tendency for heat to rise can be put to good use if you coordinate ventilation with fire attack.

    When you have darkened down the fire, and if you have shut down the stream, the heat, steam, and smoke will tend to lift a few feet off the floor, improving visibility and allowing the hoseline to advance. It will also allow any remaining hidden fire to light up so that the overhaul team can locate and extinguish it.

    The combination attack seeks to avoid disturbing this thermal balance in the event there are people within the fire building—firefighters as well as potential victims. By keeping the heat at the upper levels and the cooler air at the lower levels, the combination attack helps preserve the chances of survival. The third method of attack, the indirect method, doesn't maintain this thermal balance. Instead, it violently disturbs it, rendering this method (also called the Layman theory of indirect attack) unsuitable for interior firefighting in occupied buildings.

3. The combination attack puts the firefighters inside the structure near the seat of the fire. They are therefore in the best position to accomplish their primary goal, which is to save human life. Remember, human life, the first order of priority, far outweighs any property value. This placement usually puts the hoseline in a position to push the fire back toward its origin or out ventilation openings, which also protects the greatest amount of property as well as life.

The disadvantages of combination attack include the following:

A. The attack puts members within an extremely hostile environment. As soon as firefighters enter an involved building, they are exposed to much greater risks than they might encounter outside the building. When human life is at stake, firefighters must risk this degree of danger.

B. The combination attack requires live-fire training to develop a high degree of proficiency. Two errors are common in this attack. The first is that the nozzle man forgets to check his stream pattern and opens up on the fire in a wide fog pattern. This can be catastrophic, since the resulting steam can chase the nozzle team out of the enclosed area, possibly resulting in steam burns. In addition, the steam could kill any unprotected occupants, and it could push fire into previously uninvolved areas. This lesson will almost never be forgotten by anyone who experiences it.

The second error involves leaving the stream up on the ceiling too long without dropping the angle. This doesn't serve to cool the fuel, with the result that the stream doesn't seem to be making any progress. All the initial attack on the ceiling is supposed to do is cool the hot gases that are traveling along that area, either flame or extremely hot smoke that contains flickering tongues of fire rollover. As soon as that overhead fire is extinguished, the line should go to work on the real source of the overhead gases—the rest of the room. Although the change is visible to the firefighters, sometimes they may not recognize that the line has had its desired effect and that they must now change their angle of attack

The smoldering stage

In the third stage, the fire is no longer burning freely. There has been a serious fire within the premises that has burned up nearly all the fuel and is approaching self extinguishment, or else there is plenty of fuel left but the fire has consumed most of the available oxygen, and the fire is now being limited in size by the lack of oxygen. The second case can be particularly dangerous. In this case, there is an overabundance of heat due to the lack of oxygen (the atmosphere throughout the area is more than 1,000°F) and the building is literally filled to capacity with flammable gases, yet due to the lack of oxygen no fire is visible. This is an extremely unusual and dangerous situation, with a high potential for a backdraft explosion. With today's emphasis on energy conservation and efforts to make buildings more airtight, this is a situation that is likely to become more common. The only ingredient necessary to ignite this explosive mixture is oxygen, which can be added when firefighters enter the building to begin the attack or if ventilation is performed in the wrong location. Obviously, opening the front door to advance a hoseline for either a direct or combination attack admits oxygen to the fire area. It also exposes firefighters to a highly charged atmosphere where things can rapidly go wrong (Fig. 3–5).

Fig. 3–5 Vertical ventilation is crucial to preventing backdraft and greatly assists an interior attack.

If you suspect that a backdraft explosion is possible, the first tactic that should be attempted is to vent the highest portions of the affected area. Usually that refers to roof ventilation, which may be fine for a fire in a one-story store, or for top floor fires in larger buildings, but which would have absolutely no effect on a fire in the ground floor of a two- or three-story building. At other times, firefighters may encounter conditions that make topside ventilation impossible, or at least impractical. Cutting and pulling an adequate ventilation opening on a wooden roof takes several knowledgeable, determined firefighters, sufficient power saws, and time! On other roofs cutting a hole may not be possible or safe.

I operated at two very similar fires where this was the case. Both involved supermarkets after closing. In both cases, the stores had been closed for at least 6 hrs before discovery of the fire, and heavy smoke was pulsating from every opening. Horizontal openings were limited to front show windows, which were covered by steel roll-down security gates. As mentioned earlier, roof ventilation should be given the highest priority at incidents such as these, which had obvious backdraft potential. In both cases, however, the normal venting procedures failed to accomplish the desired results. In the first case, the building's roof had been covered with 1/8- in. steel plating on top of the standard wood roof. This is becoming an increasingly common method in high-crime areas and in high-value buildings to prevent burglars from gaining access via the roof. In the second case, roof ventilation was delayed by the presence of a rain roof—a complete new roof structure built just above an existing leaking roof. In both cases, no alternatives existed but to begin fire attack prior to roof venting (Fig. 3–6).

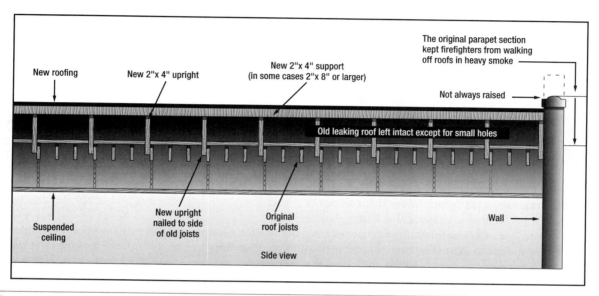

Fig. 3–6 A rain roof is an extremely dangerous alteration to an existing building. It adds undesigned weights to the structure, creates a second hidden void space that is impenetrable from below, and prevents effective ventilation of the structure below..

If roof ventilation is not possible or will take too long, consider using the indirect attack. Most potential backdrafts seem to occur in commercial buildings at night, which have had a substantial time to *get cooking* and develop to the third stage. There is rarely any need for an aggressive lifesaving effort in the fire occupancy. The indirect attack, pioneered by Chief Lloyd Layman of the U. S. Coast Guard Firefighting School for use on open-burning shipboard fires, lends itself to application at this type of third-stage fires. This is because most of the elements needed for a successful indirect attack are present at third-stage fires and are, in fact, commonly found only at this stage. These basic requirements are as follows:

1. high heat conditions throughout the area
2. limited ventilation of the fire area
3. a point on the perimeter in which to make a small opening for the injection of a 30° fog pattern into the superheated atmosphere

As you can see, high heat and limited ventilation are two classic indicators of a third-stage fire. In this case, create a very small opening in a window, door or roll-up gate and direct a 30° fog pattern through this opening, sweeping the ceiling in a side-to-side motion. Under these conditions, when water fog is injected into a superheated atmosphere, the water is almost instantly vaporized and turned to steam. In doing so, it absorbs huge quantities of heat from the fire and the surrounding areas. It also expands tremendously in volume. Each gallon of water can expand to 1,500–2,000 gal of steam when fully vaporized. This steam will rapidly fill the fire area, pushing smoke and heat ahead of it and forming an inert atmosphere even in areas remote from the point of water application. Keep the stream open until the volume of steam being expelled around the nozzle begins to diminish substantially.

If conditions permit, proceed with opening up and then advance using the combination method to extinguish any remaining fire. If resources do not permit a rapid follow-up and advance of sufficient hose streams for immediate follow-up, then it is best to leave the area sealed and let the steam soak in. This steam atmosphere must have several minutes of soaking time to have the desired smothering effect. Otherwise, if ventilation is performed prematurely, reignition may occur throughout the superheated area. In the case of the supermarkets described previously, an indirect attack was made initially and then rapidly converted to a successful combination attack. Both well-involved supermarket fires were quickly controlled with two 2½ in. handlines. Extensive overhauling and smoke removal efforts were required, of course, but personnel sustained no injuries and damage was confined to already-involved areas (Fig. 3–7).

Fig. 3–7 If the potential for backdraft is suspected in a commercial building, use of an indirect attack through a small initial opening should create a steam filled atmosphere. This will reduce the potential for backdraft when the larger opening is made for entry.

Although this indirect method of attack can work in certain circumstances, it isn't practical for the majority of structure fires for a number of reasons. The first and most important reason is its total unsuitability for use on areas that pose a potential life hazard. The conditions that make it impossible for the fire to exist also make human survival impossible. This fact, along with the soaking time required and the subsequent delay in searching for and removing occupants, poses an undue risk to human life. The second reason is the unlikelihood of finding all of the necessary ingredients present simultaneously. As all methods of attack have prerequisites, so too does the indirect attack. The four key requirements are

1. no occupants
2. limited ventilation
3. high heat
4. limited size of the potential fire area

Although there is a growing trend toward sealing buildings, fire has the ability to break through glass, thereby self-venting. If you arrive at a building to find fire or smoke venting out several openings, chances are an indirect attack may not be successful. Next comes a virtual contradiction in requirements. First, the fire must be hot enough to create the needed steam conversion. I have operated at several third-stage fires, including an after-hours nightclub, a large dress factory in a loft building, a garden apartment, and a six-story apartment house. In all four cases, the fire had subsided to the point where the heat condition had dropped below that required for either backdraft or to successfully convert a large quantity of water to steam.

On the other hand, the fire must not be so large as to overwhelm the capabilities of an indirect attack. Large-area commercial structures are unlikely locations for this tactic, since the steam-generating requirements are greater as the building increases in size. Difficulties occur in selecting points of attack on large structures, and at the same time, providing the needed confinement. If the fire has self-vented through a large number of openings, the indirect attack is not likely to work. Using a single large-flow stream to produce the required volume of steam may extinguish the fire in a small area, thus reducing the source of superheated air needed to create the additional steam for the remaining area. As in all methods of fire attack, too small a flow will have a negative effect on operations and do little to extinguish the fire. Too low a flow is likely to push fire into other areas. This fact, coupled with the delay in entering the building while the attack crew waits for the steam to work, could spell catastrophe when expected results are not achieved.

It is important to recognize that the use of the indirect method of attack alone may not prevent a backdraft explosion. In all potential backdraft situations, the application of water should, if possible, be timed to occur after topside ventilation. Although this will tend to lessen the high degree of confinement needed for the indirect method to be most effective, it will also funnel the fire gases into the proper channel, letting the hottest gases out first while the steam follows upward convection currents, extinguishing fire along the way. In addition, this coordinated attack greatly reduces the danger to firefighters from the potential backdraft.

Of course, there are going to be situations where both plan A (roof ventilation) and plan B (indirect attack) will not work for any of a million reasons. The most common situation occurs at a store fire in a middle store of a strip mall or taxpayer with a metal deck roof that makes roof cutting too dangerous and which has only large plate glass doors and show windows for access points. It is not possible to guarantee making only a small hole in an 8x10-ft plate glass window when we want to. You may inadvertently make a very large hole, admitting fresh air that could cause a backdraft. If this is the case—and we suspect backdraft is possible but we can't vent the upper areas and can't apply fog streams through small openings—then we will have to simply go ahead and let the fire do whatever it is going to do while ensuring we don't have anyone in danger's path.

In this case, order a 2½-in. line charged and positioned a substantial distance away from the danger zone (outside the potential collapse zone and out of the blast zone that could be affected by flying glass). If possible, have everyone clear out of the front of the building, position the hoseline and all personnel behind the apparatus or other parked vehicles. When the nozzle has been thoroughly bled and is ready to operate, the nozzle team signals one member who will create the initial opening. This member should be located off to the side of the involved store and out of the path of the exploding fireball and flying glass. The member should use a 6-ft or longer hook, swinging it like a baseball bat at the intended window. The nozzle man should coordinate his actions with the ventilation member, so that when the window glass goes in, so does the hose stream. In this manner, the likelihood of the backdraft is reduced. But if it does occur, our personnel are out of the most likely danger zones. After the hose stream has been operating for 30 sec or so, the likelihood of the backdraft occurring (in this area) should be over. Note, however, that a backdraft can still occur in other areas as they are opened up (Fig. 3–8).

Fig. 3–8 When creating the first opening into a sealed area, position all personnel out of the possible blast and collapse zones and have a 2½-in. line begin immediate stream application as soon as the windows are vented.

One of the misconceptions about backdrafts is that they don't occur at fires that are free-burning. That is not entirely correct. A backdraft will not occur in an *area* that is free-burning, but a backdraft can still occur in adjacent areas that have not yet been thoroughly vented. A number of firefighters have been seriously burned and even killed by this occurrence. The most common situation where this occurs is at fires on the top floor of apartment buildings, schools and other Class 3 buildings that have a cockloft that extends over the entire top floor. A serious fire in the top floor rooms fills the cockloft with superheated smoke, driving out or consuming the oxygen in this area. An unsuspecting firefighter who pulls a part of the ceiling down before roof ventilation is accomplished can be caught by a ball of fire blasting down on him from the cockloft. This is exactly what happened at a fire in Queens, NY in 1995. As shown in the photos below, fire had vented out of two windows at the rear of the exposure 2 side of this 75 x 150 ft apartment house prior to the FDNY's arrival. Firefighters searching for occupants and fire extension were seriously burned when a hole was made in the ceiling of an apartment on the front of the building, two apartments, and approximately 50 ft from the seat of the fire (Figs. 3–9 and 3–10).

Fig. 3–9 On arrival, heavy fire was found venting from the two left windows on the exposure 2 side of this building. This was a fully occupied apartment building in the middle of a Sunday afternoon.

Fig. 3–10 As firefighters searched the top floor, two apartments over from the fire apartment on the exposure 1 side, a backdraft occurred in the cockloft, blowing fire down on several firefighters, severely burning one member and forcing him to dive out the 4th window from the left (with the smoke stains above).

Some advantages of the indirect method of attack are the following:

1. It can reduce firefighters' exposure to potential backdraft situations.

2. Under the proper circumstances, this method can extinguish fires in areas where the heat condition denies entry to firefighters. Fires in the holds of ships, railroad boxcars, and shipping containers are often candidates for this method.

3. At times, the indirect attack allows a very limited crew to extinguish more fire than would be possible using the combination attack. The indirect attack can extinguish fire in areas remote from the point of water application. If a direct or combination method were used, the hoseline would have to be advanced to this remote area, usually through the interior, a slower process. In addition, ventilation support isn't as critical in the indirect attack, so these personnel can be used to press the attack. It is important to remember, of course, that this is only possible if conditions are right.

4. The indirect method often uses less water for extinguishment. When operating from an apparatus booster tank or in other situations of limited water supply, the efficiency of fog—both to absorb more heat and to produce the blanketing effect of steam—can reduce the required volume far below the amount required for a combination attack.

5. Water damage may be less than occurs using other methods. When conversion to steam is accomplished most efficiently, nearly all of the water is used for extinguishment, and there is very little excess runoff.

Some disadvantages of the indirect method of attack are the following:

1. It cannot be used in an occupied building. To do so is to concede the loss of any remaining occupants. The resulting atmosphere is untenable even for firefighters in protective clothing. Just as the steam progresses to remote areas as it extinguishes the fire, so will it affect the people in its path.

2. The presence of ventilation openings will dilute the effect of the steam. If enough windows or doors have vented prior to water application, no steam buildup is possible (Fig. 3–11).

3. A discharge of less than the critical volume can push fire ahead of the steam, an effect that may not be observed from the outside. Blowing fire into vertical or horizontal voids is particularly dangerous since the fire may be in control of so many voids by the time the situation is recognized that opening them all is impossible.

4. It isn't possible to view the interior layout until you have gained control of the situation. Unlike the direct and combination methods, it isn't possible to get an idea of what the floor plan is like by looking in below the smoke. The members will be operating in very poor visibility once they do enter.

5. Using the indirect method can increase water damage. If a company uses this method in an inappropriate situation, it may result in soaking materials that aren't threatened by the fire. Fires in mattresses and overstuffed furniture produce tremendous amounts of smoke. If an inexperienced member arrives to find heavy smoke showing from every crack yet fails to evaluate the level of heat, he may begin an indirect attack that is doomed to fail. These types of fires don't normally develop the superheated atmospheres required for successful steam conversion. The member could be operating the stream into a room that isn't even in the fire area.

Fig. 3–11 A fire that has control of the oxygen supply is not a candidate for the indirect attack method.

As you can see, fires differ drastically in *type* as well as *method* of attack. By learning to recognize each stage and selecting the most appropriate method of attack, firefighters will greatly improve their skill as well as their chances of successful rescue, fire control, and property conservation. Now let's look at the various guidelines that will influence your selection.

Choose the Proper Operating Mode— Offensive, Defensive, or No Attack

Just as the different stages of fire development can influence which method of stream application to use, other factors can affect the overall strategy used to fight each fire. The three possible choices are the following:

1. to initiate an offensive attack

2. to establish defensive positions

3. to take no action at all

Each is generally dictated by factors beyond the firefighter's control—the size of the fire, the threat to exposures, and the life hazards if personnel are committed to the operation. An offensive operation proceeds on the assumptions that sufficient resources are available to perform the needed tasks and that the threat to exposures is minimal or can be controlled. An offensive operation is interior, for the most part, although in larger fires, exterior streams can be used for the attack. Elevated platform streams are excellent for an exterior attack on large fires. Defensive operations are generally exterior, with the highest priority being protecting exposures. A fire in an occupied building obviously isn't the place for defensive tactics, due to the life hazard.

Usually defensive operations are dictated by a massive body of fire that is beyond the control of responding units. In this case, if companies cannot reasonably expect to gain control of or extinguish the fire before it spreads, they must instead position themselves to protect exposures, even if this is only a temporary measure until additional units arrive. The units should select positions where they can accomplish exposure protection and alternately hit the fire, if possible, while being prepared to shift into an offensive position when the needed resources become available. Forget the old tactic of creating a water curtain between two exposures, radiant heat can pass right through a water curtain and ignite the exposure. Exposure protection is best accomplished by coating the exposed surface with water. In this way, the water carries away the built-up heat as it flows down the surface. Water is a poor absorbent of radiant heat; thus, most of the radiant heat will pass right through a water spray. Putting water directly on the surface allows it to absorb heat by conduction. In addition, the water protects against direct-flame impingement. These two means, radiation and direct-flame exposure, are responsible for most of the spread of fire to exposures. A word of caution on stream application: avoid using high-velocity straight streams at close range. Such streams will break windows, greatly aiding fire spread. Use a fog stream or the splatter of a straight stream from a distance to keep the members farther away from the danger area.

The last option is to conduct no attack at all. This is sometimes necessary when conditions pose an undue threat to firefighters' lives, such as a potential boiling-liquid, expanding-vapor explosion (BLEVE), a potential danger at all incidents where fire impinges on closed containers of liquids. It is also being recognized as an important tactic at fires in warehouses containing poisons or pesticides, where the runoff from hose streams would spread toxins over wide areas. In this case, letting the fire burn may be the safest, least costly, and fastest means of resolving the incident. A partially extinguished fire in these occupancies is often more toxic, more enduring, and costlier than one that is allowed to burn freely. This isn't a decision to be made lightly, and the officer in command should be well advised of the possible repercussions. Generally, however, if you explain the situation to the building owner, he will agree with your decision.

When Human Life Is at Stake, an Offensive (Interior) Attack Is Mandatory

There are people who question why the FDNY committed hundreds of firefighters to the WTC Towers in the face of such heavy fire and potential collapse. The answer is simple. This was an occupied building with thousands of people trapped. That is why they called the fire department instead of the sanitation department. It is our duty to go get those people. Thousands of people survived because of the FDNY, including all but approximately 200 people below the impact points, in spite of being trapped in elevators or cut off by fuel in elevator shafts. Even after the first collapse, hundreds of people were led to safety through blocked stairways and crashing debris.

As long as you suspect that people are still alive inside a building, you will have to go in after them. A cardinal rule of aggressive firefighting is, "Don't use an outside stream in an occupied building." To do so is to push fire and its byproducts right back into the structure, where you expect to find live victims. This narrows the mode of operations down to the offensive method.

The decision to change from this mode as operations progress should be made using the same logic. "Are there any more people inside who can be saved?" As long as the answer to that question is "yes," all of the needed resources must remain committed to the task of saving lives, even though the fire may extend greatly during the time it takes to perform rescue. When the answer is "no," however, then the risk to firefighters' lives should take precedence, and fire forces should be withdrawn to safe areas. Philadelphia Fire Commissioner William C. Richmond's actions during the famous movement (MOVE)[1] siege were the ultimate example of this concept. Richmond ordered firefighters withdrawn from the danger area, even though it meant allowing almost two blocks of row houses—more than 60 buildings—to be destroyed by fire. He made the decision based on the belief that 11 occupants of the original fire buildings, all of whom perished, weren't savable. The occupants were hardcore cult members who had barricaded themselves inside one of the buildings and had fired on police officers surrounding the premises. In this case, the lives being protected were those of the Philadelphia firefighters. One year after the devastating fire, all of the properties that had been destroyed were rebuilt.

Begin Suppression as Soon as Possible

This simple guideline should be a cornerstone of engine company philosophy. The sooner you get down to putting fire the out, the sooner things will get better. Although actual extinguishment efforts may have to be delayed due to the need to perform rescue or to protect exposures, the engine company must always take the earliest opportunity to put out the fire. It makes every other fireground task go much faster and smoother. Search, rescue, removal, overhaul, and salvage—all are more easily performed when the heart of the problem has been extinguished.

Get the First Hoseline in Operation to Cover the Worst Case before Stretching Additional Lines

A major downfall of many engine company operations is that crews try to do too many things at once. As a result, all of the tasks proceed more slowly, some don't get done at all, and certain actions become counterproductive. The number of personnel on hand, the height and area of the structure, the location of the fire, and the distraction of other operations all influence the time it takes to position a hoseline. Get the first line in place between the fire and the occupants as soon as possible. If necessary, commit all available personnel to this task. Once the line has reached the fire area, the normal nozzle team will be sufficient to carry out the operation, and any excess personnel can return to the apparatus to stretch additional lines. If the building is unoccupied, place the line in a position to confine the fire.

At a fire in a one- or two-story private home, an engine company arriving with a driver and four firefighters might be expected to place two 1½-in. or 1¾-in. preconnected handlines almost immediately, if necessary. Take the same volume of fire, put it on the fifth floor of a building that has no standpipes, and the same crew would be doing well to have one line in operation 5 min after arrival. The same is true of fire in any building out of the ready reach of a short (200 ft or less) preconnect. Buildings with large floor areas, even without life hazards, demand six, seven, or even more firefighters to drag 300+ ft of hose up stairways, down corridors, and around corners. Remember, concentrate on getting that first line into operation before you split crews or start another difficult stretch with an insufficient number of personnel.

When an Attack Is Stalled, Increase Ventilation, Water Flow, or Both; if Unsuccessful, Change Tactics

Firefighters making an offensive attack should be moving forward almost constantly. Being stationary is as bad as going backward. It means that the fire is winning the contest. While the hoseline remains stationary, the fire keeps assaulting the building, undermining its structural integrity and churning out by-products of combustion. The nozzle team must defeat the enemy or be defeated. If the team stands still for too long, the fire will outflank them, spreading through hidden voids until members are forced to withdraw. The engine company must recognize the tendencies of fire and take steps to forestall its progress.

In this day of modern breathing apparatus, only two things should prevent a hoseline's advance—heat and flame. In the past, smoke did as well, but a universal SCBA policy should solve that problem. In all but a very few cases—notably in tunnels, the holds of ships, and large-area structures—heat will be the main problem.

The solution to the problem depends on the root cause. If heat is what's thwarting the advance, as opposed to a visible body of fire, the answer probably lies in performing additional ventilation, preferably opposite the advancing hoseline. On the other hand, if flame is impeding your advance, chances are that the water that you are applying to it isn't enough to do the job. Call for the pump operator to increase the discharge pressure. If practical, this works best with automatic fog nozzles. It is less efficient in fixed-orifice nozzles, such as solid-stream tips and non-automatic fog nozzles. This is a quick method to get a small additional flow. For larger flows, the most practical method is to commit a second hoseline to the same position as the first one. The two lines operating side by side should be able to advance. A good rule of thumb when committing this second line is that it should be at least as large as the original hoseline—that is, back up a 1¾-in. line with either another 1¾ in. or a 2½ in. Never back up a 2½ in. initial line with a smaller one. When in doubt, go for the larger line, especially in anything other than a residential building.

The methods described previously may be used either singly or in combination to get the hoseline moving forward. Should they fail to permit advance, you will probably have to change tactics. Something highly unusual is taking place, for it takes a lot of fire to resist a hoseline operating in a ventilated area. High winds could be fanning the fire, accelerants could be feeding it, or the original hoseline crew could be physically spent from their efforts. In any case, what you have done so far hasn't worked, nor is it showing signs of working.

You will have to try something else, or sooner or later you will be chased out of the building. The officer in command is responsible for deciding what alternative courses to follow and giving the necessary orders. Perhaps what is required is a change in the direction of the attack, from the leeward to the windward side of the fire. In other cases, temporarily withdrawing units to a safe area while an extra-heavy body of fire is darkened down with a master stream may allow the handlines to advance and complete extinguishment. In still other cases, fire that is inaccessible to firefighters may require an unusual attack, using cellar pipes, distributors, or high-expansion foam, for example. The officer in command must be flexible in making these decisions. Don't become so captivated by the visible fire that you focus solely on its location. Many times, alternate, albeit unorthodox, means of applying water will have very impressive results. For example, at several fires in apartment houses, I have seen hoselines pinned down in the hallway by flames that were blown at them by high winds. The lines were therefore prevented from reaching the door to the fire apartment. In several instances, a line was brought into an adjoining apartment, and a small hole was breached in the common wall. This line was then able to darken down the heaviest bodies of fire, thereby allowing the pinned lines to advance.

As another example, I recall operating at two commercial fires where, despite the best efforts of firefighters for more than 20 min, flames were pushing personnel back. In both cases, access to the fire areas by hose streams was extremely limited. A last-ditch suggestion was made to apply water by distributor from the roof, the only readily accessible location. In the first instance, heavy fire existed in one portion of the cockloft over a one-story,

125-ft-wide, 150-ft-deep supermarket. The fire had originated in the rear of the store, over the meat section. The ceiling over this entire 30-ft-deep, 125-ft-wide area was covered with Formica glued to ½-in. plywood to facilitate sanitary cleaning. This ceiling construction prevented ready removal from below, and a partition in the cockloft hindered access by the hose stream into this area from the remainder of the store.

Unfortunately, this partition didn't hinder the spread of fire into the rest of the store. Here, faced with a fire that was inaccessible from the four sides and the bottom, the IC rejected the suggestion to use distributors and cellar pipes from the only accessible area, the top. All of the needed equipment was readily available, since three power saws were operating on the roof. Instead, the firefighters were ordered to keep trying the same tactics that had already proved unsuccessful. The result: the fire was darkened down from above anyway by a tower-ladder stream after fire had burned away about a 50 x 100-ft area of the roof. When it was recognized that the initial, normal tactics had been unsuccessful, prompt application with just one distributor would have limited fire damage to a 30 x 30-ft hole.

In the second case, we encountered a sizable body of fire in a one-story, 50-ft-wide, 125-ft-deep windowless warehouse. Access was limited to a single door and a rolling steel overhead garage door, both located on the front wall. The first hoseline was unable to advance more than a few feet inside the front door because of intense heat. Roof ventilation operations determined that the fire was in the rear of the building, but even with the additional ventilation, the line wasn't advancing. Fire was beginning to progress toward the front, pushing the line farther back. A suggestion to apply water to darken the flames from the roof was rejected. Instead, the officer in command elected to attempt to knock several holes in the brick and cement-block walls for stream application points. This is a much slower operation than using a power saw to make a 1x1-ft hole in a wood roof. In this case, despite the operation of two elevating platform streams, two ground master streams, one ladder pipe, and several 2½-in. handlines, fire took possession of the entire structure. The collapse of the building created a convenient commuter parking facility.

These examples illustrate the need for flexibility when faced with other than routine situations. Various alternatives may be available to an officer in command, but if he fails to recognize the need for a change in tactics or select and implement a viable tactic, the operation is almost certainly doomed to fail.

When an Effective Offensive Attack Isn't Possible or Hasn't Succeeded within 20 Min, Prepare Defensive Positions

The need to confine the fire, either inside or outside the building, is a high priority in the strategy of fire attack. It is an effort made even in buildings where there is no civilian life hazard, provided that it doesn't unduly endanger personnel. Firefighters, however, must occasionally be reminded that they aren't superhuman. The most common way for them to realize this is when a building is destroyed despite their best efforts. Firefighters who have spent any appreciable length of time in the service can recall a building or two that they felt could have been saved. To be sure, errors in strategy or tactics may have played a part in some of these losses. In other cases, however, the building was condemned to die long before the fire department received the first alarm. This result cannot be justified, however, when the fire continues to spread even after the department begins operations.

Barring any unforeseeable circumstances, such as stored explosives, there is little excuse for fire to extend beyond the structure of origin. Yet time and again it happens, *after* intervention by the fire department. A large part of the problem rests with the IC's failure to conduct a continuing size-up of the fire problem. Conditions on the fireground are dynamic, changing almost continuously. So, too, must be the evaluation of those conditions.

What looked like a single-bedroom fire in the rear of a one-story home somehow managed to surprise the officers in command when fire broke out in remote areas, threatening to extend to exposures. In almost all cases, when fire does extend to exposures, the cause can be linked directly to the IC's failure to follow the basic rule outlined previously, which may be paraphrased as follows: "If you haven't whipped it in 20 min, expect it to whip you." The officer in command should know what to expect when hoselines advance—flames darkening down; changes in the movement, volume, and color of smoke; and the production of steam. If these things don't occur as expected, additional resources may be required either for examination or operation.

You must realize that the average house or apartment fire is no match for modern 1½-in. or 1¾-in. hoseline, especially if ventilation can be performed in sync with line advancement. If the fire hasn't darkened down after 10 min of hoseline operations, something is drastically wrong. If the tactics recommended previously—increasing ventilation and water flow—haven't succeeded, anticipate full involvement of the structure as the next result, for if you haven't reduced the intensity of the fire by this time, structural elements will begin to fail. These partial collapses are the danger signal that all interior forces must withdraw to safe locations, usually outside of the building. It isn't practical to perform offensive operations under these circumstances, since the fire traveling in the void spaces is nearly impervious to outside streams. The result is that the fire will have access to the entire structure, and it will claim its prize.

Incident operations become long, drawn-out affairs as outside streams hit fire that breaks out of void areas, and tons of water are poured onto the building in the hope that some of it will find its way to the hidden fire. If enough water isn't available, or if the application points are insufficient, then the burning void area as well as the original fire area will become heavily involved. The total destruction of the building under these circumstances is a real possibility. In some structures, notably those that have truss construction, even 20 min may be too long to prevent total collapse.

When the officer in command recognizes that he may be forced to withdraw from the offensive interior attack, he must immediately begin preparing defensive positions. This may require calling additional resources, even though pumping capacity and apparatus at the scene aren't overtaxed. Sufficient personnel must be available to get the outside streams into position and to charge them, allowing for an orderly withdrawal of interior lines. It isn't enough to wait for a company to come out of a building and then have them go off to stretch a defensive line, particularly when exposures might be threatened. By the time the line is positioned and in operation, the fire may have already extended to exposures.

When Forced into a Defensive Mode, Consider the Possible Effects of Total Involvement of the Structure

When an interior attack is neither possible nor practical, fire traveling in void spaces has an advantage over suppression forces. It can break out in areas quite remote from the original fire area, possibly outflanking the fire forces. If it manages to reach several nearby areas simultaneously, the entire structure may become fully involved. Depending on the size of the structure, access points for stream application, and exposures, such a conflagration can outstrip the units available to combat it. Because of this, once you are forced into a defensive mode, call for additional resources. These resources should be of sufficient type and quantity to prepare for the worst if or when all hell breaks loose (Fig. 3–12).

Recognize that radiant heat can cause extension to exposures even across wide avenues. A master stream from an elevated platform can cover a frontage about 100 ft in length and up to about three stories high, depending on type and length, *if* it can operate in front of the building. If radiant

Fig. 3–12 When the situation is deteriorating, plan for the loss of the entire structure. Look for taller surrounding buildings that will be threatened when fire vents from windows or burns through the roof, and position hose streams to protect them.

heat or the threat of collapse forces withdrawal to the flank of the fire area, its reach may only be about 50–75 ft. Ground- or pumper-mounted master streams have even less flexibility under extreme circumstances, but due to their better reach and penetration, they are far superior to handlines in similar situations (Fig. 3–13).

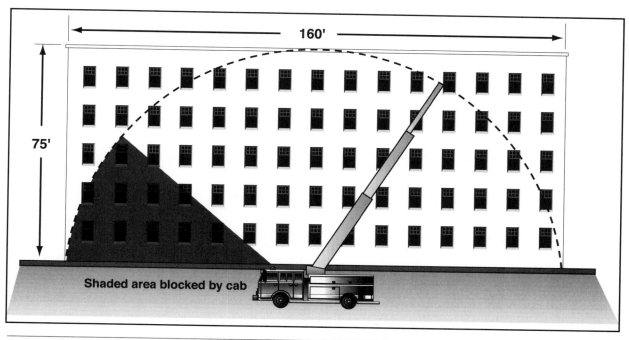

160'

75'

Shaded area blocked by cab

Fig. 3–13 Officers and drivers all need to know the capabilities of the available aerial devices. Scrub area—the area a platform can physically touch at a given distance from the building—is a key criterion when spotting apparatus at a scene.

A final guide for engine company firefighters is as follows: *When in doubt, lay it out.* If you even suspect that a hoseline will be required at a given location, go ahead and call for it. By the time you confirm your suspicions, the fire may have gained enough headway to force you into playing catch-up. As the old saying goes, "it's better to have it than to wish you had it."

Notes

1   "Nobody was Supposed to Survive," by Alice Walker, from *Living by the Word*, London: Women's Press (1988), 155–7, 159–60.)

04 Hoseline Selection, Stretching, and Placement

You have responded along with the first-due engine company to a midmorning alarm for a fire in a furniture store. As you approach the block, you see what appears to be a slight haze in the area. Pulling up in front of the building, you see none of the usual signs of a working fire. You are about to investigate when an agitated off-duty member approaches from the interior, informing you, "You're going to need a line in there in a hurry." You mentally note smoke, now visible from a doorway at the left-rear corner of the 100 x 125-ft store. You turn to your crew, a driver and three firefighters, and give a few commands that will have a great bearing on the outcome of the fire: you order a hoseline to be stretched into the building.

This chapter covers the following three interrelated topics that directly affect putting water on a fire: selecting the proper size and length of hose, quickly and efficiently stretching the line, and making sure that it is stretched to the proper position. This chapter deals primarily with offensive modes of attack, meaning interior handlines for the most part. Supplying master streams and auxiliary appliances will be covered in chapters 5 and 6.

Putting the proper size hoseline at the required place in the least possible time with the most efficient use of personnel is the job of the engine company. In many cases, all future actions hinge on accomplishing this task correctly. If it isn't, rescue and fire control may not be possible. A large number of variables affect the selection and use of a hose stream. Flow rates of 12–325 gpm are available for interior handlines. The selection of a hoseline and nozzle combination must be suitable for a given location. No single choice can be mandated for every contingency. Should the line be a ¾-in. or 1-in. booster, or a 1½-in., 1¾-in., 2-in., or 2½-in. handline? Where should it be stretched—to the fire floor or the floor above? Should it go through the front door or the rear? Should the line be stretched by way of an interior staircase, over a ladder, or up a fire escape? Should it be hauled up by rope along the outside of the building? Should the selected hose be preconnected, or should it be taken from a main hosebed? The answers to these questions will come after evaluating the necessary size-up information. Such factors as the building's occupancy, construction, height, and area, as well as the location and extent of the fire, have a direct bearing on hoseline selection and placement. The method of the stretch will be determined by such factors as the presence of standpipes, open stairwells, and usable courtyard windows.

The key to successful hoseline selection is to look at the situation briefly before taking any hose off the apparatus. Make sure that the line stretched is appropriate for the task. Far too often, firefighters have stretched an inappropriate line because "that's the line we *always* stretch." This often involves stretching booster, or red, lines into structures, a totally unsatisfactory solution. In at least two large departments, the administration is so strongly against this practice that the booster lines have been removed from the apparatus. Stopping short of such drastic measures means that the members riding on the pumper must be able to make the right choice based on the situation at hand.

Regardless of the method of attack you choose or the type of stream you employ, two criteria determine whether your effort will successfully extinguish the fire. The first is that the amount of water discharged be of sufficient volume to remove the heat being generated. The second is that the water actually reaches the heart of the fire and not be carried away by convective currents or turned to steam. These two criteria combine to determine what size hoseline will be appropriate.

The first consideration, required volume, is relatively simple. In fact, formulas that can predict the required amount of water flow have been devised based on the volume of the area and the weight of the fire load. These formulas, discussed in chapter 2, vary from 10 gpm for every 100 sq ft in a low fire-load setting to 50 gpm per 100 sq ft for high fire-load areas. This is determined by the amount of heat that the fuel can produce. Materials vary as to the amount of heat they give off. For instance, one pound of wrapping paper will give off about 7,100 BTUs, whereas one pound of styrene foam gives off about 18,000 BTUs. In this case, the styrene foam would require about 2½ times more water to extinguish than would the wrapping paper. Still, in either case, if you don't apply enough water, the fire won't go out. Simply stated, firefighters must apply enough water to absorb all of the heat being given off or the fire will continue to extend. (Water absorbs about 9,275 BTUs per gal when raised from 70° to completely vaporized, so in the case of the pound of wrapping paper, a little less than 1 gal of water would be required to cool the burning paper, while nearly 2 gal would be required for the same one pound of styrene foam, assuming complete vaporization.)

For fires in small spaces with light fire loads, a smaller line may be used. If the fire is in a larger area or if it involves materials that have a high rate of heat release, a larger line is required. The larger-size hose has another advantage over the smaller hose: the reach of the stream and the effect that it has. The ability of a stream to remain intact in the face of a high wind or a heavy body of fire is a distinct advantage, as is the ability of a stream to punch through ceiling or wall surfaces to hit hidden fire. To combat Class A fires, the water must cool the fuel to the point where it ceases to give off flammable vapors. In the event of heavy fire involvement, a stream lacking reach and penetration, while theoretically flowing the required volume of water, will have a difficult time extinguishing the fire. This is due to part of the water being turned to steam before it reaches the burning materials. In this case, the steam will likely be dissipated without having any smothering effect. Water can also be carried away by the updraft of hot air and fire-plume gases, thus being unable to reach the fire area. The use of a stream with good reach and penetration also serves to protect the firefighter, putting out fire and cooling surfaces in advance of the firefighter's entry to that area.

Factors Affecting Hoseline Choices

Present-day pumpers are usually equipped with a variety of discharge appliances, from booster lines to master streams, both preconnected and non-preconnected. How does a department choose the proper combination of lines for its area? More importantly, how does the first-arriving pumper crew select the proper line for the particular building to which they are responding? The department has the luxury of time when deciding how to outfit a pumper, since they have time to evaluate rationally all of the variables that will affect future operations. An engine crew pulling up at 3:00 AM shouldn't be hindered by decisions made in the light of day, when all of the factors could be considered logically without people screaming and sirens wailing. The ideal situation is to spec out a pumper so that it can handle a majority of the types of fires that can be expected in a given area with minimal effort and so that it will provide an effective attack on the nonroutine fires that might occur. By providing the right selection of hoseline choices for a given response area, the guesswork is reduced for the nozzle team (Fig. 4–1).

Start by examining the response area in which the unit is to operate, and determine what the majority of operations will involve. What is the most common occupancy? What is the construction type and height, width, and depth? What is the setback distance from the street? What is the water supply—hydrants, wells, drafting points, or tanker shuttle? Hose loads that work well in spacious suburban tracts with good hydrant supplies are ineffective in urban areas, crowded with large non-fireproof apartment buildings. Similarly, the hose load that works well in these tenement areas may be impractical for use in heavy industrial areas or in high-rise commercial areas. Once you have determined the true nature of the district, you are ready to start analyzing the hoseline choices that are available and to select the proper options. *No single size or length of line is a panacea for all situations.* Let's examine the possible combinations and the effects that the various factors have in determining the correct line (Fig. 4–2).

Fig. 4–1 Urban row houses require several beds of midsize lines (preferably 1¾ or 2 in.) for interior work as well as a bed of 2½ in. for larger fires and a supply line for master streams for advanced fires.

Selecting Attack Lines

The consideration of occupancy will have a direct bearing on the required fire flow rates and, therefore, the diameter of the attack line. Fires in residential occupancies are the most common fires you will encounter. For this reason, many departments gear their attack entirely in this direction. Residences, be they single-family homes, garden apartments, apartment houses, or mobile homes, have three characteristics that play an important role when selecting hoselines.

The first is the need for speed. In the United States, more than 75% of fire deaths occur in residential buildings. Undoubtedly, some of these lives could be saved if the fire could be extinguished more quickly. This requires a line that can be stretched with a minimum of personnel.

Fig. 4–2 A suburban neighborhood that consists of widely separated one story homes might use 1 ½-in. or 1¾-in. preconnected lines for nearly all their operations.

The second is the relatively low fire loading. The amount of fuel found in an average living area is much less than that found in a similarly sized commercial or manufacturing area. Thus, you can use smaller lines.

The third characteristic is the presence of dividing walls or partitions between rooms, which has two effects, one desirable and the other not so. Partitions tend to reduce the extent of involvement or at least slow the spread of the fire. In high-rise firefighting, this fact is known as compartmentalizing an area. Even a flimsy hollow-core wooden door has some containment value if it is closed. Partitions or walls allow an attack crew to approach relatively close to the involved area before stepping into the doorway to face the enemy. The disadvantage, however, is that walls can be obstacles to hoseline advancement. Anytime a hose makes two bends, a member must be positioned between the two bends to feed hose to the advancing nozzle team. If this isn't done, the line won't be able to move no matter how big the people are at the nozzle and no matter how hard they tug.

Residential occupancies are generally much more confined than other areas, so the need for a maneuverable-size hose is often more critical here than in a store or factory. As a practical matter, if a fire in a residential occupancy has passed the flashover stage, a hoseline of at least 1½-in. diameter is required, yet past experience has shown 2½-in. hose to be impractical and unnecessary in all but the largest fires. That narrows down the choices to 1½-in., 1¾-in., and 2-in. hose. Respective practical maximum flows of 125, 180, and 225 gpm can help you decide which is most appropriate for your area. A unit that only responds to single-story homes built on slabs will probably be able to use 1½-in. hose to good advantage, while units that face larger homes and those with basements or cellars may require 1¾-in. or 2-in. hose. Companies that respond to many fires in multistory apartment buildings have found 1¾-in. hose to be very suitable. A 1¾-in. line provides sufficient flow to knock down four or five rooms of fire rapidly, yet it also allows the necessary speed in stretching and the flexibility during advancement that is required in such buildings. Occupancies that have characteristics similar to those of residential buildings are likely candidates for this intermediate size of line. Hospital patient care areas and school classrooms present similar degrees of compartmentalization and light fire loading. Of course, a heavy body of fire present on arrival should indicate to firefighters that the intermediate-size lines should be left on the apparatus; it's time for the 2½-in. line. As one old sage puts it, "You don't go bear hunting with a .22."

Certain occupancies should dictate stretching a 2½-in. hose whenever any sizable body of fire is present. Most commercial and industrial properties fall into this category. The reasoning is rather simple.

- The fire loading in such buildings is usually heavier than in residential occupancies.
- The floor areas demand a longer-reaching, harder-hitting stream.
- Often there are more flammable materials around that can accelerate a minor fire (Fig. 4–3).

There has been a disturbing trend recently toward eliminating 2½-in. hose for handlines and relying solely on 1¾-in. or 2-in. hose for interior attack lines. This is a fine plan if the only fires you'll ever face are in one- and two-family homes, but it can be disastrous at fires in commercial buildings or under severe fire conditions.

A classic example involved a large *L*-shaped, six-story, non-fireproof apartment building. Fire had involved two apartments on the second floor and had entered the plumbing and structural voids prior to the arrival of suppression forces. The department arrived with four engines, three ladders, and a rescue company. They found the occupants of dozens of apartments calling for assistance. Members began immediate rescue operations and initial attack, yet it was obvious that they would need assistance. Fire, pushed by 10–15 mph winds, was rapidly spreading upward toward the common cockloft as well as into the exposed apartments above. My department was called as the seventh or eighth mutual-aid engine

Fig. 4–3 A response area that consists of large numbers of non-fireproof factories and other commercial occupancies should have multiple long beds of 2½–in. hose as well as master stream supply lines.

company. We were directed to report to the top floor with standpipe rolls to assist another company that was trying to cut off the spread of fire. Our engine officer took one look at the volume of fire showing and decided to leave the rolled up 1¾-in. standpipe packs and instead to take six lengths of 2½-in. off the main bed. This was flaked over each member's shoulders, and off we went to do battle. In the meantime, the officer had selected a 2½-in. break-apart nozzle with a 1¼-in. solid tip. He then placed an automatic fog tip in his pocket.

On arrival at the top floor, we were met by a crew that was pinned down in the hall, unable to advance because of heavy fire venting into the hall from two fully involved apartments. Our unit hooked up to a separate standpipe riser and stretched up to the location of the first line. All but two of the members on the first line left at this point to exchange air bottles. The single 2½-in. line, discharging more than 325 gpm, was able to advance to and darken down the two apartments. At this point, additional members advanced the 1¾-in. line for overhaul in one apartment, while a rolled up length of 1¾-in. and a nozzle were connected to the break-apart 2½-in. nozzle to overhaul the other apartment. There was no doubt in anyone's mind that the 2½-in. line had made the difference between advancing and being pinned down. This lesson has been learned many times over by fire officers, sometimes the hard way (Fig. 4–4).

One statement that is often heard in justification of replacing 2½-in. hose with newer, smaller-diameter lines is, "With the new 1¾-in. or 2-in. hose and automatic nozzles, we can maintain the same flow as with the 2½-in. line." This simply isn't true. The laws of hydraulics do not change because of hose or nozzle design. The larger the diameter of the hose, the more water it can carry with the same pressure drop. The larger the opening of the discharge orifice, the larger the flow of water at the same pressure. The same improvements in hose and nozzle

Fig. 4–4 A hose bed equipped for a tenement area provides two beds of 1¾ in. filled out with 2½ in., a 2½-in. attack line, and a bed of 3½ in., for supplying master streams.

designs made for 1¾-in. and 2-in. hose have been applied to 2½-in. hose. Equip your 2½-in. hose with the same grade of fine automatic nozzle that you have on the intermediate-size lines, and you have a big line that can deliver big punch, more than 325 gpm. In fact, buying new equipment isn't required at all in many cases. Simply dust off that 1¼-in. solid-tip nozzle that's been around for years, and you have just equipped yourself with a line that flows the equivalent of almost two 1¾-in. lines—and with better reach!

Reach is a definite advantage of larger streams. When surveying your response area, look for large-area buildings. The larger floor areas that you find in commercial, manufacturing, and storage occupancies translate into large fire areas. What appear to be small-to-moderate fires can rapidly extend, rolling over the heads of advancing crews. The larger flows available from the larger-size lines ensure a greater reach if both lines are operating at the same nozzle pressure. An additional benefit of the larger line is often mistaken for a disadvantage. Many departments don't use 2½-in. lines when they should because they say they don't have enough people to handle the larger line. That's simply not true, and a sure sign that they haven't been shown how to use it properly. A 2½-in. line flowing 325 gpm through a 1¼-in. tip can be handled by two people and moved by three. To apply the same 325 gpm with 1¾-in. hose requires four people to hold or move the two lines. The 2½-in. line also adds reach and striking power over the smaller lines.

A very common structure across the United States is that of one-story commercial buildings of ordinary construction, commonly called a taxpayer or strip mall. The average store in such a building measures 20 x 75 ft. A 2½-in. handline provides enough reach to begin putting water on the entire depth of the store almost as soon as you enter, if necessary. A prudent (and inexpensive) addition on apparatus is to include a 2½-in. preconnect as well as the two intermediate-size lines. This facilitates rapid knockdown when you are faced with heavy fire, and it also provides a rapid means of extending a 1¾-in. or 2-in. line that won't reach to the seat of the fire without a drastic increase in friction loss. It also eliminates the excuse of many firefighters after having stretched a 1½-in. preconnect into a fire that was too large to handle: "We thought that by stretching the preconnect, we could save time and cut off the fire."

Many fire companies, of course, find themselves confronted by buildings that are deeper than even the reach of a 2½-in. handline. Any heavily involved commercial building larger than about 20 x 75 ft will need more water than a single 2½-in. line. (For example, 20 x 75 ft = 1,500 sq ft, an application rate of .3 gpm/sq ft works out to 450 gpm required flow; 3 gpm/sq ft is the recommended flow rate for most commercial occupancies.) For this purpose, a quickly applied master stream may be sufficient. In other cases, however, the area of the building may dictate that manual firefighting is impossible once a serious fire has developed. Witness the K-Mart distribution warehouse fire in 1982 and the General Motors Plant fire at Livonia, Michigan, in 1953 (Fig. 4–5).

How Long a Line

While the occupancy, the area of the building, and the size of the fire determine the diameter of the line, the building's height, area, and setback from the street all serve to indicate the length of hoseline to use. Whether a hose needs to be stretched around a staircase as opposed to hoisted up the stairwell is another consideration. You can preplan this information. Preconnected lines of sufficient length to cover the majority of residential structures in the area are an extremely important part of a modern pumper. It often isn't practical, however, to provide sufficient preconnected hose lengths to cover every building. Avoid overly long preconnects, since the friction losses build up rapidly and result in either reduced flows or dangerously high engine pressures.

As a rule of thumb, four lengths (200 ft) of 1½-in. hose is the maximum length for preconnect if you want to flow 125 gpm, whereas you can use up to six lengths of 1¾-in. hose at 150 gpm or 2-in. hose at 190 gpm. If it is necessary to go farther than these distances, chances are that a preconnect isn't the answer. Filling out the stretch with 2½-in. hose—connecting 4 to 6 lengths of the medium hose to 8 to 10 lengths of 2½-in. hose—is an alternative. This drastically cuts the friction loss for the remaining lengths (average 3–5 psi/100 ft through the 2½-in. hose), while the lead 1¾- or 2-in. lengths maintain the necessary maneuverability inside the fire area (Fig. 4–6).

Another alternative is to stretch the first line from a main bed of 2½-in. hose to the vicinity of the fire area and then place a length of intermediate-size hose and nozzle on the tip of a break-apart 2½-in. nozzle. Having a main bed of 2½-in. hose packed with the male end up allows you to operate in the larger buildings that you are likely to encounter. You must have sufficient hose to reach all areas of every building in town, unless, of course, a given building is equipped with a standpipe that reduces the distance to the fire.

Fig. 4–5 *This apparatus serves an area with many large factories, warehouses, and flammable liquid tank farms. The rear step mounted master stream reflects the need for rapid application of large streams with minimal manpower.*

Fig. 4–6 *These beds of 2½-in. and 3-in. hose provide at least 800 ft of hose each and a break-apart nozzle to reach into the many large-area factories and warehouses in this district.*

Even relatively small commercial buildings can require 300 ft or more of hose to reach the fire. This distance again outstrips the practical length of preconnected hoselines. A fire that requires 2½-in. hose is a major operation. The fire is already past the flashover stage, and you should give priority to ensuring that you have enough water and hose to complete the attack. In this situation, the worst thing that you can do is stretch short and not be able to reach the fire. This is a terrible position to be in, since personnel will be pinned down, unable to advance and finish off the enemy. They must either retreat to a safe area, add more hose and start again, or hold in place while waiting for another, longer line to finish the job. This is a very punishing predicament, since protective clothing isn't designed to stop the transfer of heat to the body but rather only to slow the rate at which it penetrates. The longer a firefighter is exposed to heat from an uncontrolled fire, the more heat will soak through the clothing to the firefighter. The way to lessen this exposure is to put out the fire quickly (Figs. 4–7 and 4–8).

Fig. 4–7 To fight a fire in this large area factory you must stretch enough hose to cover the entire building. Notice the pumper stopped right at the entrance door to facilitate the hose stretch.

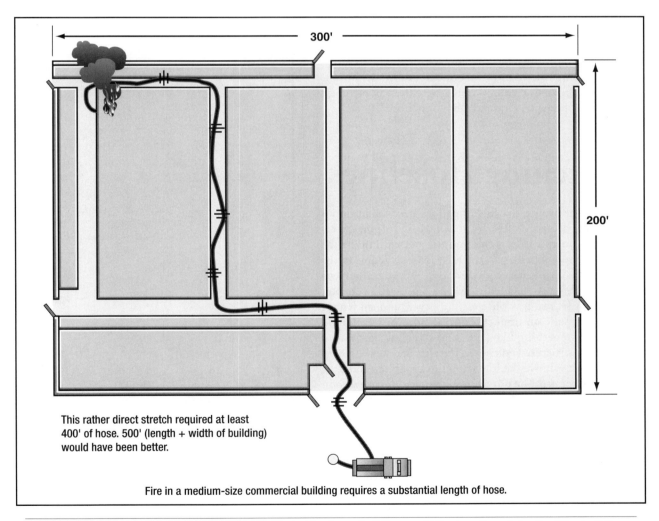

300'

200'

This rather direct stretch required at least 400' of hose. 500' (length + width of building) would have been better.

Fire in a medium-size commercial building requires a substantial length of hose.

Fig. 4–8 When estimating the amount of hose needed for large buildings, add the length and depth of the building.

PIKES PEAK COMMUNITY COLLEGE

To avoid stretching short, you must stretch enough hose to cover the entire fire area and often the floor above the fire. How much hose is enough? A good rule of thumb is to have enough hose to equal the width of the building plus the depth of the building plus one length for each floor above or below the level that the fire is on. For example, for a fire in a one-story warehouse 200 x 200 ft, at least 400 ft of hose should be available at the entrance to the building (Fig. 4–9).

Although not as critical as stretching short, take care to avoid overstretching as well. Overstretching results in unnecessarily high pump pressures and added clutter around the fire area. It also increases the chances of kinks when extra hose is flaked out in a confined area. In the previous example, if the fire were obviously concentrated right inside the door, 150–200 ft of hose would probably suffice. When the location or extent of the fire is in doubt, however, stretch sufficient hose to cover the entire structure. A good way to ensure that you have the correct amount of hose is to have the nozzle team arrive at the fire area with one or two lengths to spare. As the nozzle team nears the fire area (but is still in a safe area), one radio-equipped member can contact the members at the apparatus, letting them know how much additional hose is required. An alternative method would be to train all members to estimate properly the required amount of hose.

Stretching Hoselines

Getting the proper-size hoseline to the needed location is the beginning of a successful fire attack. How the line arrives at this location can vary tremendously depending on the building features as well as on the hose loads available to the firefighters. The presence of a standpipe system has obvious effects on the hose stretch, which will be discussed in chapter 6. Other factors are less obvious, however, and may have nearly as dramatic an impact. Routine hoseline stretching usually goes something like this: after the crew has done a size-up of the structure and chosen the appropriate hoselines, members must remove the line from the apparatus and drag it to the point of use. The nozzle man removes the nozzle and sufficient hose to allow him to arrive at the fire area with at least one length (50 ft) of hose. There are many ways—loops, folds, and rolls, for example—to provide the working length so that it is easy to carry and flake out. Whatever method you choose, it must provide a simple way for members to know how much hose they have available. They must also be able to carry the hose safely and securely, leaving one hand free for climbing stairs and ladders (Figs. 4–10 and 4–11).

Fig. 4–9 Having too much hose can also create major problems. Flake out all hose where it can be moved for advance before the line is charged.

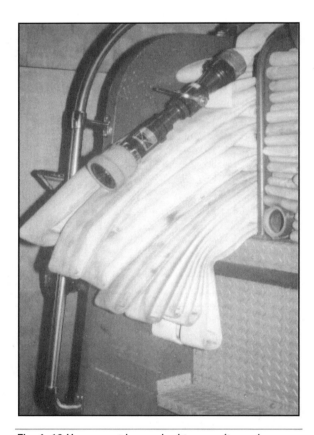

Fig. 4–10 Hose must be packed to permit nozzle team members to remove it easily and carry a predetermined amount (usually one 50-ft length) with them to the operating location.

After the nozzle man has removed his working length, he should step off two or three paces, and then pause to allow the next member to remove his folds of hose. Preconnected lines should have some type of loop or fold to allow this second member to remove the entire remaining hose quickly. Once this has been completed, both members proceed to the fire area. The second person, the backup man, should stretch his hose to provide as much spare hose as possible near the fire building, where it may be readily advanced as needed. *This is separate from the length that the nozzle man brings with him.* The backup man should not, however, bring unneeded hose inside the structure, since this would impede operations. En route to the fire area, all members must take care that the uncharged hoseline doesn't run under any doors that could close over it, restricting flow and movement. This is best accomplished with a chock. If no chock is available, use a rug or a welcome mat. This simple act can be crucial if things go wrong inside.

A good friend of mine related the following experience. The first-arriving engine and ladder at a fire in a two-story brick home were met at the front door by a resident indicating that a person in a wheelchair was trapped in the fire apartment at the end of a 20-ft-long public hall. A heavy fire condition existed in the apartment (people were jumping out of the second-floor windows) and time was of the essence. A three-member forcible entry and search team entered the front hallway, followed by a three-member nozzle team. The resident at the front door held the outer door open for the entering crew. The top of the door to the fire apartment burned through as the company prepared for the attack. Obviously, they couldn't perform a search until the hoseline had darkened down the fire.

In the meantime, some members were removing people via portable ladders, while other members attempted to reach the trapped occupant from the rear. About the time that the top of the door burned through, the smoke and heat forced the resident at the front door to back away. When he did, the door swung closed and locked. Shortly thereafter, a ball of fire blew out into the public hall, engulfing it in flame. In a matter of seconds, six firefighters found themselves trapped, driven to the floor by fire, unable to bring themselves up off the floor to force the door open (Fig. 4–12). They were saved by another incoming unit that noted the uncharged line going under the door and then fingers trying to reach under the door to pull it open. Several members received burns requiring treatment, and all were certain that they would have perished if help hadn't gotten there when it did—all because no one had chocked the door open.

Fig. 4–11 One method that permits a member to carry at least the lead length up stairs around banisters is to fold it into a horseshoe shaped loop that can be carried over one arm and flaked out rapidly.

Fig. 4–12 A charged line wedged under a door prevents any further advance of that line and can severely restrict water flow to the nozzle.

While your difficulties with un-chocked doors may be less harrowing, they can be just as much of a problem. The kinking that occurs when a line is charged under a door will severely hinder the development of proper nozzle pressure. You may not notice this while bleeding out the line, since a small flow will get past. Then, when you have begun the attack, you will have insufficient flow to darken down the fire. The charged line also works as a wedge, locking the door in position, cutting off entry or escape just as effectively as the door lock did in the house incident. Finally, if the effect on the flow isn't critical enough to halt the advance, the binding of the hoseline will be. Any spare hose on the street side of the door is stuck there. The same exact situation occurs when a line is stretched up a stairway that has a narrow space between flights or between a flight and a landing. In this case, it is much better for firefighters to stretch the line around the banister and along the stairs than to risk pinching the line. This practice obviously takes more hose than going straight up. A reasonable estimate is one length per flight of stairs (Figs. 4–13 and 4–14).

Fig. 4–13 A wide well allows the hoseline to be carried up the stairs instead of dragged around the staircases. This greatly reduces the amount of hose needed. The presence of a well must be communicated to the people stretching the line to avoid too much hose being brought into the building.

Fig. 4–14 A narrow well with a charged line wedged into it can have the same effect as wedging it under a door.

Many older buildings have open stairs with a wide well running from bottom to top. These are dangerous when fire starts on a lower floor, allowing heat and gases to rise rapidly. However, they can be useful to the hose stretch when the fire is on the upper floors. In this case, the nozzle man simply carries his assigned hose up the stairs while the backup man keeps the hose in the well. When the attack team reaches a safe area, they must secure the hose to prevent it from being pulled back down by the weight of the water. They should pull up enough hose before the line is charged so a hose strap or rope hose tool can be secured below the next coupling. This takes the strain off the coupling—strain that is trying to pull the hose right out of the coupling. Only one length of hose is required to go from the ground floor to the fifth floor when using this method. This fact must be relayed back to the street, or the member estimating the hose required will pull off four unneeded lengths if he follows the one-length-per-floor rule of thumb (Figs. 4–15 and 4–16).

Fig. 4–15 The nozzle team should carry their line in the well to save effort. Do not stretch more than two lines in this fashion in a well, otherwise they will wrap around each other and end up in a knot that prevents advance.

Fig. 4–16 Hose that is stretched up a well must be secured with a hose strap or rope to take the weight off the couplings. Otherwise, the hose can pull right out of the coupling when the line is charged.

Often, as these older buildings with open wells were modernized, an elevator was added in place of the stairwell. In such a case, the stairs are said to wrap around the elevators. If no stairwell is present, and if the fire is on the fourth floor or higher, stretching hose around the banister becomes a very lengthy, tiring, personnel-intensive operation.

An excellent alternative is to use a rope to haul up the line on the outside of the building. In this procedure, at least two members ascend to a safe area that has access to street side or courtyard windows. They carry a suitable length of light nylon rope with them and lower the rope out of the window to another member who has stretched the hoseline to a point directly below. This member secures the rope to the hoseline (a snap link or clip saves you from tying knots) and signals the members above to hoist away. They haul up sufficient hose and then secure the line in place with a hose strap as previously described. The line can then be advanced to the fire.

This method has tremendous advantages, particularly speed. It is far faster than stretching hose around the staircases, it requires fewer personnel to put the line in place, and it provides a ready means to stretch additional line. As a side benefit, it keeps the stairs clear for access as well as evacuation. Remember that the members must stage at a safe area, usually on the floor below the fire. The line is then advanced up the interior stairs to the fire floor, meeting one of the primary objectives—protecting the staircase. The first time I witnessed this tactic, I was at first dismayed and then amazed. I had recently transferred from a unit where it was unusual to encounter a building more than three stories high that wasn't fireproof or without a standpipe system. The new area I was working in was predominantly all six-story, non-fireproof multiple dwellings; some went as high as eight or nine stories without a standpipe system. This loophole in the building code caused tremendous difficulties when it came to stretching hoselines. It wasn't uncommon to commit eight or more members to get the first hoseline in position rapidly (Figs. 4–17 and 4–18).

Fig. 4–17 A staircase that wraps around an elevator shaft or other obstruction requires a long tedious stretch.

Fig. 4–18 A better approach to stairs that wrap around an elevator shaft would be to haul the line up by rope to the window on the half landing on the floor below the fire.

About 6 AM one day, we responded to an alarm for a fire on the top floor of one of these six-story apartment houses. My company was assigned to search and vent the adjoining apartments. As we entered the ground floor, I met two members of the first engine in the lobby flaking out five or six lengths of hose. As I climbed up the five flights, I was thinking what a poor engine company they must have been not to have a line above the ground floor, since they had arrived a minute or so before we had. You can imagine my surprise when I rounded the staircase from the fifth floor to the sixth to find a charged hoseline coming in the stairway window. The fire in the first room was already darkened down.

What had looked like utter confusion in the lobby was part of a well-planned and executed hose stretch. The officer and two members had gone ahead with a 1½-gal bleach bottle containing 100 ft of ⅜-in. nylon rope. The rope had a clip on each end and was easily deployed when laid into the bleach bottle. At the staging area, one end is clipped around a banister or handrail to prevent it from falling, while the whole plastic bottle is dropped to the waiting firefighter below. This method can also be used to bring a line up to a fire escape or to the roof. A word of caution on the fire escape line: this may be a fast way to get water on the fire, but it shouldn't be the first method you try. The line operating from the fire escape will push fire and gases back into the building toward any occupants—not an acceptable first tactic (Figs. 4–19, 4–20, 4–21, and 4–22).

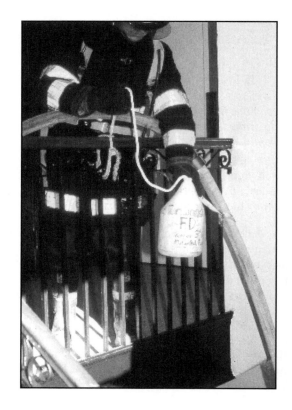

Fig. 4–19 A 1½-gal bleach bottle easily stores and deploys 100 ft of nylon rope.

Fig. 4–22 The rope stretch up a fire escape is much faster and less complicated than using the hook to hoist the line.

Fig. 4–20 An exterior rope stretch is an excellent way to quickly stretch the third and later lines to upper floors of apartment houses and other large buildings. Note that the line is being brought in to the floor below the fire then up the interior stair. This is a crucial strategic concept.

At times, however, it may be necessary to fight a fire from an unusual location such as a fire escape. This may be necessary when an extremely strong wind is pushing in from the fire escape side, driving fire toward the stairway in such a manner as to prevent entry from inside. It may be acceptable in this case to advance with the wind at your back. Situations like this may occasionally occur when there is no fire escape. In this event, the line is sometimes stretched up an aerial ladder or from a platform. While it is fine to use these devices to get personnel and hoseline into position, you should be alert to prevent attempts to turn the aerial devices into standpipes. Hoist up the line and secure it as described previously. This allows the aerial device to remain free to perform the following primary functions: upper-floor entry and egress, roof entry, and heavy-stream operations in other areas, if necessary. It should be taboo to stretch a handline directly off the end of a ladder pipe or from an outlet on a platform unless the fire is in its final stages and the line is to be used strictly for overhauling.

Fig. 4–21 Avoid the temptation to use the aerial device as a temporary standpipe. This ties it up, preventing its movement if it is needed for other critical purposes.

I have witnessed the consequences of violating this provision a number of times. In one case, the handline proved inadequate for the task and was forced to withdraw. The platform's master stream was urgently needed elsewhere to cut off spreading fire, but it couldn't be used until the handline crew had withdrawn to a safe area where their line could be shut down. At yet another incident in a nearby town, fire that had spread managed to burn the roof off of a three-story commercial building, in part because additional manpower and saws couldn't be brought to the roof by platform. Again, it was tied up supplying handlines. Both of these fires resulted in substantial losses at least in part due to tactical errors that could have been avoided if alternate means of supplying streams in upper areas had been provided.

The Placement of Hoselines

Of the three items to consider when putting a line into operation, the ultimate destination of the hose will probably have the greatest effect on the fire. If you stretch too large a line, the fire will certainly go out, but it will take longer. If you stretch too small a line, you may not be able to extinguish the fire without some help, though you might be able to confine it by taking a purely defensive position. Certain hose stretches lend themselves to different building configurations, but nearly all will get water on the fire at some time.

Improper placement of the hoseline, on the other hand, can have disastrous consequences affecting the safety of occupants and firefighters as well as determining the extent of damage to the structure. As I mentioned earlier, the fire-escape stretch may be the fastest method, since it doesn't position the hoseline at the proper location. In deciding where to position the hoseline, remember the following priorities: protect human life, confine the fire, and then extinguish it. This usually involves placing a hoseline between the fire and the victims. This is, in fact, the highest priority for hoseline positioning. Keep in mind that in unusual circumstances, a separate line may be required to protect each individual rescue. This should *not,* however, delay the stretching of a hoseline to the seat of the fire.

Take, for example, a fire on the second floor of a three-story building. Heavy fire exists in an apartment and has trapped the occupants of the apartments above, driving them to the windows to await rescue. The fire venting out of the windows below is preventing their descent via the fire escape and aerial ladder. In this case, one line should immediately be stretched to protect the rescue, while another line should simultaneously be advanced to the door of the burning apartment via the interior stairway. In protecting the rescue, the nozzle team must avoid directing their stream into the windows, which would drive the fire back into the building. Instead, they should position themselves on the side of the window and use a narrow fog pattern to blow the fire away from the exposed occupants (Fig. 4–23).

In any multilevel building, except for the need to place hoselines to protect a specific life hazard, the main priority must be to position a hoseline to protect the interior staircases. Stairways function very well as natural flues, drawing all of the products of combustion toward them. By placing the initial hoselines to protect these locations, we accomplish three things. First, we place a hose stream between the fire and any occupants trapped above the fire. Protecting the life hazard is our highest priority and must take precedence over other considerations. Second, we confine the fire, stopping its access to its most likely path of extension. Third, we protect the most common means available for bringing additional personnel and equipment to the fire area.

Fig. 4–23 The most critical function the first nozzle team must accomplish at most multistory buildings is to protect the interior staircase.

In the rope stretch mentioned previously, it is worth noting that the hoseline should be brought back into the building, either on the floor below the fire or into a stairway window at the midpoint between the fire floor and the floor below, then advanced up the final flight to the fire floor. In any case, if the initial line is driven back by heavy fire, it should only have to go as far as the staircase, where it can maintain a holding action until help arrives. In the meantime, the line protects the stairwell. A special note regarding below-grade fires: it is perfectly acceptable to stretch a dry line into a building and carry it up to the third floor or more before calling for water if the fire is above you. When the fire is below you, however, and you suspect even a light-to-moderate fire condition, you must have the line charged before starting down the stairs. If the fire suddenly intensifies, the hoseline may well be your only protection as you climb back up the stairwell—a veritable chimney.

For fires in one-story commercial buildings or outside fires of significant proportions, the hoseline should be placed to cut off the fire, not to chase it. Generally, this involves attacking from the unburned side. Get ahead of the fire and drive it back where it came from. Of course, this may initially be more punishing on the members. It means getting inside and slugging it out with the devil, not just standing outside squirting water at it. In the long run, however, it will result in less punishment, since it will extinguish the fire more rapidly with less area involved. Be alert to prevent the common mistake of taking a hoseline directly to the location where flame is issuing from a building. When fire is blowing out the front three windows of the ground floor, inexperienced firefighters tend to want to attack directly from this point. The firefighter seems almost mystically drawn to the flames. To begin water application from this point, however, can create havoc. It is going to drive fire, heat, and gases back into the structure, right where you don't want them. By taking the line through a door to the interior entrance to the involved rooms, you are in a position to blow the fire where it was trying to go anyway—right out the windows.

There are exceptions to positioning a line to attack from the unburned side. In the past, some texts emphasized that this strategy of attacking from the unburned side was the method to be used at all fires. That idea became popular in the 1960s, when SCBA, 1½-in. hose and Navy fog nozzles became popular. Before that time, when fire attack consisted of using 2½-in. hose with a solid tip nozzle on almost every fire that was beyond the capacity of a small *booster* stream, and fought without the benefit of SCBA, no firefighter in his right mind would propose dragging a ton of charged 2½-ft hose through a building full of smoke and heat when they could walk right up to it and blast it into oblivion right through the nearest window. It just didn't make sense.

Once the modern inventions of the 1950s and 1960s resulted in mask-equipped firefighters moving lighter hose with a wide pattern fog nozzle, pushing smoke and heat ahead of it through the building, it instantly became the *in thing* to do, because for the first time it was possible. While there are fires that will benefit from this attack, firefighters must be alert for those cases where a different attack will be far more beneficial. Specifically they include any fire where there is a potential life hazard. In this case, the line should be brought to a location that puts it between the fire and any trapped occupants. A particularly dangerous situation occurs when the fire involves the means of egress from the building. When a fire begins in a lobby or stairway, waste no time stretching through from a remote area. It is of the utmost importance in this case to get water on the fire as soon as possible to keep the primary route of escape open. Many fires that occur in the means of egress are the result of arson, involve accelerants, and are particularly deadly events. Speed is of the essence.

An additional exception might be made in the case of a vacant or abandoned building. If it is less punishing on the firefighters, and if the fire won't be seriously extended by doing so, by all means attack from whichever location will involve the least danger and discomfort. Fires in vacant buildings don't just happen. They are set for a variety of reasons. One of their worst effects is firefighter injury and death. The building being abandoned by its owners is not a high-priority item in the scheme of things. If some extra fire or water damage occurs within this type of structure, it is preferable to injury or even discomfort to the firefighter.

This brings up the subject of placing values on property risks vs line placement. It should be quite clear by now that the life hazard will determine the location of the line placement at occupied buildings. But what methods can point out the proper placement when no life hazard is present? Two phrases are often heard when discussing placement to cut off extension or to protect exposures. They are the following: *the most severe exposure* and *the exposure most severely threatened*. At first glance, these may seem to be one and the same, but there is an important distinction between the two that a member making the decision must clearly understand.

I responded with an engine company and a tower ladder late one bitter cold evening to stand by in the quarters of a neighboring department that was battling a multiple-alarm fire in a vacant hotel. Shortly after our arrival at their quarters, the IC requested that the pumper respond to the fire scene and that another company be called to take their place on standby. About 1 min after this pumper responded, another alarm was received for a fire in a vacant frame house of two and a half stories. I responded with a tower ladder and an ambulance with the nearest pumper coming from 10 min away. The situation looked grim as we entered the block. The original fire building was fully involved, with fire severely threatening a similar structure on the exposure 2 side (both were in the process of renovation), as well as a one-story garage in the rear. Exposure 2 was separated from the fire building by about 20 ft and exposure 4 was a similar building, separated by about 35 ft. All of its occupants had self-evacuated prior to the arrival of the fire department and so informed us. All of the buildings were of stucco-coated wood framing.

On sizing up the situation, a call was placed for "at least three more engines and another truck," but they would all be at least 15 min away. It was getting down to heavy decision-making time. Exposure 2 was obviously the most seriously exposed building, since it was already smoking. Our firefighting resources consisted of two 2½-gal water extinguishers, two 20-lb extinguisher (for Class A, B, or C fires) and eight men. Exposure 4 was further away from the fire building, but it was the closest of three buildings that led like a fuse to a large lumberyard. The stage was set for a large loss. Exposure 2 had no exposures on any of its sides, other than the fire building. Our crew entered exposure 4 and successfully kept fire from extending to this structure for more than 10 min, removing all of the combustible drapes and shades and judiciously using two water extinguishers and pots and pans full of water from the bathroom and kitchen sinks. I would like to tell you that we saved that structure, but it was not to be.

When the first engine company arrived, we were almost whipped, but we knew that quick placement of one 1½-in. line into exposure 4 and a second 1½-in. line to the yard to hit the underside of the eaves would surely save us. The engine company failed to recognize the important priorities, however. They started operating on the original fire buildings and the well-involved exposure 2 because that's where the main bodies of fire were. Unfortunately, they had expended their 500 gal of booster tank water before they realized that they had connected to a frozen hydrant. Both exposure 2 and exposure 4 were involved in fire by now, and the next engines were still several minutes away. Fire was now threatening to extend down the fuse to exposures 4A, 4B, and 4C, the lumberyard. Our tower ladder was positioned for master stream operations in the hopes of stopping extension beyond exposure 4. Pleas were relayed to incoming engines to stretch their first lines to the intake of the tower, where the bucket would put water right on the critical locations. Again, unfortunately, these pleas were ignored. The first incoming unit elected to dump their water through a deluge gun at the main body of fire on the exposure 2 side, which simply ate it up until they, too, realized that they were on a frozen hydrant. When additional units finally arrived, a reliable water source was finally located. A supply line was finally stretched into our tower, which then operated on exposures 4A and 4B, stopping the extension toward the lumberyard. The total loss was as follows: four buildings destroyed, two more heavily damaged, and one other less severely damaged.

It is likely that, if the first engine had recognized the difference between the *most severely exposed* and the *most severe exposure,* the damage would have been contained to the two buildings under renovation. It should be mentioned that, at the time of this incident, there was no incident command system in effect in this jurisdiction. The original officer in command, a lieutenant, had a great deal of time to evaluate the conditions, yet when incoming units arrived, he was powerless to tell their commanders what to do. Radio communication was readily available, but incoming units failed to heed the requests. As a result, that officer made a hasty judgment and acted as he saw fit. Each subsequent arriving unit acted independently.

There is a bit of good news to this third edition. In previous editions I had reported that the same type of thing has repeatedly happened in this jurisdiction due to a lack of an effective incident command system (ICS). Happily, this seems to be changing for the better. In 1999, I fought another very large apartment house fire in this same town where things seemed to be spiraling out of control, with various mutual aid units performing whatever actions they felt were appropriate without consulting with the IC or sector commanders. When the chief realized the extent of this lack of control, he took the most extreme action I've seen in some time. He pulled everybody out of the building and called the officers together and gave them explicit orders. Either they go to a particular position and operate as directed by that sector commander or else they'll be sent home. He then proceeded to line up each company in the order they would be deployed, and sent the sector commanders back up with one engine and one ladder each, holding the rest in the lobby staging while the sector commander deployed their first units, then called for more in sequence. The second time around the fire was brought under control in about 30 min. It was a textbook operation once it was made clear that the ICS would be followed, or else there would be consequences.

When making a decision about whether to commit the first line to interior operations or to act as an exposure protection line, a number of items must be considered. The first is the extent of the fire. Is it already so large that the initial attack lines are unlikely to darken it down before it extends? If so, the first line should be placed in a defensive position to protect the exposure, and it should be of sufficient length and capacity so that when additional help arrives, it will be able to begin offensive operations.

A second factor is the attack capability of the unit. A pumper that arrives at a heavy fire in a store with three men and only 1½-in. preconnects will be severely handicapped. If fire is threatening a nearby structure, they should concentrate on protecting the exposure and realize that there is little possibility of saving anything in the store. On the other hand, if the unit is equipped with a large-diameter supply line and a preconnected master stream, they may well be able to protect the exposure and darken down the main body of fire by themselves.

The third factor to consider regarding extension to exposures is the construction of the buildings, particularly the exterior siding. Buildings of noncombustible siding may still be exposure hazards. Brick, concrete, and steel buildings may all have windows that can allow the fire to vent to the outdoors or to enter the exposure. The degree of danger to these buildings depends largely on the size of such window openings and on their proximity to the fire building. Fire venting out of a window in a brick wall is a relatively small source of energy, since it will radiate in all directions, expending much of its energy on the brick of the exposed fire building. The most severe threat would be to a window in the opposite exposure directly in line with and slightly higher than it. Barring direct-flame contact, the most severe threat is from radiant heat, which travels in straight lines. Distance is the greatest protection from radiant heat. The energy absorbed varies with the distance from the source. With as little as a 10-ft gap between two noncombustible walls, there is little threat from exposure if the first line puts out the fire. Of course, additional personnel should be assigned to examine the exposure immediately, but the crew on the first line shouldn't hesitate to operate first on the original fire (Fig. 4–24).

Fig. 4–24 Closely spaced wood-frame buildings present a serious exposure hazard.

Fires in frame structures, though, can be a different situation altogether. Here, once fire vents out a window, it will readily extend on the building's siding. The surface fire rapidly multiplies the amount of energy being thrown off. The same rules apply to distance: the further away, the less heat received. Still, if the entire wall surface of the fire building is involved, the separation of 50 or 100 ft may not be enough to prevent an exposure from becoming involved. The closer together buildings with combustible siding are the greater the danger. A hoseline must be positioned to protect this exposure before fire extends to it. Otherwise, the initial attack crew may come out of the original fire building to find a severe fire roaring away in the next building. The speed with which this can occur is sometimes startling. Once my company was called to the scene of a cellar fire in a three-story balloon-frame apartment building 25 x 60 ft. The first line was already operating into the cellar via the interior stair, but fire was reported spreading up the walls toward the cockloft. My partner and I were assigned to take the saw to the flat roof. We joined several other firefighters already cutting holes to allow the ascending flames to vent out rather than spread beneath the roof and burn off the top of the building.

As part of routine duties, I made a fast survey of all four sides of the structure, checking for alternate escape routes, fire extension, and persons in need of assistance. I noted a similar three-story frame building, exposure 2, about 3 ft to the left of the fire building. Things were looking pretty good at this stage. Water and steam were coming out of the front cellar window as I looked over the edge of the roof. There was a little fire venting out a side window toward the rear, but I felt little concern. Obviously, the line was operating in the cellar, and in a matter of minutes they would put out that part of the fire. Our job was to locate the extension into the cockloft and vent it and the heavily charged top floor.

As the first 4x8-ft vent hole was being cut, a large quantity of black smoke started to rise from the alley between the two buildings. Before anyone had a chance to react, the smoke turned to a solid sheet of flame for the entire 60 ft depth of the building, roaring like a jet engine 30 ft above the roofline. The radiant heat from the massive inferno drove the entire roof team into retreat, a headlong scramble for the aerial ladders positioned on the far side of the structure. In the confusion, my partner was bumped by another member and nearly fell off the roof. The radiant heat was severe enough to burn the paint on our helmets. It melted the reflective stripes on our coats and gave several members second-degree burns (Fig. 4–25).

The fire spread almost instantly into all three floors of the fire building and exposure 2, resulting in a third alarm before control was achieved. The culprit: asphalt siding and narrow separation. The small fire I had seen venting out of the cellar window extended to the asphalt siding, commonly known as gasoline siding. The opposing wall was sided with the same material. Remember that two closely facing vertical surfaces are the ideal geometric arrangement for the spread of fire. The thermal radiation feedback from each wall serves to heat up the opposing surface, releasing additional flaming gases. The process is self-accelerative. Once started, such a fire will rapidly run out of control. If a single water stream of even a small caliber had been in position in that alley when fire vented out of the original window, the fire would have been controlled in the original building. Instead, two structures were destroyed and several members were injured.

Fig. 4–25 An advanced asphalt siding fire requires immediate use of a master stream.

Remember the following priorities: rescue, confine, extinguish! Additionally, remember that, while direct-flame impingement and flying brands may attract a lot of attention, radiant heat can sneak up on you, igniting structures upwind and considerably far way. To protect these threatened structures, water must be applied to whatever surface is absorbing the radiant heat energy. As with most situations, the stream best suited for this task may come from a variety of different nozzles or appliances. There is a considerable variety of nozzles and appliances, and there is a considerable variety of devices for directing or applying streams. A moment spent considering the benefits and drawbacks of each may pay dividends when you need them on the fireground.

Nozzles and Appliance

A nozzle does several things to the water it delivers. First, the shutoff portion, usually a ball valve, regulates the amount of water flowing past the opening from zero flow through wide open. In normal use, set the nozzle in the full-open position, thus allowing it to operate as designed. If it is necessary to reduce the flow to maintain control, throttle down the handle until you reestablish control. This should only be a temporary measure. The correct procedure should be to notify the member supplying the water to reduce the pressure to the correct range. Of course, if only a portion of the flow is required for fire control, crack the nozzle appropriately.

The next function of the nozzle is to increase the velocity of the water flowing through the hose (Fig. 4–26). By narrowing the area that the water has to pass through, the nozzle makes the water go faster. This is one feature that gives the stream its reach. If the nozzle opening is increased beyond its maximum optimum nozzle size for each diameter hose, the stream will start to decrease in reach. This is easy to understand. Think of a 2½-in. hose flowing 250 gpm through an open hose butt. The 250 gpm comes out of the open butt and falls to the ground in 3 or 4 ft. Now visualize that same 2½-in. line flowing 250 gpm through a 1⅛-in. solid tip. The reach should be more than 60 ft. The closer you get to the 2½-in. diameter of the open butt, the less velocity the water will have if the flow stays the same. As a rule of thumb, no nozzle should have an opening greater than half the diameter of its supply. For example, on a 2½-in. handline, the 1¼-in. tip is the largest that should be used. For a 1¾-in. line, the $^{15}/_{16}$-in. tip is about the maximum.

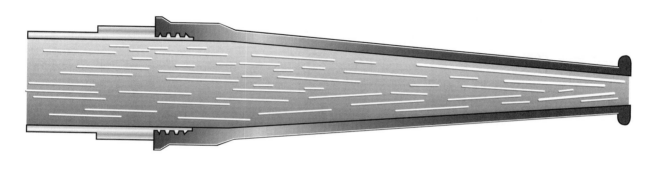

Fig. 4–26 The converging bore of a solid stream nozzle shapes the stream for maximum impact at the lowest energy loss.

It is possible to use a larger tip if the hose stretch is kept short and the discharge pressure is raised sufficiently high. For example, a 1¾-in. line that is only 100 ft long could supply a 1⅛-in. tip if the discharge pressure was 160 psi. This rule applies to fog nozzles as well as solid tips. Trying to use a 250-gpm fog nozzle on the end of four lengths of 1½-in. hose will result in a short-range stream with little striking power (the force produced by the weight of the water multiplied by its velocity). Again, the 1⅛-in. solid tip would have more striking power than the open 2½-in. butt if both were flowing 250 gpm.

Striking power is a useful factor to keep in mind when the fire is hidden, either in deep-seated piles of materials or behind building surfaces. You can make the water go faster to blast through materials by reducing the size of the nozzle opening. I have used this technique with success at rapidly moving cockloft fires as well as at fires hidden behind partitions. A stream with good pressure will strike through a plasterboard or tile ceiling. Water application is made without having to pull down the ceiling with hooks—a time-consuming process—and it doesn't have to wait for the entire opening to be made (Figs. 4–27 and 4–28).

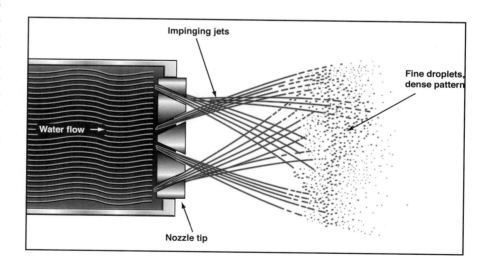

Fig. 4–27 Impinging jet nozzles create a dense fog by directing streams of water into each other.

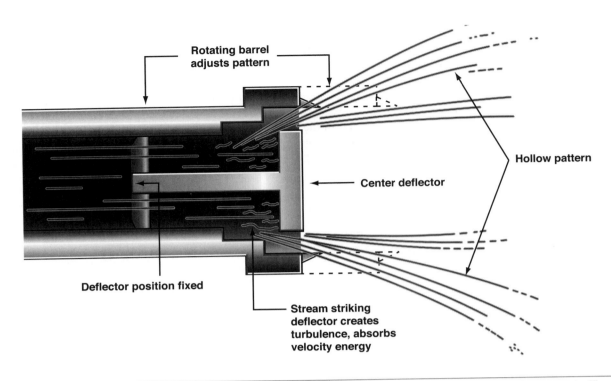

Fig. 4–28 Peripheral fog nozzles create a hollow center pattern by spreading the water out around a center baffle.

The third function of a nozzle is to give a stream its shape. This is where the greatest variety of styles is found. Some nozzles, notably solid-stream nozzles, are designed to provide a compact, cohesive stream with maximum reach and striking power. Others are designed to provide a fixed, cone-shaped spray pattern for use where it would be dangerous to use a straight or solid stream, such as around high-voltage electricity. The most versatile style of nozzle is the variable-pattern, peripheral fog nozzle, which allows you to select any stream pattern from straight through wide-angle spray simply by rotating the nozzle's outer barrel.

I don't get involved in arguments between proponents of fog nozzles over solid stream nozzles. It shows that people are not familiar with all the tools at their disposal. Each type has its place. In many districts where overcrowded tenements with tight bends and long hose stretches make up the majority of the response area, a solid-stream tip is favored. It allows maximum penetration of hidden voids. It extinguishes fire without creating suffocating clouds of steam, and it allows 50 psi more engine pressure to be used for friction loss, rather than nozzle pressure.

At a fire on the top floor of a five- or six-story apartment house, 350 ft or more of hose might be required to reach the seat of the fire. Using a $^{15}/_{16}$-in. solid tip on 1¾-in. hose, you can flow 180 gpm with a nozzle pressure of 50 psi and an engine pressure of 190 psi. The same stretch with a fog nozzle on the end would require 240 psi at the pumper. The low pressure capability of the solid stream nozzle is recognized in the regulations of FDNY. At fires in high-rise buildings where hoselines can be supplied from the standpipe system, the initial line is *required* to be 2½-in. in diameter and equipped with a solid stream having both 1⅛-in. and ½-in. stacked tips. Initially, pressures on the upper floors of these buildings may be quite low, as low as 30 psi flowing. By providing the 2½-in. hose, the unit virtually eliminates friction loss with the low flow of the ½-in. tip, which may be sufficient to extinguish or control smaller fires before the pressure is augmented by building pumps or fire department pumpers. If the flow proves too low, the line can be upgraded to a 250-gpm line by removing the outer ½-in. tip as soon as the proper flow becomes available.

Of course, the previous selections aren't the answer to every situation. I know of several cases where members tried to take these nozzles and make them the primary attack nozzles in suburban areas, with poor results. The variety of fire hazards in the suburbs is broader than in the inner city. One company with solid-stream nozzles on their preconnected lines is the first-due engine to a large bulk oil terminal with heavy flammable liquids traffic through the area. In the event of a serious Class B fire, the two preconnected lines will be a hindrance to operations; either in capability or for the time it takes to switch nozzles. A peripheral fog nozzle should be the primary choice for the preconnected lines in this area. Besides the flammable liquids hazards, many suburban homes now have propane barbecues, which demand an adjustable fog pattern for approaching a burning cylinder (Figs. 4–29 and 4–30).

Other types or classes of fires call for specialized types of nozzles, for example, foam-aspirating nozzles, piercing or bayonet nozzles, and cellar nozzles. Unfortunately, the need for these specialized nozzles isn't usually evident as the first lines are being stretched. Once these lines are operating, it can be quite troublesome to stop and change nozzles when the need is recognized. A solution is to ensure that all nozzles are of the break-apart design, with male hose threads on the discharge side of the shutoff handle. This not only ensures rapid placement of the tip, it also allows the line to continue to operate while the most efficient tip is being brought to the point of operations. There is no need to shut down the line at the pumper, since control is at the nozzle man's fingertips. It also allows you to add lengths of hose rapidly to extend a line that is too short, or to break down a 2½-in. line to 1¾-in. or 1½-in. after the main body of fire has been knocked down for overhauling.

Fig. 4–29 A 1¾-in. hoseline can easily be extended from a break-apart 2½-in. nozzle for overhauling.

Fig. 4–30 Break-apart nozzles can quickly be fitted with a foam nozzle if needed.

Types of Fog Nozzles: Constant Gallonage vs. Constant Pressure

The following two main types of fog nozzle are currently in use: the constant-gallonage and the constant-pressure nozzle. Past designs of fog nozzles included variable-gallonage nozzles as well as combination solid-stream and impinging-jet fog nozzles. While automatic nozzles have gained the lion's share of the market, some departments have decided to return to the constant-gallonage type. You should understand how each works, whether ordering new equipment or using what you already have.

The constant-gallonage nozzle is really something of a misnomer. It refers to the effect that a change in the stream position (straight through wide fog) has on the flow. Early versions of peripheral fog nozzles varied the flow as the pattern of the stream changed. By rotating the outer stream-shaper barrel, the opening around the center baffle increased or decreased. In many cases, this was a very drastic change, up to 33% of the nozzle's rated flow. Usually the lower flow was in the straight-stream position. This wasn't sufficient for an interior combination attack. Improvements to these early designs resulted in the opening around the center baffle remaining in a fixed position while the stream was changed from fog to straight. Thus, a *constant* amount of water flows out of the opening regardless of the stream position. The gallonage is constant only as long as the nozzle pressure remains constant, hence the misnomer. If the pressure drops, so does the flow and vice versa.

These nozzles are usually designed to deliver a specified gallonage when the nozzle pressure is 100 psi. This pressure results in the best patterns, as well as in providing the optimum reach and striking power. Raising the nozzle pressure above 100 psi can be counterproductive, giving only a marginal increase in flow, yet producing compounding increases in nozzle reaction. Nozzle pressures below 100 psi produce streams with poor reach and striking power and with less-dense fog patterns. The reduced flow may be insufficient for the intended purpose. Some models have a feature that permits the nozzle man to change the size of the discharge opening manually, generally by twisting a ring around the nozzle. This permits the nozzle man to adjust to varying water supplies. A problem develops if the setting is inadvertently placed at too low a flow to control the fire. This may result in a 2½-in. handline nozzle discharging as little as 120 gpm, a total waste of time and manpower, since this flow can be more easily applied by a 1½-in. line. The stream will have a good reach, and it may appear to be a good stream, but it simply may not be flowing the needed quantity of water. A good idea is always to ensure when beginning operations that the manual setting is at the nozzle's maximum flow. If insufficient water is available to supply this flow, the stream will show it. At that point, the setting can be adjusted downward until sufficient water becomes available. In the meantime, the water that is being discharged will be discharged at its best efficiency (Fig. 4–31).

A new twist to the constant-flow nozzle that is currently en vogue is the low-pressure fog nozzle. These nozzles deliver their rated flows at either 50 or 75 psi. The intent is to deliver a suitable stream at less than the *standard* 100 psi used with fog nozzles. This is desirable for several reasons, the first being to relieve the nozzle team of some of the work that they have to perform. By lowering the pressure at which a nozzle operates, the nozzle reaction is also lowered. A flow of 250 gpm at 50 psi takes less effort to hold than one of 250 gpm at 100 psi. Those who use smoothbore nozzles have known this for years.

The second reason for using low-pressure fog nozzles was made quite clear during the fatal fire at One Meridian Plaza in Philadelphia. There are times when 100 psi isn't available for nozzle pressure (Fig. 4–32). This primarily occurs during standpipe operations. (See also chapter 6.)

Constant-pressure or automatic nozzles are the most recent improvement in nozzle design. The basic concept is to use a spring to regulate the size of the discharge opening. The spring acts to counter the nozzle pressure. As the pressure increases, the spring allows the size of the discharge opening to increase, flowing more water. If less pressure is available, the spring contracts, reducing the size of the opening.

Fig. 4–31 It is counterproductive to set an adjustable-gallonage 2½-in. nozzle at 120 gpm for initial attack.

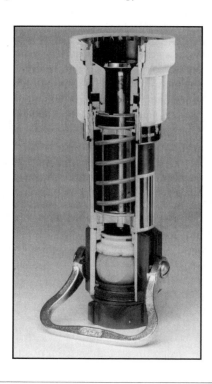

Fig. 4–32 An automatic fog nozzle has a center baffle that can move in or out, opening or closing the size of the waterway, depending on the amount of pressure that pushes against its internal spring.

The spring acts similarly when the nozzle man adjusts the gallonage on the manually selected nozzle. In theory, the spring should maintain a constant 100 psi as long as the flow past the spring is within the nozzle's design operating range. Some of these nozzles can flow from 60 to more than 350 gpm with a relatively constant nozzle pressure.

It should be noted that these nozzles don't change the laws of hydraulics. They cannot make water. A hoseline is a closed system until the stream leaves the nozzle. Whatever flow is pumped into the pumper end of the line will come out the nozzle end—no more, no less. If insufficient flow is being provided, insufficient flow will be applied. While nozzle pressure helps to give a stream its reach, it is the quantity of water in gpm that puts out the fire. With automatic nozzles, as with the manually selectable gallonage nozzles, it is possible to have a good-looking stream with sufficient reach that is flowing less than the desired gpm. Again, a 150 gpm stream from a 2½-in. line is wasteful. With non-adjustable (solid-stream and fog) nozzles, any decrease in flow is readily apparent, and steps can be taken to remedy the cause of the decrease. The reduction may not be noticed with automatic nozzles; at least not right away. The nozzle team finds that their attack is stalled or, worse, that they are being driven back. A call to the pump operator confirms that there is sufficient pressure, yet something is wrong. The culprit is likely to be a kink in the hoseline, which can act just like a gate valve, reducing the flow. If the apparatus is equipped with flow meters, the decrease will immediately be visible, and steps can be taken to remove the offending kinks. If there are no flow meters, the change may go unnoticed until the attack runs into trouble.

The nozzle man is the person best able to judge the stream. All personnel should be familiar with the accompanying backpressures at the desired flows. If the line is suddenly much less difficult to control, chances are that there is less than the desired flow. As a rule of thumb, if one person can control a 1¾- or 2-in. handline, the line isn't delivering its designed flow. It should be *work* to control an attack line. If it isn't, something is wrong.

Specialized Nozzles and Appliances

Fires in certain problem areas, such as shafts, cocklofts, and cellars, can create difficulties for those trying to put water on them. Heavy fire or unusual layouts can prevent direct entry to an area. A fire in a vertical shaft such as a pipe chase can spread quickly and pose the added danger of falling objects, which could cause injury. These and other situations may require using specialized discharge devices to hit the seat of the fire. An efficient, routine interior operation (locate the fire, force entry, confine the fire, ventilate, and extinguish) will rarely involve these specialized appliances. However, under certain circumstances, they might be the only thing that prevents the destruction of an entire building. Tools that get so little actual use must be thoroughly understood if you expect to fall back on them, *the last line of offense* (Fig. 4–33).

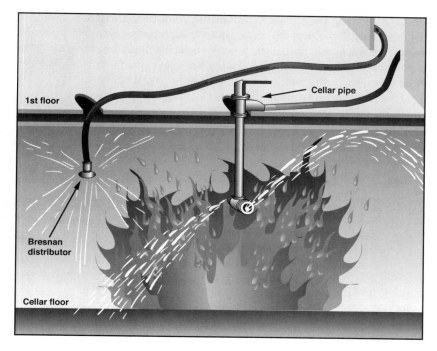

Fig. 4–33 Cellar pipes and Bresnan distributors can be useful for knocking down heavy fire when conditions prevent direct entry into the fire area.

Cellar nozzles and distributors are among the last-resort devices for applying water where entrance to an area is impossible or impractical. They have been further removed from active duty by the development of SCBA, power saws, and high-expansion (high-ex) foam, which is now used in similar situations. Still, none of these completely eliminate the need for them. Unfortunately, by the time the decision is made to employ them, many times the structure is no longer savable. To be effective, they should be used as soon as practical after the need for them is recognized. This generally happens when an interior handline attack has failed, either due to heavy fire, an inability to vent the area, or to reach the seat of the fire. If no practical alternatives exist—such as increasing the water flow, or ventilation, or creating more access points—the cellar nozzles can deliver sizable volumes of water from more tenable areas. They must be carefully monitored for effectiveness; since they may have to be relocated a number of times to ensure that the entire fire will be darkened down. Since their task is to darken down the fire rather than extinguish it, once the cellar nozzles have served their purpose, they should be shut down and the handlines advanced to finish the job if conditions permit.

There are many difficulties associated with the use of cellar nozzles. The first is that the area from which they must be applied may be untenable. The floor above any working fire is a smoky, hot location, but cellar fires can produce severe conditions. A protective handline should be stretched into the first floor location of the cellar nozzle to protect the persons cutting the access holes for the nozzles, as well as to protect the entire crew while operating the device. The next factor is the structural stability of the first floor itself. Remember that the fire has been attacking the supporting members all during the time of the initial attack and for an unknown period of time before that. Even if the conditions on the first floor aren't severe, you may be operating unknowingly over a heavy body of fire. Extreme caution is indicated when putting one of the cellar nozzles into use; you must know its capabilities. Cellar pipes, such as the Baker pipe, only apply water in one or two directions at a time. For this reason, they must constantly be manned, with personnel directing the stream for greatest effect. This can be too dangerous to be worth the risk of floor collapse. One redeeming factor is the reach of the stream from these devices, which can be 50 ft or more. It may be possible to cut a hole over a safe area and use the reach of the stream to knock down the fire.

Revolving nozzles, on the other hand, such as the Bresnan distributor, have a very limited range, a radius of only 15–20 ft. This means that the nozzle has to be placed nearly directly over the fire. As a trade-off of sorts, though, it isn't necessary to man these devices continually, since they distribute water in a circular pattern without human guidance. To facilitate using either of these devices, it is recommended that a gate valve be installed on the hoseline one length back from the nozzle so that the flow can be controlled from a position of relative safety. By placing the gate valve 50 ft back from where the nozzle is operating right over a heavy fire, the crew controls the flow without having to worry as much about the floor caving in beneath them, since they are 50 ft away from the heavy fire. Hopefully, there is not more fire directly beneath them (Fig. 4–34).

High-expansion Foam

No mention of cellar pipes and cellar fires would be complete without at least a brief discussion of high-expansion (high-ex) foam. As previously mentioned, it is replacing the cellar nozzles in many cases because of its ability to be applied from a relatively distant, safer area. High-ex is an extremely expanded (from 400:1 up to 1,000:1 ratio) solution of water and detergent. Its primary function is to fill an enclosure, thus pushing out the products of combustion and replacing them with a water solution, which will then cool the source of the fire.

Fig. 4–34 High expansion foam may be used to control fires in cellars and other locations where it is too dangerous to send firefighters.

High-ex foam has primarily been used as a defensive tool when the troops have been pushed out of a cellar and nothing else seems to be working. *Let's throw high-ex at it since nothing else seems to be working.* What's likely to happen is that the building will burn down and the high-ex will get the blame. In reality, when the high-ex was put in, it was already too late in the majority of cases. When the crews withdrew from the cellar, chances are that the first floor was evacuated as well. It's that action that permits the upward fire extension, resulting in loss of the structure. When high-ex foam is discharged into the area, an opening must be provided near the upper spaces, preferably opposite the place where the foam is being injected, to allow the heat, smoke, and flame to escape as it is pushed ahead of the advancing foam blanket.

Much like operating a fog nozzle into an enclosed area, ventilation must be provided or the fire will follow its own paths, usually up behind furred partitions, pipe chases, and the like. If crews aren't able to open these areas and extinguish this extension, the building will be headed for destruction. Of course, if the reason that the cellar has been evacuated is for fear that the first floor will collapse, then it isn't possible to operate on that floor. In such a case, the IC may be forced to recognize the likelihood of a total loss. If high-ex foam or any other method succeeds in darkening down the fire, it is a win. If not, it is a draw. The building was designed to be a loss by the architects, contractors, owners, and occupants, as well as by the building and fire prevention officials who permitted a heavy fire load to exist in an un-ventable area without automatic sprinklers (Fig. 4–35).

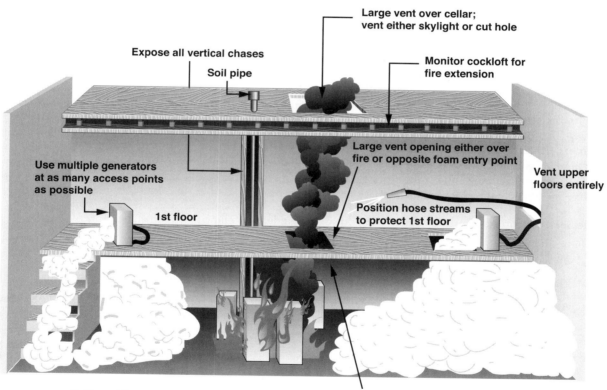

Ceiling of fire area will be full of superheated gases being driven ahead of foam. They MUST be vented.

Fig. 4–35 In order for high-ex foam to successfully control a cellar fire, several other steps must also be taken, including ventilation over the fire and stopping upward extension. If the floor above the fire is unstable due to potential collapse, these actions may be unsafe.

Several factors must be present for high-ex foam to stand a chance of extinguishing the fire. It must be applied as soon as practical after the need for it is recognized, and while the floor above is still tenable. It must be applied in sufficient quantity to do the job. Fire and heat break down large amounts of foam, and huge amounts may be going into rooms or areas where there is no fire. Just like water, if the amounts given are insufficient to cool the fuel, the fire will grow until it burns down the building or runs out of fuel. Finally, the foam must be applied where it can reach the seat of the fire. Fires in remote areas, behind partitions or stock, may not be accessible to the foam. In such a case, it may take time to realize that the foam isn't able to do its job. By the time the equipment has been moved to a new location, it is often too late.

High-ex foam, like cellar pipes or distributors is unlikely to achieve total extinguishment without substantial luck. In some cases, if structural stability seems assured, as is likely for a cellar fire in a reinforced concrete building, it may be possible and actually beneficial to use the high-expansion foam to gain control of the fire, then send crews in with handlines for final extinguishment and overhaul. This would be required for example in buildings where a thorough search was not completed before the units were forced to withdraw. In other situations such as fires in commercial buildings where all occupants have been accounted for or where the structural integrity is in doubt, the high-ex foam should be continuously reapplied, allowing for a prolong soaking time to attempt complete extinguishment without endangering fire personnel.

If it is necessary to enter a high-ex blanket, all members must be properly prepared. Submergence in foam is a very disorienting experience. During a furniture store fire, I led a crew into the cellar of an attached exposure to the fire store. The building we were in was being filled with high-ex that was being pumped into the cellar of the main store, through an open doorway we were sent to protect. The closest analogy I have for this experience is swimming under murky water, something I have done more than once as a diver in New York harbor. Light travels only a few feet in high-ex and is widely diffused through all the bubbles, making it difficult to pick out the source of fire, or even a spotlight type beam from a handlight. There is virtually no visibility, akin to the thickest smoke conditions, Even the thermal imaging camera (TIC) is blinded by the foam blanket.

Sound is also muffled, the already muffled voices from within an SCBA face piece can barely be heard 3 or 4 ft away. The foam blanket acts as a shield from any heat buildup overhead, which can be a serious problem if firefighters either break down the foam blanket through substantial movement or by use of a hose stream on a fog setting. The sudden loss of the shielding high-ex blanket, coupled with the introduction of the fog stream into the high temperature layer could result in steam burns to members exposed. Caution should be the watch word whenever members enter a high-ex blanket.

Members entering a high-ex blanket should do so only under the following circumstances:

1. The entry is for lifesaving purposes.

2. The entry is needed for fire containment purposes, and the structural stability of the area is assured.

3. All members are properly equipped as follows: SCBA donned, with personal alert safety system (PASS) device activated, operating in constant contact with either a search guide rope or a hoseline. The lead member in particular must be using a pike pole or other hook to probe each step in front of them to avoid stepping into holes, tripping, etc. Move slowly, feeling for solid footing with each step. Better yet, crawl, but be very aware that you will not sense any heat through 3 or 4 ft of high-ex.

4. Electricity to the affected area should be shut down, not so much for fear of an electrical charge through the foam blanket, but to prevent a wet firefighter from accidentally stumbling into a live electrical source or into machinery that may be running or start automatically.

5. A rapid intervention team (RIT) is on standby at the point of entry, dedicated to this operation.

6. If entering for fire extinguishment, have the handline charged and operate only on a straight stream position, sweeping the ceiling well ahead of the nozzle team's advance. This will reduce the confined heat without causing a major break down of the foam blanket until you are certain of the level of heat overhead.

Class A and Compressed Air Foam Systems

For decades, scientists have been experimenting with ways to make water a more effective extinguishing agent. In the 1960s, chemists discovered ways to reduce water's surface tension. Surface tension is the tendency of water molecules to cling together. It results in much water from fire a stream running off the surface it's applied to rather than soaking into it. With materials, such as baled hay or cotton, stuffed furniture or mattresses, that created serious problems with extinguishing efforts, requiring very extensive opening up efforts to complete extinguishment. Ordinary, untreated water applied to these materials often extinguished only surface flaming, leaving a deep-seated smoldering fire burrowing inside, to rekindle later if the material was not torn apart to expose the hidden fire.

Wet water was the description given to the several additives (primarily surfactants, a derivative of detergents) that first addressed the reduced surface tension. I was first introduced to this product in 1970, at my first hay barn fire with the Stillwater (OK) Fire Department. I was not overly impressed with it, since the barn and all the hay in it was totally destroyed despite the use of the *magic elixir*, but I had no frame of reference as to how much overhauling the agent saved us. To me the barn was the reason we were there, and it was a total loss. To the experienced Stillwater firemen, the barn was a total loss before we even left quarters, the minimal overhauling was the primary benefit of the *wet water*.

At approximately the same time, the Union Carbide Company developed *rapid water*, another additive, this one designed to improve the flow of water through a hoseline. By injecting *Polyox* as it was marketed, it was possible to flow 250 gpm from a lighter 1¾-in. line that typically can only handle about 180 gpm. Several fire departments, including the FDNY saw the potential benefit of this improvement in its ability to allow fewer firefighters using 1¾-in. hose to flow the larger volume that previously required the heavy 2½-in. hose. Rapid water injection modules and 1¾-in. hose were added to all FDNY engines, and one or two firefighters were taken off. Unfortunately, by the late 1970s when serious problems developed with maintenance of the injection modules after several years of heavy use, the FDNY was left with 1¾-in. lines that were limited to 180 gpm and two fewer firefighters per engine to stretch the 2½-in. hose when the larger flows were required.

During the 1980s, many rural departments had taken a considerable liking to the various wetting agents, but they were still largely confined to the advantages of reduced overhauling. Manufacturers continued to experiment with their formulations, adding and modifying various characteristics, introducing the concept of Class A foam, which combined the lower surface tension of the wetting agents with a bubble forming surfactant for use during forest fires. The foaming agent allowed a thicker coating of the agent to cover a porous surface, depriving the fire of oxygen, but more importantly, giving the water trapped in the bubbles more time to soak into the nooks and crannies of the burning material to root out the hidden fire inside. A similar concept has been in use for many years in the form of high-ex foam, which is used to bury burning Class A materials and provide a longer *soaking time* for the amount of water discharged. High-ex foam is actually one of the first Class A foams developed.

Like *light water*, another additive developed for use at Class B fires, primarily involving aircraft crash firefighting, early Class A foams were marketed for use with existing nozzles and appliances. Both agents were designed to be discharged from a standard non-aspirating fog nozzle. Later, air aspirating foam nozzles designed for Class B foams were used with the Class A foams, with the resulting increase in expansion ratios seen with these nozzles that add air to the solution stream. Fog nozzles generally only cause a 2:1 or 3:1 expansion of most foam solutions. Air aspirating nozzles typically create a 7:1 or even a 10:1 expansion ratio. A higher expansion ratio means a thicker blanket of foam and longer soaking times, which is desirable. But it comes at a price, since the streams do not have as great a reach as a water stream from a solid tip or even a fog nozzle, and they have no ability to penetrate through ceilings.

In the 1990s, as the *urban-wildland interface* fire problem gained recognition, many rural departments found themselves fighting structure fires in the middle of the forest, with *forest fire* equipment, including Class A foam. What was soon apparent was that this forest fire foam worked as well inside a house as it did in the backyard. A particularly useful feature of the Class A foams for this application was to coat the exterior of an exposed structure with foam as a wildfire approached. The foam helped to prevent ignition. For this purpose, it was found that the greater expansion ratios achieved by aspirating the foam solution allowed the foam to cling to walls and other vertical surfaces better than the wetter foam produced by fog nozzles. Manufacturers pushed this discovery still further by examining whole new ways to control the expansion ratio—not by adding the air at the nozzle, but by injecting a measured amount right at the pump. Thus was developed the commercial compressed air foam system (CAFS).

CAFS can be produced in several ways. One method recently marketed for wildfires involves connecting a small diameter hoseline to a gasoline-driven leaf blower worn on a firefighters back. Early apparatus-mounted units used separate power supplies to run a compressor. More modern units use a power takeoff (PTO) driven compressor operated off the main vehicle engine. All CAFS require the same basic components as follows:

- a foam concentrate supply

- an injection device to proportion the correct amount of concentrate into the water stream (for CAFS use, Class A concentrate is usually used in a range of 0.3–0.6%, or 3–6 gal of concentrate for every 1,000 gal of water used, as opposed to 3–6% for Class B foams)

- a source of compressed air that can be injected into the foam/water solution at the correct rate

Some advocates of CAFS claim a remarkable improvement in fire extinguishing capabilities over plain water. Claims have been made that a CAFS stream can knock down a fire in one-fourth of the time of a conventional hoseline using only one-third the water. This remains to be proven, although there is a great deal of anecdotal "evidence" to support claims of greater efficiency. To date, no scientific measurements exist to back up such claims. One noted CAFS expert, Dominick Coletti of the Hale Pump Company and author of *Class A Foam-Best Practices For Structure Firefighters* (1998, Lyons Publishing, Royersford, Pa.) acknowledges that the addition of a foaming agent does not change the laws of thermodynamics or the physical properties of water. According to Coletti "the critical application rates and needed fire flows for structural suppression cannot be reduced by adding surfactants to the water."

Water extinguishes by cooling. CAFS solution is 99.5% water. The tiny percentage of foam concentrate does not radically alter the properties of the remaining water. All it does is allow the water to penetrate deeper into porous material and by forming a foam blanket, remain in contact with the hot fuel on vertical surfaces and ceiling longer than plain water does.

The room on fire is producing X amount of BTUs. It will take the same amount of water or Class A foam solution to remove those BTUs. Over the past 35 years, the applied fire flows (as opposed to the theoretical fire flows) has jumped from 8–10 gpm high pressure fog guns to 150–180 gpm handlines. I have taught tactics classes for departments that use CAFS, and when I mention those numbers, firefighters often say how high that is compared to the flows they are using with CAFS, which is often in the 50–60 gpm range. They equate the 50 gpm CAFS stream as the equivalent of a 180-gpm water stream. The way I approach these claims is to review the size of the actual fire areas involved with them and compare that to any of the standard fire flow formulas. If the fire is in a 10 x 12-ft bedroom, the 10-gpm high-pressure fog gun of the 1950s and 1960s will eventually extinguish it. A 50-gpm CAFS stream will obviously overpower it. If the department does not encounter more than a three- or four-room residential fire, the equivalent of a lightly loaded 20 x 25-ft area, the 50 gpm CAFS stream will have little difficulty. If the same stream were used at a fire involving part of a single store in a strip mall—say the same 20 x 25-ft area—the 50 gpm CAFS stream is likely to be inadequate, since the fire loading of many stores is typically double or triple that of a residential area. That small area would likely take 100 to 150 gpm to control.

This theoretical discussion of CAFS use is borne out by an evaluation of CAFS by the Boston Fire Department in 1992–93. In that test, Engine 37, Boston's busiest unit at the time, was outfitted with a CAFS system for one year. Their overall evaluation was that the CAFS was equal to or superior to water as an extinguishing agent (*Compressed Air Foam for Structural Firefighting: A Field Test*—Boston Massachusetts, Technical Report TR-074, U.S. Fire Administration). Engine 37's apparatus was piped to supply two foam handlines as well as the deck gun, with flows of 70–130 gpm on the handlines.

Boston typically looks for about 150 gpm from their plain water 1¾-in handlines, so the system did show a distinct advantage for the CAFS use. But again, it was not a scientific study of capabilities, since there is no examination of whether the fire would have been within the capability of a 70- or 130-gpm water stream. It is interesting to note that after this successful evaluation of the extinguishing capability of CAFS, Boston has not elected to install CAFS on any of their new apparatus. One problem with a CAFS system is that it is a fairly complex system, which requires maintenance. Like many urban departments with a heavy workload and shrinking budgets, a system that adds another $30,000–50,000 to the price of each pumper and then adds to the annual maintenance costs is not seen as a cost effective measure. Boston experienced repeated failures of the foam bladders on this unit, as well as compressor failure, which put the unit out of service (OOS) completely as a CAFS unit.

This situation goes right back to New York City's experience with the breakdown of the rapid water units in the 1970s. The failure of such a critical component as an air compressor puts firefighters' lives at risk to a far greater extent in an urban environment, where firefighters may be operating on floors above the fire, depending on the hoseline to protect their escape routes, than they would in a wildfire situation. Until such issues as reliability of components, cost and maintenance are resolved, it is unlikely that many urban areas will make the switch from plain water to CAFS. But once these items are overcome, I believe there will be a rapid changeover. As saving the barn was not the driving force behind using wet water, the benefits of reduced labor during the always-dangerous overhauling stage alone will be a driving force behind CAFS use at structural fires.

Special Nozzles: Applicators, Piercing Nozzles, Bent Tips

A number of truly special nozzles are available to the fire service for unique applications. They are, in many cases, considered antiques, unsuitable for use by a modern, progressive fire department. A problem with our hardware has helped to foster this image, yet an objective evaluation of the fire problem and potential often reveals that these devices still have some merit, especially when used by innovative engine companies.

Applicator pipes or bent pipes were once nearly standard equipment for most engine companies, and they have been around for decades. Generations of firefighters learned to use these devices to extinguish oil fires as well as to extend the reach of the line into an involved ship compartment without being right next to it. In fact, their use is still very common aboard ships. Land-based firefighters have moved away from the old 60-gpm combination Navy nozzle, which is normally needed to connect the applicator pipe to a hose line. Instead, they have searched for more efficient, higher-volume lines for our 1½-in., 1¾-in. and 2-in. handlines. As a result, the applicators and their cousins, the piercing nozzle and the foam-aspirating tube, have become virtual museum pieces.

Many firefighters have encountered situations where the newest state-of-the-art nozzle required that the firefighter put himself in a very dangerous location to extinguish the fire. In reviewing these incidents, the question was often asked, "What could we have used to do that faster or from a safer position?" Occasionally the older nozzles were brought back into use; however, the old Navy nozzle was still needed to complete the hookup. In addition, the need for these special items often isn't obvious when the first line is being stretched. Once this line is charged, it is too time-consuming to shut it down, change the nozzle, and then go back into operation. It may also be too dangerous, not having any water to stop a sudden flare-up.

At least one manufacturer has recently recognized this problem and has taken strides to solve it. The solution is simple, again centering on the break-apart nozzle. Normally, this line will be equipped with an automatic fog tip with a built-in twist shutoff. Removing the tip instantly converts the line to an efficient solid-stream nozzle. A line of accessories, including bent applicator pipes, piercing nozzles, and foam-aspirating tubes, are available with 1½-in. hose threads on their bases. Now a line can continue to operate while a needed accessory is brought up. Then it is only shut down momentarily at the nozzle while the appliance is put on. The stream is then ready to go again with the most effective discharge head in place. Let's look at some of the applications for these older nozzles and appliances.

The situation that I believe has the greatest benefit from the use of the applicator pipe is the extreme wind-driven fire in Class 1 *fireproof* high-rise multiple dwellings. These fires occur in apartment houses and have killed several firefighters and dozens of civilians. They have defied conventional attacks and demand an alternate method of water application be found. The applicator pipe is the most readily available, effective tool to be found.

Gasoline tank trucks pose a severe threat to any and all fire departments. None are immune from potential disaster. In the past, applicators served only to shield members from the heat of a flammable liquids fire, and their use was counterproductive at times. The water they applied broke down or washed away foam blankets.

The world of flammable liquids fires has changed in recent years, and the bent applicator has become a highly useful tool. The great majority of gasoline tank trucks in the United States today are constructed of aluminum. When exposed to fire, these trucks tend to melt from the top down to the surface of the fuel. Standard nozzles produce a high-velocity foam stream that, when directed at these tanks with open tops, splashes additional fuel out of the container, worsening the situation. The applicator pipe supplied with aqueous film-forming foam (AFFF) can be used to extinguish these fires from the ground level with greater safety than having a member climb onto the truck to apply the foam. This wasn't possible years ago, when protein foam was the only choice for flammable liquids fires. Unlike protein foam, AFFF doesn't require a great deal of aeration. The low-velocity fog delivered by the applicator allows the foam to be gently applied to the surface of the fuel so as not to cause the splashing that is so common with high-velocity types. For smaller fires on top of these tank trucks, the applicator has another use, to discharge dry chemical up and over the top while the operators remain safely on the ground. In order to accomplish this, the fog head must be removed. One person directs the stream while another person discharges the dry chemical extinguisher (Figs. 4–36 and 4–37).

Fig. 4–36 A bent applicator can be used to discharge dry chemical to extinguish flammable liquid fires remaining on the top of a tank truck.

Fig. 4–37 This is a two-person operation. One member controls the discharge from the extinguisher while the other directs the stream.

Another innovative use of the applicator has been on fires involving the undercarriages of many of the heavier motor vehicles. Many buses and tractor-trailers now rely on air-bag shock absorbers to provide a smooth ride. A fire under these vehicles often results in failure of the rubber air bladders. In turn, the vehicle can drop several inches. A firefighter with part or all of his body under the vehicle can find himself pinned between vehicle parts or between the vehicle and the road. This also exposes the firefighter to chunks of burning rubber. By using the applicator, water can be applied to almost any point under the vehicle without any part of the member's body being placed in danger.

Piercing applicators could also be used to advantage if they could be placed in service easily. At a simple car fire, the task of gaining entry to the hood or trunk area is often made difficult by fire within these areas. By using the piercing nozzle, the fire can be readily extinguished before the hood or trunk is opened, thus simplifying the task of manipulating the locking devices that must often be forced. This device could also be valuable at cockloft fires and deep-seated fires in baled materials (Fig. 4–38).

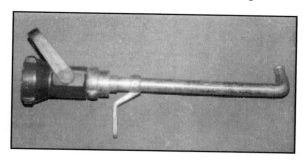

Fig. 4–38 The bent tip attached to a break-apart nozzle is a practical tool for fires in rubbish compactor chutes.

The bent tip is a device whose use seems to be restricted to New York City, where it is seen at almost every major fire as a water fountain on a pumper's discharge. This device is designed to allow stream application at a 90° angle to the hoseline, similar to an applicator. Unlike the applicator, however, it is compact, and it discharges a solid stream. In the past, it was primarily used at fires in dumbwaiter shafts (no longer a common problem) and at cockloft fires (before power saws made roof ventilation so much more effective). Presently, it could prove to be very useful at fires in compactor chutes, where the danger from falling debris and exploding containers warrants extreme care. Here is a device that allows water to be applied from a sheltered position, as well as up or down the shaft.

In closing this chapter on hoselines, let me point out that it isn't the routine but the unusual fire that causes us problems and pain. A good engine company is prepared to accomplish its main mission, to put out the fire, regardless of the circumstances that they encounter. This requires courage, skill, and dedication on their part, as well as innovative thinking and careful planning. Those who think that they have seen everything and have all the answers are in for a rude awakening one day as the devil introduces them to a whole new aspect of his work.

05 Water Supply

Water, at least for the foreseeable future, will continue to be the primary extinguishing agent for the vast majority of structure fires. Even when water is mixed with any of a variety of additives, wetting agents, Class A or compressed air foam, or Class B foam, to increase its effectiveness, it is still primarily water that is essential to form the solution desired. Obviously, having a sufficient quantity of this agent available is a very high priority for all firefighters, from ICs down to the ranks.

An understanding of water supply is crucial. A great deal of training material in the past has been devoted to the study of hydraulics. Volumes have been written detailing various formulas, conversion factors, and friction losses. It would be foolish to attempt to condense this body of knowledge into a single chapter of any text for two reasons. First, with the advances being made in the fire hose industry, the charts and formulas being published today will be outdated relatively quickly. Second, the task is an impractical one. It wouldn't fit; moreover, it isn't required. I feel that I can speak from experience. After majoring in fire protection in college, I spent seven years as a sprinkler system design engineer. For dozens of systems, I have devoted large amounts of time to hydraulic calculation, determining discharge flows, friction losses, *K* factors, and all else.

This was all done by hand, before personal computer programs for such applications became popular. At the same time, I was a firefighter, pump operator, and eventually a line officer in a fire department engine company. At no point during this time was there any need for a great deal of knowledge of hydraulics on the fireground. For the entire time that I've been associated with the fire service, however, I have seen a glaring need to improve firefighters' understanding of water supply. Does that sound contradictory? It shouldn't, if you understand the basic difference between the two subjects.

Water supply is concerned primarily with ensuring that the proper *volume* of water is available at the required location on the fireground. Hydraulics is a study of the factors that influence *how* this water gets to the fireground. To understand water supply, you need to understand the basics of hydraulics. Even more important, however, is to understand that it is the *amount* of water that is applied to the fire that puts it out, not the pressure at which it is applied. If enough volume isn't delivered, the fire simply won't go out. Of course, pressure is useful to provide reach and penetration, but it must be a type of pressure that is useful to the firefighter at that particular time.

Basic Principles of Pressure

To understand water supply and delivery systems, we must first understand water or, more accurately, how water behaves. Water is a fluid that cannot be compressed by applying pressure. This makes it very convenient to pump. Other fluids, such as air, shrink in volume when pressure is applied, only to expand again when that pressure is relieved. That makes for very complex flow calculations. Water is pretty straightforward, since 1 cu ft of it will occupy virtually 1 cu ft regardless of the pressure. That 1 cu ft of water weighs about 62.5 lbs and is equal to about 7.5 gal.

At a relatively young age, we learn that water seeks its own level. There are six additional physical facts that apply to water under any condition found on the fireground. They are as follows:

1. Fluid pressure is perpendicular to the surface on which it acts.

This is what allows us to hold water in a tank and, by use of a gauge, to determine the pressure that it is exerting. The pressure of the water at the base of the tank is the same as that acting on the side where it adjoins the base (Fig. 5–1).

2. The pressure in a confined body of water at rest is the same at all points.

If a length of fire hose is charged and then laid flat with the nozzle shut down, there is no flow. No matter where in that line you measure the pressure—at the pump panel, at the nozzle, or anywhere along the length—it will be the same. When this no-flow condition occurs in water mains, we refer to it as static pressure (Fig. 5–2).

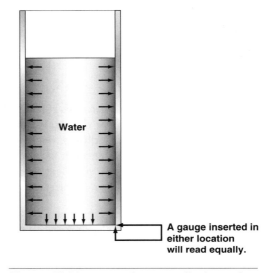

Fig. 5–1 Fluid pressure is perpendicular to the container it acts on.

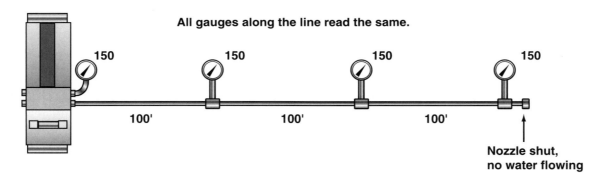

Fig. 5–2 Pressure in a confined body of water at rest (no flow) is the same at all points in the line.

3. The pressure of a fluid in an open container is proportional to its depth (Fig. 5–3).

As with all fluids, water has weight. The higher the water, the more weight acts on the bottom. When water is in an open container, the pressure at the surface is 0 psi. For every foot of height that the water rises, it exerts .434 psi of force. This is known as *head pressure*. (The weight of 1 cu ft of water, 62.5 lbs, divided by the number of square inches in the base of 1 cu ft, 144, equals .434 (Figs. 5–4 and 5–5).

Another useful fact to remember about water is that 1 psi of force is required to push a column of water up 2.3 ft. If you look at Fig. 5–5, it shows a column of water 23 ft high exerting a pressure of 10 psi at the base. But think of it another way. How much pressure does it take to push a single drop of water out of the top of an open valve of a standpipe riser located 230 ft above the pump? (See the bottom of the page for the answer.)

Since that force acts perpendicular to the sides of the container, a pressure gauge inserted at a point anywhere in the column will measure the downward pressure of the water above it.

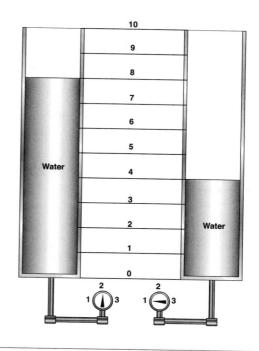

Fig. 5–3 The pressure exerted by a fluid in an open container is proportional to its depth.

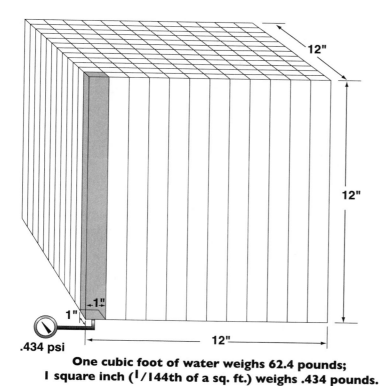

One cubic foot of water weighs 62.4 pounds; 1 square inch (1/144th of a sq. ft.) weighs .434 pounds.

Fig. 5–4 A column of water 1 ft high exerts a pressure of .434 psi at its base.

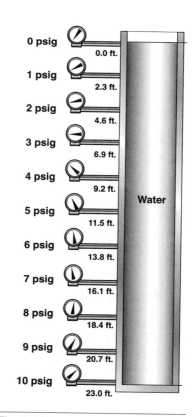

Fig. 5–5 Water pressure increase 1 psi for every 2.3 ft of depth.

4. The pressure of water in an open container is independent of the shape and volume of the container. It depends solely on the depth.

This is what allows water storage tanks to be supported on legs with only a narrow down pipe to supply the flow. This is much more efficient than building a cylindrical container if the desired objective is to provide head pressure, the force resulting from the difference in elevation. Head pressure is useful to move water from an outlet to a nozzle or to the intake to a pumper. On the other hand, if volume is all that is required, with the pressure being supplied by a pump, a low-level storage tank of cylindrical design is all that is required (Fig. 5–6).

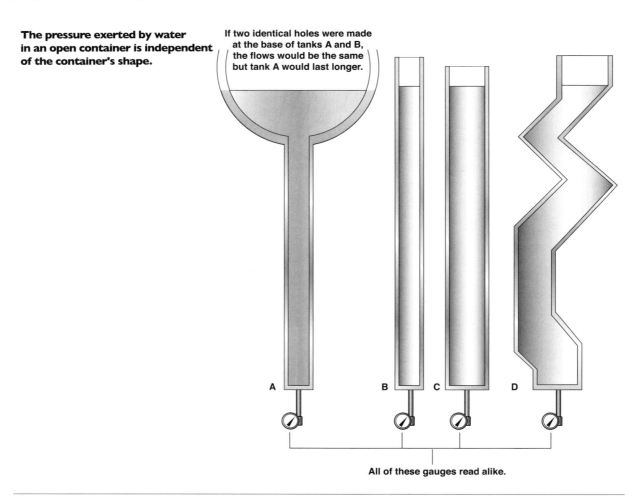

The pressure exerted by water in an open container is independent of the container's shape.

If two identical holes were made at the base of tanks A and B, the flows would be the same but tank A would last longer.

A   B   C   D

All of these gauges read alike.

Fig. 5–6 The pressure exerted by a fluid in an open container is independent of the container's shape.

5. The pressure of a fluid is proportional to the density of the fluid.

As stated in rule 3, all fluids have weight, including air. Naturally, some are heavier than others. For example, mercury is 13.546 times as heavy as water. On the other hand, it takes thousands of feet of air to create what we call atmospheric pressure. The weight of the air above us is equivalent to the pressure exerted by 33.9 ft of water or 14.7 psi. That is the average pressure pushing down on the surface at sea level, often referred to as one atmosphere.

This is a vitally important fact for firefighters attempting to draft water from any open-topped source, since it is the weight of the air above us that actually pushes the water up through the suction hoses to the intake of the pump. The pump doesn't actually suck the water up to the pump when drafting. A special pump is used to remove all of the air in the suction hose, taking the weight off the surface of the liquid, thereby allowing atmospheric pressure to push water to the pump. This one-atmosphere pressure sets the maximum height that water can be siphoned or *drafted* from a container. If the intake of the pump is 34 ft above the surface of the liquid at any time, it will be physically impossible to draft. In practice, even the 33.9 ft is an impossible goal to reach, since friction loss in the hose, strainer, and pump detracts from this maximum theoretical value. Air leaks also take their toll. In addition, the greater the flow attempted, the more resistance there is to pushing the water up the suction hose. A pump may be able to pick up a prime with a minimum flow only to find that it cannot maintain the prime when attempting to increase the flow output (Fig. 5–7).

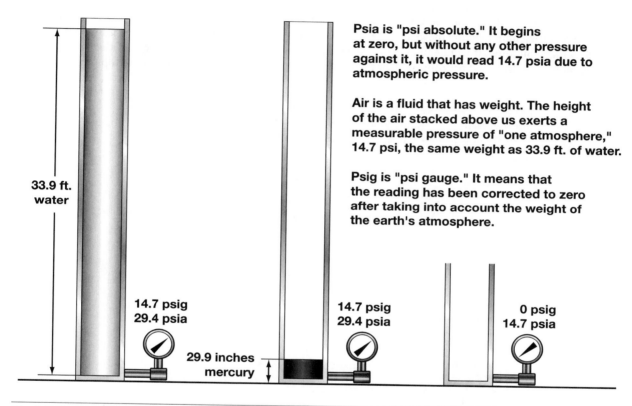

Psia is "psi absolute." It begins at zero, but without any other pressure against it, it would read 14.7 psia due to atmospheric pressure.

Air is a fluid that has weight. The height of the air stacked above us exerts a measurable pressure of "one atmosphere," 14.7 psi, the same weight as 33.9 ft. of water.

Psig is "psi gauge." It means that the reading has been corrected to zero after taking into account the weight of the earth's atmosphere.

33.9 ft. water

14.7 psig
29.4 psia

14.7 psig
29.4 psia

0 psig
14.7 psia

29.9 inches mercury

Fig. 5–7 The pressure exerted by a fluid is proportional to its density.

A final consideration at this point may in some cases be the rise and fall of the tide. Remember, the maximum drafting height depends on the level of the surface of the water below the pump. A unit that is able to draft water 20 ft below the pump at high tide may lose its prime at low tide, when the surface of the water is now 28 ft below the pump, even if the end of the suction hose is 5 ft below the surface! To further complicate the issue of drafting, compound gauges found on fire apparatus are calibrated to indicate inches of mercury for the vacuum side of the gauge. The gauge manufacturers would do the fire service a favor simply by changing the numbers to a more useful *feet of water* calibration. This would not only simplify the training but also the judgments required to draft water. If the gauge clearly showed that the maximum vacuum attainable on the gauge is the maximum height above the surface of the water that the pumper can be to draft effectively, perhaps there would be fewer attempts to draft from bridges 30 ft above the waterline or from a shoreline 35 ft higher than a quarry.

6. Pressure on a confined fluid is transmitted equally throughout that fluid.

This may be rephrased to say that, in a closed system, any application of force will be felt evenly throughout. To return to our charged hoseline with the closed nozzle, if a vehicle were to drive over the hose, it would exert an evenly distributed downward pressure throughout the line. If the pressure were to exceed the strength of the hose, the hose would burst at its weakest point, which could be hundreds of feet away. This allows fire hose to be tested in multiple sections, since all are equally pressurized.

Fire hose streams are basically confined systems. With the nozzle open, whatever water is pushed into the inlet must come out the discharge. The motors on modern fire apparatus are quite powerful devices, but they aren't capable of discharging unlimited quantities. They must overcome resistance to the water flow. That resistance can be in the form of friction loss or head pressure. Take, for example, a standard Class A pumper, rated to deliver 500 gpm at 250 psi. If the flow is out through a deluge nozzle at grade level, and if the supply of water to the intake of the pump is adequate, there will be no problem in meeting the goal of 500 gpm. On the other hand, if this pumper is tied into a standpipe riser in a 60-story office building, it will be unable to deliver any water to the top floor. The weight of the column of water 600 ft high requires at least 260 psi coming in at the base just to push it up and out through the top outlet (Fig. 5–8).

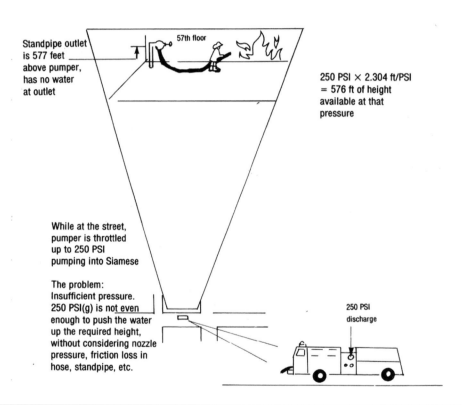

Fig. 5–8 A pumper discharging at 250 psi would not be able to push a single drop of water out of this nozzle on the 57th floor, since the nozzle is located 577 ft above the pump. A 250 psi will only move the water up 576 ft.

Resistance to flow is also produced in other ways, including physical obstructions in the line, such as a kink. Kinks can severely restrict flow, reducing the area of the hose for the water to squeeze past. But the most common resistance to flow is friction loss, which results from the interaction of the water moving through the hose. Only a fraction of this loss is actually due to the friction between the water and the inside of the hose. The remainder is due to the turbulence within the stream itself. This turbulence increases as the volume of water increases. At low flows, there is almost no turbulence, and all of the water molecules move along quite smoothly. This is called *laminar flow*. As flows increase, however, the water molecules bounce off each other and the hose with more frequency, losing energy in the process.

For each diameter of hose, there is a certain flow that, once exceeded, results in a marked increase in turbulence, and therefore, friction loss. Once this flow is reached, friction losses increase to the point where it is impractical to use that size of hose to attempt to supply any more water (Fig. 5–9). Simply stated, you can't put 10 lbs of stuff into a 5 lb bag.

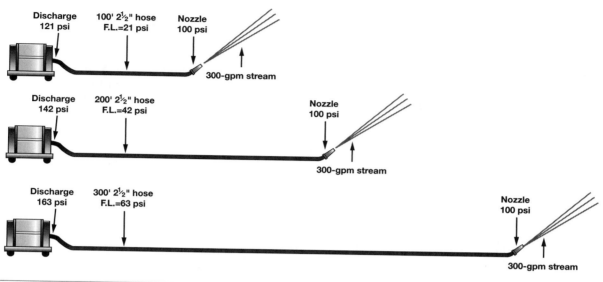

Fig. 5–9 Friction loss is cumulative along the length of the hose line. The longer the line, the more friction loss you will have, as long as the diameter of the hose and the flow remain the same.

Friction loss occurs along the length of the hose. Therefore, if the flow remains constant, the longer the hose, the greater the friction loss. This is a direct relationship—doubling the length doubles the friction loss. Friction loss is also related to the diameter of the hose—the larger the diameter of the hose, the less the friction loss for the same flow. For example, it takes about 21 psi to push 300 gpm through 100 ft of new 2½-in. hose. To move the same 300 gpm through 100 ft of new 3-in. hose only requires about 8 psi. This allows some flexibility. The 300 gpm can be pushed about two and a half times farther in the 3-in. hose using the same energy, 21 psi. In other words, if we had 21 psi available for friction loss in 2½-in. hose, we could only flow 300 gpm 100 ft away. By replacing the 2½-in. with 3-in. hose, the same 300 gpm could be moved 250 ft (Fig. 5–10).

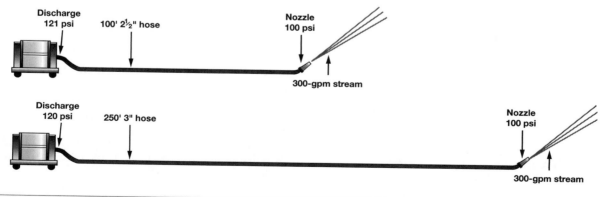

Fig. 5–10 By increasing the diameter of the hose, more water can be delivered farther using the same pump discharge energy.

An alternative to this might be to use the same energy to push more water through the hose. If the nozzle is only 100 ft from the pump, we don't need the extra energy saved by 3-in. hose. What might be more useful would be to discharge extra water. By using 3-in. hose instead of 2½-in., we can move 500 gpm that same 100 ft by using the same energy, 21 psi of friction loss—*maybe*. You see, while being able to determine friction losses is useful, it cannot tell you exactly what the supply capabilities are for any given situation.

The following other factors are involved: the capacity of the pump, the capacity of the water supply source, and the capacity of the hose. Each plays an interrelated part in moving and supplying water. When attempting to analyze the best water supply choice, you must base that decision on the best available information. That information may include a number of variables that you can't determine until the fire has begun. On the other hand, a number of items are readily known before the incident and should be known to all pump operators as well as company and chief officers. These include the rating of the pump, the discharge characteristics of the various nozzles, the capacity and length of the hoselines, and at least a general idea of the strong and poor water supplies in the area. In the case of target hazards, or areas with poor water supplies, you should make plans in advance to obtain alternate sources of supply. All of these items play direct roles in delivering water to the scene (Fig. 5–11).

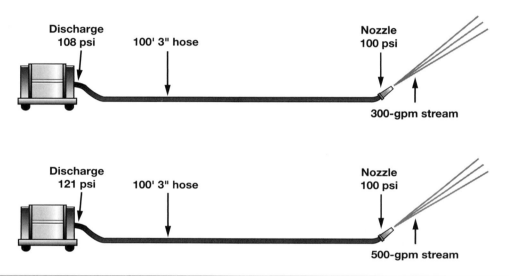

Fig. 5–11 Increasing the amount of water flowing through a hose will increase the amount of friction loss, regardless of the diameter of the hose used.

Knowledge of Water Supply

Firefighters are bound to make mistakes in attempting to deliver water without a thorough understanding of how a given system works. Sometimes there are glaring errors, such as trying to supply a 1,000-gpm master stream with a 500-gpm pumper. Most times, however, they are less obvious, like trying to supply that same 1,000-gpm master stream from a 1,000-gpm pumper when the master stream is 400 ft from the pumper and the only supply hose is 2½ in.

One serious water supply failure that I witnessed involved a fire in a wire and cable factory. On arrival, we found a serious fire in a one-story factory 300 x 600 ft. Initially, 2½-in. handlines were stretched while master streams were being positioned to protect exposures and cut off extension. Water demand soon outstripped supply in the immediate area. Mutual-aid companies were immediately brought in, and an extensive relay operation was set up involving at least six pumpers, each of which dropped from the previous engine and proceeded toward the next hydrant, moving toward a major thoroughfare. As each pumper hooked in and began relaying water, it would flow several hundred gpm. In turn, however, each pumper closer to the scene lost incoming pressure and was forced to reduce its discharge. The pumpers were drawing from hydrants that were all fed by the same 8-in. main, which was incapable of sustaining the flows to all of the open hydrants.

By moving toward the source of the original main, some seven blocks away, the relay did gain about 500 gpm more than the initial supply. Still, it took six engines and over 30 min to establish this. If anyone at the Command Post had bothered to look at a water main map (or ask a water company representative), he might have seen that there were hydrants two blocks in the opposite direction that were fed from a separate 24-in. main. The substantial increase in flow that they could have provided might have enabled the fire units to make a stand. Instead, the structure was a total loss. Firefighters must understand that having a hydrant directly in front of the fire building does *not* ensure an adequate water supply.

Terms Used in Water Supply

When discussing water supply, you must know the meaning of certain terms, as well as how they affect delivery capabilities. Among the most misleading is the static pressure available at the hydrant. If you think back to rule 4, the pressure in the water main is independent of the shape and volume of the container. It depends solely on the depth or head pressure. We know from rule 2 that when no water is flowing, the pressure is the same at all points within the container (in this case, the water main). This is *static pressure*. Static pressure is of very little use to firefighters. It isn't a pressure that is able to move water, since rule 2 no longer applies once the water is flowing. Static pressure can also be misleading to firefighters. Often you will hear members stating, "We have no water supply problems here, because we have 110 psi at all of our hydrants." They are usually referring to static pressure.

Let me illustrate with probably the best example that I've ever encountered. Once while I was designing sprinkler systems, I was called to give a price estimate to extend a system to a new addition that had been added to a row of stores. The cellar area was already sprinklered and consisted of a number of small shops that fronted a subway arcade. Each store had its own sprinkler control value and pressure gauge, each of which read approximately 75 psi. Each store was fed from a common 4-in. main that ran along the front of the row. I had to connect directly to the 4-in. main to supply the new extension, so I sought the main 4-in. control value. I searched both ends of the main, the middle, everywhere, but no valves. I asked the storeowners, but they had no idea. Whenever they needed to work on their systems, they simply shut off their individual valves.

After several hours of inspecting virtually every inch of pipe, I found the sole source of supply, a ½-in. pipe connected to the ¾-in. domestic supply. It seems that the original installer had designed everything right but had forgotten one thing—the subway tunnel and vent grating ran along between the building and the water main. A 4-in. sprinkler pipe with the necessary insulation and wrapping wouldn't fit through the available space above the tunnel. So the installer had pulled a fast one, connecting up to the much-smaller domestic line that did fit. By providing numerous readily seen gauges, accurately depicting the static pressure, he had fooled everyone into assuming that the required 4-in. supply existed. This had gone undetected for 20 years and, if not for the new extension, might have gone much longer. Needless to say, I ran away from that estimate, and a brief call to the building department put an inspector on the case.

The lesson is that *static pressure is no indicator of the volume of water available from a hydrant*. Even if a hydrant is connected to a large main with good flow pressures, it may give a poor flow due to obstructions, such as a partially closed gate valve. Only full-flow testing will detect this situation, and it is recommended when time permits (Fig. 5–12).

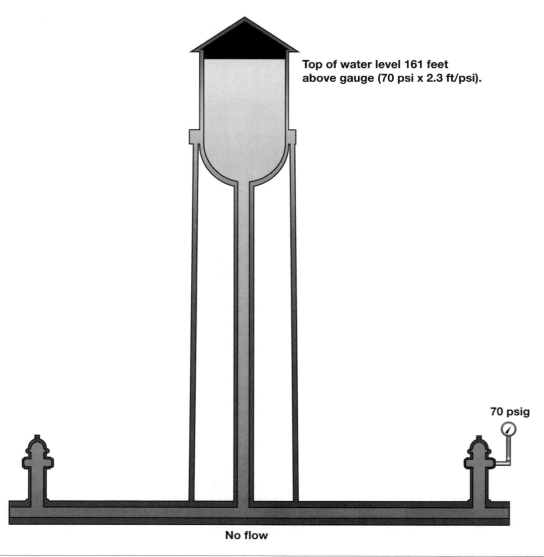

Top of water level 161 feet above gauge (70 psi x 2.3 ft/psi).

70 psig

No flow

Fig. 5–12 Static pressure is the condition present when no water is flowing in a system.

I once responded on a mutual-aid call as the company officer of an engine. We were assigned the duty of supplying a tower ladder at a serious fire in a large vacant apartment house. We were directed to take a hydrant at a corner about 400 ft from the fire. We had a 1,500-gpm engine and the ability to lay three lines of 3-in. hose at once. At the corner, we found a choice of two hydrants. Before selecting one, I had a firefighter remove the 4-in. cap and open each hydrant separately. There was an obvious difference of flow between the two. By selecting the better hydrant, we were able to supply not only a 2-in. tip (1,000 gpm) on the tower ladder, we were also able to augment the supply on a snorkel, allowing an increase in tip size from 1½ in. (600 gpm) to 2 in.

We'll discuss moving that water later. The point here is that by opening each hydrant, we were able to observe the second type of pressure, *flowing pressure,* which is much more useful to firefighters than static pressure. Flowing pressure is sometimes confused with residual pressure, but they are, in fact, separate and distinct. When we opened the hydrant, we saw how much energy was available to push the water out of the particular hydrant. It also gave us an idea of the total gallonage available from the hydrant (Fig. 5–13).

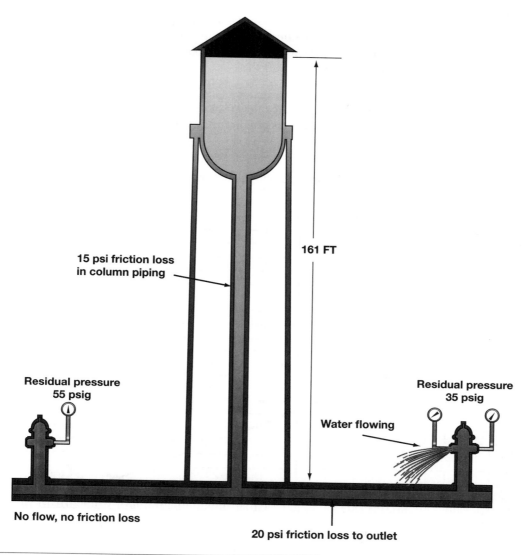

15 psi friction loss
in column piping

161 FT

Residual pressure
55 psig

Residual pressure
35 psig

Water flowing

No flow, no friction loss

20 psi friction loss to outlet

Fig. 5–13 Flowing pressure occurs at the outlet where water leaves the system. Residual pressure is measured at other points in the system while water flows.

While we were flowing that hydrant, the residual pressure was being noted on the gauges of the pumpers already in operation, which could be quite different. If a pumper directly in front of the building were connected to a hydrant that was on the same main as our hydrant, the effect on his incoming/residual pressure could be dramatic. If he were on a particularly strong main, it might not even be noticeable. Since water was flowing, any gauge on any other hydrant would be reading residual pressure, which, for argument's sake, might be 55 psi. But if a gauge were placed on our hydrant, it might read only 25 psi. That is the *flow pressure*. It results from the friction loss in the pipe from the main to the hydrant, past the hydrant valve, and up the barrel of the hydrant. The residual pressure on the other pumper's gauges may only have dropped to 35 psi, which is satisfactory.

The flow pressure being low should indicate that the pumper be placed near the hydrant rather than near the fire using an in-line hose stretch. This is a good policy to follow for all but the initial-arriving engine company, since it ensures the pumpers the fullest supply when residual pressures, and flow pressures, start to drop as a result of the increased demand for water. This pressure drop is caused by increased friction losses within the water mains as more and more flow tries to squeeze into the fire area. The minimum residual pressure that should be allowed is 10 psi, except when drafting. This ensures that a positive flow of water is coming into the pump. A spinning pump imparts energy to the water as it goes through. If there isn't enough water in the pump to supply the demand, a condition known as cavitation of the pump can occur. If it does, serious damage to the pump, including failure, can occur as the spinning pump impellers try to impart their energy to water that isn't there.

When residual pressures begin to drop, you must take action to keep the pressure from going below 10 psi. There are only two possible actions to take: reduce the discharge flow (not desirable) or increase the supply with another source (may be impossible or impractical). There are no other alternatives. A pumper cannot make water (Fig. 5–14).

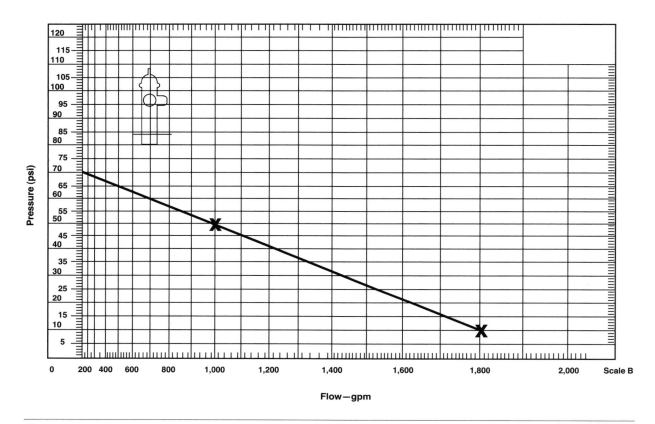

Fig. 5–14 A residual pressure graph can be useful in determining what flows can be projected from a water supply system using a variety of hose layouts. The hydrant here can supply 1,800 gpm at 10 psi, but only 1,000 gpm to feed a supply line at 50 psi flowing.

To determine the maximum flow a given supply hoseline arrangement will sustain, first determine the available flow pressure, for example, a hydrant might be capable of discharging 1000 gpm at 50 psi flow pressure. (See Fig. 5–14) A pumper may be able to discharge the same 1000 gpm at 150 psi. Next determine how much pressure will be required at the discharge end of the line. For the residual intake pressure, that is a minimum of 10 psi. Subtract this minimum end pressure from the starting flow pressure; in this case 10 from 50, and you have the amount of pressure you can afford to lose due to friction loss or elevation pressure, (if our pumper will be located uphill from the hydrant) or both. Since we only have 40 psi to work with, if we had to move this water 400 ft from the hydrant to the pumper the most we could afford to lose is 10 psi per 100 ft of hose (Fig. 5–15). (That is the way friction loss is typically expressed for fire hose, amount per hundred ft.)

	A	C	D	E	F	G	H	I	J	K
1	Flow in US G.P.M	1.5"	1.75"	2"	2.5"	3"	3.5"	4"	5"	6"
2	60	10	5	2.5						
3	95	22	11	5						
4	125	37	21	10	4	1				
5	150	54	26	13.5	6	2				
6	175		34	18	8	3				
7	200		45	24	10	4	2			
8	225		57	30	12	4.5	2			
9	250			37.5	15	6	2.5			
10	275			45	17.5	7	3			
11	300				21	8	3.5	2		
12	325				24.5	9.5	4	2.5		
13	350				28	11	5	2.5		
14	400				36	14	6	3	1	
15	500				55	21	9.5	5	2	
16	600					30	13.5	7	2.5	
17	750					46	20	11.5	4	1
18	1000						34	19	6.5	2.5
19	1250							31	10.5	4.5
20	1500							43	15	6
21	1750								20	8
22	2000								26.5	10.5
23	2500								41.5	16.5

Fig. 5–15 Friction loss for each diameter of hose increases as the flow increases, until the practical maximum flow is reached.

Consulting our friction loss chart, we can see that if we are using 3-in. hose for our supply, we can only flow 330 gpm before we exceed the 10 psi per 100 ft loss. If a pumper were to connect directly to the same hydrant, it could deliver the capacity of the hydrant at a starting pressure of 150 psi, instead of 50. So 150 psi minus 10 psi leaves 140 psi available to overcome friction loss, elevation, etc. To move the water the same 400 ft, we now would divide 140 by 4, and find that we can now expend 35 psi per 100 ft of hose for friction loss, etc. Consulting our table, we find that 3-in. hose can flow about 650 gpm at a pressure loss of 35 psi/100 ft. If we laid two lines of 3-in. hose, we could flow 1300 gpm from one pumper to the intake of another pumper 400 ft away, since this hydrant could supply that much water. Still we are not getting the maximum benefit from this hydrant, since it is capable of flowing 1800 gpm at the minimum pump intake pressure of 10 psi. There is still another 500 gpm available that we cannot use because our system of two pumpers and 2-in. supply lines creates too much friction loss. (Yes, for all you advanced hydraulic theorists, there is some *wiggle room* here, due to the pumper at the hydrant receiving its water at 10 psi. This is the most straightforward way to explain it, and the best basis for planning, since such wiggle room is often eaten up by other unforeseen problems.)

Low residual pressures are most commonly encountered when in-line pumping is used with 2½-in. or 3-in. hose as supply lines. At certain fires, in-line pumping or hydrant-to-fire lays have great advantages over reverse, fire-to-hydrant lays, notably private dwelling (PD) fires, where speed takes priority over the gpm delivered. The in-line attack can begin using preconnected lines, from handline to master stream, as soon as the apparatus pulls up to the building. In addition, any needed tools are right in front of the building and close at hand, not down the block at the hydrant. A disadvantage, however, is that maximum flows aren't available. Say, for example, that the nearest hydrant is 400 ft from the building. It is a good hydrant with 70 psi static/50 psi flowing at 1,000 gpm, and 10 psi flowing at 1,800 gpm (Fig. 5–16).

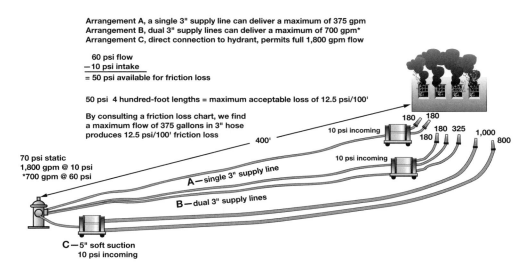

Fig. 5–16 The hydrant is only one part of the water supply system. Various hose lays and pumper positions can greatly reduce the amount of water actually delivered.

If the first apparatus, a 1,500-gpm engine, were to in-line from this hydrant, 400 ft to the front of the building, they could expect to deliver only 330 gpm using a single 3-in. supply, and 660 gpm using two 3-in. supply lines. If a 1,500-gpm pumper were to connect directly to the hydrant using a soft hydrant connection (soft suction), they would be able to flow the entire 1,800 gpm.

An alternative, then, to increase the flow to an in-line pumping engine would be to provide a means for a second pumper to connect to the hydrant and relay the extra available water to the first pumper, which has taken advantage of the speed of in-line pumping. This can be accomplished by placing a clappered Siamese in the first supply line and placing gates on the hydrant outlets. The first pumper connects in-line and initially begins receiving water at hydrant pressure. When the second engine arrives, if further supply is needed, its crew can connect it to an unused gate on the hydrant, get water coming in, and begin supplying a discharge line into the open inlet of the charged supply line's Siamese. By pumping at a pressure that is higher than the hydrant pressure, they will force the clapper valve on the Siamese to swing over, shutting off the flow from the hydrant. At this point, they are now relay pumping. By increasing their discharge pressure, they can increase the flow, *within limits*, to the first engine. Since all Class A pumpers discharge their maximum flow at 150 psi, and since the pumper at the fire scene needs at least 10 psi incoming pressure, there is only 140 psi available for friction loss (Fig. 5–17).

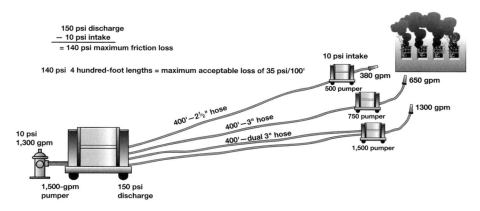

Relay pumper supplying the attack pumper.

Fig. 5–17 Adding a relay pumper to an existing hose layout can increase the amount of water available to an attack pumper at the scene.

Again, going back to the previous example of a pumper 400 ft from the hydrant, the relay pumper can boost the flows from 200 gpm to 380 gpm for a single 2½-in. supply, from 330 to 650 gpm for the single 3-in., and from 660 to 1,300 gpm for the dual lines of 3-in. However, if the full capacity of the hydrant is desired, a third line of 3-in or larger hose *must* be stretched. The problem isn't in the ability to pump the water. The problem lies in the ability to pump the water *at the high pressures required to stuff 1,500 to 1,800 gpm through long lays of small hose.*

Unfortunately, most firefighters and officers don't recognize the relationship between volume and pressure within a fire pump. For years, hydraulics instructors have drilled into firefighters the notion that the higher the nozzle pressure, the higher the flow, as long as the tip size remains constant. That is true of *nozzle* pressure where the flow is exiting through an opening, but it is *not* true of *pump* or *engine* pressures. In fact, the opposite is true. By their very design, centrifugal pumps will deliver increasingly *lower* flows as the pressure is increased. Standard Class A pumpers are rated to deliver their maximum volume of water at 150 psi. They will deliver 70% of their rating at 200 psi and only 50% at 250 psi. (To erase any doubts, simply read the specification plate on your pump panel.) In other words, a 1,000-gpm pumper will flow 1,000 gpm at 150 psi, only 700 gpm at 200 psi, and a mere 500 gpm at 250 psi.

This is contradictory to the understanding that many firefighters have of hydraulics, but it is easily explained. A centrifugal pump spins the water off its impellers, imparting energy. The impeller can do either of two things—it can give a little bit of water a strong push, or it can give a lot of water a little push. It can't do both at once. Think of it this way: a man is asked to throw a rock. He picks up a ½-lb stone and hurls it 100 ft. Then he throws a 20-lb cobblestone, but it only goes 10 ft. This fellow can do either, but he cannot throw the 20-lb rock 100 ft. To do so requires more power, just as with your pumper's engine.

Now consider the case of a 1,000-gpm pumper that is assigned to relay 1,000 gpm to another pumper 900 ft away. As the operation begins, the pumper at the water source starts pumping at 150 psi through two parallel lines of 3-in. hose. Still, the incoming supply isn't enough to feed the 1,000-gpm fog nozzle, and the stream is inadequate. The officer at the fire scene has few choices, and he radios for the supply or relay pumper to increase the pressure to the attack pumper. The cycle has just begun to worsen. The initial pumper was doing its best to deliver the needed supply. It cannot deliver more water than it could at 150 psi. Remember, it is gpm that feed the nozzle and put out the fire. By calling for more pressure, there will be less gpm discharged, thus worsening the situation. The problem is in the hose, not the pump (Fig. 5–18).

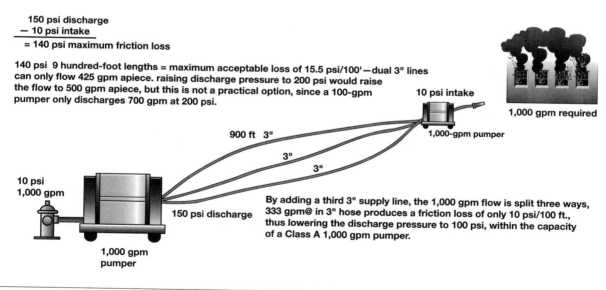

150 psi discharge
— 10 psi intake
= 140 psi maximum friction loss

140 psi 9 hundred-foot lengths = maximum acceptable loss of 15.5 psi/100'—dual 3" lines can only flow 425 gpm apiece. raising discharge pressure to 200 psi would raise the flow to 500 gpm apiece, but this is not a practical option, since a 100-gpm pumper only discharges 700 gpm at 200 psi.

10 psi intake

1,000 gpm required

900 ft 3"

3"

3"

1,000-gpm pumper

10 psi
1,000 gpm

150 psi discharge

1,000 gpm pumper

By adding a third 3" supply line, the 1,000 gpm flow is split three ways, 333 gpm@ in 3" hose produces a friction loss of only 10 psi/100 ft., thus lowering the discharge pressure to 100 psi, within the capacity of a Class A 1,000 gpm pumper.

Fig. 5–18 For every hose and pumper combination there is a maximum possible flow. When that is reached and more water is needed at the scene, adding an additional supply line may be what is required.

To flow 1,000 gpm in this layout requires both 3-in. lines to supply 500 gpm apiece. That produces a friction loss of 21 psi per 100 ft of hose. The hose length of 900 ft means there is 189 psi of friction loss in the line. Still another 10 psi is required for the attack pumper's intake residual, meaning that the first pumper will have to pump at 200 psi to overcome the friction loss and deliver 1,000 gpm. By design, however, a 1,000-gpm pumper discharging at 200 psi can only flow 700 gpm. This layout is therefore destined to be unsuccessful. There are possible solutions. The most practical at this operation, once it is already underway, might be to have a third line stretched between the two pumpers. By dividing the 1,000-gpm flow among three supply lines, the friction loss is cut to about 10 psi per 100 ft. Thus, only 90 psi is required for friction loss and 10 psi for residual, creating a net discharge pressure of only 100 psi. This supply pumper will be able to handle the task easily.

Another solution, but not practical once the relay has begun, is to replace the 1,000-gpm supply pumper with a 1,500-gpm-or-larger model. That will ensure success, since a 1,500 gpm unit is rated to deliver 1,050 gpm (70% of 1,500) at 200 psi. This solution could deliver the 1,000 gpm within the existing dual 3-in. supply line arrangement, but it would be unable to deliver its capacity (1,500 gpm) this distance. The problem isn't always best solved by a bigger pump. It might be more cost-effective to correct the hose size. The final solution, which must be put in place long before the relay begins, is to increase the diameter of the water supply line.

A partial improvement, but not a solution, is to reduce the demand from the nozzles to a point where the 700 gpm coming in is sufficient to provide a good stream. For instance, if in the previous example, the master stream nozzle were equipped with a 2-in. solid tip, the tip pressure would only be about 40 psi and very ineffective. By reducing the tip size to 1½-in., the 700 gpm could be blasted through at a tip pressure of 120 psi, greatly extending the stream's reach and penetrating power. But this may not be the answer to the problem. If the fire demands 1,000 gpm to extinguish, 700 gpm just won't do the trick and the fire will continue unabated. This is a drawback of automatic nozzles, particularly on master stream devices, where there is no way to feel the nozzle reaction. The stream from the automatic nozzle at 700 gpm *looks* as good as it does at 1,000 gpm (Figs. 5–19 and 5–20).

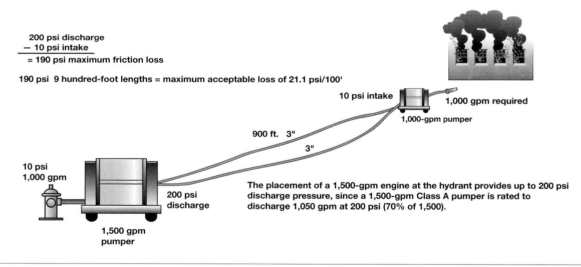

200 psi discharge
− 10 psi intake
= 190 psi maximum friction loss

190 psi 9 hundred-foot lengths = maximum acceptable loss of 21.1 psi/100'

10 psi intake

1,000 gpm required

1,000-gpm pumper

900 ft. 3"

3"

10 psi
1,000 gpm

200 psi
discharge

The placement of a 1,500-gpm engine at the hydrant provides up to 200 psi discharge pressure, since a 1,500-gpm Class A pumper is rated to discharge 1,050 gpm at 200 psi (70% of 1,500).

1,500 gpm
pumper

Fig. 5–19 Having a larger capacity pumper as the relay pumper will allow greater flows, since the larger pump will flow more at an elevated pressure.

Large-diameter Hose

As our civilization progresses, inevitably so does the scale of our surroundings. In the early 1960s, 1,000-gpm pumpers were very common, with a fair distribution of 500 and 750 gpm units. On the other hand, units of 1,250 gpm and larger were rather uncommon. Today, quite the opposite is true. Models of 500 gpm are extremely rare;

750 and 1,000 gpm models are found far less frequently than in the past, and 1,500- to 2,000-gpm models are becoming quite common.

This is not surprising when you consider the increases in the size of today's structures and the increased fire loads. Buildings having millions of square feet of floor space are being built. Gasoline tank trucks that were limited to 3,000-gal capacities in the 1950s are now being replaced by 14,500-gal behemoths. If nothing else demands higher flow, surely the vast increase of the plastics industry does. According to one estimate, the fire loading of a building with large amounts of plastics is *double* that of the same building containing only ordinary combustibles.

The problem then develops: how to move the supply made available with these larger pumping units? The answer is large-diameter hose (LDH). LDH was first introduced in this country in the early 1970s and has been increasingly accepted. As the problem discussed previously illustrates, 2½-in. and 3-in. supply lines have reached their maximum potential. They may be satisfactory for low gallonages (under 1,000 gpm), but when flows in the range of modern pumpers are being considered, lines of these sizes are inefficient as supply lines. That, then, leaves a choice of three common sizes as follows: 4 in., 5 in., and 6 in. (Note: Yes, 3½ in. and 4½ in. are available, but their usefulness is endangered by a lack of supporting fittings and more suitable larger hose.) There are advantages and disadvantages for each.

Fig. 5–20 *When the quantity of water is limited, the pressure that it is discharged at may mean a more or less effective stream. Here, the same 700 gpm is discharged through a 2-in. tip (shorter, less power-ful stream) and through a 1½-in. tip (longer, high-pressure stream).*

In the late 1970s and early 1980s, 4 in. was a size adopted by some departments as a replacement for their 2½-in. supply lines. While a single 4-in. line is vastly superior to a single 2½-in. line, it is actually only marginal in supplying large quantities of water (1,000 gpm and over). In fact, as far as friction loss is concerned, a single line of 4-in. is almost exactly the same as two lines of 3-in. hose (Fig. 5–21).

4" friction loss

gpm	500	600	700	800	900	1,000	1,100	1,200	1,250
gpm	4.75	6.84	9.31	12.16	15.4	19.0	23.0	27.4	29.7

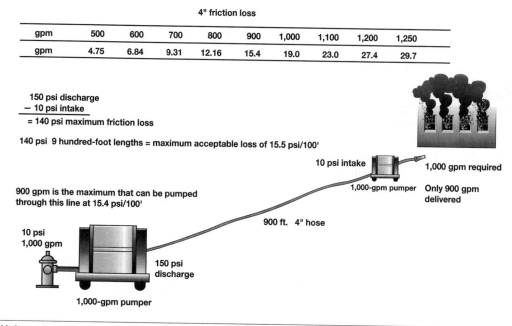

150 psi discharge
− 10 psi intake
= 140 psi maximum friction loss

140 psi 9 hundred-foot lengths = maximum acceptable loss of 15.5 psi/100'

900 gpm is the maximum that can be pumped through this line at 15.4 psi/100'

10 psi
1,000 gpm

150 psi discharge

1,000-gpm pumper

10 psi intake

1,000-gpm pumper

900 ft. 4" hose

1,000 gpm required

Only 900 gpm delivered

Fig. 5–21 *Using 4-in. hose in a relay provides only a 45% efficiency. Two 1,000-gpm pumpers only flow 900 gpm in this arrangement.*

Consider the problem of supplying 1,000 gpm 900 ft from the source. No major benefit is produced by replacing the two lines of 3-in. hose with a single line of 4 in. Supplying 1,000 gpm through a single 4-in. line demands 181-psi discharge pressure (19.0 psi/100 ft x 900 ft, plus 10 psi intake). A larger pumper, or an additional supply line, will still be required to supply this 1,000 gpm. (Note: NFPA Standard 1932 recommends a maximum discharge pressure of 200 psi for LDH. Any length more than 900 ft, regardless of the size of the supply pumper, will require an additional supply hose to keep discharge pressures below 200 psi. For these reasons, it would be a prudent move to restrict the use of 4-in. hose to pumpers of 1,000 gpm capacity or less.)

Pumpers of 1,250–1,500 gpm are best supplied with 5-in. supply hose, although dual lines of 4-in. hose are an option. One of the advantages of LDH, however, is that it allows for rapid deployment and return to service with minimum manpower. Dropping dual supply lines either takes more time or requires more people. When packing up, either twice as many people are needed or the operation takes twice as long. Inserting a single line of 5-in. hose into the previous problem opens up a variety of options. If the two 1,000-gpm pumpers are used, the supply pumper only has to overcome 59 psi of friction loss and 10 psi residual intake pressure for a net pump discharge pressure of 69 psi (6.6 psi/100 ft x 900 ft plus 10 psi residual). In areas of strong water supply, it may be possible to eliminate the relay (supply) pumper entirely. If the hydrant can support 1,000 gpm at 69 psi, there is no need for the second pumper. It can remain in service for another alarm or may not even need to be purchased in the first place, thus saving hundreds of thousands of dollars (Fig. 5–22).

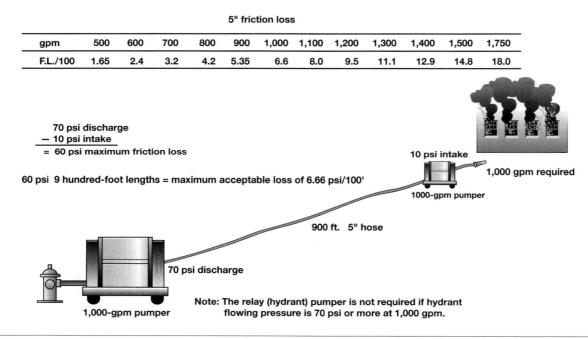

5" friction loss

gpm	500	600	700	800	900	1,000	1,100	1,200	1,300	1,400	1,500	1,750
F.L./100	1.65	2.4	3.2	4.2	5.35	6.6	8.0	9.5	11.1	12.9	14.8	18.0

70 psi discharge
— 10 psi intake
= 60 psi maximum friction loss

60 psi 9 hundred-foot lengths = maximum acceptable loss of 6.66 psi/100'

10 psi intake

1,000 gpm required

1000-gpm pumper

900 ft. 5" hose

70 psi discharge

1,000-gpm pumper

Note: The relay (hydrant) pumper is not required if hydrant flowing pressure is 70 psi or more at 1,000 gpm.

Fig. 5–22 Replacing the 4-in. hose with the 5-in. line allows the flow to increase to 1,000 gpm and may eliminate the need for the relay pumper if the hydrant supply is strong.

Another option is to use positive pumping through LDH. In this case, the pumper stops at the fire scene and drops a water manifold along with a variety of hose, nozzles, and appliances. The pumper then proceeds to the hydrant and pumps back to the manifold. The friction loss here is 59 psi, plus 10 psi for the manifold. That leaves 130 psi available for handline or master stream use, or as auxiliary supplies to other operating pumpers. If the manifold is placed near the fire, keeping discharge lines short, this should be adequate. One problem with using such an arrangement is due to hardware problems. Most such manifolds don't have individual gauges on each gated discharge. If the manifold were to be used to supply one 2½-in. handline, 150 ft long with a solid tip, and a 1½- or 1¾-in. line, 200 ft long with a fog tip, there would be a difference of more than 75 psi between the proper discharge pressures for each outlet. A lack of gauges could result in one line being dangerously over-pressurized or the smaller line being severely under-pressurized (Fig. 5–23).

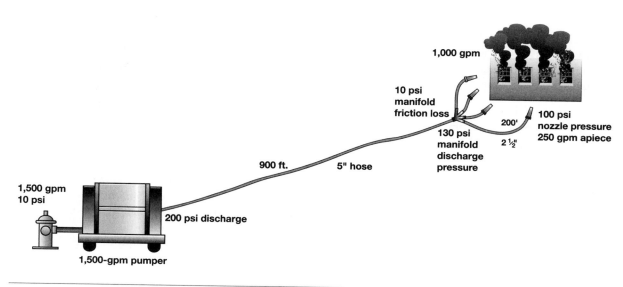

Fig. 5–23 Positive pumping through LDH is not efficient, due to the low working pressure rating of the large hose (185–200 psi) and the high nozzle pressure (100 psi) required of most nozzles.

You should seriously consider using 5-in. hose for supplying 1,250- and 1,500-gpm engines, and it can be extremely valuable as well for supplying smaller engines at long distances (Fig. 5–24). For example, it is possible to relay 500 gpm for more than two miles using a 750 gpm pumper drafting from a farm pond, relaying to an attack pumper at the fire (1.65 psi/100 ft x 10,600 ft = 175 psi + 10 psi residual = 185 psi discharge pressure).

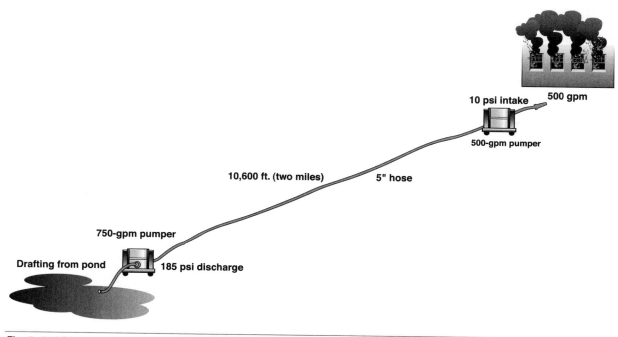

Fig. 5–24 LDH is particularly suited for long distance relays that would not be possible with smaller lines.

As great as 5-in. hose may be for flows up to 1,500 gpm (friction loss of under 15 psi/100 ft), it, like all other sizes, also has its limitations. Pumpers of 1,750–2,250 gpm are currently being marketed, and I have seen at least two 3,500-gpm models at an oil refinery. Industrial fire pumps are available to 6,000 gpm, basically as standard items. With the rapid progress being made in diesel and turbine engines, it is becoming increasingly easy to match a power plant to these large-capacity pumps.

It took years of design and building and about $1 million (in 1965!) to build New York City's Superpumper, which had a capacity of 8,800 gpm. It is possible today to put a 6,000-gpm pumper on a standard chassis for probably one-third of that price. An area requiring such massive supplies can get them at bargain-basement prices compared to 30 years ago.

The problem with having that much pumping capacity available, however, goes right back to the hose. How do you move such large volumes efficiently? Since at 2,000 gpm, the friction loss is 26 psi/100 ft, 5-in. hose isn't the answer. When you consider that a 2,000-gpm pumper is a specialized piece of equipment, likely to have to go a bit of distance to reach a water supply able to meet its capacity, you realize that long stretches may be the order of the day. That rules out 5 in. and points toward 6 in., which is currently available. All of the advantages of 5 in. apply to 6-in. hose—longer stretches are possible and greater flows are available. It does with a single line what would otherwise require two lines of 5-in.hose. Currently, there isn't as large of a selection of appliances and fittings produced for 6 in. as for the more popular 5 in., but as the demand for more of these large-capacity, 2,000- and 2,250-gpm engines increases, so too will the supply of fittings.

Where will it all end? Currently, one of the projects that I am working on is a disaster preparedness program designed to prepare for catastrophic events, including disruption of our city's water supply, coupled with potential fires. One of the systems we are looking at uses 6,000-gpm mobile pumps, discharging through 12-in. hose. This hose and pump combination could flow 6,000 gpm more than 8,000 ft—a mile and a half! Similar systems are in use in several petrochemical refineries and at least one municipal water supply system already. Let's hope that such systems are only needed for catastrophic events and not for routine firefighting. As I said, the technology is available to put even larger pumps on apparatus. If there is enough demand, 8-in. hose is a possibility, truly moving the water main from underground to the surface of the street. Hopefully, before either you or I have to drain one of those monsters, the building officials, planning and zoning boards, and the insurance and fire officials will have demanded sufficient improvements in construction, sprinklers, and separations to make such large demands unnecessary (Fig. 5–25).

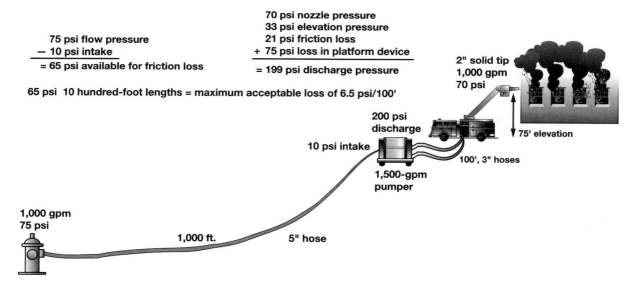

Fig. 5–25 Supply elevating platforms that do not have built in pumps requires a high flow and high discharge pressure (200 psi +). The 5-in. hose is best used as a supply line to a pumper near the base of the platform.

The use of LDH of any size should take advantage of its low friction loss and work around its relatively low maximum discharge pressure. This is best accomplished by using the hose as an in-line supply. This was graphically illustrated at one test of the capabilities of 5-in. hose. The initial objective was to determine if a single 5-in. line, 1,000 ft long, could supply 1,000 gpm to a 2-in. tip on a tower ladder. The initial arrangement had a 1,500-gpm engine lay from a hydrant 1,000 ft away to the base of a 75 ft tower ladder. Two 3-in. hoses were then stretched from the pumper to the tower. When the hydrant was charged, the pumper was able to maintain 10 psi residual and discharge more than 1,000 gpm to a 2-in. tip on the tower at an elevation of 75 ft. This required a pump discharge pressure of 225 psi, a safe pressure for the double-jacket 3-in. hose. Having proved the benefit of an in-line stretch, the pumper was then positioned at the hydrant, pumping back through the 1,000 ft of 5 in. directly into the tower. The result was a net *loss* of flow of approximately 150 gpm. How can that be? The most efficient method of water delivery is positive pumping, where the pumper is located on the hydrant. One hose manufacturer claims that this is 300% more efficient than in-line pumping.

The answer is simple. It goes back to what you are trying to deliver: volume or pressure. LDH is best suited to deliver volume. Trying to pump directly into an aerial device without a relay pump is a lost cause. All of the following factors are working against you: first, nozzle pressure (100 psi for a fog tip), then elevation pressure (to push the water up 75 ft takes an additional 33 psi). Then there is the friction loss through the hose or piping (at 1,000 gpm, it is more than 75 psi on many models). That is a total of 207 psi *at the base of the aerial*. That exceeds the NFPA's maximum discharge pressure for LDH before you even start. Here the hose's maximum rated working pressure is the problem. To flow the desired volume of water this way risks losing the entire operation due to a burst length of hose. If the nozzle is a solid tip, you are somewhat better off than you would be using a fog tip, since solid tips only require 80 psi nozzle pressure for master streams vs. 100 psi for a fog tip. In this case, it required 235 psi to flow only 850 gpm from the tower. Surely if the hydrant flowing pressures aren't high enough to support such a long in-line stretch at such a high flow, a relay pumper must be placed at the hydrant while the pumper positions itself adjacent to the aerial device (Fig. 5–26).

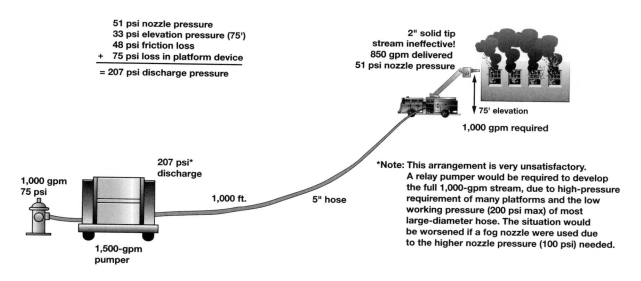

Fig. 5–26 Using 5-in. hose for positive pumping to the same platform could actually reduce the flow to the fire, while exceeding the rated working pressure of the hose.

Applying Heavy Streams

The use of aerial devices to apply heavy streams goes back more than 100 years to the days of hydraulically raised water towers. Modern apparatus have turned all this into a high art. The tower ladder is the most versatile, most highly maneuverable master stream on the fireground today. Unfortunately, it and its cousins, the ladder tower and the ladder pipe, are more often misused than used properly.

There is more to proper stream application 50 ft in the air than just squirting water, just as there is at ground level. Water functions primarily as a cooling agent. Then, to do its job, it must end up on the fuel, cooling it until it ceases to give off more flammable vapors. That means that the water must be able to hit the seat of the fire, not merely the flame that results from the burning process.

At ground level, when the fire exceeds the capabilities of a handline, a master stream is often brought into play. One of the most common reasons for doing this is that the fire has extended into the cockloft, which allows a rapidly spreading fire to extend to several areas almost at once. In rather short order, the roof decking will burn through and the IC will be faced with a lot of visible fire blowing through. A much more satisfactory approach is to put a lot of water on the fire from underneath, where it will cool off the fuel, thus stopping the production of gases and, consequently, flame. The biggest mistake is to think that by directing a stream through a hole in the roof, the fire will be stopped. This is almost never the case. This action just drives flame, heat, and smoke back under the roof, where it spreads out. Remember, the basic purpose of a roof is to keep water out. Period. It doesn't discriminate between an elevated master stream and one of the Almighty's torrential downpours. As long as part of the roof is intact, the fire will burn beneath that section, unimpeded by the stream coming through the hole in the roof (Fig. 5–27).

Fig. 5–27 Platform streams are very effective on fires in the cockloft, if applied from below!

As I mentioned earlier, a far more successful approach is to get a stream up and under the roof from below. In the event of heavy fire, the most maneuverable master stream is the one at the end of a tower ladder basket. The operators can put the basket right down on the sidewalk and simulate a ground-mounted deluge set with the ability to move rapidly from window to window or store to store as quickly as the fire is darkened down. Remember, height isn't a requirement for the use of a tower ladder master stream.

Many rapidly spreading fires in taxpayers have been stopped in their tracks by the quick application of 800–1,000 gpm from the sidewalk. For this application, the telescoping boom devices are superior to the articulating boom, since the front of the building may contain overhead power lines or other obstructions. The telescoping booms can rotate easily right into place under the wires. On a normal street, the articulating booms require that the elbow be almost straight up above the basket for the basket to be put on the sidewalk. The telescoping boom is capable of flowing more than 1,000 gpm, while ladder pipes on aerial ladders are limited to about 600 gpm on the fly ladder, the topmost section. Another definite advantage of the platform is the range of motion of the stream. The platform nozzles can cover almost every angle in front of, above, below, and to the sides of the nozzle. Ladder pipes are restricted to a very narrow range of lateral movement away from the centerline of the ladder. This difference becomes most pronounced when operating the streams through windows and other restrictive openings. When specifying a new apparatus, it pays to look closely at platform payload capabilities vs. the boom angle and the length of boom extension. The low angles involved with the low-level master stream attack may limit some apparatus from performing these maneuvers (Fig. 5–28).

As a final comment on pressures and water supply, let me point out that water, like all fluids, takes the path of least resistance, and that pressure differences always try to balance themselves out. Don't try to operate appliances requiring substantially different pressures from the same wye, water thief, or manifold if at all possible, since it is inefficient to do so.

Take, for example, a 3-in. line that is laid to a water thief on the top floor of a non-fireproof school. Again, if practical, all lines should be equipped with either fog nozzles or solid tips. Let's say that two 1¾-in. handlines are stretched, each with a fog tip and 100 ft of hose. To operate properly, they would require about 140 psi at the outlet of the device. Then a larger backup line is put into operation, consisting of 100 ft of 2½-in. equipped with a 1⅛-in. solid tip. This line only requires 65 psi at the outlet of the device. Still, the device must be fed with the maximum pressure demand, 140 psi plus about 15 psi for friction loss through the valve. This means that there is a great disparity in operating pressures and, therefore, the potential for accidents and injuries. Although it is possible to provide the correct flowing pressures to all lines by means of the gate discharge, this isn't a surefire solution, since very high pressures will be encountered while opening and closing the nozzles (Fig. 5–29).

Fig. 5–28 Not all platforms are meant for stream application into the cockloft at low angles. Obey manufacturers' operating instructions to avoid potentially disastrous failure.

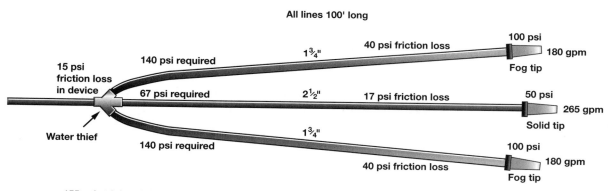

All lines 100' long

15 psi friction loss in device

Water thief

140 psi required — 1¾" — 40 psi friction loss — 100 psi — 180 gpm — Fog tip

67 psi required — 2½" — 17 psi friction loss — 50 psi — 265 gpm — Solid tip

140 psi required — 1¾" — 40 psi friction loss — 100 psi — 180 gpm — Fog tip

155 psi at inlet of device required to supply 1¾" lines, but this high pressure is more than twice the pressure required for 2½" line, making for a potentially dangerous situation. Avoid mixing nozzle types on shared supply sources.

Fig. 5–29 When supplying water thiefs or manifolds, it is best to only supply one type of device with similar characteristics, i.e. all fog nozzles or all solid tip nozzles.

Flow Meters

One easy way to take much of the difficulty out of water supply is to use flow meters, either in lieu of or in conjunction with the pressure gauges. Similarly to calibrating the vacuum or compound gauges in feet of water instead of inches of mercury, the idea is to simplify the mental calculations of the firefighter. Flow meters are calibrated to read in gpm delivered. Again, the important factor in water supply is volume, not pressure. By knowing the gallonage rating for each appliance, it is a simple matter to establish the correct flow.

In the case of preconnected lines, the pressure gauge is normally sufficient for routine operations. All factors are known in advance, from hose length to desired gpm and operating pressure. Friction-loss charts are consulted and a standard operating pressure is established, perhaps after some trials to ensure the mobility and flow of the line. It's a simple matter from this point on to provide the desired flow. Simply open the throttle until the dial reads the pressure that you were told. In some cases, a line is drawn across the face of the gauge to show where the dial should be.

But what about the unusual cases, when a kink develops in the line, severely restricting the flow? In this case, the nozzle team must contact the pump operator and verify that adequate pressure is being maintained. The operator should spot the drop of flow with a flow meter and try to find out the cause. If it is a kink and not merely that the full flow isn't required, the operator may spot the problem and take action to have it removed before the nozzle team even realizes that there is a problem. Another situation in which flow meters pay off is when hose is stretched from the main hose bed and the exact length of the hose isn't known. In this case, by knowing the type of nozzle that is to be supplied, it is a simple matter to dial up the proper flow (Fig. 5–30).

Fig. 5–30 An adjustable-gallonage master-stream nozzle should be set at the flow that provides the most efficient nozzle pressure.

In closing, let me again reiterate that *it's gpm that puts out fires.* Although it may be advantageous to use pressure to increase reach and striking power, the minimum flow must satisfy the demand. This may be accomplished by selecting a nozzle size that will still flow the desired gallonage, but at a higher flowing pressure. That does *not* imply that fog nozzles deliver more velocity than solid nozzles, since a large portion of their 100-psi nozzle pressure is wasted bouncing off the center baffle, then the pattern shaper, and finally off the rest of the water. What it does mean is that, if a nozzle can have its flow varied, and if one selection provides a higher flowing pressure, then using that selection will most likely be beneficial.

Realize that the most satisfactory method to supply large flows is to use large hose. If LDH isn't available, the next best option is to use multiple lines. (Note: By splitting the flow between two lines, the resulting friction loss is only 25–35% of the same flow in a single line. By splitting the flow among three lines, the friction loss is only 10–15% of the one-line loss.)

By preparing for the worst-case scenarios in water demand, you not only take care of the worst case, but all the lesser situations as well. At times, however, no amount of water will prevent a total loss in these situations, primarily those involving delayed alarms. Be realistic in your expectations. If it takes 45 min to put a relay into operation or to assemble the number of pumpers and discharge appliances, the fire may be over before the water reaches the nozzle. Remember, the owner had the option of installing his own water supply and distribution system but chose not to.

Sprinkler Systems and Standpipe Operations

Operations in Sprinklered Buildings

Automatic sprinkler systems have often been called silent sentinels and for good reason. They often sit unused in buildings for years on end, maintaining their unending vigil without being called on to perform. Then one day, an emergency occurs, and they spring into action immediately, usually delivering a knockout punch to the invading enemy. Fire departments have good reason to stress ensuring that a working sprinkler system is installed in all premises, since they almost totally eliminate the loss of life, both civilian and firefighter. Yet I, a firefighter and a former sprinkler system engineer, was almost killed at a major fire where sprinklers played a major contributing role. What, then, must a firefighter know about operations at sprinklered buildings?

A properly designed automatic sprinkler system is the firefighter's best ally, if it is properly used. It is neither a panacea nor, as some have suggested, a means of eliminating manual firefighting. On the other hand, the presence of an operating sprinkler system shouldn't be blamed for hindering firefighters to the point of abandoning a building.

For more than 100 years, sprinkler systems have protected lives and property around the world, achieving a remarkable success rate. The NFPA has compiled statistics on sprinkler operations reported to them, and in more than 100,000 operations, sprinklers controlled or extinguished the fire more than 95% of the time. In fact, in more than 90% of the cases reported, only one or two heads went off. More importantly, there has never been a multiple-death fire (more than three killed) in a building having a fully operational wet sprinkler system. Records of operation in Australia show an even higher success ratio.

When properly maintained, automatic sprinklers have proved to be the most effective means of protecting life and property. Of the reported causes of sprinklers failing to control a fire, human error (including fire department error) is largely to blame. (Note: Three New York City firefighters were killed in a 1999 fire in a high-rise residential building that had a partial sprinkler system that protected only the exit halls. The sprinkler system sectional valve was likely never turned on and no water ever flowed from the system. This is a fairly common problem with partial systems that are connected only to the domestic water and do not have a FDC to augment or bypass the domestic systems. I am convinced that if the sprinklers in the hall had operated, these three firefighters would be alive today.)

Fire Departments vs. Sprinklers

In past years, there have been catastrophic fires that were directly attributable to the decisions of fire departments to shut off the sprinkler system and rely on manual firefighting efforts. The officers in command of these fires have given numerous explanations for their failure to let this extremely valuable tool do its work. Such excuses include a mistaken belief that the fire was so small that it didn't require the use of a sprinkler system; the desire to limit *unnecessary* water damage; that the firefighters couldn't find the fire due to the heavy smoke, so they wanted to let the fire light up so they could find it; that the firefighters were complaining of discomfort from having to work under the spray; and the belief that a limited water supply could be more useful through fire department handlines. Although it is true, in certain isolated cases, that it may be necessary to close sprinkler valves to conserve water for manual firefighting, it is usually best to allow the sprinkler system to continue to operate until the fire has definitely been extinguished (Fig. 6–1).

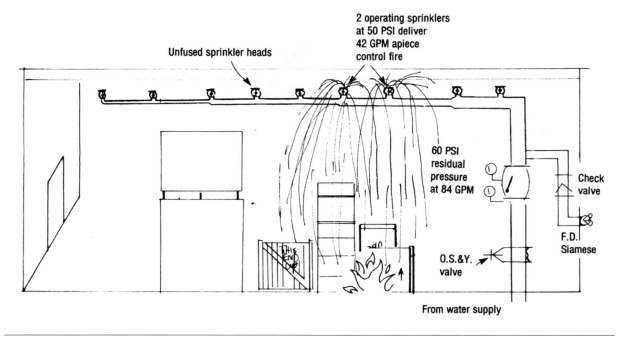

Fig. 6–1 Flowing pressure from the street mains may provide each sprinkler with more than 40 gpm apiece initially.

Only in extreme cases should the sprinkler system be shut down pending completion of a careful examination. For example, an explosion that demolishes the sprinkler piping in the fire building and the open pipes are letting large quantities of water escape with no effect on the fire, or where a rapidly expanding fire overtaxes an under-designed system. In these cases, the remaining water may be better used for exposure protection. Remember, just because the fire is under control doesn't mean that it's going to stay that way. It may be checked by the operation of hose streams and the sprinklers or a second fire may be burning in a remote area. Take away the sprinklers and you may lose control.

Fire Department Operations

Most firefighters reading accounts of such shutdowns reassure themselves, "My department would never do anything like that." Stop to reconsider what effect the operations of your engine companies are having on the operation of sprinkler systems. Every sprinkler system should be required to have a FDC. In fact, in larger buildings, more than one Siamese should be provided to allow for the larger flows required and to reduce the chance of a single Siamese being OOS. These Siameses require prompt support.

One of the first-arriving engine companies should be assigned the duty of supplying the sprinkler Siamese. Each normal sprinkler head flowing at 50 psi will deliver about 40 gpm. This is the equivalent of having a man with a booster line stationed every 10 ft, ready for the fire to start. If the first companies connect to a hydrant on the same main as the sprinkler system and start supplying handlines, they will cause the pressure available to the sprinkler heads to drop. If they drop the main pressure to a point where the sprinkler is operating at 5 psi, the heads will now be flowing only about 10 gpm each. Try to visualize the effect that reduction may have. Imagine yourself at a fire where you're just holding your own with a 2½-in. handline when the chief orders you to shut down your line and replace it with a 60 gpm, 1½-in. line. That is the effect we are having on the sprinkler system when we start to compete with it for available water. Under these circumstances, the fire may intensify and overwhelm the sprinklers causing more heads to open and the pressure to drop still lower (Fig. 6–2).

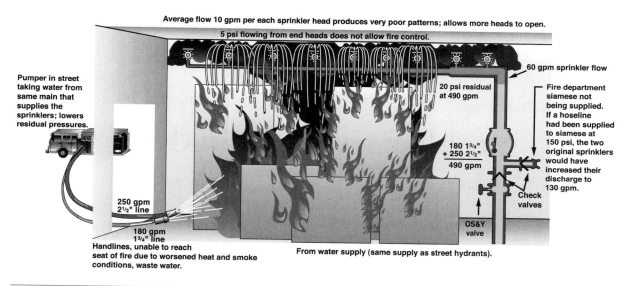

Fig. 6–2 If fire department operations lower pressure in the street main, sprinkler flow will drop. Loss of the building could result.

If, on the other hand, one of the first lines is supplied to the Siamese and backed up by a second line, we will now begin to discharge more water at a higher pressure where it will do the most good, right over the fire. A standard sprinkler operating at 100 psi flow is discharging 55–60 gpm, the same flow as 1½-in. Navy fog nozzle.

When planning to supply Fire Department Connections, you should consider the following factors: supply the system early, before residual pressures in the mains start to drop; supply the system with multiple lines of the largest size hose possible, using two lines of 2½-in. hose as a minimum.

If we plan for the worst, considering each sprinkler has to operate at low pressures because of many heads being opened, we can, for speed of calculation, figure each head to be covering an average of 100 sq ft with a flow of 20 gpm. Using these figures, we can see that, with 2½-in. supply lines of any great length, we are limited to about a 500 gpm flow, covering an area of about 2,500 sq ft (50 x 50) and feeding only about 25 heads. That's acceptable for light-hazard occupancies like schools and office buildings, but if our fire is in any type of commercial or industrial occupancy, we may be getting into trouble. Fires in these buildings may release more heat, requiring larger flows. In this case, larger diameter supply lines will be of great aid.

If possible, establish a separate water supply to reinforce the sprinklers. If the fire requires supplying more than one handline for mop-up, it would be a good idea to have a separate pumper supplying the handlines. Another pumper should feed the sprinklers, preferably with large hose, using a hydrant on a main separate from the one supplying the handlines or master streams. This pumper should keep the lines to the Siamese as short as possible and operate in volume, discharging at 150 psi.

There are several reasons for using this method. First, most pumpers today are rated to discharge their maximum rated capacity in volume at this pressure. Raising the pressure cuts down on the volume discharged (see your pumpers discharge plate on the pump panel). Remember, it is the *amount* of water we apply that puts out the fire, not just the pressure. For this reason, it is impractical to try to supply additional lines from this pumper, since they may require in excess of 200 psi, depending on type, length, and other variables.

Another reason for not raising the pressure much more than 150 psi is the condition of the sprinkler system. When newly installed (in most systems), the piping should be hydrostatically tested to 200 psi. That may well be the last time the system is ever subjected to that high a pressure. If the system has been in place for very long, it may be subject to many damaging forces, such as rust, corrosion, impact, freezing, expansion, and contraction. If we come along with our high-pressure pump and start cranking away, we may be responsible for blowing apart fittings and putting the system Out of Service (OOS) at the time that it is needed most.

Problems with Sprinklers

One problem that sprinklers produce—and with which many firefighters must be prepared to cope—is that fires in sprinklered buildings are often more smoky than those in unsprinklered buildings. It's not hard to figure out why. Fires in unsprinklered buildings often progress further than equivalent fires in sprinklered buildings. Often, such a free-burning fire, getting a good head start, will vent itself before we arrive. Sprinklers, on the other hand, detect the fire early, sound the alarm, and begin to apply water. This thwarts combustion, so carbon monoxide levels will increase initially, and the water spray tends to cool the fire gases, making them sink. Probably the biggest reason we think of such fires as smokier is that the spray pattern creates a draft similar to that of a fog nozzle, pushing smoke and gases down to the floor. These three factors—increased carbon monoxide, sinking fire gases, and smoke being pushed down—remove the fresh air layer often seen at fires. This mandates the use of masks even though the intensity of the heat is less. Using masks will allow firefighters to move in and extinguish any remaining fire without having to shut down the sprinklers first. Mechanical ventilation will often be required to move this cold smoke, especially in below-grade areas.

An infrequently encountered situation, but one that causes firefighters a lot of grief, occurs when sprinklers go off after the firefighters are already at the scene. Normally, the sprinkler water-flow alarm is our first indication of a fire. When the alarm sounds, we must determine the location, get dressed, board the apparatus, respond, and possibly force entry before we get near the area of the water spray. In the 3, 4, or 5 min that this response takes, the sprinklers are operating over the fire, cooling it and the ceiling temperatures before we arrive.

Late one night, I responded to a three-story, wood-frame multiple dwelling that had a sprinkler system that only protected the interior stairs. A line was in place on the top-floor landing as we forced the apartment door. When the door opened, we were met by fire rolling out onto the stairs. We opened the line and began our advance when suddenly all hell broke loose. We were hit by a blast of steam, hot water poured over us, and the visibility immediately went to zero. As the officer on the line, I felt that we were being pushed back by an outside opposing stream, and so relayed a plea to shut down the tower ladder. We kept the hose stream operating. In a couple of minutes, the conditions got better. The fire went out, and I learned a valuable lesson. In buildings equipped with partial sprinkler systems, you can't rely on those systems to extinguish fire in a room where there is no sprinkler head. You should immediately send a member to locate the shutoff and to stand by for orders to shut it down. Once you are certain that the fire won't extend, shut down the system and leave a man at the valve to await further instructions. It may be necessary to reopen the valve if conditions change drastically.

Another fire in a partially sprinklered building was creating havoc with traditional fire attack until the officer in charge recognized the situation. This fire was in an interior store of an L-shaped shopping mall. The stores were unsprinklered and, for the most part, separated from the mall only by open-mesh security gates. The sprinkler system protected only the central area of the mall and couldn't have any extinguishing effect on the fire in the stores. A heavy body of fire existed in the store of origin and extended to two additional stores. Roof ventilation of existing skylights was prompt, but it provided no relief, even though each one measured 10 x 25 ft. The presence of two sprinkler heads in each skylight effectively put the lid back on these natural vent points. Very little headway could be made on the main body of fire due to the severe heat and zero visibility.

As an engine officer, I was ordered to stretch a 2½-in. handline into the mall area and attempt to locate and attack the seat of the fire. While leading the line in, I became separated from the line and lost my bearings. I remained calm, located the nearest wall, and began to move along the storefronts to safety. Commercial buildings aren't like residential fires, in that you may be hundreds of feet from the nearest door or window. When forced to travel by maintaining contact with a wall, you must follow the layout of the walls. In a shopping mall type of layout, that means following along the show window return areas. To go across a 30-ft-wide store might involve traveling 80 or 100 ft or more. That eats up your breathing air and escape time. I found myself out of air in this heavily charged mall, uncertain whether I was even moving in the shortest direction toward safety or toward the fire. I wasn't radio equipped and thus couldn't call for assistance. I survived by breaking through a plate-glass store window (fortunately not covered by a roll-down gate), crawling through the less heavily charged store, and letting myself out the back. Fortunately, this store wasn't equipped with any locks on the rear that required a key. The smoke that I inhaled provoked a case of bronchitis that took two months to cure (Fig. 6–3).

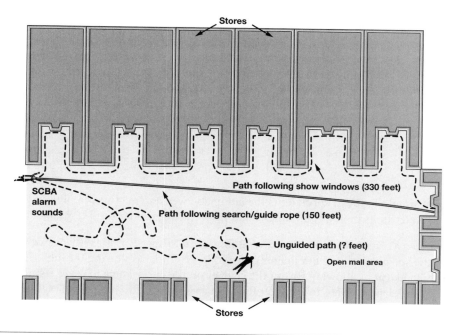

Fig. 6–3 This diagram illustrates three potential paths a firefighter might take when exiting a fire in a mall or other large structure. Following a search rope or a hose line should be mandatory. Any other option risks death. Partial sprinkler systems can compound the smoke condition.

Upon being informed of the situation, and realizing the severe threat to the firefighters, the officer in command ordered the sprinkler system to be shut down in that area. This allowed natural ventilation to progress in the fire area, permitting an attack and extinguishment. It also left the zones in the rest of the mall operating to prevent extension in those directions. The man sent to shut down the sprinklers was radio equipped and ready to open the valve if the situation deteriorated.

This brings out a key point in fighting fires in sprinklered buildings: ventilation is an absolute must, but it will often be extremely difficult to accomplish. The laws of Mother Nature are often seemingly violated. Smoke doesn't rise as expected; instead, it sometimes hangs low. That's due to the twofold action of the sprinkler cooling and pushing down the smoke. In many cases, cellar, sub-cellar, and windowless areas are sprinklered as a result of the building code. Obviously, venting in such areas will be difficult, since it will involve inside rooms of larger buildings.

What alternatives are available to the officer in command at these situations? Fog lines will often suffice in the immediate fire area, since fire and water damage have already been done. Smoke ejectors will do in relatively small areas with openings to the exterior. In large structures with limited openings or many small rooms, the building's heating, ventilation, and air-conditioning system (HVAC) may be the only possible way to move the quantity of air desired. This may necessitate the aid of building maintenance people.

A Suggested Strategy

1. Know beforehand the location of sprinkler system shutoffs and Siameses.
2. Commit supply lines to sprinkler Siameses early, providing proper volume and pressure.
3. Get handlines in place, staffed by mask-equipped members.
4. Ventilate the area, anticipating difficulties and using proper techniques.
5. Only after the fire is definitely under control, shut down the sprinkler system, drain it, and restore it to service.

Types of Systems

Firefighters may encounter several different types of sprinkler systems, each with its own operational characteristics.

Automatic wet system

The most common and simplest type of sprinkler is the automatic wet system. This is a pipe system connected to a source of supply that has water in its piping at all times, right up to each sprinkler head (Figs. 6–4 and 6–5).

Usually each system has an alarm valve that serves several functions. Primarily it is a check valve, preventing water from flowing back out of the sprinkler piping but opening to admit water that is flowing through the heads. If water is flowing, it also transmits an alarm by allowing part of the clapper on the check valve to cover a small pipe when the check valve is closed. When the clapper opens to let water flow to a sprinkler head, water also flows through this small pipe, tripping electric switches or turning the paddle wheel of a water motor alarm bell. If the water flow stops, the clapper falls back down, resetting itself and covering the pipe.

False alarms can occur if there is a large fluctuation in the pressure of the water supply to wet systems. This happens when the pressure from the surge on the bottom of the clapper is greater than the pressure in the system on top of the clapper. The clapper temporarily lifts and then resets.

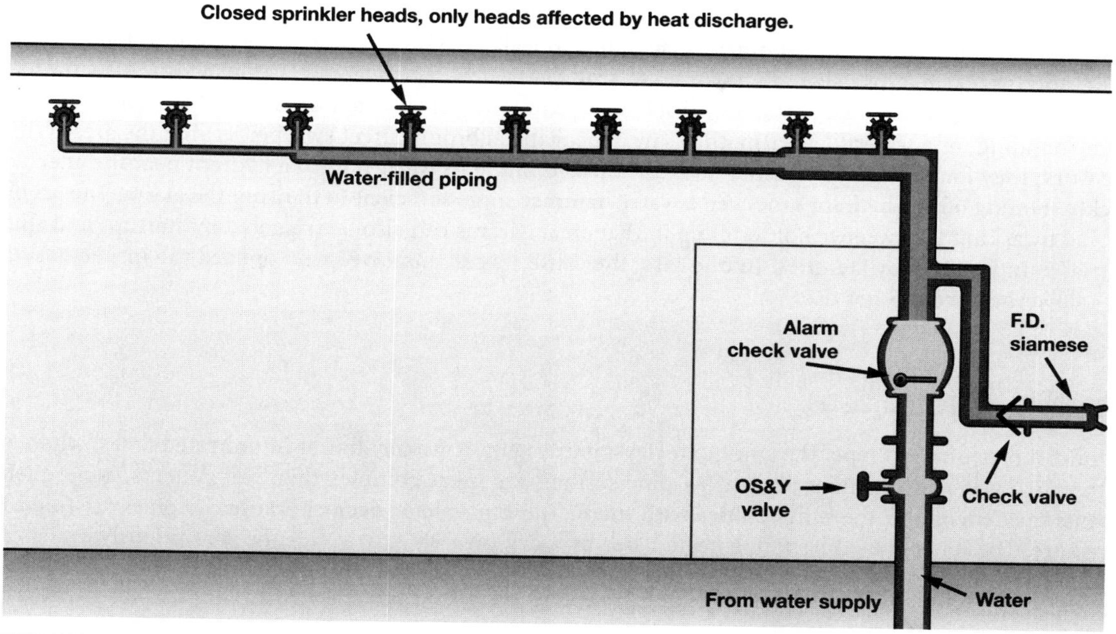

Closed sprinkler heads, only heads affected by heat discharge.

Water-filled piping

Alarm check valve

F.D. siamese

OS&Y valve

Check valve

From water supply

Water

Fig. 6–4 Automatic wet-pipe systems are the most reliable type system.

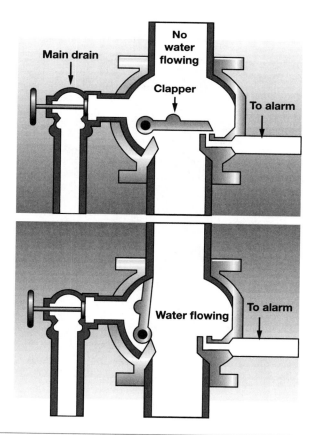

Main drain

No water flowing

Clapper

To alarm

Water flowing

To alarm

Fig. 6–5 Water flow in a wet-pipe system lifts a clapper (check valve) that sounds an alarm. The clapper will close if water flow stops or if a pumper pumps into the FDC.

If false alarms recur at the same location, you should root out the problem. First, notify the building owner that repeats won't be tolerated. Sprinkler contractors can adjust the alarm piping so that water will flow for a longer period before it sets off the alarm. This delay can be for up to 1 min (very few surges would last this long). Also, try to find the cause of the problem. One chronic problem was solved when firefighters noticed a pattern. Whenever they responded to the same building for a waterflow alarm, they saw a mechanical broom street sweeper leaving the area. With a little detective work, they found that the sweeper had just finished filling its water tank at a hydrant near the affected building. Quickly shutting off the hydrant produced a water-hammer surge sufficient to transmit the alarm. The problem was easily solved by asking the sweeper not to fill up in that area. Alarms can also be triggered by startups and shutdowns of industrial pumps at nearby factories. In one case, the cause was an employee who opened the inspector's test valve to water a flowerbed every other day.

Automatic dry (or dry-pipe) systems

The next most common type, the automatic dry-pipe system, is usually found in unheated areas, where freezing poses a threat to water in sprinkler pipes. Dry-pipe systems are more complex than wet systems, consequently, fire departments may encounter more difficulties with them. These problems occur because the pipes are filled with air under pressure. The water is held back in a heated area by a dry-pipe valve. Air must be continuously available to this system, either from a compressor or bottles and regulators. Otherwise, the air pressure may drop to the point where the valve trips and sends an alarm (Figs. 6–6 and 6–7).

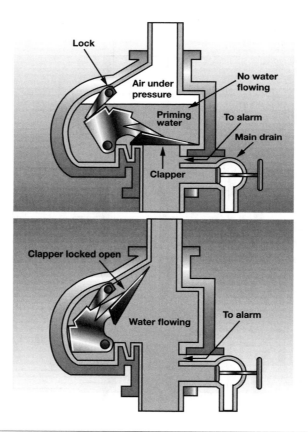

Fig. 6–6 Automatic dry-pipe systems have a clapper that should lock in the open position when the valve trips. The dry-pipe valves clapper is larger than a wet-pipe clapper, to allow a low air pressure (with priming water seal) on the top to hold back a higher water pressure below.

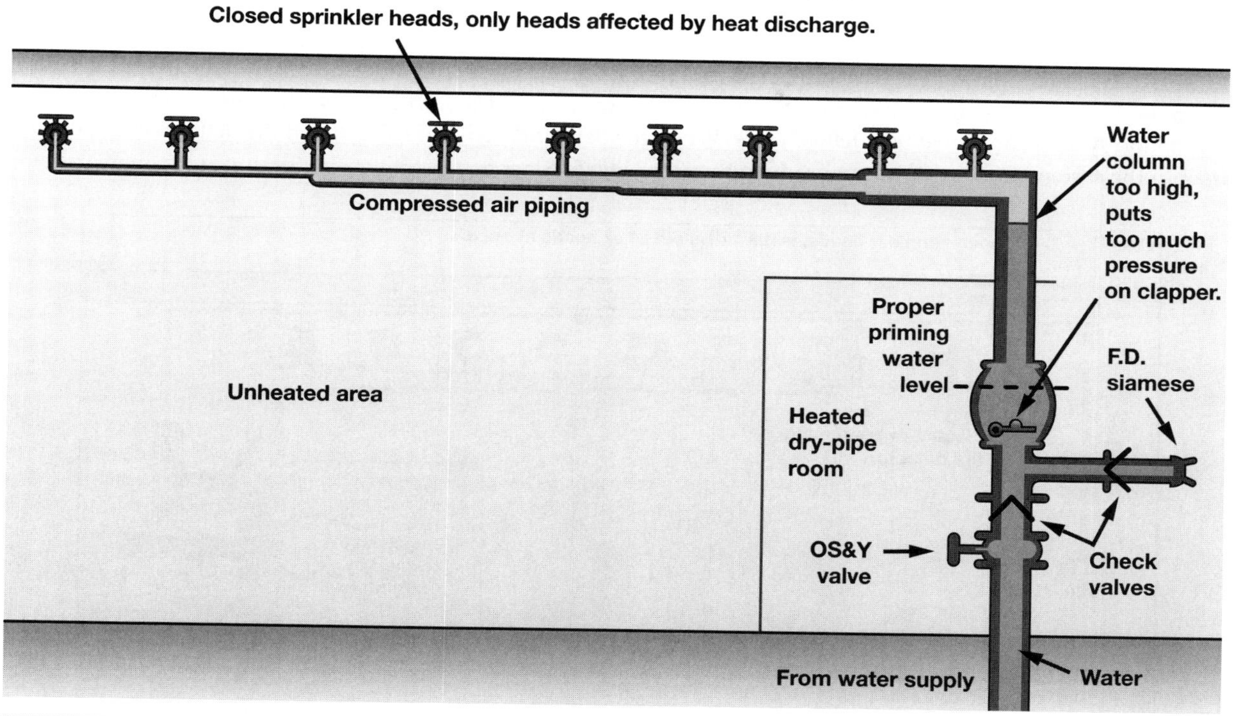

Fig. 6–7 Dry-pipe valves must be located in a heated room, to keep the supply pipe and valve from freezing. If too much water is allowed into the dry-pipe system, the weight of the water can prevent the valve from opening during a fire.

The dry-pipe valve functions similarly to the wet-alarm valve, with a couple of exceptions. In the dry-pipe, there is a clapper to hold air in the system. This clapper varies from a wet-system clapper in that it has a much larger top than bottom surface, allowing a much lower pressure on the top to hold back a high water pressure on the bottom. For most valves, the air pressure on top will be about 30–40 psi, which will hold back up to 100 psi. This is called a *differential dry-pipe valve.*

Another difference of the dry-pipe valve is that its clapper is equipped with a lock-open feature. If the pressure surges continually, lifting the clapper and pushing in a little more water each time, the system will eventually fill with water, which could freeze and destroy the system. A less obvious problem, called *water columning*, occurs when the water above the clapper is so high that its weight will hold the clapper closed, even after a sprinkler has opened and all of the air has bled out. When this happens, no water will discharge on the fire. A lock is installed on the clapper to prevent this so that if a flow equal to one sprinkler head occurs, the clapper is locked open and an alarm is sent. A dry-pipe system won't reset by itself, nor can the alarm service reset it. The owner should arrange for the services of a qualified sprinkler contractor.

When firefighters arrive in the middle of the night at a dry-pipe valve trip, where there is no fire present and no water flowing, they are faced with some difficult decisions. If they leave the system on and silence the alarm, they face two problems. If the system trips for an actual fire, another alarm won't be received. Also, if the system freezes, it can't flow water on the fire, and the ice in the pipes and fittings can literally blow the system apart.

If firefighters shut off and drain the system, what will happen if a fire occurs? A possible tactic is to shut and drain the system, and to order the owner (preferably in writing) to provide a watchman's service throughout the building until the system is reset. The watchman should be taught how to reopen the system in case of fire. If there is no threat of freezing, firefighters can leave the system on as a wet system, again with a watchman's service to sound any necessary alarm.

Deluge systems

These consist of a piping system connected to a water source. Open sprinkler heads are attached to the piping throughout the area. These heads have no heat-sensing elements, but are merely nozzles. The area is also protected by a fire detection system connected to a deluge valve. When the detectors sense fire, they open the deluge valve, and water flows into the system and out all of the heads in the area (Fig. 6–8).

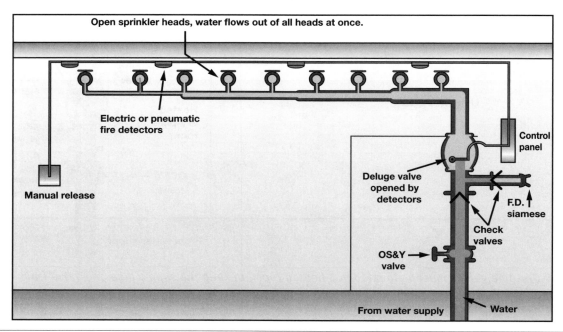

Fig. 6–8 A deluge system has open sprinkler, which allows water to flow from all at once, flooding the area.

Systems of this type are used to protect areas such as aircraft hangars and flammable-liquid loading docks, where an extremely rapid spread of fire is expected. Fire departments responding to a building protected by a deluge system must be prepared to augment the supply of very large volumes of water. Once the fire has been extinguished, they must use great care in shutting down any control valves, since the large flow can easily create a dangerous water hammer if the system is closed too quickly (Fig. 6–9).

Fig. 6–9 Deluge systems are often found as protection for flammable liquid facilities, such as this truck loading rack. This system injects foam into the piping to discharge through special foam/water sprinklers.

Pre-action systems

Another special type of system, the *pre-action system*, consists of an automatic detection system coupled to an automatic dry-pipe system. Pre-action systems are often installed over very expensive electronic equipment that could be damaged by a water leak. The piping is usually filled with compressed air and connected to an automatically operated valve that exhausts the compressed air only if the detection system (heat- or smoke-activated) sense fire. This lets the dry-pipe system fill with water when the detectors sense fire, which is usually at a stage before the sprinkler heads would have fused. The pre-action system saves time in getting water on the fire, since it doesn't require that all of the air bleed out through a sprinkler before the dry valve trips. It differs from a deluge system in that it uses closed sprinklers, so water discharges only over the fire area (Fig. 6–10).

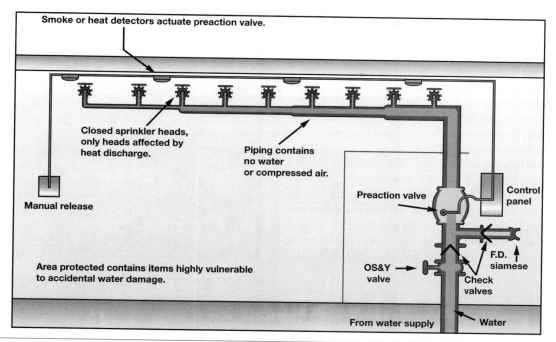

Fig. 6–10 Pre-action systems are usually located over very expensive, water sensitive equipment. They combine a fire detection system with a deluge type valve that supplies closed sprinkler heads.

The most recent innovation in sprinkler systems uses heat detectors, not only to activate the system, but also to shut it off when the area has cooled. By using closed heads to localize the flow, this system can turn the water on and off again if fire rekindles in a remote area. The primary purpose of this system is to limit water flow to just that needed for extinguishment, thus reducing water damage. Its water-conservation feature makes it an excellent choice for a system that has a limited water supply, such as a pressure tank. It may be found where prevention of water damage is a must, as in computer rooms, tape-storage areas, electronic switch rooms, and telephone switching centers. Firefighters operating in these areas should be aware of the possibility of sprinklers suddenly activating, engulfing them with water, steam, and smoke while they are at work. This can also occur if the structure has on/off sprinkler heads on a standard wet, dry, or preaction system. These heads have a heat-sensing element built right into them, and it will cycle on and off continually. They don't generally require any chocking or replacement by fire department personnel.

Nonautomatic systems

This type of system is sometimes confused with the automatic dry-pipe system. The first nonautomatic system goes back more than 100 years, when fires in commercial cellars and sub-cellars of large urban areas defied firefighters' efforts to descend. Fire codes soon called for sprinkler systems to be installed, but the building owners complained that it would cost too much to connect to a water supply. The compromise was to install a system with no water supply connection other than the fire department Siamese, so firefighters wouldn't have to descend into those hellish areas. This ignores the fact that the fire was often so intense by the time it was spotted that it had opened every sprinkler in the area, creating a tremendous water demand; otherwise, it had destroyed the piping, making the system useless.

The first attempts to provide some type of sprinkler system in these areas involved so-called perforated pipe systems, which consisted of rows of pipes with small holes (⅛ in. to ¼ in.) drilled every few inches along the top and sides. These pipes were run along the ceilings of the occupancy, connecting to Siamese connections at the street level. They functioned much like a modern deluge system, in that water came out of every pore when the system was supplied by a fire hose. Some of these systems, more than 100 years old, still exist in older cities (Fig. 6–11). They shouldn't be relied on for fire protection, but they may serve to indicate the location of the fire by the amount of smoke pouring out of the Siamese!

A variation of the nonautomatic system was developed after a fire at the New York Telephone Switching Center in 1975. At that incident, conditions in sub-cellar areas were so severe that more than 700 firefighters using more than 1,050 air cylinders had to work 60 hr to bring the fire under control. Telephone service for much of Manhattan's business area was knocked out for days. As a result of the incident, FDNY and New York Telephone planned the installation of a so-called special suppression system in many telephone-switching centers. This is a custom-designed nonautomatic system using 365° temperature sprinkler heads connected to piping designed to withstand 1,500°F for 15 min An essential part of the system is a fire alarm that notifies the fire department immediately. The system remains nonautomatic to protect the delicate electronic switching gear from accidental water leakage (Fig. 6–12).

Fig. 6–11 Perforated-pipe Siameses are the very first nonautomatic systems. They are very old and unreliable.

Combination sprinkler and standpipe systems

In yet another move to reduce the cost of fire protection systems (and thereby improve the chances of more of them being installed) combination standpipe and sprinkler systems have been developed. These systems were a logical way to extend sprinkler protection into existing unsprinklered high-rise buildings in the early 1970s, after a spate of serious high-rise fires demonstrated once and for all that high-rise office buildings must be sprinklered for the safety of their occupants. In these buildings, which already were required to have a standpipe system due to their height, it was a very logical move to allow the sprinkler contractor to tap into the existing standpipe riser, rather than have to run a brand new riser the height of the building. This permitted the owner to sprinkler older buildings at a reduced cost. I said that in these old buildings, it made sense to allow this; because we were getting additional protection we wouldn't have otherwise. In the old buildings, we also had tremendous capacity in the standpipe risers already. They could easily flow 500 gpm, and many were capable of much more.

Fig. 6–12 Many telephone switching centers were equipped with a special suppression system, a specially designed nonautomatic system, for their cable trays.

In New York City it is common to find 8-in. diameter risers in the taller buildings. Since the sprinkler systems were only required to flow anywhere from 150 to 200 gpm in a light-hazard office floor, the standpipes could easily supply this additional demand. The problem is that new buildings are being built this way, with hydraulically calculated piping that removes all the reserve (think *wiggle room*) out of the system. Examine the photo in Figure. 6–13 carefully, and you will observe that the supply pipe for this combination system is only 2½-in. pipe that comes into a *T* connection. The Fire Department's 2½-in. hose valve connects to the bottom of the *T* while the supply to the sprinklers ascends from the top. What do you think will happen to the pressure in the sprinkler system the moment the fire department opens the standpipe valve?

Professor Frank Brannigan describes such engineering as *state of the art*. He goes on to explain that state of the art means "we don't really know what it's going to do, so we're trying it out on you." I know what it's going to do, but apparently people who have never crawled down a smoke-filled hall feel it's not that dangerous. Just like state of the art pressure-reducing valves (PRVs) at One Meridian Plaza!

Size-up at Sprinklered Buildings

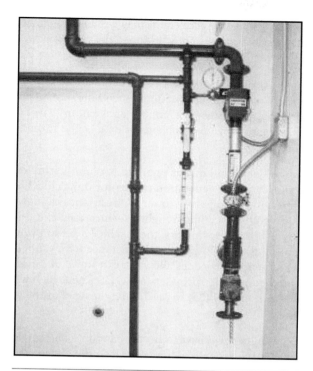

Fig. 6–13 The latest state of the art design is a combination standpipe and sprinkler system. Note: The 2½-in.pipe feeding both the 2½-in. fire department hose outlet and the sprinkler system. What happens to the sprinkler operation when the fire department charges their handline?

In sizing up a fire, we must consider auxiliary appliances, such as sprinkler, standpipe, foam, and other fire-suppression systems. We must first determine whether or not an auxiliary system is present. In some cases, the presence of certain systems, such as a standpipe in a high-rise building or a foam system at a bulk oil storage yard, will be assured by building or fire codes.

The presence of automatic sprinkler systems, however, is more difficult to forecast. Many very large buildings aren't sprinklered, and many smaller buildings are fully protected. Often the method of alarm receipt, *i.e.*, *valve alarm* or *water-flow alarm*, for example, will indicate the presence of a sprinkler system. Some exterior signs that a building is protected by a sprinkler system are the presence of a Siamese connection, the sight or sound of the water motor alarm gong, and running water discharging from pipes coming through the wall of the building. (**Note:** A Siamese connection may look the same for both sprinkler and standpipe systems. There should be a marked metal plate on or adjoining the Siamese to indicate its purpose. Some jurisdictions mandate color-coding Siameses to simplify the task of identifying which system each one serves.) Of course, the best way to determine whether or not a sprinkler system is present is by prefire inspection. This also familiarizes us with its location and type of controls and lets us check its operational status.

Upon arrival at a fire in a sprinklered building, we should first determine whether or not the building's sprinkler system is operating, since this will indicate our course of action. This is usually easy to do if the building is occupied. We will often be met by building personnel who have investigated the cause of the alarm and can lead us to the problem.

If a serious fire is found in a building that is protected by a sprinkler system, you should immediately call for assistance. In many departments, sprinkler alarms only receive a minimal response, based on the premise that greater than 95% of all fires are controlled by one or two sprinkler heads, and an even larger percentage of responses involve no fire at all but are minor malfunctions such as pressure surges or low air pressure in dry-pipe valves. Even if your department sends a full response to sprinkler alarms, you should expect that some of the responders have become complacent about these type responses and may be holding themselves back a little, waiting for a report from the scene, fully expecting to hear it's another *false* alarm caused by a malfunction. Another more pressing reason for needing help though is the simple fact that you're likely to need it!

Since most sprinklered buildings are commercial or industrial in use, you've just found yourself at a serious commercial fire, which is far different from the residential fires you usually fight. Even if you do fight a good number of commercial fires, a fire in a *sprinklered* building where the sprinklers are OOS is likely to be much more difficult than the norm—and potentially deadly. If you find a serious fire in a sprinklered building, generally the system will have been turned off. That could be due to an arsonist closing the valve, meaning accelerants may well be present. Most firefighters think you can treat a fire in building that has a sprinkler system that is not working in the same way we do a building that was not sprinklered in the first place. That could prove to be a tragic mistake.

When the installation of a sprinkler system is being contemplated, especially in new construction, building codes often permit *trade-offs*. In other words, because the sprinkler system will make the building much safer, the building code allows the owner to install less costly and often more dangerous materials in a number of situations, since *the sprinklers will take care of any fire problems*. Many of these items are not obvious problems as long as the sprinklers work, but can create havoc if the sprinklers are OOS.

Examples of some of the conditions that are allowed in a sprinklered building as opposed to a non-sprinklered building include the use of materials with higher flame spread and/or smoke generation ratings for things like floor, wall and ceiling coverings (carpets, wallpaper and the like). Some codes also permit lower fire resistance ratings for structural elements and walls, and a particularly dangerous issue for unsuspecting firefighters—increased exit travel distances. What that means is that if the building were being built without sprinklers, the distance a person would have to travel to reach an exit is set at a particular distance, say 100 ft, which might be the distance for a school or an office. If the building is built with an automatic wet sprinkler system many codes allow the exit travel distance to be doubled, to 200 ft. Now picture yourself fighting a serious fire because the sprinklers are OOS when the low air alarm on your SCBA begins to sound. Suddenly you have to crawl an extra 100 ft to reach safety instead of the way the building without sprinklers would have been designed.

If we find a fire in a building where the sprinklers are not operating, we will have to stretch handlines and conduct a manual attack, but we should also make every effort to get the sprinklers into action as well. Immediately begin supplying the FDC and monitor the effect that this effort achieves, if any. Be sure to alert the interior forces that you are about to start water to the system. Remember the impact that the sudden operation of the sprinklers will have: a large steam/smoke cloud and red hot water will descend on the members. They may need to temporarily retreat to a stairway or other refuge outside the immediate fire area. The advantage of getting the sprinkler discharge where it is needed though—right over the seat of the fire— usually far outweighs any small delay this causes. Note that in the case of large buildings that have several sprinkler *zones* or separate systems, the FDC will join the system below the main outside stem and yoke (OS&Y) control valve. If this valve is closed, pumping water into the FDC will not feed any water into that system or zone. Pump operators supplying the FDC should note whether discharge and intake pressures fluctuate when the gate valve feeding this line is closed. If there is no fluctuation in pressures as the valve is closed and there is no water flowing into the system, a closed sprinkler valve may be the culprit.

As soon as possible, send a reconnaissance team equipped with forcible entry tools including bolt cutters and a portable radio to the sprinkler control valve location. If the sprinklers are not operating, chances are that the valve is closed. This recon team may be able to open the valve and restore protection. Again, all interior units should be warned of this impending action, and should be directed to report the effects of sprinkler operation to the IC. They should also report any situation where water does not flow from the heads in the fire area. Other valves may be closed or there could be a large break in the piping remote from the fire area.

If there is a large break in the piping, you may as well shut down the control valve and save the water for the manual attack or for exposure protection. I would keep the lines supplying the FDC charged if there is any possibility that other portions of the system are operational, as this may assist in confining the fire to the section of building that is already involved. Be sure to remind the recon crew that part of their job is to verify the actual position of the control valves, since a closed control valve may be an indication of arson. They may have to testify in court as to what position the valves were in when they arrived, and any attempts they made to open them. If valves are located and opened, this crew should stand by at the valves awaiting orders to close them if it suddenly causes severe operating problems or if large breaks in piping are discovered that render the system inoperable.

If the building isn't occupied, and no signs of an actual fire are detected, our actions will depend on our size-up. First, look for any obvious signs of water flow, such as a ringing water motor alarm, a steady flow of water from drain pipes near control valves, or water running out from under doors or down stairwells.

Consider the following scenario: at 1:30 AM, you receive a water-flow alarm for a carpet warehouse. A central alarm service has notified the building's owner, who is responding with keys, but he won't be there for about half an hour. When you arrive on the scene, you note a Siamese and a water motor alarm gong in front of the building. The gong isn't ringing, and there are no external signs of a problem. To force the main entrance to the building, you would have to break through a heavily secured steel-clad door. As you pass the alarm gong, you notice a small puddle on the ground and water dripping occasionally from a pipe nearby. You check with the alarm company and learn that it is no longer receiving the alarm signal.

Your choices of action include the following:

1.   Force entry to the structure and locate the cause of the alarm.

2.   Hold some or all of the units and await the owner.

3.   Return all units to quarters.

Prefire planning can pay dividends here if you made note of the type of sprinkler system that's in the building. A wet-pipe system, subject to pressure surges, will sound an alarm and then reset when the pressure stabilizes. If this building has a wet system, a pressure surge is the likely cause, but further investigation is necessary. Immediately make a rapid survey of the perimeter of the building. Look for any signs of fire, smoke, or water within the building, and locate the least damaging means of entry. Send a member to the Siamese connection to check for water flow. This is done by feeling the pipe for vibrations and putting your ear to it to listen for the sound of flowing water.

If nothing indicates an immediate problem, you may elect to release all units except the first engine and ladder, which will be able to handle the situation. On the other hand, if there is any sign of fire, smoke, or water flow, you are justified in making immediate forcible entry by the most advantageous means. You must force your way in and make a complete examination of the premises to determine the cause for alarm. Often a member will be able to climb a ladder and enter through an elevated area, such as a roof or window. (It will be difficult for thieves to gain access here after you leave.) A minimum of two members can then make their way down and unlock the door from the inside to allow a thorough search. In smaller buildings, they can complete the investigation alone and report their findings to the officer in command.

A few words of caution about this method: first, when entering a building like this, be especially alert to hidden hazards like open shafts and automatic-starting machinery. Also, beware of security dogs and nervous night watchmen, especially armed ones. Make a lot of noise before you set foot inside. Call out in a loud clear voice, "Fire department, is anyone here? Finally, advise the owner (preferably in writing) of any damage incurred while forcing entry so that he may have it repaired.

That leads us to ask, "Why shouldn't we wait for the owner with the keys? After all, the system has reset itself." Aside from the discomfort and inconvenience to the people left there at 1:30 AM on a miserably cold, rainy night, there are some very important reasons for not waiting. First, we aren't positive that there isn't any cause for alarm. Any number of things may have stopped the obvious signs of water flow, such as defective or vandalized water motor gongs, valves in the alarm line left in the off position, or debris stuffed into drain lines. In these instances, water will still be flowing on the fire or, in the case of accidental discharge, on material subject to water damage (Figs. 6–14 and 6–15).

Fig. 6–14 A gravity-fed water source offers very limited pressure to the top floor sprinklers.

Fig. 6–15 A pressure tank offers the advantage of increase pressure for the top floor. Both are limited duration supplies that must be augmented immediately through the fire department Siamese connection.

In some cases, though, our failure to act may allow an insignificant fire to intensify. This happens primarily in systems that have a limited water supply and depend on the fire department's support of the Siamese to ensure total extinguishment. Many structures have systems like this, from rural nursing homes to inner-city warehouses. The common denominator is their limited supply. Rather than having a large connection to a public water main, these constant-pressure systems are fed either from a gravity tank on the roof or a pressure tank, which is much like a giant, pressurized fire extinguisher. Depending on the hazard, the capacity of these tanks varies 2,000–5,000 gal or more. They are generally designed to provide 30–90 min of supply. If this supply runs out and the fire isn't totally extinguished, the fire department must step in, supply the Siamese, and overhaul. Otherwise, the fire will multiply unchecked.

At some arson fires, the water flow is shut off after the heads have activated. This typically means that, as the arsonist fled the scene, he realized that the sprinklers had darkened down his handiwork, and so he shut off the outside stem and yoke valve on his way out.

One final reason not to delay our investigation is that some very valuable equipment may be exposed to damage. Once when I was working as a sprinkler contractor, I was called at 9 AM to reset a sprinkler system that had tripped at about 6 AM For three hours, a 1,500 gpm automatic fire pump had been running with no water flowing. The churning of the impellers had heated the water to the point where it did major damage to the pump itself, costing tens of thousands of dollars to repair. The damage occurred even though the fire department had responded. (Since the firefighters found no fire, they elected to leave the scene.) Another time, I found a fire pump that was actually spurting steam under pressure from around the packing glands and the pressure relief valve.

Locating and Operating Controls

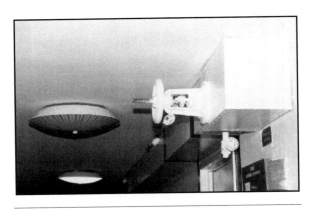

Fig. 6–16 An OS&Y valve controlling sprinklers in this floor of a multistory building. All members should note the position (open or closed) of such sectional valves as they pass by.

As soon as we have determined that a sprinkler system is present and functioning, we should locate the system controls and station a two-person team there. They should be able to identify the chief components of the control valves and know how to operate them. The control valves are generally located near the outside signs of the sprinkler system (water motor gong, Siamese, drains), but that isn't always the case. Some buildings have several zones, with valves located at the ceiling, 15–25 ft above the floor and accessible only by ladder. The best policy is to seek the assistance of the building maintenance people. If they aren't available, you may fall back on your prefire plan or knowledge gained from previous visits to the building (Fig. 6–16).

In the past, if the previous two methods failed, we used to be able to look at the sprinkler piping and trace it back to its supply rather quickly. The design of the system was almost always known as a tree system. The farthest branches were the smallest and, as the pipe diameter increased, we reached the trunk and finally the root of the system. In recent years, however, the hydraulic design of sprinkler systems has changed drastically. Sprinkler systems are now often designed as loops of equal-diameter supply mains surrounding the building, cross-connected to form a grid. Locating the supply valves by following the pipe can be a much more complex, time-consuming affair. It is far better to spend a little time before, rather than during, the fire to locate the controls.

The most common type of sprinkler system control is the OS&Y valve. It is usually found at the base of the sprinkler riser near the front of the building, where the water main enters. This is an indicating-type valve, in that you can see what position it is in by looking at it. The OS&Y is equipped with a spoked handwheel that is turned to open or close it. By turning it counterclockwise, a center screw stem rises and sticks out beyond the valve, indicating that the valve is open. By turning the handwheel clockwise, the stem recedes into the valve, indicating that the valve is closed. We can tell whether the valve is open or closed by the amount of the stem protruding. A quarter-inch or less means that the valve is closed; a fully open valve should have about as much stem showing as the thickness of the pipe above or below it. (For example, a 6-in. alarm valve should have about 6 in. of brass stem protruding below it.)

The task of locating the controls is simplified for us when the sprinkler contractor or designer locates the main shutoff outside the building. This usually makes it easier to locate, and it can most often be operated on command by other than attack personnel. Two devices, called the *post indicator valve* (PIV) and the *wall indicator valve* (WIV), greatly resemble each other and serve the same purpose—to control the flow of water. In some cases, a single valve may control the entire system, or a separate PIV or WIV may be found for each zone. Both valves show whether they are open or shut by means of a sign within the stem of the valve reading *open* or *shut*. The PIV is normally found with its operating handle secured by a padlock in the open position over the operating nut. To close this valve, we remove the padlock, take off the detachable handle, and place the end over the operating nut, much in the way that you would operate fire hydrant. Then, turn the PIV clockwise to close it (Fig. 6–17).

The WIV also has a sign in the stem, but instead of having a separate handle, it is often permanently fitted with a spoked handwheel to operate it. Unlike the OS&Y valve, however, the handwheel doesn't raise or lower a stem. The only indication of the valve position is in the sign's window. It is generally chained and padlocked open to prevent unauthorized closings. To close the WIV, we simply remove the chain that secures it and turn the valve clockwise (Fig. 6–18).

Firefighters should also be able to recognize and operate the 2-in. main drain valve. This valve is fairly easy to spot among the mass of piping around a sprinkler control valve. Quite often, it is labeled *main drain,* and its physical features distinguish it from the other valves. The main drain is usually a 2-in. pipe, the second largest pipe beside the large supply main. All of the others are usually much smaller. In addition, the valve is of an unusual design—it is often an angle valve, with the feed and discharge at 90° angles to each other. The name of the valve indicates why and how we use it, since a sprinkler system may contain several hundred gallons of water within the piping. After we have shut off the control valve, this water remains and will continue to pour out through any open sprinkler heads. Firefighters can speed up the process of salvage and overhaul by draining this water out through the drainpipe (Fig. 6–19).

Fig. 6–17 A PIV. Note the **open** sign in the window.

Fig. 6–18 A WIV. Note the chain and lock to prevent tampering.

Restoration of Protection

One often-debated aspect of operations at sprinklered buildings is the fire department's role in restoring the system to service. Some departments prefer to chock flowing sprinklers and not touch any other valves or fittings. This approach presents an advantage in that protection for the rest of the building isn't compromised. It is often unsatisfactory, however, due to different sprinkler styles, high ceilings, low visibility, and numerous other factors.

Fig. 6–19 All members should be able to locate the main drain on an alarm valve. (It's usually an angle valve.)

Some departments take a hands-off attitude to this problem. Once the fire is out, they claim their work is done. This is a fine approach if building personnel are on the scene and are notified of the urgency of restoring protection. At 3 AM on a holiday weekend, however, it may be quite some time before a representative of the owner shows up. The alternatives are to leave the building unprotected, to leave a unit at the scene awaiting the owner, or to get a wrench and sprinkler head from a supply that should be mandatory on the premises, and replace the fused head (90% involve only one or two heads).

There are so many styles and so many temperature ratings that your department doesn't want to be liable for making a mistake and installing the wrong type, you say. Some decorative heads may require special wrenches to replace, and these should be left to the qualified contractor. For the most part, though, the worst that could happen would be to install the wrong temperature and the wrong style. This should be a temporary measure until a contractor arrives and corrects your choice. In the meantime, should a fire occur, the premises are protected. Even an incorrect head will fuse eventually and apply water to the fire. It will also re-notify the fire department and possibly prevent the total loss of the structure.

All sprinkler heads are marked as to temperature. They are also color-coded to indicate the temperature, if they are not chrome-plated.

All sprinkler heads are also marked as to their intended placement as follows: upright, pendant, or sidewall. You don't have to be a fire protection engineer to replace a fused head. Simply remove the old one, take it to the spare head cabinet, match it to the same style, and replace it. You should mark the location of the replaced head with a piece of bright cloth or banner tape on the pipe. Mark the OS&Y shutoff valve with similar tape and a note to check the head. Finally, in all cases, direct the owner (in writing) to have the system examined and the heads inspected by a qualified sprinkler contractor.

Standpipe Systems

Standpipe systems are sometimes confused with sprinkler systems, usually by very inexperienced firefighters who haven't seen each type of system in action. Although both will usually have a FDC that looks very similar (and certain other similarities, such as style of control valves), the purposes of the two are distinctly different (Fig. 6–20). A sprinkler system sounds an alarm and actively attacks the fire But a standpipe system is a very passive device, in some cases being no more than a vertical pipe requiring fire department pumpers even to supply water to it. Before you can discuss operations at a building equipped with standpipes, you must understand the types of systems that you might encounter, as well as the classes of systems within each type. As with sprinklers, the type of system relates to the water supply. NFPA Standard 14, *Installation of Standpipe and Hose Systems*, describes the following five types of standpipe systems: automatic wet, automatic dry, semiautomatic dry, manual wet, and manual dry.

Fig. 6–20 Sprinkler and standpipe Siameses should be clearly labeled to indicate what each supplies.

Fig. 6–21 A manual dry standpipe on the exterior of a building. Most of these are extremely old and should be considered unreliable.

NFPA 14 was first adopted in 1912 and remained largely unchanged as far as fire department operations from standpipes was concerned until 1993. NFPA 14 has been extensively revised after the catastrophic fire at One Meridian Plaza in Philadelphia in 1991. At that horrible high-rise fire, the standpipe system failed due to improperly adjusted pressure-regulating valves. Three firefighters died because of someone's mistake. Still, part of the problem rests with the tactics selected by fire departments. NFPA 14 was written reflecting the fire tactics that were commonplace in 1912, using 2½-in. hose and solid-tip nozzles. When departments decide to use 1½-, 1¾-, or 2-in. hose and fog nozzles, they violate the hydraulic design of standpipe systems. The standard has been revised, but the vast majority of standpipe-equipped buildings were built under the pre-1993 standards. In this text, I will try to highlight the tactical aspects of pre-1993 and post-1993 buildings (Fig. 6–21). The problem is, when you pull up at 3 AM to fire out four widows, can you tell them apart?

As previously described, a basic standpipe system is as simple as a vertical pipe, with valved outlets on each floor to connect the hoses. Ask any firefighter who has ever stretched hose around the staircase for six flights how valuable a standpipe is, and you'll surely get a hearty endorsement. The system described previously is a manual dry-standpipe system. The only time there is any water in it is when the fire department pumps into it. In some cases, these systems may be found adjoining the front fire escape balcony rather than inside the structure. Another common location to find manual dry standpipes is in buildings under construction, where no heat is yet provided. That explains the lack of water in areas with freezing temperatures.

There are several inherent problems with this system. Since there is normally no water in the system, the piping may be unable to withstand the working pressure due to unseen corrosion over many years. Often, this is only discovered when an attempt is made to use the system, which is obviously too late. Even if the system can take the pressure, you can expect delays in developing sufficient pressure. Since the valved outlets are often located in public areas within buildings, the valves are often opened, either intentionally or unintentionally. Manual dry standpipes are often found in public parking garages, where they are subject to vandalism. Since there is no water in the system, it is easy for thieves to steal the brass outlet valves rendering the entire system OOS. Since no water comes out, no apparent harm is done. Yet, when the fire department attempts to supply the system, water pours out all of these open valves, preventing the development of a proper operating pressure and causing unnecessary water damage in remote areas. In addition, valuable personnel must be sent to shut down all of the open valves at a time when manpower is likely to be critical. In general, the installation of manual dry standpipes should be discouraged (Fig. 6–22).

The manual wet-standpipe system is a slight improvement over the manual dry type. It has a small water supply, which reduces the problems of corrosion and open valves, but it is too small to supply an adequate stream. Firefighters must know about this in advance and supply the Siamese connection with a pumper before a hose stream is put into operation. Manual wet systems should be discouraged in favor of automatic wet systems.

The preferred type of standpipe system is the automatic wet system. Wet standpipes, just like wet sprinkler systems, have a source of water under pressure right up to each hose outlet. This not only prevents the problems just mentioned, it also speeds the application of water to the fire.

Where freezing weather makes the installation of wet standpipes impractical, a better alternative to the normal dry standpipe is the semiautomatic dry system, which has a deluge-type valve connected to a set of manual pull boxes at each hose station. To charge the system from the domestic supply, the user pulls the box, which sounds the alarm and trips the deluge valve. Although this isn't as desirable as the wet system, it is superior to the manual dry (Fig. 6–23).

The fifth type is the automatic dry standpipe, which is nearly identical to a dry automatic sprinkler system in that the piping is filled with compressed air and is connected to a dry pipe valve. Opening the hose outlet valve allows the air pressure to drop and water to fill the pipe and, eventually, the hose. These systems aren't common due to the maintenance problems that they present, and they are not desirable from a fire department point of view because of the need to bleed the pressurized air out through the nozzle.

The following sources of water supply are similar to those in automatic sprinkler systems: direct connection to city mains (both with and without booster pumps), gravity tanks, and pressure tanks. It is important to know the source of supply to the system, both in volume of water available and in pressure. Where the supply is from either a gravity or a pressure tank, the firefighter has only a limited time either to put out the fire or to get a pumper on line to augment the supply. In the case of a 5,000-gal fire reserve, a single 2½-in. handline will exhaust the tank in 20 min. At least initially, the pressure at which water flows is perhaps more critical than the quantity available.

Fig. 6–22 Manual dry standpipes are often subject to vandalism. The brass outlet valve was stolen from this outlet in a parking garage.

Fig. 6–23 Automatic dry standpipes may be found in some areas that are subject to freezing, such as in this parking garage. To operate this system, you must first activate the key switch to the left of the pipe.

The pressure in a gravity tank-fed system is directly related to how far below the top of the water level the line is placed. If a gravity tank is located 20 ft above the roof of a 100 ft-tall building, there will be about 50 psi available at the first-floor outlet. This pressure results from the height of the water above. Every foot of water height exerts .434 psi on the bottom of the pipe. That is the reason the tank must be raised above the roof: to ensure at least a minimum pressure at the top-floor outlet. For every 10 ft above the roof that the tank is raised, the top-floor pressure goes up 4.34 psi. (Note: For speed in estimating elevation pressures, a figure of 5 psi per floor may be substituted for 4.34 psi per 10 ft.) Now, do you see a problem with connecting to this standpipe with 1½-in. hose and a fog nozzle that requires 100 psi? Although many building codes specify a minimum pressure to be maintained at the top-floor outlet, very few require 150 psi or more, which is what it takes to supply a 120-gpm flow through 150 ft of 1½-in. hose with an average fog nozzle. In fact, some codes allow as little as 15 or 20 psi at the top floor (Figs. 6–24 and 6–25).

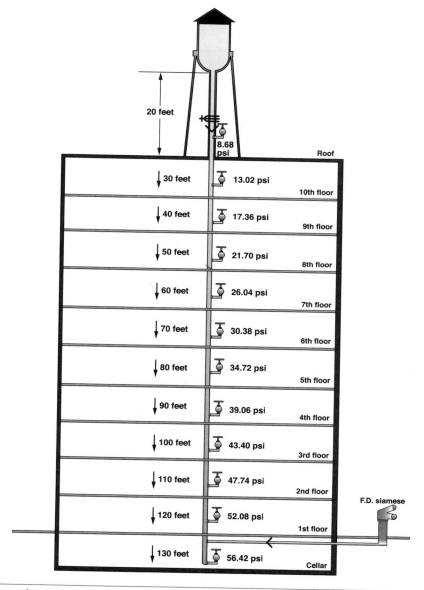

Fig. 6–24 A gravity tank as the only automatic supply for a standpipe system is very undesirable, since the low pressure at the upper floors prevents placing a hose stream in operation until the fire department Siamese connection is properly supplied.

Fig. 6–25 A six-story building with 75 psi at a ground-floor standpipe may have only 45 psi (static) at the sixth floor.

Ah, but you say, "We don't allow any gravity tanks in my jurisdiction. Everything is either fed directly off a city main or assisted by booster pumps." Then consider the following scenario: the static pressure in the street is 100 psi. What is the flow pressure at the fourth-floor outlet? While 100 psi might be fine for a lower-floor fire, particularly if the line is kept relatively short to minimize friction loss and if a solid-stream tip is used, what is the result of trying to use a fog nozzle on the 4th floor, or worse, the 8th or 10th?

Booster pumps are supposed to be a solution. They may be in some cases, but you must know their capabilities. Many are governed to supply a designed flow at a pressure that the fire department considers rather low, if the fire is on an upper floor. In fact, relief valves or other pressure-control devices are required to ensure that the maximum pressure at the top-floor outlets is *no more than 65 psi (100 psi after 1993), where 1½-in. occupant hose is provided).* Now, what do you know about the flow pressures of your area's standpipe systems? The only way to find out about them, short of flow testing each one (which is not a bad idea), is to analyze the water supply. With a cap gauge (and a bucket to catch the inevitable spillage), measure and record the outlet pressures on the highest and lowest outlets. Then you can realistically determine a course of action. There is no point in relying on pressure or flow that isn't available.

"But what about the FDC and the pumper?" you ask. They should, in fact, be the primary source of water supply. A very high priority must be placed on augmenting the domestic supply. Unfortunately, in many locales, all that is required is a single FDC with two 2½-in. inlets, regardless of the size of the structure. (After 1992, NFPA 14 requires two remote Siameses in high-rise buildings *unless* the fire department permits otherwise. *Don't!* For that matter, demand three or four.) This may be totally inadequate in a heavy-fire situation, especially if 2½-in. hose is used to supply the connection. The demands can quite rapidly exceed the amount of water that can be stuffed into such a small opening.

I recall operating at one six-story apartment building fire (Class 3, ordinary construction) where six lines were being operated off of the three separate standpipe risers. Only one Siamese connection, feeding all three risers, was available. At the height of the fire, the Siamese blew apart because of the excessive engine pressures being used in trying to keep up with the demand for more than 1,200 gpm. All of the attack lines immediately lost all but a trickle of water from the domestic supply, and the fire made considerable headway before alternative sources of supply were developed. A good practice is to have supply lines stretched to pump *into* the first-floor hose outlets of each riser, in addition to the Siamese, whenever a serious fire demands the use of a standpipe system (Fig. 6–26).

Classes of Systems

Within each of the types of systems described previously, there may be a number of different classes of standpipes available, as designated by NFPA 14. The primary distinction in this system is the intended users of the standpipe. Class I standpipe systems are those designed primarily for use by fire department personnel or others trained in handling heavy fire streams. The outlets on Class I systems have 2½-in. threads and are mainly used to allow fire department personnel to use their hose and nozzles. Building occupants aren't normally expected to be able to handle such a flow so, most of the time, no hose is provided for occupant use. Occasionally, however, high-hazard occupancies, where it would be reasonable to expect a fire brigade to operate, may be equipped with 2½-in. hose for occupant use.

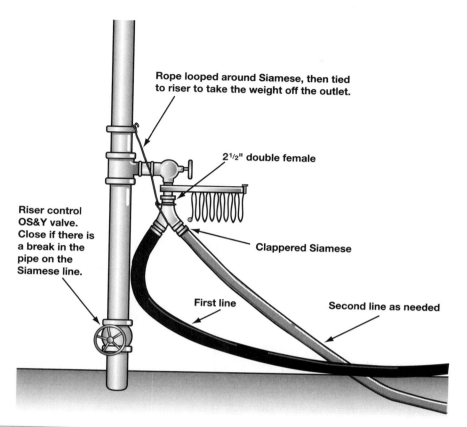

Rope looped around Siamese, then tied
to riser to take the weight off the outlet.

2½" double female

Riser control
OS&Y valve.
Close if there is
a break in the
pipe on the
Siamese line.

Clappered Siamese

First line

Second line as needed

*Fig. 6–26 It may be necessary to pump into the first or second floor hose outlet valves for several reasons, includ-
ing a defective fire department Siamese connection, broken piping, and to augment supplies. Note the riser control
valve below the first floor outlet.*

Class II systems are designed for use by building occupants to control minor fires until the fire department arrives. As such, they are provided with 1½-in. hose, equipped with either an open tip or fog nozzle. The usual design flow for this system is only 100 gpm and, as such, shouldn't be relied upon by fire department personnel for serious interior fires. In fact, as a matter of policy, the fire department shouldn't rely on any building hose for their hose attack lines. Such hose is rarely, if ever, inspected. It is almost never tested, and it is often missing or damaged. Instead, if a major fire develops, 2½-in. handlines may have to be hand stretched.

Many jurisdictions permit Class II systems in buildings that are fully sprinklered, or in residential and other low-hazard occupancies. The assumption is that the lines will readily extinguish the small fires likely under these circumstances. That's fine as long as the fire knows that it's only allowed to get to a certain size and then stop. But when it doesn't, the firefighters stuck with a Class II standpipe can be in big trouble. Take the case of the sprinkler valve that is closed in the fire area or the apartment fire that is being driven back into the building by strong winds. Both can easily require 400–500 gpm or more.

An obvious answer to the previous dilemma is the Class III standpipe system. It is designed to allow a department to use heavy hose streams as well as providing first-aid hose for occupant use prior to the arrival of the department. This can be done in several ways, including separate valves for 2½-in. and 1½-in. outlets, as well as the more common practice of placing a 2½-in. by 1½-in. reducer on the 2½-in. outlet valve. This last method is a compromise that introduces difficulties. The NFPA recommends that hose for occupant use be located outside of the stairwell on the fire floor. This is to prevent a charged line from holding the stairway door open, thus filling the stairway with smoke while it is being used for evacuation. This is much more likely when untrained people are operating the hoseline than if the fire department is operating it.

The fire department should evaluate what effect opening the stairway door will have on the stairwell. If it will endanger occupants of the stairwell, the attack must either be delayed until the stairwell can be cleared of occupants, or another point of attack must be found. The untrained building occupant is unable to make that kind of critical judgment; their immediate thought is to put out the fire. When faced by heat and smoke, occupants should be expected to abandon the hose and flee, leaving the standpipe hose to chock the door open. This isn't to say that hose for occupant use is totally worthless. In fact, there have been many instances where it has greatly reduced damage and even saved lives. Many times, however, it has been because trained people just happened to have been nearby.

It is important to note that these pressure-restricting devices don't function to limit static pressure. When the nozzle is shut, substantially higher pressures build up within the hoseline.

Another problem with the use of the 2-in. outlet to provide Class III 1½-in. hose connections is the requirement that automatically protects the untrained operator from excessive pressures. Systems that have more than 100 psi at any hose outlet could be dangerous to the untrained user. As such, some type of device is required that will restrict the pressure to a maximum of 80 psi (100 psi after 1993). That's fine for the untrained civilian with the ½-in. tip on his 1½-in. first-aid hose, but it obviously presents a problem when the fire department wants to hook up 1¾-in. or 2½-in. hose with fog nozzles, and wants 125–160 psi to supply it. (Now NFPA 14 permits up to 175 psi at 2½-in. outlets but still allows 100 psi as the minimum pressure.)

There are several styles of these devices available, and firefighters must know how to handle them if they are to use the system to its fullest capacity. The first style is a simple orifice plate installed between the reducer and the 2½-in. outlet. A hole is drilled through a 2½-in. circular brass plate. Based on the static pressure at the valve and knowing the size of the nozzle at the end of the hose, it is possible to determine how large a hole is needed to allow only the desired flow through the plate. In effect, the plate creates friction loss to reduce the flowing pressure on the downstream side (Figs. 6–27, 6–28, and 6–29).

Another similar device, known as a vane-type Pressure Reducing Valve (PRV) consists of two sets of overlapping holes on a movable plate. As an adjusting stem is turned, the holes can be moved from being totally

Fig. 6–27 Can you spot the pressure reducer in place on this standpipe outlet?

Fig. 6–28 Here is the pressure reducer and the 2½ x 1½ in. reducer removed from the outlet.

Fig. 6–29 Looking through the vane-type pressure reducer in the full open position, you can see how little of the waterway is actually available for water to flow through.

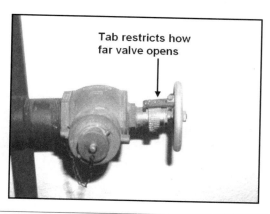

Fig. 6–30 Some PRVs only open part way unless bypassed for fire department use.

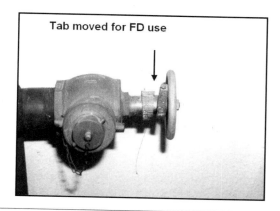

Fig. 6–31 This valve is bypassed by pushing the steel tab out of the way so the valve can be opened all the way.

Fig. 6–32 This style PRV operates with a fire extinguisher-type pin in place. The fire department must pull the pin, prior to opening the valve, to be able to open it fully.

non-aligned (closed) to totally aligned. Even when the holes are open, however, there are severe restrictions in the stream, amounting to more than one-half of the outlet area. As a precaution, before connecting any fire department hose to a standpipe outlet, probe inside the threads with a spanner wrench or other small tool to make sure there are no obstructions. Usually, if there is anything within about 3 in. of the thread, it will be some type of restrictor that should be removed. A member who manually controls the outlet valve should maintain fire department control over outlet pressure. This member should be in radio communication with the nozzle team to coordinate flowing pressures. In effect, it places the pump operator at the fire floor.

A third type of pressure-reducing device, with an entirely different design, is becoming common. These are combination outlet control and pressure-reducing valves, called *pressure-reducing valves*. They are designed to allow only a set maximum pressure through the valve, regardless of the flow (from zero flow to the maximum designed, usually 100 gpm). They must be bypassed for fire department use. This can involve removing a pin similar to that of a fire extinguisher. After the pin has been removed, the valve operates as a normal gate valve. All members must be aware of how to operate each and every style of valve that they may encounter. Otherwise, it may not be possible to establish the needed fire flow (Figs. 6–30, 6–31, and 6–32).

During the disastrous fire at One Meridian Plaza in Philadelphia, the PRVs were of a type that couldn't be bypassed, and they required special tools to adjust. They had been set at too low a setting to provide an effective fire stream through the 1¾-in. hose and fog nozzles that made up the department's standpipe kits. Another problem with the PRVs, as opposed to the simple pressure-restricting devices, is that they incorporate moving parts that can fail. During one survey by a major metropolitan city, more than 75% of the PRVs failed to provide a satisfactory fire stream. These valves require periodic maintenance and flow testing. This should be done at least every three years, preferably every year. If possible, their use should be discouraged in favor of the more reliable, removable pressure restrictors or adequately designed pressure zones (Figs. 6–33, 6–34, and 6–35).

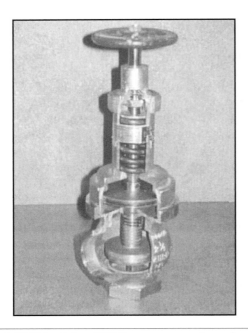

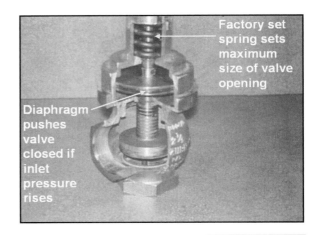

Fig. 6–34 These spring-operated PRVs are dual-chambered. Increasing pressure at the inlet flows through the center stem to the upper chamber, which in turn pushes the valve further closed, reducing the outlet pressure.

Fig. 6–33 This style of PRV is a very complex valve that has been shown to be unreliable unless flow tested to capacity on a regular basis. Most standpipe outlet valves are rarely or never used.

Operations

Fires at buildings that are equipped with standpipe systems tend to become major events for a number of reasons, the first of which is the sheer size of such a structure. You simply don't find standpipes in one- or two-family dwellings and other small structures. Building codes usually specify the requirements for a standpipe system based on the size of the structure—for instance, any building more than 75 ft in height or more than two stories and 20,000 sq ft per floor. These are big buildings. If they aren't sprinklered, the potential fire area is quite large.

Besides the difficulties of locating the fire in such a structure, there is the added difficulty of venting a building that may be out of the reach of ladders, or an area that is so far from windows that horizontal ventilation is useless. Add to this the time factor, often called *reflex time*, that it takes to respond to the building and set up for operations, and the fire has some obvious advantages. The firefighters need some weapons to help them cut down these advantages.

The first item needed is a *plan*. Obviously, operations involving the use of a standpipe will be different from other routine operations in which hose will be stretched from some hosebed directly to the site of the operations. Therefore, the first item in the plan is to determine which

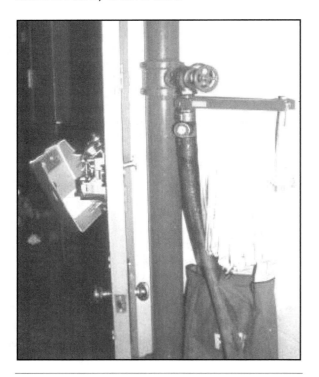

Fig. 6–35 In New York City, a pressure gauge is connected to the length of hose at the standpipe outlet to allow the control member to know when he or she has established the proper flowing pressure to the line.

buildings are equipped with standpipes and which are not. In most cases, this should be done as part of prefire planning. In the case of an urbanized area with a great number of relatively large buildings, however, this may be rather difficult. Of course, some buildings, by their very size, will obviously be standpipe buildings: the 12-story college dormitory, for example, or the 20-story office building. Other buildings may give the impression of being large enough to meet the local requirements for standpipes only to have responding units find that a handstretch is needed. That delays fire attack, since the members have geared up for one operation and may have gotten all the way into the building before realizing their mistake.

There are many loopholes in building codes that can confound even the most knowledgeable firefighter.

One evening, my engine company responded to a fire on the sixth floor of a nine-story fireproof housing project that was set back approximately 250 ft from the street. It was an extremely hectic summer evening with many alarms receiving a one-engine and one-ladder response. We pulled in front of this building and saw no evidence of fire. The ladder company quickly made its way to the fire floor while the rest of us started toward this building with standpipe lengths. Fortunately, as they ascended the staircase, the truck company was sharp enough to advise us that there was no standpipe. We rapidly dropped the standpipe operation and began to handstretch a line as the truck company reported smelling wood smoke in the stairway. Sure enough, on arrival at the fire floor, the truck company found several rooms involved, with the wind keeping the smoke inside rather than letting it vent.

The building had completely surprised us. Here was a nine-story building without a standpipe. We were all aware of our jurisdiction's requirements for standpipes in all buildings more than 75 ft, as well as certain other requirements. We were reasonably certain that this building required the use of our standpipe kit, yet it did not. The builders (government-subsidized) had bypassed the code through a technicality. Simply put, the code requires standpipes in all buildings more than 75 ft above the surrounding grade. By setting the building back from the street, recessing it into a low point in the land, the roof of the structure was no higher than 75 ft above the nearest street grade. That the nearest street was more than 100 ft from the building and 20 ft above the ground floor didn't bother the designers in the least. They had saved the builders the cost of putting in a standpipe (probably less than the cost of any other item in the building, from trash compactor to wall paint).

The point I would like to make is this: you must visit each structure in your area to observe the standpipe system. You must know whether you can use the occupant hose to contain a developing situation. You must also know whether the water source can supply your lines and, if not, how the fire department can supply the needed water.

The second item to consider at fires in buildings with standpipes is the location of the fire. Will you use the standpipe at all? Just because the fire is in a standpipe-equipped building doesn't mean that the first hoseline should be stretched from the standpipe.

In many cases, notably fires on the ground floor, it may be better to stretch the first hoseline off the apparatus than the standpipe. Normally, the routine handline stretch will be much faster, especially if preconnected lines are used as opposed to standpipe lines. You should evaluate the location of the fire in relation to the access. Will the preconnect reach the seat of the fire? Will using the standpipe mean that the attack team must go past the fire to use the standpipe? Are the stairs in line with the entrance or at opposite ends of the hallway? Above the second floor, the choice is usually simple. The standpipe is preferred due to the time saved, the lessened friction loss, and the amount of hose required to reach the upper floors.

Once the decision is made to use the standpipe, the attack crew should advance with the needed equipment to the point from which the attack will begin, usually the staircase nearest the fire. Some preparations are in order, however, before committing a hoseline to the attack. First, you must verify the location of the fire and ensure that the length of the hose from the standpipe is sufficient to reach the seat of the fire. That may be accomplished by a quick survey by members of the ladder company, if available, or by the officer in charge of the hoseline. In many large buildings, there are many staircases, each of which may contain a standpipe riser.

It is crucial for a number of reasons that the attack crew locates and selects the riser closest to the seat of the fire. The most important reason is to ensure that the line will reach the fire. In multi-riser buildings, the systems are laid out so that each point on the floor is no more than a certain distance from the nearest standpipe—up until 1993 the NFPA said within reach of a 30-ft-long hose stream from the nozzle of a 100-ft line. Theoretically, only two lengths of hose would have been required to reach any fire on the floor *from the nearest outlet*.

Actual fire conditions, however, seldom approach the theoretical. In New York City, the code has been modified to allow a point to be within reach of a 25-ft stream from a hoseline 125 ft long. By reducing the length of the stream, this code recognizes the low operating pressures commonly encountered at first. A 30-ft reach is optimistic. By allowing buildings 125 ft of reach, while the fire department carries 50 ft lengths of hose, the FDNY ensures that the attack crew that arrives carrying three lengths of hose will have sufficient hose to move around obstructions such as furniture that wasn't considered when the standpipe layout was being designed. Note that in the 1993 edition of NFPA 14, the amount of hose required to reach the most remote areas of each floor have been greatly *increased*!

The amount of hose required to operate on the fire floor can be much greater than the two or three 50-ft lengths described previously, depending on several variables found in building codes. As I mentioned, the pre-1993 version of NFPA 14 called for each point on a floor to be within 30 ft of a 100-ft hoseline. After the tragedy caused primarily by an ineffective standpipe system, NFPA 14 was modified to allow builders to spread the standpipes even further apart! Now NFPA 14 required that each point be within *150 ft plus the height of the staircase to the outlet on the stair landing below* the fire floor in non-sprinklered buildings (since new outlets are being encouraged to be located at the half landing between floors rather than at the floor landing directly adjacent to the exit door.)

In other words, at least four 50 ft lengths of hose will be required. The distance increases to 200 ft, plus the height of the stair landing, meaning at least five lengths of hose in a sprinklered building. I can virtually guarantee that none of the members of the NFPA committee on standpipes ever tried to hump two lengths of 2½-in. hose plus a standpipe bag up 20 flights of stairs in bunker gear and a mask, but they felt compelled to make our job harder for us in order to get support from builders and the real estate industry. That's an absolute outrage, especially coming as a direct result of the revision caused by the loss of three brave firefighters from Philadelphia!

Another serious problem with many codes including those used in federally subsidized building programs, is a provision that allows the maximum distance limitations outlined previously to be met by placing the standpipe outlets outside the protection offered by the enclosed stairways, out in the public hall. What that accomplishes is saving the builders the cost of another staircase, but it means that firefighters have to carry even more hose in order to hook up in the hall of the floor below and stretch up the staircase where they should have been in the first place, unless they're suicidal enough to try to hook up out in the public hall on the fire floor. I hope they're smarter than to try that. Maybe if we explained the conditions these changes force us to work under to the press every time we have a loss of life in a fire and how the safer designs might have made a difference except for the greed of the real estate industry, we might be able to reclaim some of the safety precautions that were designed in back in 1912.

Selecting the Attack Stair

When attempting to locate the seat of the fire and select the nearest standpipe, the attack team on the fire floor may be operating in zero visibility. Intelligence gathering in a clear area will pay great dividends. Information received with the initial dispatch, or interrogation of fleeing occupants, can be extremely useful. For instance, if the initial report indicated a fire in apartment 5J on the fifth floor, the attack crew has an idea of where they're going. But if they open the door from the stairway into the public hall on the fifth floor and are met by a solid wall of black smoke and heat, they may be totally lost. The answer is to recognize patterns in building design and construction, particularly in multistory buildings.

For reasons of economy, floor plans of multistory residential buildings are usually very similar from floor to floor above the ground floor. Each floor requires much the same utilities as follows: water, sewer, HVAC, gas. It is cheaper to run a common riser straight up to serve each floor than it is to run separate services. Therefore, features tend to be stacked, one directly on top of the other. By dropping down to the floor below the fire, the attack crew can get the layout before entering the smoke-filled hall. Select the stair that is closest to apartment 4J as the attack stair. By counting the number of doors you feel from the stair to apartment 4J, you can gauge your progress down the smoke-filled fifth-floor hall toward apartment 5J.

This survey of the floor below can often be carried out by the officer of the attack team while the members are assembling the necessary equipment. If practical, a quick survey of apartment 4J itself would also prove beneficial. Realize, however, that this pattern does not work for office buildings or other commercial occupancies, other than locations of elevator lobbies, stairs (usually), and toilets (Fig. 6–36).

In addition to having enough hose, selecting the closest stairway and standpipe is important for other reasons. Who wants to crawl down 150 ft of heavily charged hallway if another stairway would leave you only 50 ft from your objective? Remember, too, that once the hoseline has been advanced through a stairway door, that door becomes blocked open, turning that stairway into a chimney. That staircase can be heavily charged during a serious fire, eliminating its use as an evacuation route. (For further discussion of the evacuation and attack stairway, see chapter 15.) It is usually better to use staircases remote from the fire to bring down civilians.

Fig. 6–36 Selecting the most appropriate attack stair is critical. You don't want to crawl down an extra 150 ft of fire-filled hall or into the teeth of a wind-driven fire.

That chimney effect (not to be confused with stack effect) has a direct bearing on the operations and layout of the attack crew. Since smoke, heat, and possibly fire may be drawn toward the open stairway door, it is important to be sure that the area isn't rendered so untenable as to interfere with such vital operations as operating the standpipe outlet control valve or the advance of spare hose. High heat or fire entering the stairway at the fire floor could prevent a member from standing up to operate the outlet valve, which is usually at head height. For this reason, it is recommended that the hoseline actually be connected to the standpipe outlet on the floor below the fire. In addition, this places all of the spare hose down low, out of danger of being burned.

In fact, in fire-resistant buildings, spare hose should always be flaked out on the floor below if any there are any indications of a serious fire. This allows the crew to make all of the needed connections in good visibility, usually without requiring SCBA. If it is necessary for any reason to retreat temporarily to the staircase, backing the hose down that staircase will be far simpler. Connecting hose on the floor below does have its drawbacks. An additional length of hose is usually required, and if the fire is in reach of three lengths from the fire floor, four lengths are required from the floor below. In addition, extra effort is required to advance the hose up that staircase as the attack progresses. The advantages of safety and speed, however, are clearly superior to the drawbacks.

One thing that firefighters should *never* do is hook up to the standpipe on the fire floor if the outlet is located outside the stairwell in the public hall. Several times I have witnessed the confusion that results when smoke and heat enter the hallway before the hoseline is ready to advance. At one incident, the engine company was thrown back in a scramble for their lives when fire blew out into the hall as they were hooking up. A member of the ladder company was caught in the hall and burned to death.

A key part of the standpipe operations plan is the proper equipment and sufficient personnel, all of which must arrive at the fire area. An easy way to ensure this is to assemble them in advance as part of a standpipe kit or bag. One major item is hose of sufficient length, rolled or folded so as to be easily portable. As outlined previously, the need for four lengths of hose is quite common and due to NFPA 14 revisions, five or more lengths will increasingly be required in sprinklered buildings (Figs. 6–37, 6–38, and 6–39).

Fig. 6–37 A standpipe kit with fold-up hose should be readily available for areas that have buildings with standpipe systems. It should include at least three lengths of folded 2½-in. hose and low pressure nozzles.

Fig. 6–38 A break-apart nozzle with a St. Louis adapter between the shutoff and the fog tip. The St. Louis adapter is a compact solid-stream tip that can be kept in place behind the fog tip, ready for quick use in case low pressures are encountered.

Fig. 6–39 A standpipe kit should contain all the tools needed to operate from the systems you may encounter.

The diameter of the attack line from a standpipe is a critical element determining the likelihood of success of the attack. The year that the system was designed is a key determinant of the hose and nozzle requirements. Many *progressive* departments have taken to equipping their standpipe kits with lightweight 1½-in. hose with plastic fog nozzles in an effort to create a package that one man can readily transport. This policy can create havoc, a greater demand for manpower and equipment, and greater fire damage if the small-volume stream is unable to quickly extinguish the fire. In that event, additional people will have to bring up additional larger-size hose (which may not be ready for quick transport as the standpipe rolls are) and hook up to outlets farther away while the initial line attempts to hold the situation in check. A better option would be to ensure that the needed volume of water is available right at the start by initially providing the proper size hose, this means 2½ in.

Remember that NFPA 14 was originally developed when fire departments typically used 2½-in. hose and a solid tip nozzle. These pre-1993 systems are designed to supply a *maximum* of 65 psi at the top floor hose outlet. That is sufficient to supply a 1⅛-in. solid tip flowing 250 gpm through 100 ft of 2½-in, ½-in. hose. That's all it is designed to do. If your department decides 2½-in. hose and solid tip nozzles don't work for you, then you have to do something about the pumps and piping in the standpipe systems. Stretching three or four lengths of 1½-in. or 1¾-in. hose to a fog nozzle is doomed to failure under these circumstances, since the nozzle pressure and friction losses built up by this combination of appliances requires approximately 175 psi at the outlet.

The good news is that after 1993 the NFPA recognized this weakness and boosted the *maximum* allowable pressure at a 2½-in. outlet to 175 psi, so it is possible that you *might* be able to operate the 1½-in, or 1¾-in. streams described previously under the right circumstances. The bad news is that even in the 2000 Edition of NFPA 14, the *minimum* pressure required at 2½-in. outlets has only been increased from 65 psi to 100 psi. I have always planned on having the minimum, that way if I find that I have more it is a pleasant surprise. To plan on having the maximum is foolhardy, it's not going to happen, since designers almost always design to satisfy the bare minimum.

It is interesting to note that even under the new standards, the 100 psi minimum pressure is barely enough to supply a solid tip nozzle through 2½-in. hose, since the length of hose required to reach all areas has increased from 100 ft to either 200 or 250 ft, depending on whether the building is sprinklered or not. Four lengths of 2½-in. hose flowing 250 gpm requires 80 psi at the outlet valve, while five lengths requires 90 psi. Any nozzle that requires more pressure than the 50 psi of the solid tip is likely to have a serious pressure problem, even in the newest buildings.

In the past I only used to advocate 2½-in. hose for commercial and office building fires, believing that 1¾-in. or 2-in. would be sufficient in residential buildings due to their lighter fire loading and greater compartmentation. This is no longer the case. A series of catastrophic fires in residential high-rises in the 1990s has shown that even a one room fire in a Class 1 building can exceed the capacity of the smaller lines. Four firefighters and more than a dozen civilians have been killed in these severe fires in New York City alone. Bring the 2½-in. hose in the first place, since you can't tell from the street when the smaller hose won't be able to do the job. Remember the line about going bear hunting with a .22.

Another tool that is required to operate effectively from a standpipe is a nozzle. By now, the need for a low-pressure capable nozzle should be obvious; 50 psi is about all you're likely to have to work with as nozzle pressure, especially on the upper floors of a building.

While some fog nozzles are now being billed as *low-pressure*, or *dual-pressure*, you need to find out what that means before lugging it up to the 10th floor. I am partial to the solid tip for standpipe operations. They operate at as low as 40 psi nozzle pressure, and offer another important advantage: they readily let pass rust and scale that often flows from a standpipe, foreign objects that can clog a fog nozzle. (**Note:** The FDNY mandates that the first line operated off a standpipe shall be 2½-in. with a solid tip for good reason—they've experienced all the pitfalls many times.) This is not to say that a fog stream will not be useful. At times the fog stream may be used to vent a fire area after knockdown. The fog stream can be provided under these circumstances by including a fog tip with 1½-in. base threads in the standpipe bag. All nozzles used in standpipe operations should be of the break-apart design, with 1½-in. hose threads located just beyond the shutoff, so that the fog tip or a length of 1¾-in. hose can be placed in operation from the 2½-in. line when required. (The use of the 1¾-in. line should only be permitted after the main body of fire is definitely knocked down, for overhaul purposes.)

Other necessities in the kit include a spanner wrench, any necessary hose thread adaptors (some standpipes have iron pipe or other threads on the outlet and require adaptors), as well as a 14-in. or 18-in. pipe wrench. The need for the spanner wrench and adaptors is self-explanatory, but the pipe wrench bears explanation. At times, the operating wheel of the outlet valve has been removed, whether by vandals, the arsonist, or building management to prevent unauthorized openings. This is of little concern to the firefighter. All he knows is that there is no operating wheel present. The pipe wrench will also help open a stuck valve or one in a tight location. The practice of carrying a spare handwheel in the kit is probably a futile effort. By Murphy's Law, if you carry nine different-size wheels, the one you'll need is the 10th size. The pipe wrench is adjustable to fit all 10 and any in between. The 18-in. size weighs a little more, but companies that do a *lot* of standpipe work prefer it, since it is the smallest wrench that has a jaw opening large enough to fit around a 2½-in. coupling. Very often this extra leverage is needed to remove a stuck cap, reducer, or pressure-reducing device.

A very useful accessory that should be added to the kit is a number of wooden door chocks. Normally, the line will have to be advanced through at least one door. Keeping three or four light wooden chocks in the kit ensures their presence, preventing the problem of the line binding under the door. Equally useful are door-latch search markers. At most fires in standpipe-equipped buildings, search will be a major part of the operation. Duplication of effort must be avoided in order to speed up the search for life.

At the same time, a member who has entered a potentially hazardous area should be certain that the door to that area won't latch closed behind him, either locking him in or delaying his escape. The door latch search marker serves a dual function, and it can be quite effective. The basic principal is simple. With a common spring-loaded, self-latching door, any obstruction between the latch and the keeper will prevent the door from locking. The latch strap is designed to slip over the doorknob on both sides of the door, thus covering the locking mechanism and preventing locking. The dual function is served by a department policy. When a member puts such a strap on the door, it means that a search is underway in that area. The next unit arriving on the scene would know to check with the people inside that area and then proceed to the next area to begin their search. Afterward, the strap would be released from the occupancy side of the door and left hanging from the hallway or public side, thus indicating that the primary search has been completed in that area and that units should proceed to the next area to be searched (Figs. 6–40 and 6–41).

Additional items needed on the fire floor before the start of operations include the following: forcible entry equipment (flathead axe, halligan tool, and *K* tool as a minimum) and preferably a hydraulic forcible entry tool, handlight, and a portable radio (two would be preferable). Another useful tool for these buildings is a 2½-gal water fire extinguisher. This extinguisher is extremely valuable at controlling developing fires such as mattresses and stuffed furniture while the hoseline is being prepared and may even eliminate the need for a hoseline if used in the early stages. One firefighter can bring it and a 6-ft pike pole for venting windows. It wouldn't be unreasonable to expect a team of four firefighters to be able to advance this equipment to the sixth or seventh floor on foot. One member can carry two lengths of hose, another can carry one length plus the standpipe equipment bag, while the third and fourth provide the forcible entry tools and handlight. Especially where four lengths are needed, the four-person team will be useful, but as we know, a four-person team is

Fig. 6–40 A door latch marker in place prevents the door from locking behind the search team, and it also signals later-arriving units that members are already operating in this area, and they should proceed to other, unsearched areas.

Fig. 6–41 A door latch marker in place only on the hall side tells members that an area has already been searched.

the minimum number allowed by Occupational Safety and Health Administration (OSHA) standards. What we must ensure is that the team arrives with at least the minimum tools described previously. As the hoseline is advanced, the safety team will have to be augmented by other responders.

The attack team should proceed by the most appropriate means to the floor below the fire. If the fire is on the seventh floor or below, the members should climb the stairs instead of taking the elevator. If the fire is above the seventh floor, the members may want to take the elevator. (For further information on using elevators, see chapter 15.)

Once at the floor below, the two members with the forcible entry tools and the water extinguisher can proceed to the fire floor and evaluate the conditions, then locate the fire and the best route to it. The remaining members can begin flaking out the hose and connecting the lengths to prepare for advance of the line. The advantage of having at least two portable radios is evident here. The members doing the size-up on the fire floor can relay instructions more readily. Also, once the attack has begun, the attack team can radio any requests back to the member manning the standpipe outlet.

Before a member connects a hoseline to any standpipe system, the valve discharge opening should be inspected visually, or inspected by cracking the valve open to clear the outlets of any items that are stashed inside. In many areas, these outlet valves have become favorite hiding places for drug dealers. A standpipe located in the staircase provides a storage space out of view of the general floor area where the dealer and customer can make their transaction. Once the hoseline is connected and charged, any items in the valve will be forced through the hoseline up to the nozzle.

Additional manpower, when it is available, may be used to transport a variety of additional equipment and perform truck functions. All members should be aware, however, of the importance of getting the first hoseline in operation and of the lengthy delays likely to occur in getting this line operating. In addition, since the majority of buildings that are equipped with standpipes are of Class 1 construction (fireproof), the threat of interior fire spread and the danger to life is greatest in the immediate fire area. If the first line begins operation as quickly as possible, the remaining tasks, such as search and ventilation, will be simplified. Once enough personnel are available to ensure the rapid advance of the first line, additional personnel should be committed to search the fire area and then stretch a second line, if needed. Searches will also be required of all stairways that serve the fire floor, from that level all the way to the roof, since these vertical shafts become flues for fire gases, and people trying to descend are overcome. This is especially sometimes true of the attack stair, since the hoseline will block open the door on the fire floor. Elevators must also be accounted for.

The need for pumpers supplying Siamese connections should by now be obvious. A good rule of thumb is to have a separate pumper feed each Siamese connection if more than one is found. This is common in buildings with separate sprinkler and standpipe systems, as well as at larger buildings where several Siameses may be found for each system. Supply hose to Siameses has been known to be slashed by shards of falling glass. Having more than one supply has prevented injuries and a total loss of water.

Another recommendation is to use the largest size hose available for this purpose. Simply because the FDC is fitted with two 2½-in. females doesn't mean that 2½-in. is the hose to use. The idea is to deliver the maximum volume to the system. Right on the other side of the wall, the pipe is likely to increase to 5 or 6 in. in diameter. (Note: Two 2½-in. lines aren't the same as one 5 in.)

In unusual circumstances (where the Siamese is frozen or damaged due to vandalism, or where the check valves within the structure are stuck, or where the riser control valves are shut down), it may be necessary to bypass the Siamese connection entirely. In this sense, the standpipe system has an advantage of having auxiliary inlets to the system at the lower-floor hose outlet valves. In the event that obvious damage to the Siameses is present or the system isn't receiving adequate pressure through the initial lines, secondary supply lines should be stretched to the hose outlets on the first or second floor. Connecting a 2½-in. clappered Siamese connection to the outlet with a 2½-in. double female simplifies the task of adding additional supplies. Note that this tactic won't work if these lower-floor valves are outfitted with pressure-regulating valves, since they sense and react to increased pressure on the *normally downstream* side of the valve by shutting down (Fig. 6–42).

Fig. 6–42 This 5-in. line is pumping into the first floor hose outlet. Be aware of the pressure limitations of LDH. Also consider nearby building standpipes as a possible source of water to feed an engine in the event a hydrant that you have already committed to is defective.

Indications that the original Siamese supply lines aren't doing their job range from the obvious to the subtle. A call for more water or pressure from the hoseline crew when the outlet valve is wide open is obvious. A pump operator who is committed solely to supplying a Siamese may notice that the pump suction housing is becoming warm to hot when touched. This indicates that the pump is churning the water with little or no flow. An alert pump operator can detect this condition before the heating takes place by keeping an eye on his intake and discharge gauges. By noting the hydrant's static pressure before opening his gates to the discharge and comparing it to the intake pressure with water flowing, he will have an idea of the volume that is being supplied. A large decrease in intake pressure indicates a significant flow. No change in intake pressure indicates little or no flow. This can be verified by observing the effect that closing the discharge gate has on the discharge pressure.

With any significant flow, closing the discharge gate should increase the pressure reading on the discharge's gauge. Good starting discharge pressure for standpipe system operations is 100 psi plus 5 psi for each story above grade. If the hoseline is to be equipped with a fog nozzle, this pressure should be increased to 150 psi plus 5 psi for each story above grade. This will provide a suitable nozzle pressure with an allowance for friction loss through three lengths of 2½-in. hose, the standpipe, and 100 ft of 3-in. supply hose from the pumper to the Siamese. The use of smaller diameter (1½-, 1¾-, or 2-in.) hose may require an increase in this starting figure, since they usually have higher friction losses than does 2½-in. for their *design* flows, *i.e.*, 1¾-in. hose at 180 gpm equals 40 psi per 100 ft friction loss; 2½-in. hose at 250 gpm equals 15 psi per 100 ft.

For additional operational points regarding standpipe systems, see chapter 15.

Table 6–1 Chart of pump pressures for high-rises.

Fire Floors	Solid-Tip Nozzle	Fog Nozzle
Floors 1–10	150 psi	200 psi
Floors 11–20	200 psi	250 psi
Floors 21–30	250 psi	300 psi
Floors 31–40	300 psi	350 psi
Floors 41–50	350 psi	400 psi

Add 50 psi for each additional 10 stories.

Table 6–2 Summary of NFPA 14 requirements.

	Pre-1993	Post-1993
Maximum psi, top floor	65 psi	175 psi–2½-in. outlets 100 psi–1½-in. outlets
Minimum psi, top floor	65 psi	100 psi–1½-in. outlets 65 psi–1½-in. outlet
Minimum flow, top floor	600 gpm–1 riser 750 gpm–2 risers	600 gpm–1 riser 750 gpm–2 risers
Maximum flow required at top floor, depending on the floor area of the building	2,500 gpm for standpipe only	1,250 gpm for both standpipe and sprinkler

07 Ladder Company Operations

This chapter deals with the general functions that, over the years, have come to be known as truck duties. This by no means implies that a ladder truck is a required part of these functions. Instead, it merely separates these functions from the primary effort of the nozzle team, namely, attacking the fire. Some people call these duties support functions. Taken literally, they are. Since, just as a house cannot stand without a foundation, fire attack cannot be successful without these truck duties being performed. Unfortunately, many people in our line of work seem to feel that since these tasks don't actually involve slugging it out with the flames, they aren't the primary duties of a fire department. Nothing could be further from the truth!

The reason that we have all held up our right hand and been sworn in is to protect life and property. This mission may take some unusual forms, and it is often the job of support personnel to carry them out. Among the tasks that we will consider here are Laddering, Overhaul, Ventilation, forcible Entry, Rescue and Search, salvage, and control of Utilities (LOVERS U). Not all of these items will be required at every fire, but at some incidents, the tasks become more complex and are subdivided further.

Developing a plan in advance that embraces all of these tasks allows you to assign apparatus, manpower, and equipment. A good aggressive interior attack is the result of a balance between the attack team and the supporting cast, the ladder personnel. In areas that don't have ladder trucks, it is still possible to have this arrangement if the members recognize the tasks that must be done and provide the people and tools to perform them. For instance, I worked in one very busy neighborhood where there were four single-engine company houses. The alarm assignment always had two ladder companies assigned to a structural fire. Yet, in these engine companies of five firefighters (including the operator) and one officer, one of the enginemen was always designated to bring the forcible entry tools and a water extinguisher immediately to the fire floor with the officer. These companies realized that, if there were a fire across the street from the firehouse, they might be operating by themselves for two or more minutes before a ladder truck got there. So they arranged a plan for this contingency by assigning a member to bring the forcible entry tools and water extinguisher if the ladder company was delayed. If they were to arrive simultaneously with or after the truck, the member would assume his normal engine duties.

Conversely, just having two gigantic aerial devices pull up to a building fire doesn't automatically ensure that the necessary truck work is going to be performed. A plan of action is needed that specifies what areas are to be covered, by whom, what tools must be taken, and what, when, and why certain actions must occur. All must know and understand the plan.

Having a plan does many things. First, it formalizes thinking, making people consider in advance of what must be done. A plan also assigns a degree of priority to each element. By putting it in writing and distributing copies of it, a plan lets each member know what is happening. Moreover, a plan establishes accountability for one's actions, since it should establish what particular players should do. Although it is difficult to determine all of the variables that may be present on the fireground and to establish a procedure to cover each one, it is far more difficult to determine these variables as you pull up to the incident. Having a plan makes life far simpler for everyone, and it makes any operation go much more smoothly. By nature, the operational plan may have to be quite general, but it should at least provide the following specifics: the number of people to be committed initially to each of the major areas of responsibility, the tools to be provided at each area, and the general scope of duties.

For example, let's say your area primarily consists of single-family homes of similar design, since they were all part of a postwar development. As part of the past 20 or 30 years of experience, your department has gained some insights into what works and what doesn't in these buildings. This should be reflected in the operational plan. You probably know that there is a high life hazard in the bedrooms, so you must launch a concerted search effort in these areas as soon as possible. At the same time, you know that the aerial device is almost never used, and that roof ventilation is also rarely required. Part of your plan might state that the officer and one member from the jump seat will form an interior search team and proceed immediately to the fire area, while the chauffeur works with the remaining member to try to reach the bedrooms from the outside, either to search them or to vent for the advance of the hoseline.

We know from past experience that the following three basic tools will be required at almost all structural fires: a forcible entry bar of some type, such as a halligan tool, a flathead axe for driving in the forcible entry bar as well as chopping, and some type of hook or pole to extend our reach, pull ceilings, open walls, and the like. By assigning tools in advance, we ensure that all of the basic tools are provided at the location, without omission or needless duplication, every time they are required. Of course, any plan you make must be realistic. If you only expect 4 people, don't assign tools for 14.

As a rule, when deciding the number of personnel to be assigned to an area, always try to use the buddy system. For interior teams especially, that means at least two members should operate together. As you will see, that is the bare minimum just to carry the most basic tools needed at most fire areas. For less hazardous areas, such as performing ventilation outside the fire area, a single member may be assigned, but he should be paired up for accountability and tracking purposes with another member working in roughly the same area. For instance, in a three-member truck company, the interior team (called the forcible entry team, for lack of another name) must be two members at minimum. The ladder unit chauffeur should be expected to do more than just bring the rig to the scene, yet he shouldn't be wasted as just another body inside. He can operate on the perimeter of the building, still available to place the aerial in service if need be, as long as he maintains contact with someone else on the outside, either the officer in command (OIC), the engine chauffeur, or another firefighter—just as long as someone knows where he was operating and when he was last seen. Of course, compliance with OSHA's *two in, two out* requirements means that before a member enters any immediately dangerous to life and health (IDLH) atmosphere, that member must team up with another firefighter and report his/her position, unless human life at risk demands immediate action to save a life.

You must consider several factors when planning assignments that mean a member may be operating alone for even part of the time. The first is the experience level of the member. Someone who isn't thoroughly trained and who hasn't demonstrated the ability to act independently or to protect himself shouldn't be given this assignment. The second factor is the availability of portable radios. The member expected to be operating alone should at all times be assigned a portable radio with which he may inform others of his position and, if necessary, call for assistance. Normally, a member responding with the pumper will always be operating in the vicinity of other firefighters. This isn't always the case with people who have been assigned truck duties. For this reason, the plan must cover the majority of cases likely to be encountered in a unit's response area, pointing out actions that should be taken.

In addition, a mutual understanding of the needs of each task is highly desirable. The respect that it breeds for the other guys provides a good incentive for a member to complete his assigned task as quickly, thoroughly, and safely as possible. In other words, in many departments, both career and volunteer, a member joins and is assigned either to a ladder or an engine company. Barring transfer, that is where he remains for his entire fire service career. After 10 or 20 years, he may have very little idea of what the working conditions are like for the other guys. The engine man who is taking a severe beating from heat and steam may have no idea how difficult it is for a truckie to climb over two fences, bluff his way past a watchdog in the backyard, then lower and climb down the vertical drop ladder on the fire escape, just to vent three windows. Nor does he care how difficult it was. All he knows is, "This smoke ain't lifting at all." At the same time, a member of a ladder company may have little idea of how vital his task is to the interior members unless he has experienced the benefits of his actions on the inside, when it's really "ripe."

One way to avoid this problem is by assigning people to the ladder companies only after they have spent some time, say two years, in an engine company. This may be a difficult task in some volunteer organizations, particularly where individual companies recruit their own members, but it can be accomplished. To be sure, the benefits in firefighter safety alone are worth the effort. Often, due to the wide variety of tasks that a ladder company is expected to handle, the members of a truck crew may have to operate without an officer's direct supervision. This requires an experienced firefighter who can make correct decisions on his own; one who can operate in the fire area without the protection of a hoseline and not get himself in a dangerous position. You don't expect that of a member with only six months or a year in the department.

Ladder Company Functions at Structural Fires

No matter what type of apparatus arrives at the scene, certain basic actions must be taken. Going back to the basic sequence of actions—locate, confine, extinguish—the first action to take is often considered a truck duty. First, find out where the fire is before committing a hoseline to the location. Generally, if a single pumper arrives first, they aren't going to wait for a ladder truck to show up before taking this step, but at times it is an absolute must. At fires in buildings with large floor areas, such as malls or high-rises, prematurely committing a pumper or a hoseline can have a disastrous effect if the line is stretched to the wrong location. Just because there is smoke showing doesn't mean that that's where the fire is. Of course, if there is going to be a considerable delay in the arrival of truck personnel or if the building has a small floor area, the engine crew may be much better off initially sending one of their members, armed with the proper tools, with the officer.

Let's take a look at an actual plan, see how it works in real structures, and then consider how it might be adapted for use in almost any situation. The first item to be decided is the number of people that will be part of the plan. At very large buildings, in some cases, there may be assigned positions for more than 25 truck personnel. If, however, the number of personnel responding is only 12, then there is no sense planning for 25. The idea is to outline the responsibilities and activities of the initial responding units at the most common type of structural fire, then to make the plan flexible enough to work in the rest of the buildings in that area. Certain buildings or occupancies may occasionally pose an unusual situation, requiring a change of tactics, but in general, these are the exceptions and should be well known within the department.

Ladders 4

An excellent example of a ladder company operations guide is FDNY's tactics bulletin previously known as *Ladders 4*, and now officially known as *Firefighting Procedures, Volume 1, Book 6- Private Dwellings*, but still called *Ladders 4* by at least one grumpy old deputy assistant chief who can't forget studying the old book dozens of times. This manual covers nearly all aspects of ladder company operations at one- and two-family house fires. *Ladders 4* is an expanded, modified version of *Ladders 3*, the original tactics manual for ladder company duties in multiple dwellings (more than two families). *Ladders 3* explains such factors as venting for fire, venting for life (see chapter 8), and other items that a firefighter must understand before making tactical decisions in any structure. For our purposes, though, since private-home fires are the predominant fire problem nationwide, *Ladders 4* is a more appropriate example of operational plans. (The concepts of *Ladders 3* will be explained in chapter 13.)

Since its inception more than 140 years ago, FDNY has responded to many thousands of fires of almost every type. As with any fire department, things haven't always been letter perfect. Mistakes have been made and will continue to be made as new situations are encountered. Still, FDNY has realized that, by formalizing the game plan and experience of its most knowledgeable members, they can develop plans that work for their most common structures. This has been possible sometimes because people tried things that didn't work but finally found what does work after encountering similar situations time and time again. They then wrote their experiences down so that the newly appointed firefighters could begin to learn about these items without having to do it the hard way, through improvising and on-the-job training, which is very costly in terms of property and lives. When many people think of New York City, they think of the high-rises of Manhattan, a real fire problem. But there are four other boroughs that also make up New York City, and these have tens of thousands of one- and two-family homes. These are the proving ground for *Ladders 4*.

The ladder company operations in this plan are based on having five firefighters and an officer respond with the apparatus. Because of circumstances relating to job descriptions in FDNY, the officer isn't assigned actual physical tasks such as forcing doors, pulling ceilings, or the like. His main duty is like that of an orchestra conductor, supervising the rest of the team to produce the maximum benefit and safety for all. The assignment of manpower may thus seem a bit askew to other departments. The plan assumes that there is a working fire in an occupied private house, that an engine will arrive, and that the six-man truck crew will be augmented by another six-member truck crew later into the fire. It places great emphasis on saving human life even though no obvious life hazard may be visible—and rightly so. One thing that has been learned over the years is that you cannot assume that there is no one inside. Considering that the brain can survive without oxygen for only a few minutes, there is no room for error in getting to victims as soon as possible.

Realizing that most people who become aware of a fire will attempt to flee via their normal path of egress, an interior team is an absolute must. They must be equipped to force entry for the hoseline, expose hidden fire, vent as they move, and report their progress to the IC. Thus, the following three basic tools must be provided: forcible entry iron, flathead axe, and hook. A portable radio and a powerful flashlight (in addition to the personal flashlight carried by all of the ladder personnel) should also be considered necessities. This position is of the utmost importance, and this is where the officer should be. In *Ladders 4*, the officer and two firefighters make up the interior search team, and in addition to the three basic tools, they also bring a 2½-gal water extinguisher for containing mattress and other incipient fires, or for knocking down extension on the floor above. The officer carries the handlight and a small personal tool for venting windows and other tasks. By providing strong leadership at this position, he can have a great influence on the outcome. He can assign one or both members to a task, remaining free to evaluate changing conditions and extend the search without waiting for the initial task to be accomplished. The primary duty of this interior team is to search the immediate fire area and the means of egress. In addition, they provide the needed forcible entry, opening up blind spaces and venting to ensure rapid advancement of the hoseline. That leaves three members to perform other duties.

The remaining members are further divided into two teams. The chauffeur and one member work together at the rear of the structure, using a portable ladder to gain access to any upstairs bedrooms for search and ventilation. That leaves one member to operate alone, not normally a good idea. However, by assigning this member specific duties and a specific location, the risks are minimized. This member, known as the roof man, is to work at the front of the building, where personnel are normally observable and may be assisted if entry into front bedroom windows is necessary. His assignment is to perform venting, entering, and searching (VES) of the front second-floor bedrooms from the exterior. VES is highly useful for getting to possible victims located above the fire. By assigning this man to the front of the building, he can often take advantage of porch roofs as stable work platforms, as well as safe areas of refuge if forced to make a hasty retreat.

In developing this plan of action, FDNY took many factors into consideration, primarily based on past fire experience, and they established a list of priorities that are reflected in the members' assignments. The interior team is the largest team, and it has the experience of the officer directly available. This is logical, since the primary duty of the officer is to protect his people. The firefighters on the inside are in the greatest danger, trying to get close to the fire as well as above it via the stairs. Under conditions of high heat and dense smoke, the task of the interior team, to conduct search and rescue, is quite formidable. For this reason, three members are initially given this task (Figs. 7–1 and 7–2).

A high priority is placed on search of the means of egress and the bedrooms, since the area of highest life hazard—regardless of the time of day—is the bedroom. Statistics nationwide, as well as for New York City, bear this out, and it is easily explained. An alert human being will flee a fire if he is physically able. Persons who are asleep aren't alert, and thus cannot flee. This includes many people who work nights and sleep during the day, as well as many older folks who nap each day. Young children in cribs, the temporarily disabled, and older bedridden folks will all be found in bedrooms the great majority of the time. The odds are very high that, if you are going to have to rescue a victim from a dwelling fire, it will be from the bedroom. The plan also recognizes that fires most frequently start on lower floors, severely exposing the interior stairway and rendering it useless as a means of access until the fire has been darkened down. By having the three remaining members perform VES, access into the critical bedrooms is made immediately. By working from the ladder or roof into a window, the members aren't exposed to the severe interior conditions for as long. They are also in less danger of having their escape route blocked by fire. If conditions worsen, they are only a matter of feet from the ladder.

The plan dictates the tool assignments for each member based on his task and area of responsibility. For the outside team, the chauffeur provides an axe while his partner provides a hook and a halligan tool, thus supplying the needed tools to force any rear doors as well as to vent windows. The roof man is assigned a hook for venting and a halligan for prying.

Fig. 7–1 VES is used to put searchers into rooms where victims are trapped beyond the fire.

Fig. 7–2 Ladder company members performing VES enter a bedroom from the porch roof.

Notice that there is no mention so far anywhere in this plan of roof ventilation. This is a lower-priority duty than the life hazard. In practice, it usually isn't needed and, if required, is assigned to additional personnel, probably the second ladder company.

Now, before you say, "That's fine for New York City, where they have two six-member ladder companies responding, but it won't work here," stop and examine the important points. The first is that an interior team, two members minimum, must begin search and assist the attack team in the interior. The second point is that roof venting isn't initially required, but horizontal venting, which greatly speeds the advancement of the hoseline, is required. The third is that, by properly selecting a window for entry, you have a very good chance of entering the location of any missing occupants. Instead of six members, this tactic can be used with almost any number. Of course, the fewer people available, the longer it will take to complete all of the tasks. Still, by assigning different avenues of approach, the chances of success are multiplied, while the chances of being delayed or even prevented from reaching the victims located above the fire are reduced.

"What positions should I assign?" you ask and, "What tools should the members carry?" The answer is that the positions must cover the most vital areas first, providing all of the needed tools to that area before people are assigned to other areas. The plan outlined previously is for use on one- and two-family houses. It requires modifications for use in other occupancies. These modifications are often best determined at the local level based on conditions observed in the field beforehand. As a rule, an interior team should always be assigned and equipped with a halligan tool, a flathead axe, a pike pole, lights, and a portable radio at minimum. Obviously, if the fire is inaccessible from the outside (above the reach of ladders, for example) or if it is in a windowless building, there is no need for an outside team. Any additional people might be assigned to the floor above the fire with a similar complement of tools. In the case of one-story buildings or top-floor fires in multistories, the outside team might be assigned to the roof with power saws and hooks. The idea is to devise a plan that will put the needed personnel, armed with the proper tools, at the locations where they are required.

You say, "I'm with a volunteer department where manpower varies with the time of day. How do I assign the duties of six men to four people on the rig, and what happens when I do have six men?" The answer is to establish duties and assign positions based on priorities. Rather than attempting to assign duties to people, assign the duties to the riding position. Establish a policy whereby riding positions are filled in order (they usually are anyway, in fact). Say that, after the chauffeur, if only one member shows up, he should ride in the cab and act as the officer. If only two members show up, the next one sits in the right jump seat and teams up with the officer as the interior team. By providing a brief written tool list and short job description at each riding position, as well as mounting the needed tools right there, each member will know the responsibilities of that particular position. As additional members board the apparatus, they are automatically assigned their tasks. The positions must, of course, be assigned in order of importance, so that when only three show up, the first priorities are taken care of before someone begins work on an item of lesser importance. The entire idea is to make all of the critical tasks known to everyone and to get the needed people and equipment in place to complete them (Figs. 7–3 and 7–4).

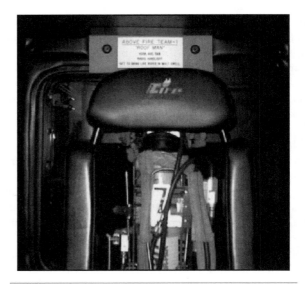

Fig. 7–3 The duties and tools of each member should be clearly marked at each riding position.

Fig. 7–4 Mount the proper tools at each riding position to ensure that the basic tools arrive at the incident.

What are the required ladder company tasks, and what are the priorities? Going back to LOVERS U, you should examine the possible timing and priority of each item. Some of the items will shift in their priority, depending on the situation. For example, salvage operations at a vacant-building fire will be very low, whereas salvage may be the highest priority at a fire in an art gallery. Obviously, search and rescue will usually be the first priority, but at a fully involved structure where there are flames shooting out through the cracks between the bricks and every other opening, the lifesaving effort may be all but forgotten.

Laddering

"What do the members of a ladder truck do?" *Laddering* is usually the most obvious answer to that question. However, laddering is a task that should be done for a specific purpose, not simply to put ladders against a building. I've known chiefs who demand specific ladder placements regardless of the situation. One demanded a ladder be placed on the front of every fire building. It didn't matter that the fire was on the 16th floor and that the search crew was short-handed, as long as there was a ladder on the front of the building. Another had a policy mandating that at least three sides of every house must be laddered. The ladder crew spent the first 5 min after their arrival putting up ladders that weren't used. They weren't even positioned properly to be of any use as escape routes—they were just up to satisfy the chief. That's a damn shame.

Portable ladders have a variety of uses, such as providing access, rescue routes, escape routes, ventilation points, and the like. They may replace aerial devices where access to an area can't be made by vehicle. The proper placement and use of a portable ladder in the early stages of a fire can often prevent loss of life, cut off fire extension, and make the firefighter's task easier. All firefighters should be familiar with their department's complement of ladders and remain alert to put them to use where they are needed. Regardless of whether you are responding with a pump, engine, or rescue unit, chances are that sometime in the not-too-distant future, your job will be affected by the use of a portable ladder. If used properly, a ladder can simplify the task. Improperly used, it can be a handicap.

The following factors affect the selection and placement of a portable ladder: reach, weight, nested or stored length, material of construction, and the number of personnel required to place it into operation. By selecting the proper ladder for the task, you improve your unit's efficiency, increasing the speed with which a task is accomplished.

I recall operating at a fire in the attic of a well involved one-story home. Roof ventilation was requested and an incoming ladder truck was assigned to the task. Horizontal ventilation of the main floor and search were also needed, since the fire had begun in a rear bedroom. I was questioning the absence of both until I walked to the side door of the home and saw the reason. The officer and four members had gotten so wrapped up in cutting the roof that the entire crew was committed: one man carrying the power saw, three men carrying a 35 ft extension ladder, and the officer busily *supervising* this affair. When discussing the wisdom of this selection later, the members readily admitted that the task could have been accomplished just as well, and more safely, by one member with a 16-ft ladder and an axe. The members were unable to explain what had prompted them to use the longer ladder. The only explanation that I've been able to come up with is a lack of training and experience in using portable ladders.

As with almost everything else in the fire service, purchasing decisions made 20 years ago still affect today's operations. Standard operating procedures set 20 years ago and hose loads on a pumper purchased 15 years ago can both be changed without too much difficulty and expense. Ladders ordered five years ago or tomorrow are likely to be around for a while, though, and replacing them can be costly. Still, as with most other items, if they aren't properly selected for their anticipated function, they may be nearly useless.

Factors Affecting Ladder Selection

Of all the factors affecting ladder selection, whether for purchasing or use, the single most important item must be its length. The strongest, lightest, safest ladder in the world is of absolutely no value if it is unable to reach the objective. In many cases, departments are lulled into a false sense of security by the presence of aerial devices. They begin to think that longer-length portables aren't required. Before deciding to eliminate the 45-ft ladder or even the 55 ft from your department's inventory, you should survey your area to determine whether it's needed. As a certain Fireman Murphy might say, "If there's one room that our ground ladders won't reach, sure enough, that's where they'll be needed, and you can bet that *I'll* be the one in the room needing the rescue." Yes, I know that it takes five members to raise these monsters (six for the wooden ones), but if there are no alternatives, what would you do?

I know of one area on an oceanfront where there are literally hundreds of rooms beyond the reach of any portable ladders and are inaccessible by aerial devices. That may be acceptable in buildings of fire-resistive construction, but not in frame or ordinary-construction buildings without sprinklers or fire escapes. I am certain, in this era of manpower shortages, that this situation is repeated across the country. Just remember, it could be you up on the fifth floor needing an escape route when the longest ladder on the rig is 35 ft.

At the same time, the ladder complement should be consistent with the conditions encountered in the response area. I have seen ladder trucks carrying two 45-ft, two 35-ft, and one 28-ft ladder, as well as 12-ft and 14-ft straight ladders, thus surpassing the NFPA requirement for 168 ft of ground ladders. Yet these were totally useless on the fireground in their own first-response area because they were unsuitable. When selecting ladders for length, the first consideration should go to extension ladders. A 35-ft extension will replace a 28-ft, a 24-ft, and, in a pinch, a 20-ft ladder, since the nested length of a two-section 35-ft is 20 ft. The ability to adjust the ladder to the required working length is a major plus for the extension ladder. The alternatives are carrying many more ladders of various lengths (one of the main reasons in the past for tillered aerial ladders) or to use an incorrect ladder at an improper angle. An added benefit of the extension ladder is that the less experienced member will be more likely to select the proper ladder, or at least a ladder that will do the job.

Selecting a length for the task is more complicated than it might at first seem. It depends to a large extent, of course, on the floor at which the task is to be performed. As a rule of thumb, residential buildings can be estimated as 9 ft from floor to floor. Increase this estimate to 12 ft from floor to floor in commercial buildings. Add to this an average windowsill height of 3 ft above the floor, plus a window height of 4 ft, and you can begin to estimate the required length of your ladder (Fig. 7–5).

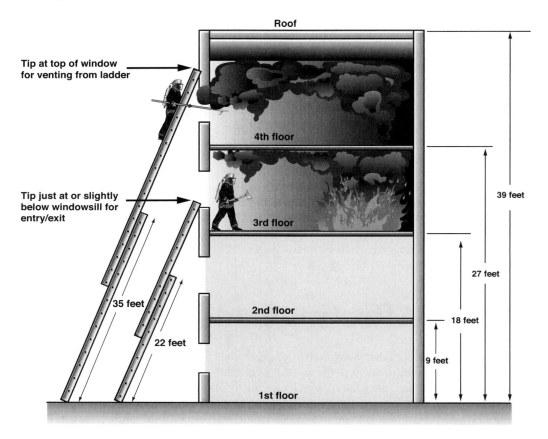

Fig. 7–5 Select the right length ladder for the job based on the task and the floor to floor height of the structure. In residential buildings this is typically 9 ft.

Consider, for example, the structure illustrated previously. A quick size-up should indicate to the firefighter that the second-floor windowsill is about 12 ft above grade, while the third-floor windowsill would be about 21 ft above grade. This is good basic information, but it isn't all the information needed to select the proper ladder. The members must also have an idea of the intended use of the ladder. If the task is window entry, rescue, firefighter escape, or hoseline advance, a shorter ladder may suffice than if ventilation or overhaul of the exterior need to be performed.

Proper Climbing Angle

The proper climbing angle for a portable ladder is about 70°. This puts the climber in a position to walk nearly erect while holding on to the rungs or beams. This 70° angle eats up the vertical height of the ladder, since the ladder must also cover a horizontal distance. For this reason, a 12-ft ladder won't reach the windowsill of the second floor shown in the following diagram. At a 70° climbing angle, about 6% of a ladder's total length is lost to this factor. The remaining 94% is called the working length, the height that a ladder will actually reach above grade. To reach the second-floor windowsill requires a 13 ft reach.

To ensure the proper climbing angle, it isn't necessary to go through any geometric formulas to determine when the ladder is at the proper climbing angle. Instead, a quick estimate of the working height, divided by four, gives you the approximate location to place the butt of the ladder away from the building. Be sure to take any eaves, balconies, or other overhangs into account. Set the butt down at this location, then hoist and lower the ladder into position. Before proceeding upward, check the climbing angle by standing on the bottom rung and reaching out for the rungs at chest height. If you can touch them with your arms fully extended without having to lean in or out from the ladder, then the ladder is at a good angle. The idea is to achieve a comfortable, stable position rather than meet some preset figure that requires a protractor to measure.

Proper Tip Placement

As previously mentioned, the task for which the ladder is being used will determine the proper location of its tip. If a firefighter is to ventilate a window by breaking it, the best position for the tip of the ladder is even with the top of the window frame, just off to the upwind side of the objective. This allows the member to ascend to a height from which he can break the glass without being in its path as it falls. In addition, it keeps his hands up high enough so that glass won't likely slide down the handle of the tool. At the same time, the member shouldn't be too high, since the reason we are venting is to allow heated gases, smoke, and flame to escape. These will usually be at the upper level of the window. Too high on the ladder, and the member could be caught in fire driven at him by shifts in the wind.

For most other uses of portable ladders, the proper location of the tip will be just at or slightly below the objective. Take the task of entering or exiting through a window. By placing the tip just at or below the windowsill, the entire width of the window is kept clear for the firefighter. Any other placement will impede entry or exit. Some sources recommend extending the ladder one or two rungs into the window with the thought that a firefighter looking for a route of escape will be able to see the ladder tip and move to it quickly. There are some flaws to this theory. The fact is that a vented window is much more readily found under fire conditions than the ladder. By completely clearing the window of glass and sash where the ladder is place, much more will have been accomplished than merely placing the ladder. This sends a signal that this is the escape window, and it also helps achieve superior ventilation, easing conditions.

Conditions must be rather severe for a firefighter to require the portable ladder for escape, and chances are excellent that the firefighter will have an SCBA on his back. Considerin g that the average residential window is only 30 in. wide, and that a firefighter with an air pack on his back is nearly 20 in. wide, there just isn't enough room to fit a 16-in. wide ladder in the window opening. In addition, placing the tip of the ladder at the sill allows the member to climb onto it without rising up into the greatest area of heat or even fire (as do other methods), further reducing the chances of injury (Fig. 7–6).

Materials of Construction

Presently, the vast majority of ladders in the fire service are made of aluminum. Wooden ladders have largely been replaced, since aluminum versions are lighter and require less maintenance. The light weight of aluminum allows one person to handle most ladders up to 20 ft in length, both straight and extension. Two men can handle a 35-ft aluminum extension ladder if they butt it against the building while they raise it.

Aluminum has its drawbacks, however, the most common being its electrical conductivity. In the past, fire-fighters have been electrocuted and many more injured when aluminum ladders were used in the vicinity of live electrical lines. Composite ladders, having beams of nonconducting fiberglass and rungs of aluminum, offer at least a partial solution to this problem. Electric companies have used these composite ladders for years, but they haven't really caught on with the fire service. Their weight is only slightly heavier than that of all-aluminum ladders, and their load-carrying capacity is comparable. Of course, all ladders—wood, aluminum, and fiberglass—can conduct electricity if they are coated with water. Always use extreme care when operating near power lines.

Fig. 7–6 Any ladder that a firefighter might be required to use as an escape route should be positioned with the tip just at the windowsill.

A drawback of both fiberglass and aluminum is their lack of resistance to heat. The process used in manufacturing aluminum ladders requires heat-treating. Subsequent exposure to high heat can seriously weaken a ladder's strength without any outward changes in shape. The only visible evidence of this loss of strength may be a subtle discoloration similar to bluing. If you detect this, or if the ladder experiences any direct exposure to flame, take it OOS until it can be properly tested. NFPA standard 1932 requires a heat-indicating label on each section of new aluminum and fiberglass ladders. The label will change color when exposed to dangerous temperatures (300°F).

Other relevant sections of NFPA 1932 deal with load-carrying capacities and testing, as well as record keeping. A design load of 750 lbs has been selected for most ladders, since it usually takes two firefighters to remove a single unconscious person. Assigning an average weight of 250 lbs for each person, including fire apparel and SCBA, explains the working strength. This loading is up sharply from previous weight-carrying capabilities. This change, coupled with another requirement designed to facilitate safety and stability during any activity (but particularly rescue) results in ladders that are heavier than in the past. The second feature common to all ladders purchased since 1984 is an increased width of the sections. Old ladders had sections as little as 12 in. wide. These were quite cramped when attempting to assist a victim down, and they were less stable than current ladders, which are now at least 16 in. wide at their narrowest point.

As with any tool, failure to use a ladder properly can lead to injury or death. A good working knowledge of the ladder parts, how they work, their capacities, and proper usage are a must before anyone is sent out to ladder a building. Each member must know the nomenclature of the ladder, and each should be trained in the department's standard methods of carrying as well as raising it.

During training sessions, attempt to keep the situation as realistic as possible. Make all personnel wear full clothing, climb with tools and hose, and perform real-world operations, such as climbing through windows and onto the roof.

Avoid the temptation to impress new people with the great strength of the ladders. A large proportion of ladder failures occur during training. Sometimes ladders are purposely overloaded to show how much they can take. Then, when they fail, the fire departments can't understand why. One possible reason is the tendency of some departments to relegate older equipment to the training ground while maintaining the best for frontline service. This practice, combined with overloading and unusual activities sometimes undertaken to illustrate lessons, can lead to ladder damage and possible failure. One way to head off this danger is through annual service testing, by testing whenever a ladder appears to have suffered damage, and by regular inspections of all ladders.

Guidelines for the Safe Use of Ladders

1. Use the proper length and type of ladder. Don't attempt to improvise except to save a life. It is far safer to go back and get the proper ladder.

2. Use the ladder at the proper angle. Too steep an angle causes the member to climb with great difficulty and puts him or her in danger. The ladder may pull out from the wall. Too shallow an angle places greater stresses on the ladder and makes climbing awkward.

3. Position the ladder with the objective in mind. Be aware of how the ladder is to be used. Avoid shifting it after it has been placed, and don't try to make do with a ladder that has been positioned improperly.

4. Keep *all* ladders away from electrical lines. Be aware of any lines in the vicinity when carrying, raising, and climbing ladders. If it is necessary to carry a ladder in the vertical position for a short distance, at least one member must keep the tip of the ladder under constant surveillance. (Note: Always retract an extension ladder before trying to move it.)

5. Do not position ladders directly in front of ingress/egress routes or where fire is likely to vent. Such locations pose accident risks as people jostle the ladder or objects are dropped from above. Obviously, you wouldn't want to have to climb through fire if the ladder were being used as an escape route. Still, ladders are often placed in front of windows of rooms that are well involved. Move ladders off to the upwind side, if possible.

6. Do not overload the ladder. Ladder failure isn't common, but it can occur, and following loading guidelines can usually prevent it. The old standard was one man for every 10 ft of ladder length. The new standard is 750 lbs maximum for all ladders, except folding and attic ladders.

7. When climbing, *always* have one hand on the ladder. Since firefighters are usually carrying tools, this poses a problem. Still, the firefighter must at all times ensure that he has one hand free to hold onto the ladder. He can accomplish this by passing the tools to someone else, by securing them to his body with a sling or by lowering a light rope and hoisting them up. If you are carrying a tool in one hand, position both hands behind the beam, with your free hand maintaining constant contact with the beam, ready to pull you into the ladder in case of a mishap. You should also position the hand carrying the tool behind the ladder, with your wrist cocked against the beam (Fig. 7–7).

Fig. 7–7 Carrying a power saw on a sling allows you to use both hands for the climb.

8. When working on a ladder, always use ladder locks or a ladder belt to secure yourself. Nobody plans to fall off of a ladder, yet people do each year. With all the activity on the fireground, as well as the weather and fire conditions, it's no wonder. If glass falls the wrong way or a hose stream sprays through an opening, the firefighter may be caught off guard. That's where the safety precaution comes into play, keeping the firefighter safely on the ladder.

9. Don't overextend yourself. If necessary, reposition the ladder. One of the easiest ways to fall from a ladder is to try to overextend your horizontal reach. Even if you are belted into the ladder, you could get into trouble, since the sideways motion places great twisting stress on the ladder, possibly causing the ladder to spin. The simplest way to avoid this is to reposition the ladder, since even the person footing or butting the ladder could be caught off guard by the strength of this force.

10. Tie off the base of the ladder. It is a good idea, if possible, to secure ladders to the building or some other substantial object. This, however, is rarely possible. With extension ladders, you should secure the halyard to the bed section once the ladder is at the proper height. This acts as an insurance policy for the ladder locks. As an alternative to securing a ladder at the base, it is sometimes more practical to secure the tip. In the event that it is necessary to leave a ladder unattended, it should always be secured at the tip, since tying it off at the base won't prevent it from being blown over sideways. This is especially important with ladders of lighter weight. I once watched an unattended ladder at a training session blow over toward a hose crew that was attempting to advance a line through a doorway. The door took the brunt of the blow, which averted injury. After things had calmed down, the ladder was again raised, and it promptly blew down again before the safety officer ordered it taken down.

11. Before stepping off any ladder, be sure of the stability of the area you're about to enter. Even if you can see the surface, use a tool to probe the area for strength. This might seem obvious when visibility is poor but even under good conditions, your eyes cannot determine the strength of supports that might be exposed to fire (Fig. 7–8).

Portable ladders are very useful tools, allowing us to reach areas otherwise made inaccessible by fire, structural collapse, or architectural design. When used with forethought, they may be the key to an entire operation. Success doesn't demand a huge ladder truck carrying hundreds of feet of ladders. It is often possible to provide the proper selection of ladders for an area right on the pumper. But whatever ladders are available, the firefighter must know how to use them all properly, even while under duress (Fig. 7–9).

Fig. 7–8 Stepping off the ladder may be this firefighter's last act if he doesn't check the stability of the roof first.

Fig. 7–9 Note the 35-ft extension ladder raised from the 10th floor setback to the 14th floor balcony at this high-rise. Creative uses of portable ladders can help overcome serious problems.

Aerial Devices

Aerial devices add to a unit's capabilities, but they shouldn't be viewed as replacements for portable ladders. Aerial devices are commonly used for search, rescue, ventilation, access to the roof, applying heavy streams, and as platforms for overhauling or inspecting an area.

Certain tasks lend themselves to being performed by specific kinds of devices. For example, ventilating windows on the front of a building is rapidly accomplished using an aerial ladder, whereas removing an unconscious victim is far easier with an elevating platform. Roof access is often required for ventilation purposes. Even though an elevating platform delivers firefighters to a given location, keeping them less physically drained than those who must climb an aerial, many platforms also have limitations. For serious fires in larger buildings, the initial roof crew may not be sufficient to complete all of the needed tasks. To supply the needed reinforcements, an aerial ladder, or ladder tower, is superior to an articulating boom (snorkel) or to a tower ladder (Figs. 7–10 and 7–11).

The difference between a tower ladder and a ladder tower is more than a semantic one. The tower ladder is a platform attached to a boom with a built-in waterway and a sideless escape ladder. This ladder is unsuitable for routine climbing up or down because of the lack of substantial sides. The ladder tower is a platform attached to the end of an aerial ladder with a built-in waterway provided. This device, designed around an aerial ladder, provides for ready climbing. This may make it a good choice for an area that has many larger structures and only one aerial device available. As you will see, however, a better situation is to have two aerial devices respond to working fires in larger structures, if possible. If one is an aerial ladder and the other a platform, the best attributes of both are at the IC's disposal.

Fig. 7–10 A tower ladder has only a sideless escape ladder, which should only be used in an extreme emergency.

Fig. 7–11 A ladder tower consists of a basket attached to an aerial ladder.

Ladder towers typically have a lower platform weight-carrying capacity than tower ladders, and many have serious restrictions on how their master streams may be operated at lower boom angles. Be sure to follow the manufacturer's specifications in this regard, or severe damage and potential injury or death may be possible. Tower ladders on the other hand offer unrestricted master stream use at any angle or extension, making them the most versatile master stream appliance on the fireground. Their platform capacity of up to 1,000 lbs is also unaffected by boom angle, extension, or master stream use.

When training members to position aerial apparatus, the type of apparatus and the task to be performed must both be considered. Aerial ladders and telescoping platforms require a clear line of sight from the turntable to the objective. Articulating boom platforms may be able to reach over some obstacles, but they pose an additional problem, in that the boom elbow projects opposite the platform, possibly striking obstacles on the other side of the truck (Figs. 7–12 and 7–13).

Fig. 7–12 An articulating platform may be able to reach over obstructions, but you should keep it (and all aerials) at least 10 ft away from wires.

Fig. 7–13 Aerial ladders should always be extended at least 5 rungs over the roof line to make them easier to find in darkness and smoke, as well as easier to mount and dismount onto the roof.

When raising an aerial to the roof, the ladder should be extended so that at least five rungs project above the level of the roof. This allows for greater visibility of the escape route when conditions on the roof start to deteriorate. In addition, the extra projection allows the member to maintain a firm handhold at normal standing position while mounting or dismounting the ladder. The ladder itself should remain an inch or two out from the building. This avoids placing a strain on any parapet walls and allows the truss construction of the ladder to perform as designed. When the member's weight is applied to the ladder, the beams will just rest on the edge of the roof. When positioning an aerial, it is very important to make sure that the ladder is placed as square to the building as possible. If the ladder is placed at an angle to the building, one beam of the ladder will contact the building first.

As the climbing members' weight approaches the tip, the ladder will twist until the other beam also rests on the edge of the roof. This is dangerous for two reasons. First, the twisting is a severe torsional load that can lead to ladder failure. Second, I vividly remember climbing a nearly fully extended 100-ft aerial positioned so that the ladder was at an angle to the building. As I neared the tip with a power saw slung securely over my back, the ladder suddenly slid about 6 ft along the parapet. This was caused by the twisting motion described previously. As a ladder twists, it pulls the other beam to the side that it is twisting toward. The building at this incident was a six-story school about 250 ft long. The chauffeur of that aerial could easily have rotated the turntable a little bit further and squared off the ladder to the building. Due to his impatience, two firefighters and I were put in severe danger. Not only had the sudden movement nearly dislodged us and our tools, but if the ladder hadn't been stopped from sliding by a projection of the parapet, it is quite possible that it would have kept going until it cleared the building, putting a severe shock load on the ladder, possibly one that it wouldn't have been able to withstand. The way to avoid this is to position the apparatus so that the turntable is in line with the objective, then to rotate the ladder until it is perpendicular to the building (Fig. 7–14).

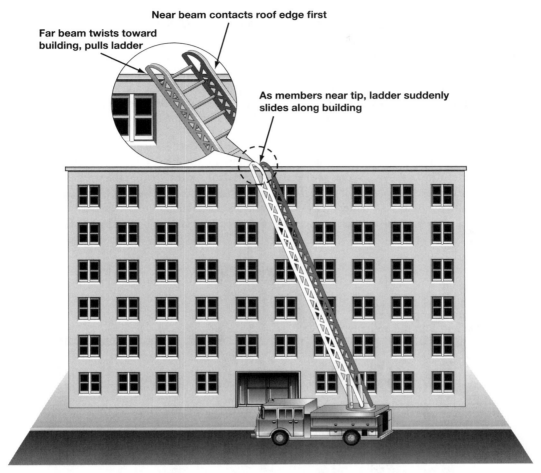

Far beam twists toward building, pulls ladder

Near beam contacts roof edge first

As members near tip, ladder suddenly slides along building

Ladder, not square to building, twists and slides.

Fig. 7–14 Make sure aerial ladders are raised square to the object they will contact. Failure to do so could lead to ladder failure.

When placing a platform to the roof, the member in the basket can place it in the best location. If the roof has no parapet wall, it is best to place the basket over the building so that it is just resting on the roof. This will allow the firefighters to step directly through the platform gate onto the roof. If the building has a parapet of moderate height (3–6 ft), place the top rail of the platform basket just slightly above the top of the parapet. This makes it easier for members climbing in or out of the basket. For unusually high parapet walls, those more than 6 ft, it is best to relocate the platform. If that isn't possible, you should gain access to the roof directly from the bucket, perhaps by using an attic ladder secured to the top sliding section of the boom. One special caution when talking about telescoping booms with their baskets on the roof is to make certain that the basket has been raised up above the roof or parapet before retracting the boom; otherwise, the basket will strike the building. Damage to the building and the apparatus is bad enough, but it is also possible to knock down parts of either onto people below.

Vent, Enter, Search

A primary function of all aerial devices is to allow firefighters access to the upper floors of a structure via the exterior. When fire on a lower floor is blocking escape from the upper floors, the aerial device is a must. It allows us to get above the fire, to search for and cut off any vertical extension, and to search for and remove any victims. It involves gaining access to the desired floor level, forcing entry to the area (usually by breaking a window, thereby venting the immediate area), and then conducting a search for fire as well as for occupants.

This method is often recommended more than an all-interior approach, since the escape route is always on the outside of the building, which isn't as likely to be cut off by extension. In this case, the elevating platform is a clear choice more than an aerial ladder. The basket is a much safer place to work from when venting the window. If a victim is found, especially an unconscious or disabled person, it is much safer, faster, and more efficient to place him into the basket than it is to use the ladder. Finally, the basket is a better area of refuge. If conditions suddenly worsen inside the building, multiple searchers can quickly pile into the basket much more easily than they can clamber onto an aerial ladder (Fig. 7–15).

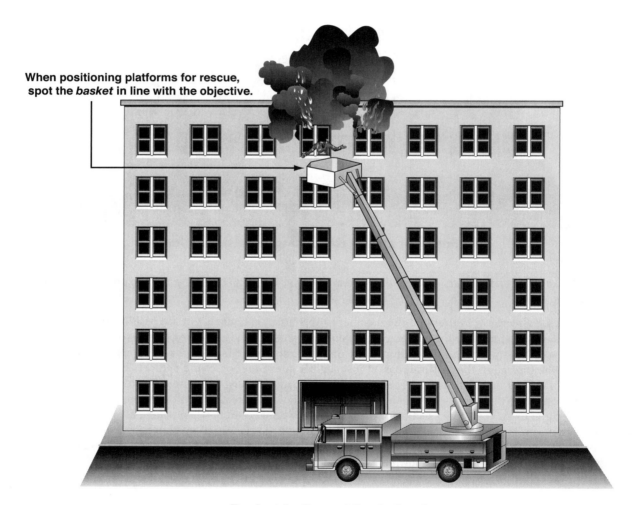

When positioning platforms for rescue, spot the *basket* in line with the objective.

Basket in line with victim for rescue.

Fig. 7–15 When positioning elevating platforms, stop the apparatus with the basket in line with the victim.

Positioning a basket to a window requires a different technique from placing it on the roof. First, most platforms have a master stream nozzle mounted on the center front of the basket. This can impede access to the basket at the center. On the other hand, the front corners of the basket are relatively unobstructed. Picture a person hanging out of a third-floor front window, screaming hysterically for rescue from the rapidly approaching fire. As you arrive at the scene with the platform, you must decide where to spot the rig. The best position is as close to the building as possible, with the basket in line with the window. This assures the shortest distance to travel, and it places the front corner of the basket at the window. Place the basket so that the top rail is level with the windowsill, making it easier to enter the basket and keep the people in the basket away from most of the smoke and heat emanating from the window. Unless the lineup is letter-perfect, you should forget about the gates in the basket. They are too narrow, and trying to align them with the window takes too much valuable time. In addition, many of these gates only open inward, reducing the space available inside the basket.

The positioning of an aerial ladder to a window is similar to positioning it on the roof, in that the rig should stop where turntable is directly in line with the window or objective and the ladder is perpendicular to the building. The tip of the aerial should be just below the windowsill, if possible, leaving most of the window free for access. Remember, the average residential window is only 30 in. wide, whereas the tip of most aerials is 20–24 in. wide (Fig. 7–16).

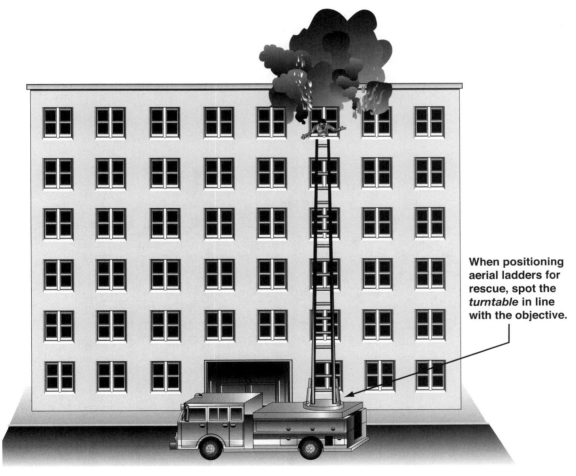

When positioning aerial ladders for rescue, spot the *turntable* in line with the objective.

Aerial positioning with turntable in line with victim.

Fig. 7–16 When positioning aerial ladders, stop the apparatus with the turntable in line with the victim.

One final word on aerials and windows: Aerials make excellent ventilation tools, especially when manpower is at a premium. The aerial operator can do initial venting of almost all of the windows within reach of the ladder from the turntable. All the operator has to do is to put the tip of the aerial through the top pane of glass and then lower the ladder, breaking the sash and bottom pane. The operator should take care not to overextend the ladder, nor to attack any steel-framed windows, such as casement windows, which could damage the ladder. In addition, the actions cited previously (extend through, then lower) should be the only actions permitted. Never allow the ladder to be rotated into a window, since this can damage the ladder and/or result in the ladder getting wedged in the opening.

This brings up the topic of scrub area, which may be defined as the amount of surface area that may be physically contacted by an aerial device. Certain types of aerial devices have greater scrub area than others. Aerial ladders and telescoping booms have greater scrub areas than articulating booms of comparable length. Generally, the greater the working height of a device, the more scrub area it can cover. Scrub area determines how wide and how high an area an apparatus may be expected to cover for access, egress, platforms, and the application of streams.

The positioning of the apparatus has a direct bearing on the scrub area. The farther away the apparatus is from the building, the less area the end of the stick can cover. At the same time, spotting too close may reduce the useful scrub area, since the apparatus may be too close to swing the ladder or boom to the lower floors, where it may be needed most. The particular *perfect distance* for an aerial device to obtain its maximum scrub area varies depending on the length of the device and the number of telescoping sections. More telescoping sections of boom for a given length means that each section is shorter, thus a four-section 75-ft boom has a greater scrub area than a three-section 75-ft boom and more scrub area on lower floors than a three- section 85-ft boom. For articulating booms, scrub area and spotting distances are determined by the length of the bed section. Chauffeurs must be trained to judge this distance from a building so as to get the most out of the apparatus. This involves setting up the apparatus at actual structures and comparing the effects of various distances. In addition, ICs must be aware of the capabilities and limitations of the devices available to them (Fig. 7–17).

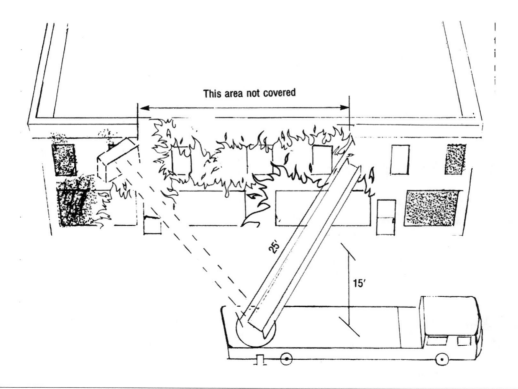

This area not covered

25'

15'

Fig. 7–17 Apparatus should not be positioned closer than the shortest nested length of the aerial device if the ladder is to be used on the lower floors.

I once responded on a mutual-aid basis with a 75-ft, four-section tower ladder to a fire in a two-story, 75 x 100 ft commercial building. One telescoping platform was already working on the front of the building and an aerial was set up on the left (exposure 2) side. Heavy fire had possession of the entire second floor. Present were several members and chief officers from another department situated closer to the fire than my own. They were somewhat perplexed as to why their new three-section 85-ft platform was bypassed in favor of the 75 ft. It soon became clear, however, when the 75-ft tower ladder was placed in a small rear parking area and proceeded to put its master stream into operation.

In conjunction with the platform in the front, operating into successive windows, the fire was rapidly swept from one end of the building to the other. It was obvious that with their longer boom section, they wouldn't have been able to place their basket to five or six windows. They would have been forced to rise up over the building, skip over to the other side, and begin again. When placing an apparatus for operation on the lower three floors of a building, the driver must be cognizant of the vehicle's profile and how it affects the movement of the aerial device. On midship-mounted devices, especially those having enclosed cabs, the absolute worst approach would be to pull nose-first toward the building if you intended to reach the lower floors, since the cab would block the ladder or boom from descending to those levels.

When positioning midship-mounted devices, the best scrub area is obtained by placing the apparatus just off and parallel to the objective. Angle the cab about 15 or 20° away from the fire building. This will get the cab out of the way for the lower angles without swinging the back in the way. Rear-mounted devices suffer from pulling nose-in, as do the midship trucks, but often not as severely, since they are usually built with a lower-profile cab to allow for street clearance. The best scrub area for rear-mounted devices is obtained by backing into position, but this can be impractical due to time constraints as well as the need to keep the street clear for other apparatus (Figs. 7–18 and 7–19).

Some guidelines for apparatus positioning apply to all types of aerial devices. The first is that they should be positioned for maximum benefit of the device. It is fairly common on the fireground to see aerial devices blocked out of the proper position by engine companies. This shows poor training. Pumpers carry hundreds of feet of hose, whereas aerials are of fixed lengths. Surely a pumper could be spotted 30 ft in one direction or another to allow the aerial a more advantageous position.

The second guide is to spot the apparatus with the fire's anticipated progress in mind. Generally, as you approach a building, slow down and observe the conditions. If any obvious conditions demand your locating at a specific position, such as a victim at a window, go ahead and spot as described previously. In the absence of any specific requirements, however, position the apparatus just past the corner of the building that you approached first. This allows the apparatus to be driven forward if the need to reposition it suddenly presents itself. That is much simpler than trying to back up into position. When positioning for access over the fire, try to spot upwind of the fire so that heat, smoke, and flame will be carried away from anyone operating above (Fig. 7–20).

Fig. 7–18 The scrub area of a midship-mounted aerial may be partially blocked by the cab of the apparatus.

Fig. 7–19 Proper positioning of this tower ladder allows it to clear the utility pole to operate on either side.

Fig. 7–20 If nothing demands a specific position on arrival, spot the turntable 15 ft past the edge of the building you pass as you arrive.

The next two items are safety rules and, as such, should be broken only under the most extreme conditions. First, keep all aerial devices 10 ft or more from overhead power lines. The metallic frame of aerial devices coupled with the direct contact between their stabilizers and the ground produces an excellent path for electric current.

Remember, direct contact with the charged object isn't even necessary. High voltages can arc considerable distances across open air. Conditions on the fireground often make it easier for this arcing to occur, given all the water flying around, creating a conductive film on surfaces that wouldn't otherwise present an electrical hazard.

The second safety rule is never to extend or retract any telescoping device, aerial ladder, ladder tower, or tower ladder while a person is climbing on the ladder. The shifting rungs have caught several firefighters' feet and legs, causing severe crush injuries. If conditions are such that the people on the device are exposed to severe conditions, rotating the turntable, or raising or lowering the ladder, should provide relief. To allow extending or retracting, however, is to invite injury.

You can avert tragedies by following these rules.

Overhaul

A well-done, efficient job of overhaul can save much hard work, prevent embarrassment, and go a long way toward improving the image of a fire department. A sloppy job wastes manpower, can upset the citizenry, and may even result in additional fire loss.

What is overhaul? What distinguishes a good job from a bad one? How much overhaul is required? An experienced firefighter should have the answers to all of these questions. Overhaul, in a broad sense, may be considered any action taken to expose hidden fire and ensure its extinguishment. The purpose of overhaul is to guard against rekindling of the fire after the department has left the scene. It will be necessary to do some damage to expose hidden fire, but officers shouldn't hesitate to authorize overhaul for fear of incurring the wrath of the occupant. The alternative to not opening up is worse.

A rekindle is often more serious than the original fire. The area has probably been vacated, the windows have been opened, and holes have been made in the walls and ceilings. All of these allow fire to spread rapidly if it goes undetected. Additionally, occupants may be slower to call the fire department, thinking that the smoke may be part of the aftermath of the original fire.

On the other hand, an efficient job of overhaul shouldn't involve a department in a house-wrecking contest. Nearly all of the openings that are made should be justifiable to the occupant. Once he sees charred material in an exposed area, it is fairly easy to explain the holes in the walls.

Where to open

Overhaul may be performed in the following two phases: precontrol and postcontrol. Precontrol over-hauling is done while the fire still has the upper hand. As such, it is often performed under very difficult conditions of heat, steam, and visibility. There is additional pressure to get ahead of the fire and cut off its extension. Pulling ceilings to expose a traveling fire in an attic or cockloft is an example. At this stage, manpower is often thinly spread and speed is of the essence, so openings may have to be made based on an officer's judgment of likely spread rather than any definite indication of fire.

If an opening is made that doesn't show fire, it means that the firefighters have gotten ahead of it and can now move back toward the source. Conversely, if fire is found, then the firefighters may have to drop back to position a line to cut off further extension.

Knowledge of a building's construction will often tip off a firefighter about where to open up. In almost all non-fireproof buildings, once fire has heavily involved a kitchen or bathroom, it is essential to examine the floor above because of the presence of pipe shafts accommodating water, sewer, and vent pipes. Fire will climb rapidly up the shaft once it enters this kind of void space (Figs. 7–21 and 7–22).

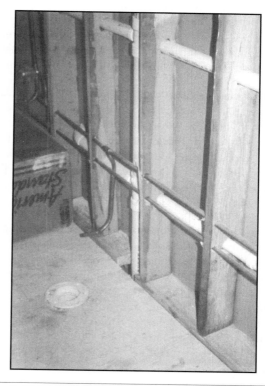

Fig. 7–21 Plumbing voids often interconnect vertically and horizontally,...

Fig. 7–22 ...creating mazes for the extension of fire.

The same is true of many older brick-and-joist buildings that are built around a skeleton of steel I-beams, also known as channel rails. The enclosure around these steel columns is often combustible, being framed out with wood lath. It is more difficult to pinpoint the location of these columns than it is to find the kitchen sink or toilet, so fire often goes undetected for some time. It is vital to check these and any other shafts for fire extension, since they provide built-in highways for extension from the bottom of the building right to the top. When checking down them, don't forget the base of the shaft, since embers do drop.

After these vertical voids, the most important areas to examine are horizontal voids such as cocklofts. Although horizontal spread is slower than vertical spread, it can still be very rapid. Fire may spread from floor to floor around exposed pipes, but firefighters can see it. Fires in voids such as cocklofts are concealed from view and pose a greater potential for trouble. Thorough examination of these areas must be made as soon as possible.

It used to be common to find horizontal voids only on the floor directly below the roof, *i.e.*, the top floor. This is no longer true for a number of reasons. First, lightweight parallel-chord trusses to support floors are commonplace. The area between the ceiling and floor above is now one open truss loft. Second, ceiling heights have been lowered for reasons of energy conservation. Usually this is evident in older buildings that originally had 9-, 10-, or even 11-ft ceilings. Often these buildings have been modernized, leaving the wooden floor joists exposed above and a suspended ceiling hung several feet below. Many times, these voids are interconnected with vertical plumbing voids, making any fire that extends into any of these spaces a nightmare for firefighters (Fig. 7–23).

The next location that must be examined is the floor directly over the fire. A quick look at the fire area on the way past will help firefighters find the likely points of extension. Lacking this information, firefighters should first examine any poke-throughs, such as radiator supply pipes, then examine the baseboards of walls. Again, knowledge of building construction can assist in picking out likely areas of extension.

For example, a cellar fire in an old, wood-frame house would dictate examining the exterior walls, since there is a possibility that the house was built of balloon-frame construction. A cellar fire in a brick and wood-joist apartment house would require greater emphasis on examining the interior partitions, since the exterior walls are brick with a plaster coating. Pipe chases and steel columns are found in the interior partitions (Figs. 7–24 and 7–25).

Fig. 7–23 A serious problem with lightweight truss construction is the horizontal void it creates at each level—virtually a cockloft on each floor.

Fig. 7–24 Balloon framing allows fire to travel at exterior walls. A fire originating in the cellar will rapidly extend...

Fig. 7–25 ...past the first floor, shown here, to the attic.

Sensing fire

Postcontrol overhauling is a less hurried, more detailed examination for the slightest bit of remaining fire. Because time isn't as critical, the openings should be made in a more thorough manner, paying more attention to reducing unnecessary damage than to speed. Generally, openings made at this stage should be justifiable both to firefighters and the property owner. Remember, the heat is off, literally and figuratively.

To decide what a justifiable opening is, firefighters must use their five senses, the first being their sense of sight. Obviously, if firefighters see fire in a partition or piece of furniture, they must root it out. Some less obvious visual signals include smoke, blistering paint from heat within a wall or container, and even the color of the object.

As a guide, normal Class A combustibles that are showing a white or gray ash on the surface are still hot enough to reignite. Dousing them with water turns this ash a solid black. This rule of thumb applies equally to almost all Class A materials, even though no open flame or glowing embers are visible. The next time you go to a brush or rubbish fire, take a look at this phenomenon (Fig. 7–26).

Fig. 7–26 TICs are tremendous assets for search as well as overhauling.

Thermal Imaging Cameras (TICs)

TICs are probably the greatest advance in overhauling since the invention of the halligan tool. The TIC allows a firefighter to view a visual image of heat, letting him move directly to its source. TIC are a great adjunct to firefighters for both searching for life as well as searching for hidden fire. In the past, the prices of these devices put them out of the reach of many departments, at least for use on every ladder truck, but recent advances in technology as well as their ever increasing popularity have helped to reduce prices to a more affordable level. In some cases, such as the state of New Jersey, advocates of the fire service have been able to convince the people holding the state purse strings that an investment in the devices would pay itself off in terms of reduced property damage, lives saved and injuries reduced as to make it worthwhile for the state to buy every fire department in the state it's own camera! I wholeheartedly agree.

Two personal anecdotes describe how valuable these units can be in terms of opening up and limiting damage during overhauling. When I was the captain of Rescue Co. 1, we had three TICs of various manufacturers that we used a lot, several times a day on average, at everything from working fires to hazmat releases. We were often special-called just for the camera, since we were the only unit in all of Manhattan that had them.

One morning we were special-called up to East 116th Street in East Harlem. The 12th Battalion had been on the scene of an odor of smoke in an occupied tenement for about 45 min, searching but unable to locate the source of what was obviously wood smoke. The building was a four-story, Class 3 multiple dwelling, with lots of void spaces that fire could hide in. It was doing just that. The two ladder companies on scene had used their normal examination techniques, looking for discoloration, feeling walls, opening baseboards, and poking small holes in ceilings but had still not been able to locate the source. Fearful of leaving a smoldering fire only to return later to find a full blown blaze, the chief special called Rescue 1 from our quarters on West 43rd Street. Within 15 sec of walking into the top floor apartment, Lloyd Infanzon, our TIC operator, pointed to a hot spot on the wall near the top of an interior partition. A halligan hook was swung into the wall, pulling the lath and plaster off, revealing an electric wire which had shorted where it passed through a wall stud, igniting the 2x4. The fire had smoldered so long that the entire top 3 ft of the stud had burned away, as had most of the double 2x4 top plate. The inch-thick plaster on the old tenement wall had confined the fire to the two bays involved, but the top plate was nearly burned through. In another ½ hr or so, the fire would have penetrated that last barrier, and spread into the cockloft, fed by a new supply of air. We would have had a serious fire. Instead, the damage was limited to a few hundred dollars for new wiring and some carpentry repairs.

The second case did not save any property as such, but it did save a lot of effort and allowed the IC to leave knowing he wouldn't face a potentially catastrophic problem later at night in an occupied hotel. Late one evening we responded to the Marriott Marquis Hotel, a 50-story Class 1 hotel with an atrium that runs up the center of the entire 50 stories. Units reported smelling wood smoke on a wide range of floors from 20 to 40. The problem here is obvious, to cover such a large area takes a lot of people, and we don't want to create a panic in the hotel waking up 20 floors of sleeping occupants. The situation was compounded by the fact that the smell was very inconsistent—it would come and go. This is usually a sign the smoke is being moved by things like the HVAC, or elevators traveling in shafts. The HVAC mechanical room was thoroughly examined with negative results, and the elevators were not a factor, since they were exposed in the core of the huge atrium, and thus couldn't create the *piston effect* they do when traveling in an enclosed shaft.

Rescue 1 was directed to assist in the search for the elusive source of smoke. The 9th Battalion Chief suspected the source might involve one of the thousands of potted plants that ringed each of the 50 floors around the edge of the atrium, since there is very little other wood in a Class 1 building. Again the TIC proved its worth when a member on the 15th floor balcony spotted a source of heat 12 stories above him. "Hey Cap, I think I found it," he said, to the astonishment of all around him. Here was a firefighter looking up at the bottom of a planter hanging over the edge of a balcony more than 100 ft above him, detecting a fire smoldering in wood mulch that two firefighters only 4 ft away from the fire could not detect! We had to guide them to the exact spot by watching them through the camera. Now, every ladder, squad and rescue company in the FDNY is issued a camera, and a newer, lighter model is being contemplated for issue to all 51 Battalions so they can observe the larger thermal picture while the companies continue their separate operations.

Hearing is the second sense that firefighters will probably use to determine when and where to open up. If firefighters crawl into a smoke-filled area and don't see any visible fire, the snap, crackle, and pop that accompanies combustion will often be the homing device, leading them to the right room or wall. A word of caution here: just because firefighters don't *see* flame doesn't mean that they don't have a problem. On the contrary, that they can hear the noise at all means that they have an open-burning fire. Call for a charged hoseline immediately. Poking just one hole into a ceiling or wall at that stage can allow fire to blossom out uncontrollably into the room.

Lacking any obvious outward signs of fire, such as crackling or open flame, and lacking a TIC, firefighters must often rely on their sense of touch. This is tricky, and it requires both caution and experience. The ignition temperatures of common combustibles are in the range of 400°F. Human skin begins to develop burns at a little more than 120°F. There is a large gap between the temperature required to produce open combustion of Class A materials and that required to injure someone's skin tissue. Indiscriminately grabbing objects in a fire area can produce very serious, painful burns. Firefighters performing overhaul operations should be very careful about grabbing or feeling objects or surfaces that they expect to be hot.

When considering whether to open a wall or surface, a good guideline to use, if no other indications are present, is the 15-sec, 2-min rule. Even if a TIC shows an indication of heat, use some discretion, since a TIC will show objects with less than 1°F of temperature difference to be *hot* to an inexperienced user. A hot water or steam pipe in a wall can cause that sort of difference. Feel for real heat if there is no urgent need to open up, no crackling noise, no smoke pushing, and no increase in temperature. The firefighter places his ungloved hand on the suspect surface. If warmth is present but there are no other signs, the firefighter might not want to open the area. Generally, if a firefighter can leave his bare hand on the surface for 15 to 20 sec without undue discomfort, there is no immediate need to open up. Just to be sure, though, that same firefighter should perform the same evaluation approximately 2 min later. If the wall feels the same, or if it has gotten warmer, then that area should be opened. But if the area has cooled to the touch, it probably won't have to be opened if no other problems are indicated.

Smell is the fourth and least used sense that firefighters can use in overhauling. Knowing that the overhauling stage produces the most carbon monoxide at a fire, and that all other toxic gases will have cooled and lost their buoyancy, the smarter firefighters will wear their SCBA well into the operation, which obviously hinders their sense of smell.

In any moderate or larger fire, the nose soon loses its ability to detect all but the most obnoxious aromas. After prolonged exposure even to these pungent odors, this ability is lost as well. By wearing SCBA, firefighters preserve their sense of smell in addition to their lungs. Well into the overhaul stage, after the building has been ventilated, this sense can be useful in distinguishing between steam and a smoldering ember.

Common sense is the final sense that firefighters should use in overhauling. Fire has a nasty habit of doing unusual things. It is the firefighter's sworn duty to protect property as well as life. By using knowledge gained through experience, firefighters learn to open things *that just don't look quite right,* thereby averting potentially disastrous rekindles. If firefighters must make a mistake, they should make it on the side of safety. It is better to keep a family up all night while firefighters search for hidden fire than it is to let them go sleep at night and have the neighbors find it in the morning—when it's blowing out both doors and three bedroom windows.

Opening up

If the knocked-down fire was a localized one (caught before flashover), say a mattress or a couch fire, opening up can be confined to the immediate vicinity of the object and the object itself. Learn to distinguish between smoke and soot stains and actual char. Objects that are charred will have to be doused with water and then opened up. A sharp knife helps in overhauling mattresses and stuffed furniture.

Items such as mattresses and couches should usually be removed to the outdoors for overhauling. This eliminates a large source of smoke from the fire area and safeguards the premises from reignition if the job isn't as thorough as it should be. It is very difficult to ensure complete extinguishment in mattresses and stuffed furniture. Fire burrows deep within and, often without even a wisp of smoke showing, can remain hidden for hours, only to break out later. It takes a lot of opening up and a good deal of water to make sure that the fire has been completely extinguished.

Three statements of caution when transporting these items outdoors for final mop-up include the following.

First, be extremely careful about how the job is done. The items should be wet thoroughly and all visible flame extinguished. If possible, roll the mattress or cushions into a bundle to reduce the surface areas exposed to fresh air. Make sure that the firefighters transporting the items wear full protective clothing, including SCBA. These materials frequently burst into flame as they reach the door or window and hit fresh air.

Second, never allow a firefighter to remove a partially extinguished item via the elevator. There aren't too many worse places to be than on an elevator when a mattress suddenly flares up.

Third, avoid throwing smoldering debris out windows into areas that are inaccessible or out of sight of members outside. They may be overlooked until units have left the scene and reignite later, exposing the building from the outside. Soak all charred debris to ensure extinguishment.

If the fire room has been subject to flashover, with a substantial amount of damage to the entire area, the firefighter can simplify the overhauling task by making the openings in logical sequence. Most interior finishes provide a reasonable degree of fire-retarding value. Plasterboard as well as plaster on wire lath and even wood lath will hold back a lot of fire if they are intact.

When seeking out the hidden fire just after a room is knocked down by the hose stream, the first openings should be made near man-made openings through the plasterboard or plaster, where the fire is most likely to have penetrated first. Open up around ceiling light fixtures, pipes that protrude, and electric outlets and switch plates. All of these are man-made holes, often covered cosmetically by a thin plate of wood, plastic, or sheet metal. If fire is found in these areas, the adjoining bays between joists must also be examined, because fire can easily follow the openings made in the joists for utilities. Firefighters must continue opening up successive bays until they find bays without evidence of charring. Additionally, each bay must be checked in both directions until it shows clean wood. As long as you keep finding charred wood, keep pulling. Be alert for intersecting vertical openings where fire can climb up and out of your view (Figs. 7–27, 7–28, and 7–29).

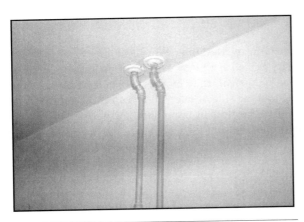

Fig. 7–27 Piping that penetrates a floor, wall, or ceiling provides a ready path for extension. This should be one of the first areas that you open up.

Fig. 7–28 Outlet and switch plates are other likely spots to find hidden fire.

Fig. 7–29 Fire had just penetrated this wood lath and plaster ceiling when fire crews extinguished it. The fire extended through the inch-thick plaster at the light fixture in the center of the photo.

When pulling ceilings, safe practices involve wearing full protective clothing and SCBA, making sure that the area below is clear of fellow firefighters and positioning yourself properly to do the work.

You should have a means of egress in mind when you make the openings. Plasterboard, in particular, has been known to come down in full 4 x 8 ft sheets, although nearly all materials can come down in one large piece given the right circumstances. If you suspect extreme damage to the ceiling construction, it is a good idea to make the first openings from the doorway. This way, if anything goes wrong, personnel won't be caught under a falling ceiling.

A particularly difficult type of ceiling to pull is the *tin* ceiling. Although this style of ceiling was originally installed as a fire retardant, it quickly proved to be more of a hindrance to firefighters than to fire. Consisting of embossed sheet metal nailed to the ceiling joists or other supports, it provides a great deal of resistance to pulling. The sheet metal also resists conventional hooks and pike poles when attempting to poke through it. The most effective method of opening such ceilings is to begin at a seam between the sheets or other openings, such as at a light fixture or around a pipe and then to pull back along the length of each sheet. The problem is that, with 3 ft of smoke between you and the ceiling, locating this seam can be nearly impossible. When visibility is poor, feel for the seams at the walls or locate any poke-throughs, such as pipes or light fixtures that may be expanded. The halligan hook has proved to be the most efficient tool for this purpose, with its uniquely shaped head allowing you to get a purchase at the edges or the seams. At top-floor fires, it is usually easier to push down tin ceilings from the roof than it is to pull them from below (Fig. 7–30).

A final consideration in overhauling is to know when to stop. There are certain cases when a fire officer should be willing to accept a rekindle. A fire that has damaged a building so severely that to enter it for overhaul would pose an undue risk of death would be a likely candidate for this, particularly if there are no exposure hazards. There is no point in injuring a firefighter by overhauling a building that will be knocked down as soon as a demolition contractor gets a check.

Fig. 7–30 Tin ceilings are difficult opponents. Use man-made openings around pipes or light fixtures to gain a purchase.

At one incident involving a vacant building piled to the roof with junk tires, an engine company was dispatched every 3 hrs to spend 10 or 15 min flooding the remains. This went on for about five days before enough of the structure was demolished to make entry for mop-up safe. Maintaining a watch line, a crew continuously at the scene, would have had little effect, since heavy equipment is often required. Additionally, it would have exposed the firefighters to the elements for no practical gain. Sometimes firefighters become so engrossed in the battle with the enemy that they lose sight of the fact that there is nothing left to save.

That brings up the subject of hydraulic overhauling, or using the hose stream to wash down entire areas. Although this is a tempting option, you must consider the damage that accompanies it. Runoff water is an obvious factor. Although some runoff water may be present due to actual fire control operations, excess water during overhaul is very unprofessional. In most situations, if a material is only smoldering, opening around it and using a little water will be far superior to not opening and trying to drown it out. Hydraulic overhauling may sometimes be required in occupied buildings where fire has charred abutting wooden structural members and it would be impractical to remove the pieces. This may be the case where fire has burned the roof joists and the ridgepole of a peaked-roof building. In this case, using a small-gallonage nozzle at a relatively high nozzle pressure can drive a stream into the cracks between the joists and the ridge or other crevices. Take advantage of hydraulic overhauling at fires in vacant buildings, where water damage is nonexistent and the dangers to firefighters are substantial. Not only is it safer for firefighters, it also conserves their strength for duties at the next call at an occupied building, which may only be minutes away. In addition, hydraulic overhauling can often speed up the return to service of the units at the scene.

Obviously, hydraulic overhauling isn't a customary duty of ladder personnel, but it is mentioned here because it is an option in overhaul. In addition, the discussion of water damage leads us right into the next topic, salvage.

Salvage

Salvage is a very broad topic that encompasses nearly every action taken on the fireground by both the engine and ladder personnel, reducing the damage from all perils—fire, smoke, water damage, extinguishment, and weather. Salvage begins at the moment of arrival, if it is made a conscious part of each firefighter's attitudes. The act of *try before you pry* prevents unnecessary damage to doors, while a nozzle man who shuts down the line when he doesn't see any fire reduces needless water damage. Sometimes firefighters must be educated to the need for salvage and the simplicity of some actions, not to mention the public relations benefits that accrue from performing a job neatly.

At times, firefighters view a heavily damaged room and its contents as only so much refuse. In fact, many items within that pile of debris may be useful, even critical, to the owner regardless of the degree of damage. It isn't a satisfactory approach merely to toss all of the contents of a given room into the yard during overhaul. Each member must be instructed to view the building as if it were his or her own home and to consider himself uninsured for the loss. What is of value and what isn't may seem quite different when viewed in this light. Particularly in a poor neighborhood, a firefighter's judgment may not be as discerning as that of the owner. The firefighter may feel that an item isn't worth saving, but to a person whose sole possessions are now little more than the clothes on his back, even scorched items may have great potential value. Personal effects, family photos, documents, and many other things may be irreplaceable. Some simple acts can go a long way toward reducing unnecessary damage.

When time permits, before pulling ceilings, protect as much material as possible by covering it. In a bedroom, throw the blankets back and pile items such as stereos, personal computers, and TVs on the bed, then pull the blankets back over them. This will protect such items against most of the plastic dust that will fill the room once overhaul commences. Another fast method is to open the top drawer of any bureaus and dressers, sweep in the contents from the top, and then close the drawers again. This prevents both plaster damage as well as water damage unless the area is deluged. Of course, salvage covers are extremely useful, and the techniques of one-man and two-man throws are valuable, but they require time and often extra manpower, neither one of which may be available. A shower curtain pulled from a bathroom, though, provides a fast, easy way to protect sensitive items from water damage, at least until the covers arrive.

Another useful tactic is to open a small hole in a ceiling, thereby allowing accumulated water to run off—water that might otherwise soak in and cause the entire ceiling to fall later. If possible, when making openings for examination, make them where they will be the least damaging. For example, if you suspect fire to be in a partition between two rooms, one side may be an open wall while the other might be full of kitchen cabinets. Take the time, if you can, to see the other side before you start pulling the cabinets off of the wall, even if it means going into the adjoining apartment. Not only will this reduce the overall damage, it will also make your work easier and faster.

Even taking up hoselines can have an effect on damage. Take couplings to the outside before breaking them apart and draining the line. In a high-rise building with a standpipe, you might be able to drain the lines into a sink or tub before uncoupling them. If not, the stairwell is usually a better place for the water as opposed to letting it trickle down through six floors' worth of occupied areas. Before leaving, turn the premises over to the owner or his representative. Make the responsible person aware of such items as broken windows and holes in the roof so that he can arrange for the needed repairs. If left uncovered, such items may lead to more serious damage from the weather. It is an excellent idea to present the owner with a copy of *After the Fire! Returning to Normal*. This is a 14-page booklet distributed by the U.S. Fire Administration, developed with the assistance of the Hollywood (FL) Fire Department. It contains many excellent tips on salvage and other actions to be taken and includes a brief explanation as to why it seems the fire department *created* damage. Also mention to the owner any actions that the department undertook to control the utilities, either to prevent or reduce damage. Be certain to mention whether a sprinkler or other vital system is still shut down.

Control of Utilities

The shutting down of sprinkler systems and other utilities is often assigned to ladder personnel—and logically so. These controls are often in locked rooms, requiring the use of forcible entry tools. In addition, members of the ladder crew are more likely to be equipped with portable radios than are firefighters riding on the engine. Portable radios speed up the shutdown process, especially when you encounter multiple controls. A member at the scene of the problem can radio the member at the controls, letting him or her know when the right switch has been thrown or the right valve closed. More important in the case of a sprinkler system shutdown, however, is to have the member remain at the control position in radio contact, standing by to open the valve again should control of the fire be lost or fire be discovered in remote areas.

Proper procedures include sending two members to shut down the utilities, since this operation often involves forcing entry into a locked area. A second person should be sent even if forcible entry isn't required. Shutting down utilities often means sending members to areas that are remote from the main activity. A member working alone is vulnerable to everything from electric shock to getting mugged. A second member, even if unable to assist directly, can at least call for help.

The members sent to cellars to shut down utilities must be equipped with masks, even if the fire is on the first or second floor. I have seen members emerge from cellars nearly overcome by carbon monoxide from a fire on the top floor and cockloft of a two-story commercial building. They didn't realize the seriousness of their condition until it was almost too late. They had been operating in a cellar that had only a light haze of smoke, so they didn't bother with their masks. Others have died in similar situations before realizing that they needed their masks.

When performing the actual tasks of shutting down utilities, members should have at least a basic familiarity with the mechanics of the operation. If possible, send someone with the relevant expertise. Many firefighters have backgrounds in plumbing and electrical work. Use this experience. A person familiar with piping can usually tell the difference rather easily between pipes for gas, water, steam, sprinklers, waste, and the like. This not only speeds control by being able to trace the lines back to the source, it may also at times prevent problems from developing.

Electricity can pose a severe danger to firefighters on the fireground and may require utility personnel to handle safely. Although linemen's gloves, sleeves, and mats are all useful tools for lifesaving purposes, you shouldn't rely upon them. Always assume all wires to be live, high-voltage conductors. Immediately request the response of the utility company. The fire department can use some routine disconnects to remove power prior to their arrival, but if you are unsure, wait for assistance. Of course, some situations can't wait, and immediate action is required.

My company once responded to a serious fire in the rear of a six-story apartment house. On our arrival, fire had possession of the entire rear of the building, top to bottom, but it was being held out of the front of the building by a favorable wind, a fairly fire-resistant wall, and three aggressive engine companies. Members attempting to reach the upper floors found the interior stairs untenable above the third floor and so resorted to the front fire escape to gain access. When the members climbed onto the fire escape, they felt a slight tingling sensation as though it was charged with electricity. They reported this to the IC, who directed the next incoming unit (mine) to eliminate the electric charge. We could see a thin wire stretched from a building across the street to the fire building, anchored off at the second-floor fire escape balcony. Two members of our team were dispatched to kill all of the juice in the building, while the rest of us made our way to the top floor to pull ceilings. Again, as each member scrambled out onto the fire escape, we felt the tingling, but we didn't consider it a severe problem. Since the power to the building was by now turned off, it seemed obvious that the line secured to the fire escape was the source.

The IC chose to wait for the utility company since, in his opinion, it only appeared to be a telephone line. Entire companies had ascended the fire escape to the upper floors without incident when the unexpected occurred. As I was pulling ceilings on the top floor, I heard a savage scream that didn't let up. Looking out the front window, I saw a member frozen in place between the fire escape and the basket of a tower ladder. The member had been about to step out of the basket when he grabbed the fire escape for support. The current had paralyzed his muscles where he stood, rendering him unable to break its grasp. A quick-thinking member inside the building was able to knock his arm loose, causing the paralyzed member to fall free. Fortunately, he fell backward into the basket rather than forward to the ground.

It seems that, while the members stood on the fire escape, they were part of the circuit and didn't offer any path to ground, similar to birds on a high-tension wire. When climbing off or onto the fire escape, the members were on wooden windowsills, offering a poor path to ground, hence they only noted tingling. When the member touched both the fire escape and the rail of the basket, however, his body offered a connection to an almost perfect ground, down through the boom and its stabilizers. The current was coming through a line illegally hooked up to the building across the street to supply squatters. Their length of wire had shorted to ground on the fire escape, delivering 220 volts to anyone completing the path. The fire escape was ordered evacuated until properly protected personnel could cut the line.

Not all electricity can be handled the same way. A very capable fire officer described to me how he and his entire crew were almost killed one night at what seemed to be a relatively routine fire. The company arrived to find a smoky fire involving an electrical transformer in the basement of a large hotel. After ensuring that the offending unit contained no polychlorinated biphenyls (PCBs,) they entered the area and attempted extinguishment with dry chemical extinguishers, but the fire kept lighting up again, fed by a live line coming off a bus bar on the wall. A large knife switch was evident on the wall, and the members felt that if they pulled it with a dry wooden pike pole, they could stop the flow of current to the transformer and thereby allow extinguishment. The officer relayed this plan to the chief, who rejected it in favor of waiting for the utility company to arrive. All of the personnel were withdrawn to the stairwell.

On their arrival, the utility company was apprised of the plan, and their faces immediately went ashen. They told the members that, had they pulled the knife switch under load, a tremendous arc would have incinerated the entire room and its occupants. The knife switch was only a safety device to ensure that power wasn't accidentally turned on while workers were operating on the lines. To open the knife switch safely, the load inside the building would first have to have been shut down in smaller portions. The moral: if you're not dead certain about what you're doing, don't do it or you may wind up dead. There are many times when the situation is beyond the control of the fire department, and good intentions don't make up for improper knowledge or equipment.

08 Forcible Entry

Not long ago, people in many parts of this country left their homes unlocked. Shops were secured by simple night latches, and few suburban or rural firefighters ever had any difficulties with forcible entry. That was something that only the big-city firefighters worried about, and even there it was often only in the high-crime ghettos. This began to change in the 1950s, as Americans became more security-conscious. New locking devices began to appear, offering more resistance to forcible entry, requiring firefighters to develop new forcible entry tools and procedures. Firefighters who once relied on a springy claw tool now depend on a stout halligan and a flathead axe or sledgehammer. In addition to such brute-force methods, progressive firefighters should have other means of gaining access to their objectives, since no one method is likely to resolve all situations. Alternatives include the through-the-lock method, hydraulic equipment, and cutting devices such as torches and saws. The better fire crews are those who can recognize the most efficient method and put it to work right away. To search, rescue, and extinguish, you must first be able to get into the structure (Fig. 8–1).

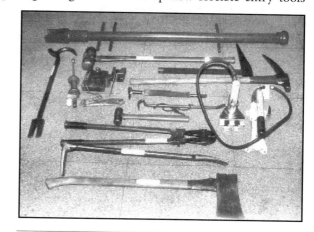

Fig. 8–1 A wide variety of forcible entry tools is required to ensure success under all conditions.

When you are confronted with a situation requiring forcible entry, take a moment to size up that situation, then choose the method that is the most appropriate. Among the deciding factors are the need for speed, the type of door and lock assemblies, the tools and manpower available, and the degree of damage that entry will create. The need for speed should be the overriding concern, whether to save a life or cut off fire that is rapidly extending. In certain cases, however, the type of door or the lock assembly will be the deciding factor, since certain types are best opened using specific methods. In some cases, there is only one way to get past a specific door. In almost all cases, the degree of danger posed by forcible entry should be consistent with the danger and damage from the hazard, usually fire. It is a very poor public relations effort to destroy a $1,000 door to shut down a faulty oil burner. Conversely, don't waste the time and effort of going through the lock when the fire is burning away the top half of the door as you work! The last factor, and the one that should be of the least concern in most cases, is the availability of personnel and resources. Two people should be able to force the vast majority of doors using only a halligan-type tool plus a flathead axe or sledgehammer, especially by employing a few select techniques.

Regardless of the type of door you encounter or the method you choose, a few basic rules should always apply.

Rule #1

Try before you pry. It is inexcusable to have a crew go through the entire forcible entry process only to find out that the door was unlocked all the time. Forget about the embarrassment that this causes—think of the time it wastes. Imagine someone lying there on the inside needing your assistance while you fool with the door! In one case, the failure of a crew to check the status of a lock hampered an arson investigation. The members had smashed in an adjoining window without bothering to check the door first. When the arson investigators arrived to investigate this obviously suspicious fire, several owners were on the scene who hadn't been there prior to the arrival of the fire department. One of them simply went over and opened the door without being challenged, thus setting the stage for a claim that the fire could have been started by anyone. Unfortunately, the inability of the firefighters to testify to the status of the door at the time of their arrival didn't strengthen the prosecution's case.

Rule #2

Don't ignore the obvious. Look for the easiest way to get in. I once followed a ladder truck onto the scene of a working house fire in a suburban area. Fire was coming out of five windows on the Exposure 2 and 3 sides of the second floor. On their arrival, an officer and a forcible entry man were met at the front by a chief who led them to a back door and told them to open it up. These three men, equipped with a halligan and an axe, were unable to gain entry for 5 min. They were finally let in by a member who had come in through the front door. The saddest part of this episode is that nearly the entire top half of the wooden door was made up of 8 x 10-in. glass panes. It would have been a very simple act to have broken one of these panes and reached in to work the lock. The members had been so wrapped up in the thought of forcible entry that they had put their blinders on. Their tunnel vision had been focused solely on forcing the door with the method that was most familiar to them (Figs. 8–2 and 8–3).

Fig. 8–2 Don't ignore the obvious!

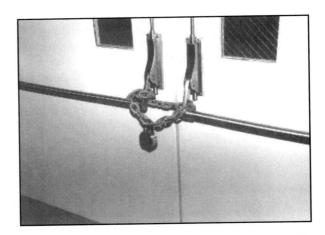

Fig. 8–3 Panic hardware chained together on the inside can prevent firefighters from escaping if they have entered by another means.

Rule #3

If possible, use the door that the occupants normally use to enter or exit the premises. Not only does this put you into their most likely exit paths, it also makes your job much easier. Firemen in ghetto areas have used this rule for years. It originated in the old-law tenements of New York City, which were commonly built with a railroad-flat layout of rooms. Each apartment had two doors to the interior public stairway. The door closest to the head of the stairway was usually the door most often used, generally opening into the kitchen. The second door opened into a front room, either the living room or a bedroom. Since most people didn't want guests entering directly into the bedroom, this door often wasn't used. With the creeping crime problem, these doors were often nailed closed or barricaded from behind. Attempting entry through them is a very draining and time-consuming task. The same principle applies to all other buildings as well today as it does to the old-law tenement. If we can find the door that the occupant normally uses, chances are that it will be less heavily secured than other doors, and we can duplicate the actions of the occupant.

Take, for example, a school or place of assembly having 8, 10, or more sets of exit doors. These doors must remain unlocked while the building is occupied, but at night and over the weekend, the panic hardware is often chained and padlocked from the inside. This creates an extremely difficult entry problem. The staff can rig every door in the building that way except one: the one they'll first reenter on Monday morning to open up. If we know in advance which door that is, our task will be greatly simplified.

Rule #4

Maintain the integrity of the door (Fig. 8–4). In other words, keep the door intact. One of the least desirable methods of forcible entry is just to walk up and smash open a plate-glass door or window, even if it is obvious that there is a serious fire within. By breaking open the glass, you have removed any control you had over the oxygen supply to the fire. In serious cases, this sudden inrush of fresh air can set off a backdraft explosion. In somewhat less serious cases, the fresh air may intensify a smaller fire. Even if a charged hoseline is in place, breaking through the door isn't the best idea, for if the line loses water (perhaps from being cut by shards of glass), the door can't be closed again to cut down the oxygen supply. At interior doors that expose stairways or other vertical openings, it is critically important to leave the door to the fire area intact, since the closed door may be all that keeps fire from blowing up the stairway being used by escaping occupants or firefighters. The members at the door to the fire area are fully responsible for the safety of everyone on the stairs above the fire. They hold the lives of those people in their hands. If for any reason the hoseline cannot contain the fire within its original area, this team *must* be able to close the door to prevent fire from blowing up the stairwell, exposing people above.

Fig. 8–4 Maintain the integrity of the door!

Forcible Entry Size-up

The forcible entry size-up begins with the alarm information, as does our fire size-up. The time of day often indicates the need for forcible entry. A business that is open to the public usually requires less forcible entry during working hours than it would after closing. In addition, the type of occupancy can also tell you which way the doors open. About 99% of residential doors open inward, whereas the opposite is true of commercial establishments and places of assembly. Knowing this can help you review the method of attacking the door, which end of the tool to use, or for that matter, *what* tool to use. Although the flathead axe is a great complement to the halligan tool in most buildings, it does have some drawbacks. Its 6-lb head doesn't give it a great deal of driving force when it is used as a hammer. When going into Class I (fire-resistive) buildings, the axe can be replaced by an 8- or 10-lb sledgehammer. There is no need to chop wood in these buildings, and the added impact of the sledge is very useful on substantial steel doors or a cinder block wall.

The forcible entry size-up should continue as soon as the building comes into view. Determine the location of the fire, any visible victims, and the door to be used for entry (usually the front door). While approaching the door, look around. Does there appear to be fire in close proximity to the door? If so, prepare for the worst. What type of door is it? Wood, glass, aluminum, or steel-clad? Wood gives, making the task easier. What is the jamb set in? Brick, wood, or some other material?

My first assignment as a New York City firefighter was in an area with several large housing developments. The walls between the public hall and the apartments were made of two layers of ⅝ in. plasterboard on metal studs. Although all of the doors were steel clad, set in metal jambs, it was sometimes possible to chop a hole in the plasterboard and reach in to open the door. At times, this proved faster than trying conventional forcible entry. At one fire I worked there this information was crucial. Fire had blocked the entrance door to one apartment where three young girls were sleeping in a back bedroom. After forcing the door to the fire apartment and finding his path blocked, Lieutenant Tom Dunphy, aware of the construction involved, led his forcible entry team into the next apartment over and chopped a hole through the plasterboard wall, removing the three children safely before a hoseline was operational from the standpipe (Figs. 8–5 and 8–6).

Fig. 8–5 A pattern of bolt heads coming through a door often indicates...

Fig. 8–6 ...the presence of drop-in bars, like those in the old-time forts.

Look at the locks. How many are there and where are they? This will tell you where to place your tool. In some cases, seeing several substantial locks may tell you to find an easier way in. Look for the presence of hinges or stop molding. If you see hinges, the door opens out toward you. If no hinges are visible, then the stop molding will be present or the frame will be rabbitted. Be aware of how the door opens and what that means. For instance, if you are operating in the smoke-filled public hallway on the fourth floor of an apartment building, and you encounter a door that opens toward you, expect trouble. Normally, apartment doors open *inward*. Doors that open out to the hall include janitors' closets and elevators. If you encounter an outward-opening door under these conditions, be sure that you aren't about to crawl headfirst into an open elevator shaft!

Doing the visual size-up of the door takes only a matter of seconds, and you can usually do it as you approach the door. Once you actually reach the door, remember the first rule; *try before you pry.* Before grabbing the doorknob and pushing the door open, however, make sure that everyone is ready to proceed. There can be a lot of fire behind a door with very little visible evidence on the outside. Make sure that everyone is out of the way, off to one side of the door and at least crouching, if not kneeling.

Keep your glove on and bend your hand back while tugging on your coat sleeve to expose a little of your wrist. Using this bared skin, feel the doorknob and then the top of the door. By using this method, you gain a lot more than feeling the same areas with your bare hand. The first is speed—you don't have to stop and put your glove back on to proceed. Second, you are exposing a nonessential part of your body. Burns to the hand are difficult to heal, since hand motions greatly hamper proper healing. The wrist is far less mobile. The doorknob may be a better place to check for heat, since it usually has a direct metallic, conductive connection from the fire side to the outer side. If the doorknob is hot, then it is time to attack. *Now* grab the knob and try the lock. If it's locked, give the door a shake to determine where the locking devices are and which are actually locked.

You now have an indication as to where to begin your work. You should also have an idea as to how strong the resistance will be. Most normal locking devices have a key cylinder located about 1–3 in. from the edge of the door. If you notice any cylinders located farther than this from the edge, it indicates that the owner has made some serious attempts at security. You may want to use a different method of attacking the door (through the lock, power tools, or hydraulic tools), or you may want to look for an easier way in. A pattern of bolt heads coming through a door indicates the presence of a drop-in bar, similar to those once used by the cavalry to lock the front gates of forts. If none of these indications are present, the task of forcible entry should be relatively routine.

Conventional Forcible Entry

Conventional forcible entry is often called *the brute-force method*, since it relies on applying great force either to bend or snap part of the door or lock assembly. As such, the person applying the force must know how to get the most out of a lever and fulcrum. The longer the lever, the greater the force that can be applied. Of course, too long a tool will be very awkward in many of the confined areas that call for such operations.

By far, the most efficient combination of manual forcible entry tools is the flathead axe paired with the halligan tool. The halligan provides four different means of getting a bite on the work. Additionally, it is invaluable for opening floors, walls, and roofs during overhaul. Since the development of power saws, the flathead axe is used primarily as a hammer to drive the halligan into spaces where it can get its bite. Using a halligan and axe (often called a set of irons) is therefore a two-person job, one to position and hold the halligan, the other to swing the axe. The more experienced person should be the one on the halligan, placing the tool in position, being in overall control of both tools, and directing the other person when to swing. When operating this way, make every attempt to have the two people work from opposite sides of the tool. This way, if the axe misses its objective, it won't carry through to strike the irons man. The actual mechanics of forcing the door depend on whether the door opens in or out (Figs. 8–7 and 8–8).

Fig. 8–7 While one member carries the irons, using them is a two-man job. The other member of the forcible entry team should bring a 6-ft hook, thus ensuring the basic triad of tools is present at every structure.

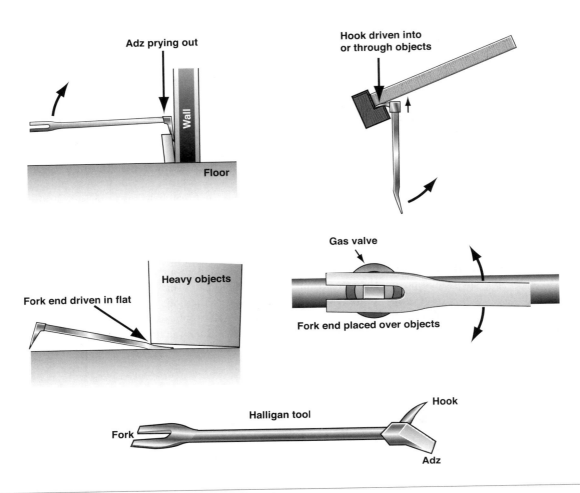

Fig. 8–8 The halligan tool is a tremendously versatile tool.

Inward-opening doors

To breach an inward-opening door, place the fork end of the forcible entry tool about 6 in. either above or below the lock, depending on whether the members are standing or kneeling. If there is more than one locking device, insert the tool midway between them. This prevents the tool from striking the bolt of the lock as it is being driven in, which will stop penetration. When using tools with a pronounced bevel on the fork, position the fork so that the outer curve is toward the door and the inner curve is toward the jamb. This helps ensure that the tool isn't driven into the door, and it provides greater leverage when it is time to pop the door. In addition, positioning it in the reverse manner tends to let the tool back out when force is applied (Fig. 8–9).

Initially, place the tool nearly parallel to the door at about a 15–20° angle from the door. This allows the tool to be driven in between the door and the molding or rabbitting. The axe man must be sure never to swing without specifically being told to do so, since the tool will have to be continually moved outward as it is driven in further and further, until it approaches a 90° attitude toward the door. This is necessary to avoid driving the tool into the jamb. Since it is the person on the halligan who decides when to move the tool, there is a distinct possibility of injury if the axe man decides to swing on his own.

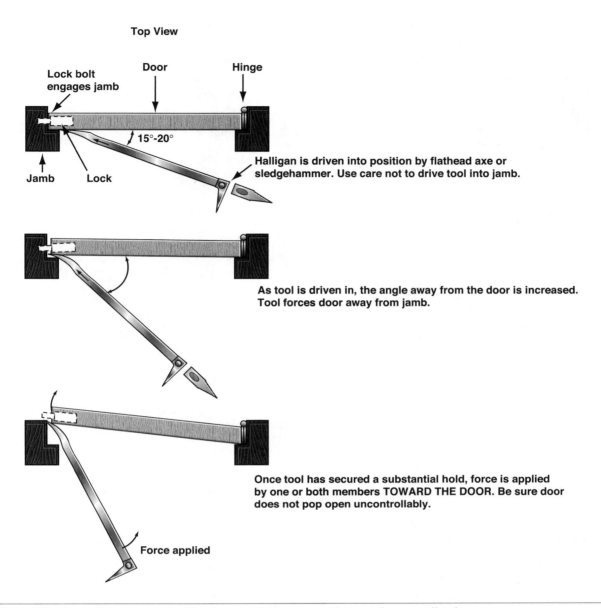

Top View

Lock bolt
engages jamb

Door

Hinge

15°-20°

Jamb   Lock

Halligan is driven into position by flathead axe or
sledgehammer. Use care not to drive tool into jamb.

As tool is driven in, the angle away from the door is increased.
Tool forces door away from jamb.

Once tool has secured a substantial hold, force is applied
by one or both members TOWARD THE DOOR. Be sure door
does not pop open uncontrollably.

Force applied

Fig. 8–9 Forcing an inward opening door is a multistep process that requires coordination.

As the tool is forced inward, the wedge shape of the fork will spread the door and jamb. Eventually the space will be wide enough to allow the lock bolt to clear the jamb or for the frame of the rim lock to tear through. In either case, this doesn't solve the problem. The task remains of pushing the door open. This will usually be possible when the fork end has been driven in 1 or 2 in. past the end of the door. Simply begin pushing the tool back toward the door. This makes the doorjamb act as the fulcrum and the halligan as a lever. In stubborn cases, both members may be required to push on the tool. Remember; use the far end of the tool to exert the most force. Before applying the force, make sure that someone has control of the door so that it doesn't pop open uncontrollably. You can do this by using a hose strap or a short rope, or simply by holding the doorknob. Otherwise, any fire behind the door might blow out and engulf the members (Figs. 8–10 and 8–11).

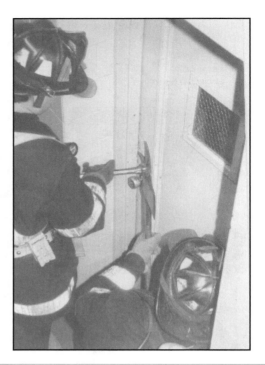

Fig. 8–10 Forcing an inward opening door.
Note: The members are on opposite sides of the
halligan tool, so that if the man swinging the axe
misses in the darkness, his follow-through does
not strike the member holding the halligan.

Fig. 8–11 An alternative method of forcing an inward-
opening door is to drive the hook or claw end deep
into the jamb as a fulcrum. Pushing the fork end
toward the door will force the door open.

Another method that is sometimes useful is to put the adze end of the tool between the door and jamb. Then, push the forked end toward the door. This might be useful if you lack a driving tool, but it isn't always successful, since the adze often slips out from the jamb. By a similar method, drive the claw end into the jamb with an axe until the claw has a substantial bite, then force the forked end toward the door.

Another alternative method, *when other methods have failed,* involves attacking the hinge side of the door. Use the same procedures described previously about 6 in. above or below the estimated position of the hinges. This will pull the screws out of the door or hinge, allowing you to push the door inward. This should be tried only as a last resort, since it violates Rule #4, to maintain the integrity of the door. If anything goes wrong, such as a delay in getting water, there is no way to close the door again to control the situation. For the same reason, it is a poor idea to knock out the center part of a door and crawl through, leaving the frame intact. Note that, if you *are* forced to breach the hinges, you should start with the top one first, since heat, smoke, and fire will subsequently be venting out. It is better to crouch below this and work on the bottom hinge than it is to stand in the line of fire from below.

Outward-opening doors

The vast majority of commercial doors open outward, since exit codes require that doors open in the direction of travel. The principles used to force them are much the same as for inward-opening doors. Spread the door away from the jamb, then apply a force in the direction of travel. The method used may vary, however. Since these doors are often found inside recesses in walls, which would inhibit movement of the iron if the fork were used, the recommended method is to use the adze end of the tool (Fig. 8–12).

Outward-opening doors recessed in walls do not permit fork end to exert any outward leverage. Before any force is applied to door, the halligan strikes the wall. Moving tool toward door cannot open it; it opens out!

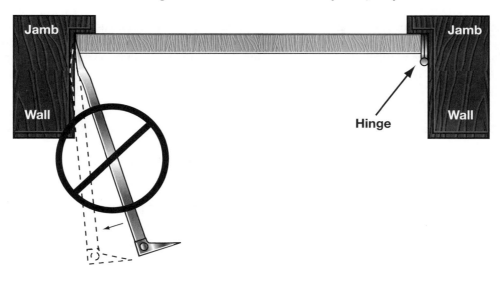

Fig. 8–12 The fork end should not be used to force an outward-opening door that is located in a recess.

Using the correct method, with the *adze* driven in between the door and jamb on the lock side, the tool projects back across the door itself. By pulling out on the end of the tool, you apply force to pull the door out. Position the adze the same as for the fork on the inward-opening door, 6 in. above or below the lock. Since the adze end can't be driven in more than the thickness of the door (it hits the rabbitted jamb), the wedge effect isn't as great as when using the forked end. This may be readily overcome by prying up or down on the opposite end of the tool, which will spread the door away from the jamb. Then the end of the tool may be pulled out, thereby opening the door.

If you encounter difficulty with the lock side of an outward-opening door, the hinge side may sometimes be forced. Be aware that you will violate the integrity of the door and won't be able to close it again (Figs. 8–13 and 8–14).

Fig. 8–13 When forcing an outward-opening door, drive the adze in between the door and jamb, with the fork end extending toward the hinges. Pull out on the fork end to open the door.

Fig. 8–14 You may be able to force the hinge side, but the door will no longer be closable.

Through-the-lock Forcible Entry

Through-the-lock forcible entry is relatively new to many firefighters who aren't aware of, or convinced of, its capabilities. Since the appearance of certain high-security locks is a new phenomenon in many areas, some fire departments just haven't had sufficient exposure to this method. Pioneered in the 1960s by Captain George Sunilla of FDNY (probably after some particularly difficult job of forcible entry), the method centers around duplicating the action of a key in the lock. Although many locks are sophisticated devices with complex arrangements, a lock's key cylinders are very simple mechanical devices. If we can get a cylinder out of the way, we can duplicate the action of the cylinder and open the lock just as though we had a key—and relock it as well, if desired.

Through-the-lock entry often provides the fastest, least damaging, and sometimes the *only* way in. It is by no means a cure-all, but it is an absolute must in any modern fire department's arsenal. You should follow the same basic steps of door size-up. Certain situations indicate the use of this method, while other situations will point you toward conventional forcible entry or some other means. Through-the-lock entry is usually indicated by a light fire condition, specific types of doors, or specific types of locks. Of course, the personnel using the tool must be familiar with the entire procedure for the various types of locks that they encounter (there are only three major types).

Those who remain unconvinced of the effectiveness of this method are those who have either seen it fail or take an unusually long time to work. This is almost always directly traceable to the members' unfamiliarity with the procedures to be followed. They just didn't know what to do (thus its other nickname, *thinking-man's forcible entry*). The same firefighters who hesitate to use through-the-lock on buildings wouldn't think twice about using it more than any other method to get into the trunk of a burning automobile. By punching the lock cylinder out of the way and inserting a screwdriver, they are doing the same thing (Fig. 8–15).

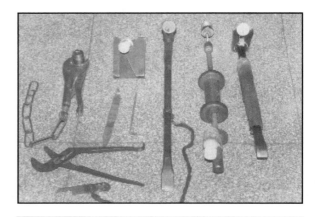

Fig. 8–15 A variety of lock pullers and key tools are shown with the lock cylinders as they would be applied.

A variety of tools may be used in through-the-lock forcible entry, with varying degrees of success. Captain Sunilla used a commercially available nail puller, but other devices, including slide hammers, the *K* tool, and even the halligan tool, can be used. By far, the most efficient way to remove lock cylinders is to use the *K* tool, which was specifically designed for the task. To use this tool, position the jaws on the cylinder, then tap them firmly into place with an axe. The orientation of the tool isn't critical and may be adjusted to fit the situation. What is important is that when the halligan is inserted into the stirrup on the back of the tool, the force being applied continues in the same direction as when the tool was being tapped on. The jaws of the *K* tool are tapered into a narrowing channel so that if force is applied in the same direction, say from the top down, then the cylinder will be pulled deeper into the notch. On the other hand, if the tool is applied from the top and the force is upward, then the tool will likely slip off the cylinder. Other devices are effective as long as they take a bite on or behind the cylinder and provide leverage to force the cylinder out of the lock (Figs. 8–16, 8–17, and 8–18).

Fig. 8–16 To pull a cylinder with the K tool, place the K tool over the cylinder...

Fig. 8–17 ...then tap it securely onto the cylinder with an axe...

Fig. 8–18 ...then insert the adze of a halligan and apply a downward force to pull out the cylinder (if the K tool was placed on from the top).

The real heart of the operation begins once the cylinder has been removed. The first step is to pick up the cylinder and examine it. This will indicate which type of lock you are facing, which tool to use next, and even where to use it. This is obviously *key* information! The next tools to be used are known as key tools. They perform simple functions, duplicating the action of the back of the lock cylinder. When you turn a key in a cylinder, it makes an attachment on the back of the cylinder turn as well. This simple rotational action is what operates virtually all locks. With the cylinder out of the way, we can replace the cylinder with a similarly shaped piece of metal and turn it, thereby performing exactly the same action.

The simplest type to conquer is the rim lock, which is mounted on the edge or rim of the door. This lock functions almost exactly like the one on the trunk of a car. On examination, you will notice a flat shaft, similar to a screwdriver blade, sticking out the rear of the cylinder. You can insert the flat key tool or a screwdriver through the cylinder hole into the slot in the lock, then rotate it to operate the lock. The latest generations of rim locks have been designed with an added security feature designed to defeat this method. A spring-loaded plate resembling a guillotine is built in so that, as soon as the cylinder is removed, the plate slides over the slot. This presents no real problem to forcible entry, however. Simply insert the claw of the halligan tool into the cylinder hole, strike it with an axe, and knock the lock right off the door (Figs. 8–19, 8–20, 8–21, and 8–22).

Fig. 8–19 Key cylinders come in the following three styles: square shaft, flat shaft (screwdriver), and cam.

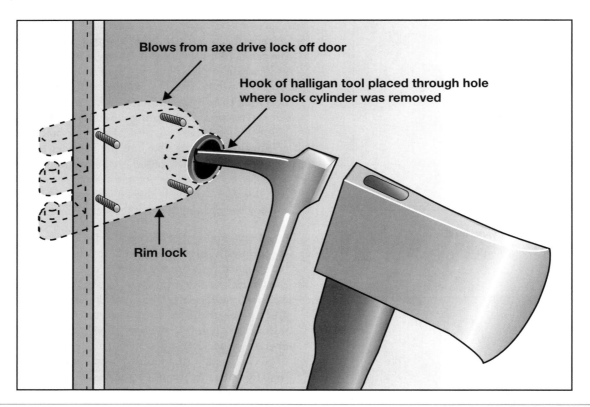

Fig. 8–20 An alternative approach when speed is essential is to drive the hook of the halligan tool through the cylinder opening.

Fig. 8–21 This should knock the rim lock right off the door.

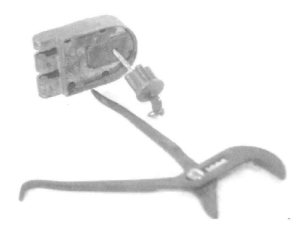

Fig. 8–22 A rim lock and cylinder with a key tool. A flat screwdriver-type blade will reproduce the action of the cylinder.

On occasion, usually in apartments in high-crime areas, you will find a square shaft instead of a flat shaft when you pull the cylinder. If you do, you have encountered a police lock. Do two things immediately. First, thank your lucky stars that you decided to go through the lock. Second, place the claw of the halligan through the cylinder hole and whack it right off the door.

A police lock is an extremely effective lock found only on inward-opening doors. It consists of a boxlike mechanism mounted permanently to the edge of the door by four wood screws. A removable ½-in.-thick steel bar extends down to a recess in the floor, bracing the door in the locked position. I have seen people work for more than 5 min trying to force it by using the conventional method. Whatever force is applied to the door is merely transmitted to the floor via the steel rod. Removing the cylinder and striking the box mechanism itself allows you to apply the force to the four wood screws rather than to the bar. Using this method, the lock should be out of the way in less than 1 min, and you'll be less fatigued. The previous method assumes that a serious fire is present or entry must be gained regardless of damage. If conditions require that less damage be done, the rotating action of the cylinder can be duplicated by using a $^5/_{16}$ in. sq piece of stock tapered to a point within about ½ in. In this case, the lock may still have to be replaced, since pulling the cylinder sometimes damages the box. However, it is a fairly inexpensive lock (Figs. 8–23, 8–24, 8–25, 8–26, and 8–27).

Fig. 8–23 Police locks present a severe challenge to conventional forcible entry methods. The police lock is only found on inward-opening doors. A steel rod braces the door closed.

Fig. 8–24 The rod is held in the closed position inside a steel box screwed to the door.

Fig. 8–26 The door can then be opened, as the bar rides up in the channel.

Fig. 8–25 When you operate the mechanism with a key or a key tool (square shaft), the bar slides over, in line with the opening.

Fig. 8–27 The mechanism and cylinder of a police lock with a key tool. A square shaft is needed to duplicate the action of the cylinder.

Mortise locks

The next major family of locks is the mortise locks, so named because they are recessed (mortised) inside the door itself. Firefighters seem to have the most difficulty with these locks because of unfamiliarity. The basic action of the cylinder is the same—rotational. When you pick up the cylinder, you will note a little pear-shaped cam nearly flush with the back of the cylinder. When it rotates, this cam activates the lock mechanism. This cam has a very limited area where it can perform work—about 180° of its path. That narrows down the area that the firefighter has to look to find the objective. In this case, there is no slot in which to insert a key tool. You must use a key tool with a 90° bend to locate the spring-loaded catch, push it down, and move it into the open position.

The design of these cylinders results in their greatest force being in the straight-down position, where most of the work will be done. Imagine that a given cylinder has the face of a clock painted on it, and that the 6 o'clock position is at the bottom where the key goes into the keyway. The place where the cylinder does its work will be between the 5 o'clock position and the 7 o'clock position. That is where the spring-loaded roller inside the lock is—there is no point in looking anywhere else or using any other type of key tool. The only way that lock is going to open is by working at either of those two positions. It is important to keep in mind that the 5 and 7 o'clock positions are in reference to the keyway at 6 o'clock. On some doors, the locks are mounted sideways at the top and bottom of the door, meaning that the keyway will either be at the 3 or the 9 o'clock position. The catch in the first instance will go from 2 o'clock to 4 o'clock or from 4 to 2; in the second instance, the catch will move from 8 o'clock to 10 o'clock or from 10 to 8 (Figs. 8–28, 8–29, 8–30, 8–31, and 8–32).

A situation that would indicate the use of through-the-lock forcible entry would be when you are investigating what is assumed to be an emergency of a very minor nature, such as a food-on-the-stove or other small fire. The damage done by going through the lock is usually easily repairable, often without the services of a locksmith.

Fig. 8–28 A mortise lock is recessed (mortised) inside the door.

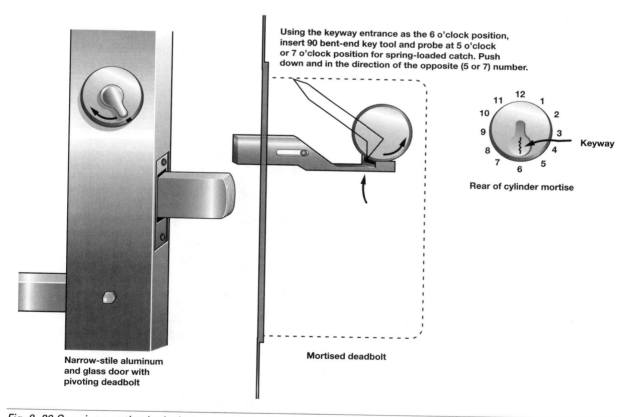

Using the keyway entrance as the 6 o'clock position, insert 90 bent-end key tool and probe at 5 o'clock or 7 o'clock position for spring-loaded catch. Push down and in the direction of the opposite (5 or 7) number.

Keyway

Rear of cylinder mortise

Narrow-stile aluminum and glass door with pivoting deadbolt

Mortised deadbolt

Fig. 8–29 Opening mortise locks in a narrow stile aluminum and glass door with a pivoting deadbolt.

Fig. 8–30 A mortise lock with its cylinder and a key tool. Note the cam at the 4 o'clock position within the lock mechanism.

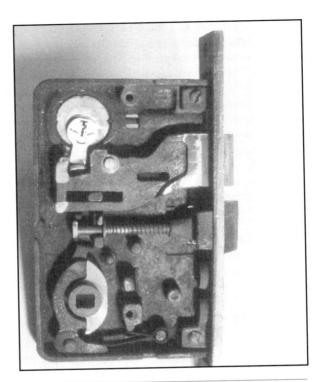

Fig. 8–31 As the cam is rotated down to the 6 o'clock position, the deadbolt begins to withdraw.

The presence of police locks (usually not detectable in advance), fox locks (plainly visible), or pivoting deadbolts (plainly visible) also indicate this method.

Most of the storefronts erected since the 1960s have aluminum framing for the plate-glass windows and doors. The doors are called narrow-stile doors (a *stile* is the frame of a door, not a style or type) and are almost always equipped with a pivoting deadbolt lock. This lock, a type of mortise lock, has a bolt throw of 1¼ in. or more to engage the jamb. There is no way to spread the door that far away from the jamb without destroying the door frame and breaking the glass, plus occasionally breaking the plate-glass show window alongside. In addition, if you do decide to break the glass, there is an aluminum push bar on the inside at waist height that forces you to crouch and crawl all over the shards. This push bar can be removed, if need be, by hammering straight in on it, thereby stripping the threads off the aluminum bolt. This should be the last choice of entry methods, however, since it is dangerous and takes more time than through-the-lock entry. It forces members to crawl and drag hoselines over slippery shards of plate glass, as well as creating an uncontrollable inrush of fresh air.

Fig. 8–32 The deadbolt is fully withdrawn as the cam completes its travel past the 8 o'clock position.

I recall responding to a row of stores at about 3:00 AM and being met by the patrons of a bar who had smelled smoke. You can guess the state of the patrons of a bar at 3:00 AM and their great desire to show the firefighters exactly how to do their job. A quick smell of the area indicated that food on the stove was the likely problem, and a search in the area revealed a luncheonette two doors down with a slight haze in the vicinity of the kitchen. We sent one member for the *K* tool while the crowd instructed us on how to get in, informing us as to how inferior we were to the neighboring city's fire department, who "would have smashed their way in by now, before the whole place burns down." When the *K* tool arrived, I called on the crowd to hold down the noise, since "we need perfect silence for this next operation." The members pulled the cylinder and opened the door in 30 sec. They walked to the back, turned of the stove, and carried a large pot of burned ham to the outside. Then they turned around and relocked the door, much to the chagrin of the dispersing crowd (Figs. 8–33, 8–34, and 8–35).

Another method of forcπing a narrow-stile aluminum and glass door may be possible. Insert the adze of a halligan tool between the two pieces of the top hinge. Pry them apart, and then lift the door out of the bottom hinge and the lock. You will have to knock the door-closer device from the top of the door, but this can be done relatively quickly.

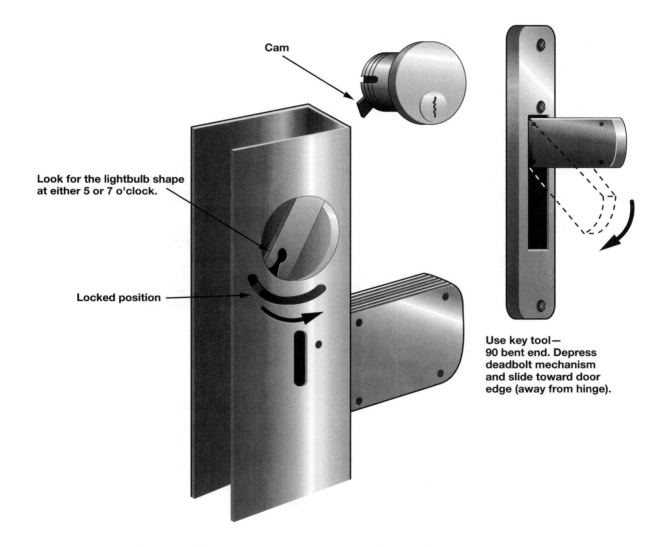

Cam

Look for the lightbulb shape at either 5 or 7 o'clock.

Locked position

Use key tool— 90 bent end. Depress deadbolt mechanism and slide toward door edge (away from hinge).

Fig. 8–33 A pivoting deadbolt style lock, commonly found in narrow-stile aluminum and glass storefronts.

Fig. 8–34 A narrow-stile aluminum-and-glass door with push bar and pivoting deadbolt is the most common commercial door in the nation.

Fig. 8–35 When looking into the cylinder hole of the narrow stile door, look for the *lightbulb* shape, and depress the spring loaded cam found right at its base.

Another device that indicates the use of through-the-lock forcible entry is the fox lock. This device is readily identifiable by a rectangular steel plate in the center of the door that shields the lock cylinder. In addition, there will be two more sets of supporting bolts near the edge of the door. This is a very difficult lock to force conventionally, since it consists of two ¾-in. steel bars extending as much as 2 in. into the doorjamb. This lock is most often found in commercial applications, usually on a steel door with a steel jamb set in a masonry wall, making for a very solid barrier. I've watched crews work for up to 10 min before finally forcing such a door. While the through-the-lock method may take 3 or 4 min, that is still better than 8 or 10 (Figs. 8–36, 8–37, 8–38, 8–39, and 8–40).

The first step in attacking the lock is to remove the shield that hides the lock cylinder. This is accomplished by placing the adze of a halligan just behind the top of the plate, directly over a set of the bolts that hold the plate on. Using an axe or a sledge, drive in the adze behind the plate, shearing the top bolt. Continue down and shear the next bolt below it. Then shift the adze to the top bolt on the other side and shear it also. Don't bother with the fourth bolt—simply spin the plate out of the way. This operation should take no more than about 2 min. This

Fig. 8–36 A fox lock from the outside. Note the lock cylinder in the middle of the door.

Fig. 8–37 The fox lock from the inside.

Fig. 8–38 Shearing the plate of the fox lock. Beware of the heads of the bolts flying off as they are sheared. Wear eye protection!

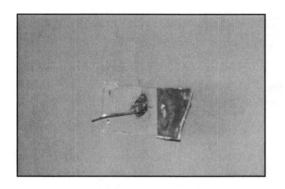

Fig. 8–39 This crew chose to shear only two bolts, and bend the plate back out of the way. (Shearing the third bolt and pivoting the plate is often simpler.) Note the square shaft key tool still in the lock.

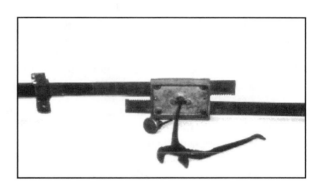

Fig. 8–40 Opening the fox lock mechanism with a key tool. Rotate toward the lower set of bolts.

gives you access to the cylinder, which may now be pulled using a lock puller, a Sunilla tool, or the adze and/or claw of the halligan. Since the steel plate is bolted flush to the outside of the door, the lock cylinder must be recessed slightly within the door. This renders the K tool ineffective in pulling the cylinder, since there is nothing for it to grab. What is needed is a tool that can be driven in slightly behind the top of the cylinder and then tapped down and behind it before you pry it out.

Once you pick up the cylinder, you will note a square shaft projecting from the back. Using your tapered square-key tool, turn the shaft to open the lock. It may take as many as two full turns to retract the bars completely. The direction that you turn the key may vary from door to door, and there is a trick in this. I once responded to a working fire at a one-story warehouse that covered an entire city block. The building was a windowless, poured-concrete structure with only one small exit door on each of three sides. My partner and I went to vent and search the rear and found the door secured by three fox locks! We advised the IC that entry to the rear would be greatly delayed, but we were told to keep trying, since units in the front were making very little headway. We set about removing the plates and pulling the cylinders. My partner then began unlocking the locks while I prepared to don my mask and tie off a search line. When he had finished all three locks, he pulled on the door only to find it still very well secured. I stopped what I was doing and went over to see what he had done. He had, in fact, turned all three locks. What he didn't know was that the owner had purposely locked two locks and left the third one unlocked so that, if anyone did pull the cylinders, they would unlock two and lock one as they turned the key tool The puzzle would be to find out which one was locking. The solution is simple if you know it. If not, you can sit there a long time trying different combinations, both clockwise and counterclockwise. The secret is that the unlocked position will always be found by turning toward the lower set of supporting bolts, the bolts that keep the bars in place.

As with any other method, through-the-lock forcible entry should be used when appropriate. Of course, there will be times when it is inappropriate. One indication of this is when visibility is seriously affected. If you can't see to locate the slot or even the cylinder, it will be nearly impossible to use through-the-lock. A condition of zero visibility often exists in fires in apartment buildings and offices, either on the fire floor or, more commonly, on the floor above. Conventional forcible entry is just as difficult under these circumstances, since you need a high degree of coordination to drive in the tool. A great number of injuries occur under poor visibility as the axe misses its target and strikes flesh. You need a method that relies less on visibility than those already given (Fig. 8–41).

The most effective solution developed to date involves the use of hydraulic-powered devices. There are now several of these on the market. Typically, such a device consists of a specially designed set of jaws connected to a hand pump. These tools are becoming increasingly popular as the task of forcible entry becomes more difficult. Once you have tried them, it is easy to see why.

Fig. 8–41 One popular type of hydraulic forcible-entry tool (HFT).

The rescue company I was assigned to as a firefighter was among the first units in New York City to try one of these devices, and we gained considerable experience with them. One evening about two months after we had gotten the tool, we were called to a multiple-alarm fire in Harlem. I was assigned the HFT and a 6-ft hook as part of an interior search team. We arrived to find a vacant six-story brick building with fire already in total possession of the second, third, and fourth floors and spreading rapidly. Special calls were sent for additional tower ladders, and the operation shifted to a defensive mode.

Our unit was assigned to stand by in the event something unexpected happened, such as an extension to exposures. We were located at the staging area when the captain of one of the ladder companies noticed the tool hanging from a strap on my shoulder. He asked about its uses and was promptly given a series of rave reviews by all of the members of the company. By the time our unit was dismissed, he had gotten the full rundown on its capabilities—8,000 psi working load, 5 in. jaw opening, 25-lb weight. He seemed impressed by our enthusiasm, but skeptical as to its actual need. He was used to working with excellent forcible entry teams and wasn't totally convinced that this device was as superior as we claimed. That was soon to change.

That night we responded to several other working fires up and down the avenue. There was obviously a berserk individual running around Harlem with gasoline and matches. He had worked the neighborhood for the better part of a year, selecting a street for a given night's activity, then bouncing back and forth along it, setting fires on it and the adjoining side streets. This night, the fires were being set along Adam Clayton Powell Boulevard. At about 6:00 AM, we went back to yet another working fire in the area of the earlier fire. This time things were different. Fire was reported on the fourth floor of a six-story partially occupied tenement, with people trapped and jumping. As we came into the block, our officer reported via radio to the chief in charge to get our assignment. Before the chief could reply, the truck captain to whom we had shown the tool earlier called the chief and asked, "Get the tool up to the floor above.

We've got four tough doors and we're not in any of them yet." We immediately proceeded to the fifth floor to find his forcible entry team still working on the door to the apartment directly over the fire apartment. As we arrived, he directed his people to let us try this new gadget (Figs. 8–42 and 8–43).

Fig. 8–42 Insert the HFT between two locks, then pump.

Fig. 8–43 The HFT forces nearly all inward-opening doors. The stronger the door, the better the results.

I inserted the tapered jaws between the door and the jamb and began pumping. In the time it takes to read these next three lines, the door popped open. The members were still unconvinced that they hadn't really done 99% of the work before our arrival only to see us *take the credit*. The captain led me to the door to the adjoining apartment and said, "Let me see that again." Again the tool went to work, and the door popped open in short order. Now he was hooked. We popped the last two doors with the same results, then the captain disappeared into the smoke to search and ponder what he had just witnessed. One man in under a minute had just forced four doors like the one that had held up his forcible entry team for at least 2 to 3 min. He became another convert, and that very morning he was on the phone trying to arrange for his company to be issued an HFT.

Speed is just one advantage. The 8,000-psi working pressure ensures success against even the most difficult locks, including fox locks, Multilocks (four-way locks), deadbolts, and even drop-in bars. The HFT works especially well in conditions of poor visibility, since all you have to do is to locate the jamb long enough to insert the jaws, then stand back and pump. Moreover, because the tool generates the brunt of the force, firefighters who use it can enter a fire area much less fatigued than can a member who has to beat on a door with a set of irons for several min. SCBA air supplies last longer, members perform better, and injuries are avoided (Fig. 8–44).

Although the advantages far outweigh the disadvantages, there are drawbacks to HFT. These tools aren't for use in every situation. You shouldn't use an HFT when damage is a concern but speed is not, such as responses to automatic alarms when no one is home. In such a case, it would be better to select a less damaging means of forcible entry—perhaps through-the-lock or by raising a ladder to an unsecured window. Also, these tools weren't designed for automobile extrication work and shouldn't be used for such.

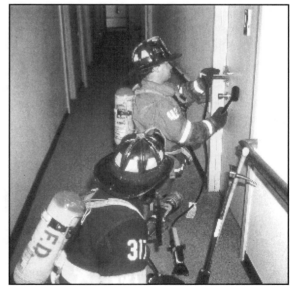

Fig. 8–44 HFTs are great tools when multiple doors must be forced in a hurry. Always bring the irons as a backup though. "The axe always starts."

The appearance of a new product should always prompt a review of the department's SOPs. Any new product should work well within the SOPs, and you should consider the HFT in this light. For instance, a knowledgeable firefighter should be able to recognize the proper method of forcible entry for the situations he encounters. Variant construction requires various entry practices. Speed, damage, and safety are always important considerations. Many methods have earned a place in the scheme of fire attack, but the HFT tool has proved to be effective when a serious situation involves heavy smoke, rapidly spreading fire, and personal risk. The tool works best on solid doors set in steel doorjambs. The stronger the door, the more useful the tool. It works best on doors that open inward, as is the case with almost all dwellings.

The tool does *not* work well with doors that open out or on flimsy or hollow doors set in wood jambs, since these don't give the tool anything to push against. In these cases, opt for conventional means. At 25 lbs, the older versions of the HFT weighed about the same as a 2½-gal pressurized water extinguisher, whereas newer, self-contained models weigh only 9 lbs.

Problems do arise when companies must decide who will carry the HFT. In many departments, the first ladder personnel send a two-person team for forcible entry and interior search, rescue, and ventilation. One person carries the flathead axe and a halligan-type tool, while the other carries a 6-ft hook and a water extinguisher if a line is not charged. All of the four following items can be vital: the hook for venting and pulling ceilings, the irons for forcible entry and overhauling. The extinguisher can darken an extending fire, prevent flashover, or permit a member to get close enough to a room to pull the door closed. Who carries the HFT? The newest versions of these devices weigh less than 10 lbs and can be slung over the shoulder, allowing the member carrying the irons to bring the HFT as well with little extra effort.

How do firefighters feel about the HFT? Like the oxyacetylene torch, a lock puller, or a separate power saw equipped with a steel cutting blade, they consider it a tool not to be used every day, but reliable when needed, aiding the search for life and helping to reduce injuries. They have made some interesting observations along the way.

During its initial trial period in New York City, Rescue 3 made skeptics into converts. Rescue 3 responds to all working fires in the busy areas of Harlem and the Bronx, working alongside or above some of the best forcible entry teams in the department. Rescue 3 was reproached for being too *quiet*, disappearing into the smoke and then coming out, entry achieved, less winded than their comrades. No one heard the familiar clang as the flathead axe pounded on the halligan tool. Companies in the area unfamiliar with the operation of the HFT grew curious. To allay any suspicion, one firefighter suggested that the man carrying the tool also carry a tape player so that he could produce the requisite sounds of pounding and cursing from the darkened hallways.

On a serious note, however, a potential problem has arisen in these days of shrinking budgets and reduced manpower. Using the HFTs, ladder company members have greatly increased the speed with which they gain entry, whereas the hose stretch is often slower than it was in the past, due to the reduced manpower of engine companies. As a result, members of ladder companies are being exposed to uncontrolled fire for longer periods while they search for the fire's location and victims. Pay attention to your surroundings to avoid being cut off or overrun by extending fire!

The Multilock Door

Short of a bank vault, one of the most difficult forcible entry challenges you could ever encounter is the Multilock door. This door, lock, and jamb assembly comes as a complete package and is usually installed in masonry walls. The lock is a mortise type of unusual design. The cylinder is shielded with an oval plate located in the center of the door. It acts to push or pull four steel rods in or out of all four sides of the door—left, right, top, and bottom. These rods measure approximately ½-in. in diameter and project as much as 1½ in. into the jamb. Using conventional forcible entry is extremely difficult on this type of door, since each of three sides of the door must be totally separated from the frame before the door will yield enough to open. Using HFTs makes the task less punishing physically, but it can still take a great deal of time, since each side must be attacked in succession (Fig. 8–45).

Through-the-lock forcible entry is also ineffective against the Multilock. The cylinder is designed to disconnect from the rods if it is removed, destroying the operating mechanisms in the process. The rods then remain in place, keeping the door locked. If faced with the Multilock door, the first step is to determine whether the lock is in fact locked. In many cases, such a door is equipped with other spring-loaded locks that lock as the door is closed. Since the Multilock requires someone to turn a key to engage it, such doors are sometimes left unlocked if the occupant only intends to run a brief errand or expects to be continually passing in and out. It is simple to determine whether the lock is engaged if the door opens toward you. Simply look for a steel rod between the door and the jamb. On tight-fitting doors, or doors that open into the occupancy where the jamb stop hides the edge of the door, you will have to slip something like a knife blade or a thin sheet of plastic under the door (an old credit card, for instance) and slide it along the edge to feel for the bottom rod. If there is no evidence of the rod being engaged, either conventional or through-the-lock forcible entry on the secondary lock is in order.

If it is apparent, however, that the Multilock is engaged, you have the following three choices: find the key, find another way in, or destroy the $1,500 door. Assuming there are no other alternatives, the fastest means of gaining entry is by using a circular saw with a metal cutting blade. Cut a triangular-shaped hole in the lower half of the door on the side between the Multilock cylinder and the doorknob. This allows you to reach in and unlock the Multilock, as well as any additional locks on the edge of the door. The Multilock usually has a handle on the inside, although some Multilock doors

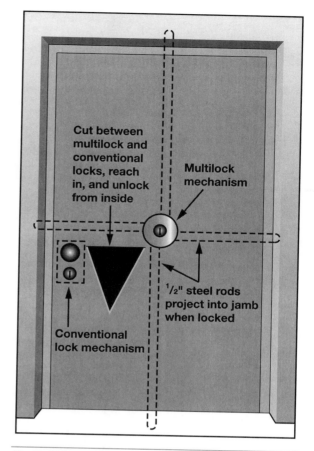

Fig. 8–45 It may be necessary to cut your way through a Multilock door. If so, cut in the lower quadrant closest to the doorknob and other locks.

require a key on both sides. If this is the case, crank up the saw and cut a 10–12 in. sq around the Multilock plate, then reach into the middle of the door and manually retract all four rods. As you can imagine, this should be a last-ditch effort to gain entry after all else has failed. It is often faster and less expensive to repair if you break through a cement-block wall than if you force this type of door.

Forcing Metal Gates and Roll-up Doors

Security gates are increasingly a problem for the fire service. Once confined primarily to high-crime neighborhoods, they have proliferated with the rise of smash-and-snatch thefts. You can now expect to find them on at least certain occupancies in almost any neighborhood where the display windows hold high-value, easily carried merchandise. Before you brush this off as not being a problem for you, survey your district early on a Sunday morning when most businesses are closed and the gates are locked. Many of the gates are difficult to detect when the occupancies are open, since they are often rolled up or pulled to the side in the stored position (Figs. 8–46, 8–47, 8–48, and 8–49).

Gates may be classified into the following three types: manual, mechanical, or electrical, depending on their means of operation. Manual and mechanical gates are the most common as well as the simplest to force. Electrical gates pose difficulties and may have to be cut open. This will be discussed later.

Fig. 8–46 Several manual roll-up gates with many locks indicate the need for additional forcible entry assistance and will delay operations.

Fig. 8–47 The square projection at the top left of this roll-up gate indicates it is a mechanical gate.

Fig. 8–48 This padlock secures a hinged angle iron cover that shields the chain hoist operating mechanism.

Fig. 8–49 After removing the padlock and opening the angle iron, remove the chain and use it to hoist open the gate. Secure the chain when finished to prevent the gate from accidentally closing on firefighters below if heat weakens the assisting springs on the hoist.

The method used to force manual and mechanical gates depends on what is holding the gate closed—the locking devices. You may find either a built-in slide bolt or a gate key, secured with up to six padlocks. On a mechanical gate, the raising/lowering chain is usually shielded by a piece of angle iron, secured by padlocks. A basic knowledge of padlock types and methods of removal is usually the key to forcing roll-up gates. Remember, however, that removing the padlock doesn't finish the job. You must remove either the gate key or the slide bolt before you can open the gate. Members often experience pulled back muscles when they forget that little tidbit in the heat of the moment and try to yank the gate open.

Padlocks fall into the following two categories: low- and high-security. Low-security models have a small bow (less than ¼ in. diameter) and are not casehardened. They aren't usually found on security gates since they are easily forced, thus defeating the gate. If you do encounter them, they may be cut easily with bolt cutters, or you may force them with a halligan or a claw tool. More than likely, anyone who goes through the trouble of installing a security gate will fasten it with the more expensive high-security locks.

High-security padlocks have bows of casehardened steel at least ⅜ in. thick. They are usually fastened much more substantially than with a simple chain. Instead, a formidable hasp is usually welded to the frame. Casehardening, a metal-treating process, gives the steel great compressive strength, which makes it very difficult to cut with bolt cutters. You must use at least 36 in.-long cutters, and you may still have trouble biting through. It is often necessary to use two people or to brace one handle of the cutters against something substantial, such as the floor. The lock often damages the jaws of the bolt cutter, so the jaws may not be able to cut the third or fourth locks if you encounter multiple locks. Clearly, another means must be available for such a situation. Power saws and cutting torches are highly effective; however, both are expensive items that may not be present on all first-arriving apparatus. And, as the saying goes, an axe always starts.

Fortunately, there are alternatives. The same processes that give casehardened steel its great compressive strength also give it relatively poor shear strength, which is the resistance a material has to being pulled in opposite directions. Firefighters can take advantage of this by driving a wedge between the two sides of the bow, thereby splitting the bow open. A commercial tool known as the *duckbill lock breaker* is available for this purpose. Basically, this tool is a steel wedge connected to a padded handle. It is driven in using a flathead axe or, preferably, a sledgehammer. It may take several blows to shatter the lock, but it works. Similar results are available without resorting to this relatively costly item. A miner's pick may be used instead. Available at any hardware store, you can cut down one side of the pick head to provide a striking surface for the sledge. It works just as well as the commercial model (Figs. 8–50 and 8–51).

Even when this simple equipment isn't available, all may not be lost. Depending on the strength of the securing hasp, you can often shatter the lock by the same shear effect by using basic forcible entry tools. Place the forked end of the halligan over both sides of the bow of the lock, then twist the tool. This forces the two sides of the bow in opposite

Fig. 8–50 *The duckbill lock breaker is a commercially available item.*

Fig. 8–51 *A miner's pick acts as a wedge, providing shearing force against the shackle of a padlock.*

directions around the hasp. This method takes strength and leverage, and the longer the tool, the better. Unfortunately, it isn't guaranteed to succeed. At times, if the hasp isn't very strong, it will twist around the lock, making it difficult to pull apart. This method does work most of the time, but if the hasp starts twisting, all that can be done is to continue twisting until the hasp itself breaks free. A 36-in. pipe wrench with a 3-ft cheater bar will serve the same purpose. Considering that you would normally use this method if no other alternatives were available, you might do well to take action right away than to wait for a truck with all the special tools that is still 3 min away (Figs. 8–52 and 8–53).

Fig. 8–52 If all else fails, you may be able to shear the hasp off the door by twisting the padlock with the fork of a halligan.

Fig. 8–53 Cutting a padlock with a power saw is the most effective technique, if the saw is readily available and starts! Know the alternative methods described here.

The fastest means of forcing multiple locks is by using power tools, either an oxyacetylene or methylacetylene propadiene (MAPP) gas torch, or a circular saw with an aluminum oxide blade. Both tools have their own advantages, lending themselves to particular situations, and both are very worthwhile investments. When cutting with either tool, it is normally necessary to cut through both sides of the bow, since nearly all high-security locks are designed to lock both shanks in place. When using the power saw, you can make the operation much smoother by having a second member hold the lock still while you apply pressure with the blade. This can be done safely by grabbing the lock with a vise grip attached to a short length of light chain. A saw is limited in the height to which it can be used safely.

Generally, any locks above shoulder height are best cut with a torch. The light, compact cutting outfits that are available today are excellent for this purpose. They can be used from a ladder, if necessary, since only one hand is required to operate the torch head after it is lit. On some gates, you will find that extraordinary steps have been taken to shield the lock and hasp from tampering. It is usually best in such cases simply to cut the U-shaped channel iron above and below the lock so that the entire piece may be removed. By attacking the locks first, we maintain the integrity of the gate so that, after the emergency, the premises can be re-secured simply by installing a new padlock.

A word or two is in order here about technique when using saws to cut steel. Most firefighters get little practice in this operation, using saws most often to cut wood. They are used to revving up the saw fully before sinking the blade into the work. This is a proper procedure for wood, but it is *improper* for cutting metal. The saw operator should place the blade on the work and then slowly increase the rpms of the saw until it is operating in the cut. Striking a metal surface with a blade spinning at high speed is likely to result in the blade rapidly sliding sideways along the work, injuring anyone in its path (Figs. 8–54 and 8–55).

The American 2000 gate lock is a unique type of lock resembling a shiny steel hockey puck. The locking mechanism and hasp are entirely concealed behind the body of the lock, leaving no place to pull, cut, or pry using conventional tools. Some success is possible using that 36-in. pipe wrench and cheater bar to twist the lock and hasp off the frame. A more reliable method is to cut the lock using a metal-cutting power saw or torch. (**Note:** This lock is made of a very strong alloy that gives off a bright white flame and white smoke when cut with a torch. Do not breathe this smoke. The cut must be made at the correct location, otherwise nothing will be accomplished. Cut across the lock, two-thirds of the height of the lock, away from the keyway. Then the entire lock may be removed.)

The need for at least two saws on the first-arriving units, one of them equipped at all times with a metal-cutting blade, should be readily apparent. In these situations, there will often be an urgent need to perform roof ventilation prior to or concurrent to forcible entry. This may be the only way to prevent backdraft. If either function has to wait for the saw blade to be changed, there will be an intolerable delay in performing critical tasks. Firefighter and civilian injuries and deaths, plus fire extension and excessive property loss, are all possible results.

Once the locks have been removed, the gate will be operable. For manually raised gates, remove the gate key or pull back the sliding bolt, then raise the door slowly. Don't attempt to throw up the gate rapidly, since you may have missed a locking device and may encounter unexpected resistance. For mechanical gates, removal of the locks gives you access to the hoisting chain. Use it to raise the gate. It may be necessary on some models to engage a clutch first. This is done by means of a second chain or cable hanging alongside the first.

Fig. 8–54 The American 2000 series lock, shown (top to bottom) with the lock unlocked with the keyway projecting; cut apart to reveal the recessed locking pin; and with the lock locked.

Fig. 8–55 By using a wrench and a cheater bar against the 2000 series lock, you may be able to twist the hasp off the gate.

Cutting the gate itself should be the last resort, since a gate is an expensive item to repair, and cutting it has many drawbacks. The first disadvantage is that it takes longer to make 12 or 15 ft of cuts than it does to cut two or three locks. The second is that almost all cuts only open part of the door or gate, and the remaining portions hinder ventilation and access. Additionally, it won't be possible to secure the premises after operations have been completed.

The following two conditions usually indicate the need to cut the door: either a heavy fire condition where it is important to get water on the fire quickly (and damage to the door is not a concern) or a door that cannot be readily opened by normal means, such as one that is damaged or electrically operated. Although it may be possible to operate an electric door by means of a mechanical override, this usually requires you to be inside the building already. If this is practical, certainly do so, but if it isn't, and if additional ventilation or access is needed, then you must cut the door (Fig. 8–56).

There are two styles of cuts that you may use. The first is the inverted *V*, a relatively fast way to make an opening in the door. Start the cut at the highest point that you can safely operate the saw, usually about head height. Make the cut diagonally to one side, as far down as you can possibly go. Begin the second cut back at the top, making certain to cross *fully over* the original cut, continuing down along an opposing diagonal. The resulting triangular section of the gate will now fold either in or out, allowing access to the area.

Making this cut is a quick way to get water on a rapidly spreading fire, but it has many limitations. Such an opening is too small to allow a member easy access to advance a line or for emergency escape. It only provides a limited amount of ventilation, and if a locked door is located behind the gate (quite common); it is very difficult to force the door, even if the door is right in line with the hole in the gate. If it isn't, you still haven't gained access. Sometimes you can expand this cut to the entire width of the portal. If the gate is surface-mounted, the remaining sections of slats can be slid out from the slat above by grasping them below the top

Fig. 8–56 If you must cut a gate for fast-stream application, use the inverted V or teepee cut. Make sure the top of the two diagonal cuts intersect at the top.

of the cut. Remove a slat on each side of the cut in this manner. This will free the remainder of the gate to be pushed up (manual doors only) while the bottom slats are pulled out. Again, this method can only be expanded if the gate is surface-mounted. If the door is recessed, the cut may only be partially expanded unless you make the cut dead center in the door. Otherwise, as you pull the slats, you will run out of room, striking the side of the building with the longer slate before it is fully disengaged. One drawback to the inverted *V* or teepee cut is the possible presence of small tabs at the ends of the slats that interlock with the steel guide rails that comprise the frame of the gate. These tabs, sometimes called wind tabs, may be found on every other slat, every third slat, or in some cases, on all of them. They serve to prevent the slats from being pulled in toward the center of the door, thus keeping firefighters from expanding the cut.

A different style of cut is required whenever you encounter a recessed door or a door with wind tabs. The 3-cut method works well and is the preferred method of cutting all doors that you intend to enter, since you cannot tell by looking at a door whether it will have wind tabs or not. Make the first cut as close to the edge of the door as possible, from the maximum height straight down to the minimum. Make another cut, the same as the first, as close as possible to the opposite edge (usually within 6 or 8 in. due to the body of the saw). Then, make a third cut in the center, from the top of the other cuts and downward about 2 ft. The resulting slats will be short enough to be pulled out before they reach the obstructing sidewalls.

At times, especially on wide doors, the weight of the hanging sections below can bind the slats that support them, making it difficult to remove them. You can avoid this by making additional short (2 ft high) cuts approximately every 5 to 6 ft. These shorter sections are easier to pull, and the operation can be further assisted by members supporting the sagging area on the low side (where the slats have already been removed). A key point to remember is that all of the cuts must be made through the same slat. Also, it helps when removing the slats to push in on one side while pulling out on the other. Once the first slat has been pulled out a little bit (using vise grips, channel locks, or by driving the claw of a halligan through the sheet metal), the operation can be assisted by a member with a flathead axe tapping only slightly more on the back of the slat. This method requires the same amount of cutting as the inverted *V* cut, but it does guarantee clearing a larger area for access and ventilation (Fig. 8–57).

Obviously, proficiency in forcible entry is an absolute must for all firefighters. If you can't get into a building, you can't put out the fire. Knowing a variety of methods gives you something to fall back on when Plan A doesn't work. As security considerations increase due to ever-rising crime statistics, the firefighter's task of gaining entry will become more difficult. It is only through the application of technology and knowledge that we can hope to keep pace with this challenge. There are already doors and locks that so drastically delay forcible entry that they seriously affect fire operations. There may come a time in the future when certain buildings or occupancies are constructed like giant

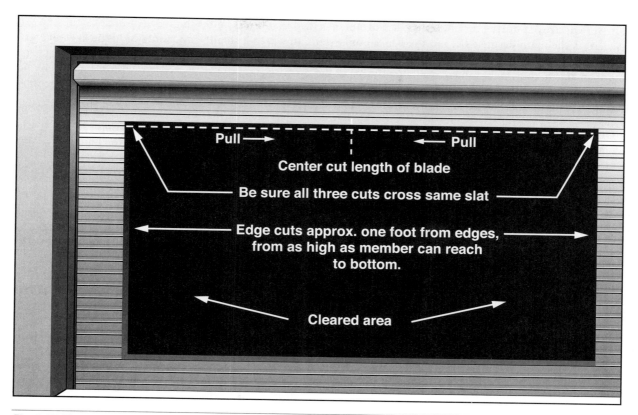

Fig. 8–57 The 3-cut method offers many advantages in terms of firefighter safety over the teepee cut.

bank vaults. Inevitably, there will also come a time when a fire unit pulls up in front of such a structure and is faced with the task of gaining entry (Fig. 8–58).

The *exothermic torch*, or burning bar, is a specialized cutting tool that burns at more than 5,000°F. It consists of a special hollow steel rod that burns at high temperatures when ignited while pressurized with pure oxygen. The rods can burn through casehardened steel padlocks quite easily but are quickly consumed themselves. The tool is a very handy device when it is necessary to cut through very thick steel such as large I-beams, train rails, or ½ in.-or-thicker plate, as is found on ships. Besides consuming large quantities of oxygen as well as steel rods, the device has one major drawback: it throws a very large jet of sparks and flame on the downstream side. Of course, this isn't a serious problem if the area behind the steel obstacle is already well involved with fire.

The best way into any locked building, of course, is by using the key. In one situation that I know of, a large number of high-security doors were installed in several buildings that were the scene of numerous fires and emergencies. After quite a few of these very expensive

Fig. 8–58 The exothermic torch can slice through heavy steel in seconds. Adequate fire protection is a must when using any torch.

doors were destroyed during forcible entry, the owners decided that it was better to give the fire department a set of keys. That is the main premise behind the Knox-box system, in which a small vault, containing the keys to a given building, is located on the building's exterior. The fire department is then given the keys to the vault so that, in the event of an alarm, they may gain entry efficiently. All of the premises that use such a system within a given city have their vaults keyed alike, so that the fire department only requires a single key. Until the time that this becomes a universal feature of every structure, however, firefighters will continue to have a need for forcible entry.

One situation where the keys are provided for you involves what are known as *series locks*. These are often found on commercial establishments. They consist of a set of locks accessible only from the inside of the building, with the keys arranged in duplicate on different locks on different doors. When opening and closing the store, the locks must be manipulated in series. The key from one lock is withdrawn from a lock as it is locked, then inserted into the next one in the sequence. The key from that unit is then withdrawn to go into the next one, and so on. Without having the first key to begin the sequence, however, the remaining doors cannot be opened or their keys withdrawn. In this case, they must be forced conventionally (Fig. 8–59).

It is only by continually evaluating the forcible entry problem in your response area and by applying all of the available technology to that problem, that you will be able to meet the challenge when the time comes. For a preview of what you will face in the future, visit your local locksmith. He will be glad to show you all of your potential adversaries. Then, try to figure out how best to defeat them. You don't have to take a course in locksmithing to do this. Follow the guidelines I've outlined here. Remember, if you can remove the cylinder intact, you can normally duplicate the action of the key. Locks that engage the jamb more than 1 in. or at multiple points are great challenges to conventional forcible entry. Other times, special tools are an absolute *must*. Learn to recognize the bigger challenges, and plan alternative methods in advance.

Fig. 8–59 The forcible entry techniques you choose must keep pace with the challenges you face. Otherwise, you may find yourself unable to perform any of your mission.

09 Ventilation

As many firefighters have discovered over the years, there is more to firefighting than just getting into the building. The real trick lies in being able to remain there long enough to complete all of the necessary tasks. The fire is doing everything in its power to prevent us from carrying out our mission. It is filling the structure with smoke so dense that it hampers or even eliminates visibility. At the same time, it drives temperatures up to the point that unprotected skin burns and a single breath will sear the throat, closing it off. All the while, the fire is extending, sending out tongues of flame, looking for easy paths to travel. Like a rampaging army, it attacks the building, constantly in search of more fuel and oxygen. It builds, driving all life before it.

Both contestants in this battle need the same raw material with which to fight—oxygen. Whoever controls the oxygen supply wins the battle. Fortunately, the good guys have a great advantage: there is an unlimited supply of oxygen on their side, and they can even take small amounts of it with them as they mount their offensive campaigns against the enemy. With their large reserves, they can sometimes afford to give the enemy some, drawing the main body of the enemy's forces in that upward direction where it will expend its energy without harm to the friendly forces. Of course, they must use care in tempting the enemy, for if this supply is allowed to fall into his hands at the wrong place, he will use it to grow stronger, possibly with explosive speed, driving the suppression forces back in retreat and inflicting great casualties. The commander who masters this art of warfare known as ventilation has a great force on his side. Used properly, it can be the deciding factor in whether the building is saved or becomes a smoldering pile of debris (Fig. 9–1).

Fig. 9–1 Coordinated roof venting prevents mushrooming and slows horizontal fire travel.

The following changes have affected the fire service in recent years: reduced staffing, energy conservation measures producing tight-building syndrome, increased awareness of property conservation, improved fans, and heavier fuel loads of plastic, producing toxic smoke. All of these factors affect one of our primary firefighting tactics—ventilation.

Reasons for Ventilation

Ventilation can be defined as the process of removing toxic products of combustion and replacing them with fresh air. It can be, but I like my definition of ventilation better. It keeps me focused on the critical nature of the process. To me, ventilation is an ongoing battle between the fire and the firefighters for control of the building. The fire is constantly pumping the building full of deadly toxins and flammable gases. The firefighters must wage war against the fire, using the right weapons at the right time to succeed.

There are two main reasons for performing ventilation as follows: venting to allow attack teams to enter and operate within the structure (venting for fire), and venting a specific area to provide fresh air for breathing and to improve visibility while searching (venting for life). Unfortunately, the distinctions are lost on many, as is the need for ventilation in the first place. The universal use of SCBA has some people thinking that ventilation is less important than it was in the past just because personnel on the fireground can breathe. There are problems with this thinking. First, it totally ignores any victims who might still be in the building. Second, it assumes that the firefighters' masks will always protect them. This isn't always so.

Several firefighters have been killed in recent years when they experienced mask problems in contaminated areas. They have run out of air, become entangled, and had facepieces pulled from their faces. At the scene of one such firefighter fatality, the process of ventilation had been undertaken with many windows on the upper floors opened by the interior forces from top and bottom, textbook style. The glass damage was minimal, and the firefighter was dead. The third problem with ineffective ventilation and SCBA is the buildup of heat that results when venting isn't prompt and effective. The mask protects you against smoke but offers nothing against heat.

When distinguishing between venting for life and venting for fire, the key factor is the timing of the ventilation. Venting for life should obviously begin as soon as possible after the life hazard is recognized. In the process, the fire may intensify somewhat, but the venting should be such that it draws the fire away from the life hazard. For example, in the crowded multiple dwellings that are prevalent in New York City, one member from each of the first two ladder trucks to arrive is assigned to proceed immediately to the roof and perform ventilation over the interior stairway and any vertical shafts. These members perform this task routinely, without orders, at every working fire. The goal is to draw the fire out through the top into the open air where it can do no further harm, rather than allowing it to mushroom on the upper floors and spread into those apartments. Immediate vertical ventilation has saved hundreds of lives by drawing fire away from the victims. It should be a primary tactic at most low-rise, non-fireproof multiple dwellings.

Venting for fire, on the other hand, is normally delayed until resources are in place to attack. For example, suppose that you pull up at 2 AM to find heavy smoke emanating from the second-floor apartment in a four-story apartment house. One member, armed with a 10-ft pike pole, approaches the building just as you are informed that the first pumper is having difficulty because the nearest hydrant is OOS for repairs. In this case, you should order horizontal ventilation withheld until the charged hoseline is in place at the door to the fire area. Once the attack has begun, you can order the windows in the fire apartment opposite the hoselines advance to be vented to allow the products of combustion to be blown ahead, thereby speeding advance.

Types of Ventilation

The venting of windows described previously is called *horizontal ventilation*. In the scenario described, it functioned as venting for fire. Horizontal ventilation can also be performed to assist the life-saving effort, but you must know the consequences of doing so. For example, if a person is reported to be in a room adjacent to the fire area, venting the windows will allow an influx of fresh air to someone whose life depends on it. Still, if you don't immediately take steps to remove the victim, you may worsen his plight by drawing fire toward the vented window. Generally, horizontal ventilation for a life hazard must be coupled with an immediate rescue effort. Either a member must enter the area and remove the victim, or a hoseline must be brought in to protect the victim where he is trapped.

The most important concept to remember when discussing ventilation is that it must be of sufficient volume to win the battle with the fire. The damage that results from our efforts should be commensurate with the amount of damage that the fire is causing, as well as the extent of the life hazard. If the fire is of a low intensity (smoky mattress or food on the stove), you should open the windows, not break them. Mechanical ventilation is often beneficial in moving this cool smoke, and it poses few of the dangers that are present when using fans during working structural fires. Window venting is the most suitable type of ventilation for most fires in houses and similar-size structures that have high window-to-room-size ratios. Window venting is very quick and relatively easy to perform, and it can be targeted to specific rooms. When interior forces are able to operate in the building, they can evaluate the conditions before deciding to vent the windows. In the past, a guide I used was, "If you can stand up long enough to manipulate the locks, then raise the windows. If you can't stand up that high or for that long due to the heat, then it's time to take out the glass." Although this is still a good rule of thumb, there are a few additional points to make.

First, you must have an idea of the progress of fire control and search efforts. For example, if the fire seems to be under control and the primary search is complete and negative (no victims found), you may want to go ahead and raise the windows instead of breaking the glass, even though the area is moderately warm or even hot. Conversely, very dense smoke conditions together with a strong potential for either fire extension or locating a victim justify breaking glass even in the absence of extreme heat.

The second factor to consider when deciding whether to break glass or not is that conditions will probably worsen before they get better. Venting not only provides fresh air to trapped victims, it also improves conditions for firefighting, leading to earlier control. Firefighters will also be less subject to disorientation or becoming trapped if hoselines are operating on the fire. Breaking glass provides twice as much window-opening area as raising double-hung windows. Since these are the predominant windows in the majority of residential buildings, firefighters must be prepared to break glass if conditions require. The lives of occupants, as well as firefighters, hang in the balance (Fig. 9–2).

Fig. 9–2 Venting windows can be done to speed the hoseline's advance, or to reach a victim trapped in the room behind the glass. Coordination of this effort with the fire attack is crucial to success.

Breaking glass, of course, isn't something that can be reversed once you have done it. Firefighters must time their ventilation efforts properly. Venting too early will allow the fire to extend, while delaying ventilation will subject firefighters to unnecessary punishment from heat. Generally, venting for fire should take place just as the hoseline begins its attack. This is best coordinated via radio. A member of the attack team should give the word to *take the windows* just before the nozzle team opens the nozzle. If radios aren't available, the person creating the vent should wait until the sights and sounds of hose-stream operation (steam and fire knockdown) confirm that the line has indeed begun operating. One exception to this sequence is in the case of tightly sealed areas, typically glazed with double- or triple-pane, energy-efficient windows. In this case, the sudden admission of oxygen as the attack crew enters can have catastrophic results. Sudden development of the fire or even backdraft are possible. When you encounter energy-efficient windows, undertake window venting as soon as possible while the attack team is in a safe area, preferably behind a closed door. If this isn't possible, it may be desirable to delay venting until after the hose stream has thoroughly cooled the fire area.

Other factors that influence the choice of vertical or horizontal ventilation, or both, include the size and location of the fire, the construction of the building, and the effects of the weather, particularly wind. Horizontal ventilation is often preferred at minor to moderate fires. A mattress fire will produce large quantities of smoke, but there is rarely any need for vertical ventilation. In fact, vertical ventilation may not be very effective, since the low heat levels of an early-stage mattress fire don't make the smoke rise very rapidly. Horizontal ventilation is often faster and easier to perform than vertical ventilation because it takes advantage of man-made openings (windows and doors). In addition, it can often be performed from ground level or from a one-man portable ladder. Vertical ventilation usually requires a longer ladder—and thus more manpower—to get to the roof.

Usually horizontal venting is also less costly to repair than vertical venting, unless there is also a man-made opening at the level of the roof. Glass is far cheaper to replace than a section of roof covering. Certain roofs can indicate the type of ventilation required. Peaked roofs, found on many private homes, are usually much more easily vented by windows in the gable ends or dormers than by cutting the roof itself. Similarly, flat roofs of poured concrete construction would defy all but the strongest venting efforts. All of this isn't meant to lessen the importance of vertical ventilation; rather, it is only meant to help define its place and importance (Fig. 9–3).

Vertical ventilation is often a deciding factor in whether the fire is stopped or continues to expand. Burning materials produce hot gas, which takes up huge volumes of space, hundreds of times the volumes of the original burning material. This gas is naturally trying to rise. When it cannot, it spreads out horizontally. If you can create a way for this mushrooming gas to go straight up and out of the structure, the gas will take a large part of the heat energy with it rather than allow it to remain inside. Horizontal venting of large quantities of this hot gas is hindered by the fact that, as it flows along under the ceiling toward an open window, it radiates heat in all directions. This means that combustibles are being preheated. It means that firefighters are being heated as well.

When an opening of proper size and placement is made in the roof under these circumstances, the relief is dramatic. Horizontal fire extension is slowed dramatically, and visibility and heat conditions improve rapidly as fresh air is drawn in at lower levels to replace the hot gas that is exiting out the top. Of course, not every fire demands or even allows for vertical ventilation. Relatively minor fires with little heat buildup, as well as those in the lower portions of structures, remote from vertical openings, probably won't benefit from opening the roof. Those fires in attics and cocklofts, as well as the floor directly below the roof, however, will be greatly aided by this tactic (Fig. 9–4).

Fig. 9–3 A serious fire in the top floor or cockloft demands roof ventilation if the fire is to be extinguished from the interior.

Fig. 9–4 Fire venting through the opening in a roof is a **good** sight. It means that the spread of fire under the roof is being slowed.

All of the ventilation that we have discussed so far has been performed through the cooperation of the enemy itself, fire. By heating up the products of combustion, the fire has made its by-products rather mobile. All we have to do is open doors and windows, and the smoke will readily pour out. This is called *natural ventilation*, since it is based on the natural action of heated gas. Occasionally, however, this natural process is insufficient to remove the products of combustion rapidly enough to allow firefighters to finish their tasks without difficulty. In the case of a smoldering mattress or a low-heat fire in the basement, the smoke may be so cool that normal ventilation won't work. The smoke will hang down low, keeping its deadly cargo of carbon monoxide with it. Similar situations arise when fully developed fires are knocked down by a sprinkler system but aren't fully extinguished. Both of these cases are prime candidates for the use of mechanical ventilation to move gas that wouldn't otherwise be displaced easily.

Mechanical ventilation can involve hoselines, portable fans, or building ventilation systems, depending on the resources available and the fire situation. Venting can be done either by sucking contaminated air out or by blowing fresh air in. Both methods accomplish the following objectives: to remove all of the contaminated air and replace it with fresh air. Certain pros and cons, which should be known to all, apply to both methods.

Venting using a fog or spray stream is one of the earliest and simplest means of mechanical ventilation. It is by far the least demanding in terms of manpower and, if done properly, can be very effective. This method relies on the fact that a spray of water entrains air with it, carrying the air out in a stream. Several factors affect how much air, and thus smoke, the stream can move. One is the volume of water flowing, since each tiny water droplet brings along a limited amount of air. The more water flowing, the more air is moved. The second is the velocity of the stream, which is a function of the pattern of the stream and the nozzle flowing pressure. Again, the higher the nozzle pressure, the more air that is moved. The pattern that moves the most air from an area is one that is tight enough to move the stream and the gas out of the area yet wide enough to entrain large amounts of air. Generally, this is in a range of about a 30° fog pattern (Fig. 9–5).

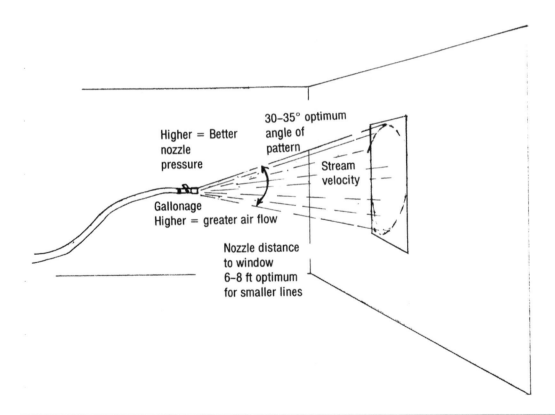

Fig. 9–5 The use of a fog pattern to assist ventilation is a very useful technique. It can also be performed with a solid tip nozzle by only opening the shutoff part way.

The final factor is the geometry of the opening and the stream. By standing back 6–8 ft from the window or door and adjusting the pattern to just cover the opening, you will create a venturi effect, which will add to the efficiency of the stream itself. At times, this 6–8 ft positioning won't be possible or practical. In the case of a well-involved commercial area of medium size, to get 6–8 ft away from the window might mean that you have to pass by a lot of smoldering material. When you begin your ventilation efforts, the fresh air being drawn toward the nozzle can cause this smoldering material to burst back into flame, drawing all the heat and smoke right past the nozzle team. In this case, it is better to give up some efficiency and remain farther back, using a narrower stream to reach the window or door. If practical, keep the fire between the nozzle and the vent point until you are sure there is little danger from this fanned rekindle.

This danger of fanning into life an otherwise smoldering fire is a problem with all types of mechanical ventilation, since the volume of air moved is generally much larger and occurs much faster than if you rely on natural ventilation. Using a hose stream for this purpose, therefore, carries a big advantage in that the personnel operating the line are in the best place to detect the fire and to use the hoseline to darken it down. Of course, there are drawbacks to using a hose stream. It can't be used at minor fires where water damage is a concern. In areas of poor water supply, the drain of using part of the available water to move smoke may mean running short of water for fire control. Nor should the hose stream be used in areas of below-freezing temperatures where the resulting ice will create fall hazards.

Smoke ejectors, blowers, exhaust fans, and other portable devices suffer none of the drawbacks of the hoseline except the same problem of a forced draft over smoldering embers. Problems can develop regarding the availability of the power supply, manpower requirements, storage space, whether explosive gas is present, and other circumstances. Still, these can usually be overcome. Enhancing the conditions by moving smoke that much faster is well worth any inconvenience.

Negative vs. Positive Pressure

Mechanical ventilation offers firefighters a great opportunity to channel products of combustion where we want them to go. The various types of smoke-removal devices offer a choice of modes of operation, either sucking smoke out of the building or blowing fresh air in. For the first 20 or so years of mechanical ventilation, their most common use was as exhaust devices, sucking smoke through the mechanism. This has proved to be a rather ineffective method. First, fans never fit an opening correctly, so some makeshift method has to be used to block off the openings around the sides of the unit. The penalty for not doing so is a churning effect, limiting the efficiency of the device. Churning results in some air from the outside being drawn into the building to be blown out again through the fan. This means that the fan isn't using all of its capacity to move smoke, since some percentage is wasted moving clean outside air.

Other problems result from debris, curtains, and other objects being drawn against the intake screen, thus blocking the flow. Another disadvantage is the location of a negative-pressure fan, hanging in or near doorways that must be used by personnel. Because they blow outward, combustible gases are drawn across the motor. These fans are often of explosion-proof design when they leave the factory, but after five years of use, I wouldn't want to bet my life on one still being so.

By using these devices the other way—as fans blowing fresh air into the structure—you can eliminate many of these difficulties and also improve the efficiency of the airflow. The idea was originally tested as a way of reducing stack effect, *i.e.*, drawing heat and smoke toward staircases, and thus the points of attack in high-rise fires. Experimentation has proved that, by putting a positive pressure in a stairwell, the flow of heated gas can be reversed, enabling attack crews to gain a foothold on the fire floor. The Los Angeles Fire Department (LAFD) expanded on these experiments and has proved the usefulness of the concept in many types of structures other than high-rises (Fig. 9–6).

Fig. 9–6 Position positive-pressure fans to cover the entire doorway.

Advantages of Positive-Pressure Ventilation

Positive-pressure ventilation (PPV) has decided advantages over negative pressure in certain respects. The first is efficiency. With PPV, the fan is set up outside the structure 8–12 ft away from the desired door opening. The airstream from the fan is aimed to fully cover the opening. Larger doors in commercial structures may require several fans. The moving airstream entrains more air along its sides as it flows, acting as a venturi to bring in still more air as the stream passes through the opening. This resulting movement of air into the structure pushes smoke-contaminated air ahead of it. Just as when advancing a fog stream, you must provide an outlet for this gas opposite the entranceway to speed the process. A fan blowing in can move almost twice the volume of smoke as compared to the negative mode, especially if churning occurs (Fig. 9–7).

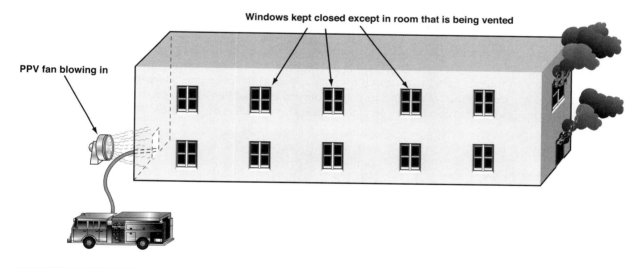

Fig. 9–7 Venting a large structure using PPV should proceed in stages. Do not open all openings simultaneously, or the fan may not be able to move a sufficient volume of air to pressurize all the openings.

Another way of looking at it is that the same job can be done in half the time. Experiments conducted by LAFD show that best results can be obtained by clearing one section of a structure at a time by first opening the windows in the room to be cleared, then moving on to the next room, closing the window or door to isolate that room. In setting up the fans, you should take any natural wind into account, working with rather than against it. This method works best with minor to moderate fires where smoke, rather than fire, is the greatest difficulty. With advanced fires that have self-vented, the effect of the fan's airstream may be reduced or overpowered by the fire's convective currents and outside air movements, due to the large number of openings spread around the building. Mattresses, stuffed chairs, and other extra-smoky fires, such as those partially extinguished by sprinkler systems, are prime candidates for positive pressure. Since PPV fans are set back from the doorways, they don't clutter the access routes. Also, they have a ready ability to remove flammable vapors safely. Natural gas or propane, for example, may have collected in areas away from openings. Positive pressure can offer the assistance they need to get them moving out, without adding the dangers of using a potentially nonexplosion-proof fan.

Disadvantages of PPV

As with any firefighting technique, PPV has some drawbacks, the most serious being the danger of fanning smoldering fire into a serious conflagration. The best way to prevent this is to have a charged hoseline in place and ready to operate. If the fire is in a concealed space, however, it may not be possible to expose it and darken it down fast enough to prevent considerable extension. Another possible situation involves separate fires in remote areas, as is common in arson incidents. In this case, the hoseline may be in place at one fire, while the other blossoms with the effect of the draft, undetected until it has gained considerable strength.

The effect of blowing fresh air from a fan onto a fire is the same as when it is directed out of a bellows onto a blacksmith's hearth. The fire intensifies. The fan mustn't be allowed to blow the fire toward any victims or firefighters on the opposite side. For this reason, PPV mustn't be used if firefighters are entering the building from any points other than where the fan is operating. This requires great coordination if VES is to be employed.

PPV requires some coordination between the members. Personnel should follow the proper sequence of opening and closing windows as each room is cleared, otherwise the effort won't be as effective. If there are many openings in close proximity to the door being used for PPV, the smoke-filled area may receive little ventilation as the fresh air simply recirculates to the outdoors. Naturally, in buildings where there is a suspected life hazard, the principle of venting for life demands that the windows be taken out. If a department becomes too infatuated with PPV, they may find that their desire to keep glass intact will have an adverse effect on any victims inside who need fresh air to survive.

A basic tenet for an interior firefighter's search for a life hazard is to *vent as you move, provided the fire won't be extended by the venting*. That is exactly the opposite pattern desired to make PPV work effectively. Thus, the decision to use PPV should be made only after considering all of the factors that can affect the situation, including the following:

- The life hazard
- The extent of the fire
- The availability of hoselines
- The degree of confinement
- Environmental factors in the vicinity of the fan (*i.e.*, dust, powders, weeds, or other materials that might be drawn into the fan, damaging it or impeding operations).
- The available equipment and the power supply

If the situation seems to lend itself to PPV, it can make your job much simpler. But remain aware of the conditions in the building, and be ready to order PPV halted if it appears to be having a negative effect on operations.

Vertical Ventilation—Taking the Lid Off

Once fire achieves headway, it builds up pressure of its own within the structure. At times, this pressure can be quite substantial, and it is responsible for creating the advancing flame front through the fire area as additional gas distills from the burning fuel and expands. The natural tendency is for this heated gas to rise. When restrained from rising, it spreads out horizontally—the *mushroom effect*. There are several ways to prevent this horizontal spread. Build a fire-resistive partition, position a countering hose stream, or let the gas continue on its upward journey, where it wants to go anyway. This is best accomplished in multistory structures by making an opening directly over any vertical arteries that the fire is exploring. The most common vertical artery is the staircase. By providing an opening of sufficient size at the top, the fire can continue skyward without extending horizontally.

Firefighters should recognize the buildings where this is possible and, in the event of a serious fire, make all efforts to open this area as soon as possible. Many multistory buildings have a skylight over the staircase. By either removing the skylight or breaking the panes, you can quickly provide ventilation over this critical vertical artery. In breaking the skylight, remember that other personnel will most likely be ascending the stairs. Don't drop the entire housing through the opening, and give the members below some warning of the falling glass. Besides calling out, "I'm taking the skylight!" on the radio, break a single small pane of glass first, and then pause for a few moments. The falling glass will signal the members on the stairs, and they can take steps to protect themselves. They should keep their heads level, looking neither up nor down. They should also hug the wall as they ascend, keeping their hands close to their bodies and off the banister (Figs. 9–8 and 9–9).

Fig. 9–8 Venting the skylight and door over a stairway will rapidly clear the interior stairs and halls.

Fig. 9–9 Remove the door from its hinges to prevent it from closing, and then it can be used as a ramp to help reach the roof of the bulkhead to vent the skylight.

Sometimes staircases continue past the top floor, ending in a small bulkhead with a door leading onto the roof. This door must also be opened to provide additional ventilation, but more importantly, to check for any overcome occupants who may have attempted to flee to the roof. Other likely vertical paths of fire are pipe chases and dumbwaiter shafts. Dumbwaiters are small, hand-operated mini-elevators that were common in multistory residential buildings built in the late 1800s and early 1900s. Their purpose was to lower the tenants' trash down to the basement for collection. Although most are no longer in use, having been replaced by compactors and incinerators, the shafts may still be present, and any fire that extends into them quickly finds another highway to the top. Often, dumbwaiter shafts were equipped with a miniature bulkhead at the top, complete with a door and a skylight. Again, if there is fire or heavy smoke in this shaft, horizontal extension will be greatly delayed if you take the lid off the top (Figs. 9–10 and 9–11).

Fig. 9–10 A skylight over a dumbwaiter shaft should be examined for fire and vented if heat or flames are found.

Fig. 9–11 This bulkhead, complete with skylight, and these soil pipes are prime items for the roof team to examine.

The second most common vertical avenue is the pipe chase. Commonly found behind kitchen and bathroom walls, these shafts extend right up to the underside of the roof boards, where a single vent pipe extends through the roof. This pipe, also called a soil pipe, provides an excellent means of locating these vertical arteries. Simply scan the roof for the soil pipe, and you'll gain an idea of the location of potential extension. By feeling these soil pipes for heat, a member can estimate whether or not fire is about to enter the cockloft or attic. Even if the fire appears confined to a lower floor, you should take immediate steps to fight it on the top floor or in the attic if the soil pipe on the roof feels hot to the touch. A hoseline must be stretched, the top-floor ceilings must be pulled, and, if fire is present, the roof must be opened.

Due to the numerous interconnections with other blind spaces, a serious fire within a building's voids is a severe threat to the very existence of the structure. Fires within these spaces are often extremely smoky because the oxygen available to the fire is limited. Firefighters will have great difficulty locating, opening, and attacking these fires because of poor visibility. The answer is total ventilation. This includes all windows and other horizontal openings as well as roof ventilation, particularly over the top of the vertical voids. Anything less, and it is likely that the firefighters will be unable to remain in positions to cut off further fire spread and will ultimately be driven out of the structure.

Venting Flat Roofs

Flat roofs are generally found on larger buildings such as stores, factories, schools, and apartment buildings. In the past, these roofs were normally much stronger than peaked roofs, since they were designed to support greater live loads. A great number of these older buildings are still around and are what I will call the old-style flat roof. Generally, the following are two types of these old-style roofs: the standard flat roof, where the main roof joists are right at the roof level, with the roof boards nailed directly to the joists, and the inverted roof, where the roof boards are nailed to a framework of 2×4s, raised several feet above the main roof joists. The style of roof that has become dominant in the flat-roof industry in the last 20 years, at least in the East and the Midwest, is the metal deck roof. This is a drastically different type of roof from a fire point of view as compared to the old-style roofs, and it will be discussed separately later in this chapter (Figs. 9–12 and 9–13).

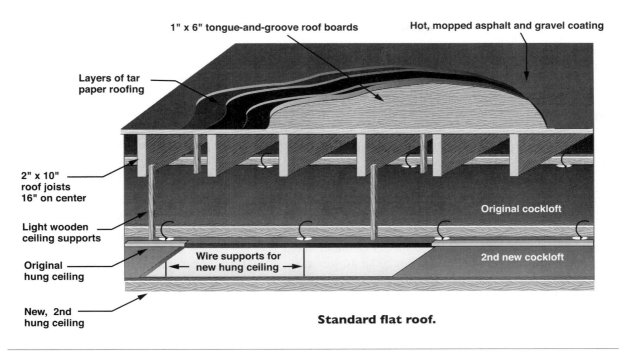

1" x 6" tongue-and-groove roof boards

Hot, mopped asphalt and gravel coating

Layers of tar
paper roofing

2" x 10"
roof joists
16" on center

Light wooden
ceiling supports

Original cockloft

2nd new cockloft

Original
hung ceiling

Wire supports for
new hung ceiling

New, 2nd
hung ceiling

Standard flat roof.

Fig. 9–12 A standard flat roof has its main roof supports (joists) at the roof level, with the roof decking nailed directly to the joists, any suspended ceiling if present is hung below the joists.

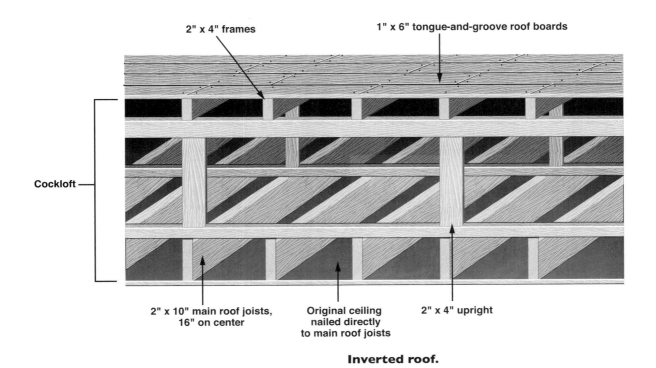

2" x 4" frames

1" x 6" tongue-and-groove roof boards

Cockloft

2" x 10" main roof joists,
16" on center

Original ceiling
nailed directly
to main roof joists

2" x 4" upright

Inverted roof.

Fig. 9–13 An inverted roof has its decking raised several feet above the main roof supports of light wooden supports (2x4s).

Old-style roof construction

Older flat roofs have wooden beams of 2×10s, or even 3×12s, as the main support beams. Typically, these are spaced 16 in. apart and can carry nearly the same loads as the floor joists below them. These timbers are limited in length to about a 20-ft span between supports, most often a masonry bearing wall or a steel I-beam. Nailed directly to the roof beams is the roof decking, usually 1×6 tongue-and-groove planking. This is generally covered by several overlapping layers of tarpaper acting as a moisture barrier (in some very old roofs, a thin layer of tin was also laid on top of the tar paper). During construction, the entire roof surface was then mopped with hot tar to seal it. In some cases, gravel was then spread over the hot tar to act as a retardant to the spread of flames. This style of roof has a limited life span, 20–30 years, and several such layers may be found in very old buildings, making the tar paper up to 3 or 4 in. thick. This is a very substantial roof, and it is relatively safe to work on, even over a heavy body of fire.

Generally, before there is any danger of the supporting joists succumbing to fire, the roof boards (1×6s) will have burned through first. Thus, catastrophic roof failure of old-style roofs is not an initial concern. Of course, if the roof boards are seriously weakened, it is possible to step through the deck, plunging a firefighter's foot or leg into the flames. Old-style construction with exposed joists can be found in older buildings such as garages and warehouses, where heating and appearance aren't primary concerns. However, for many types of occupancies, such as stores, schools, and apartments, exposed joists are unacceptable both for reasons of esthetics and energy conservation. In such cases, a ceiling has typically been added, usually a wooden framework suspended several feet below the roof from light wooden hangers. A void space known as a cockloft is thus created between the ceiling and the floorboards. This space serves as an insulator from both the sun's heat in the summer and the winter's cold. Fire in either the cockloft or on the occupied floor may be readily vented by cutting a hole directly over the fire and pushing down the ceiling. This action is absolutely critical once fire has entered the concealed space, for there is no other means of venting this area.

A working fire on the top floor of any large-area flat-roof building requires an immediate commitment of at least four men equipped with two power saws, two 6- or 8-ft hooks, a halligan tool, an axe, and a portable radio. Preferably, an officer experienced in roof ventilation should also be assigned.

Immediately upon reaching the roof (via ladder or platform, never through the interior stairs), the members should vent any vertical shafts they find, such as bulkhead doors and skylights. Simultaneously, a member can lean over the parapet and vent the top floor windows in the fire area with a hook. Be sure whenever venting windows in the fire area that the wind won't be blowing inward and fanning the flames. A member on the roof can easily evaluate the direction of the wind. *After* this initial ventilation of the stairs and windows, the members can get involved in the more time-consuming operation of cutting open the roof. If you perform these operations in the reverse sequence, you will likely delay fire control by hindering the advancement of the hoseline (Figs. 9–14 and 9–15).

Fig. 9–14 The roof team should vent top floor windows to assist the interior operations. One method that works well is to attach the fork end of the halligan to a light rope. Lower the tool to the appropriate length first, to measure the rope...

Fig. 9–15 ...grasp the rope firmly at this point, pull the rest of the rope back up, then throw the halligan straight out. Be sure to hang on to the rope! The rope will act as a pendulum, and pull the tool into the window.

For a number of years, firefighting texts have advocated that the proper size hole for a fire in a large-area flat-roof structure is 8×8 ft—a very substantial hole. Unfortunately, most texts give the firefighter little guidance on how to create such an opening. The result is a group of firefighters who have little idea how to go about it. The shortcoming is rather common, since fires in these structures are relatively few, and there is little opportunity to practice such activities.

The first step in making any ventilation opening is to locate the proper site, usually as close to directly over the fire as is safe. (**Note:** For this text, whenever I refer to "directly over the fire," I mean as close to that area as is safely possible, considering the conditions on the roof, the danger of collapse, and other such factors.) The seat of the fire is most often determined by observing its condition, either as the members make their way to the roof or once they have arrived there. Fire venting out windows is a good indicator, whereas smoke may not be as reliable. Other means of locating the proper site include communicating with interior forces by radio, feeling the soil pipes as described earlier, observing the surface of the roof for bubbles and steam, and by means of examination holes. Examination holes are small, rapidly made penetrations through the roof to look for fire. Two types are commonly, efficiently made with a power saw as follows: either a triangular opening, or a short, narrow slit that is only the width of the blade. This latter is called a *kerf cut*.

The kerf cut is the simplest, fastest type of examination hole possible. It is also easily repaired if the fire doesn't severely damage the roof. It is made simply by plunging the saw through the roof and then pulling it out. This creates an opening about ¼ × 10 in. Although quickly made, it often isn't a very positive indicator of fire, since the opening is so small that it doesn't allow you to view the areas to either side. If there is fire directly below the hole, it will show through, but fire on either side may not be visible, especially during daylight. A slightly more time-consuming method of making examination holes is to make a triangular-shaped opening where you suspect you might find fire. Simply plunge the saw through the roof three times so that the edges of the cuts overlap. The resulting 8- to 10-in. triangle will readily fall into the structure, unless you have cut directly over a joist. Still, you can easily knock out the piece with any tool. This type of cut improves visibility to the sides, and the hole may also be used to insert a distributing nozzle or pipe, such a Bresnan distributor, to knock down fire if the handlines are unable to advance.

Once you have chosen the proper site, plan the cut and make all of the members in the vicinity aware of the operation. Power saws of any type are deadly weapons under the wrong conditions. The first item to consider is the means of escape from the roof if things go badly. Make the cut so as not to be between the members and their means of escape, especially on smaller roofs. Fire venting out of a hole can seal off even a relatively large path. Be sure you have at least two ways off of each roof. If there's only one, make sure somebody arranges another (Fig. 9–16).

I can recall being stuck on a roof by heavy fire that cut off our escape routes. A shift in the wind had blocked our intended path, and the chief in charge was hesitant to call for another ladder truck just to provide us with another *fire escape*. From his point of view, the fire on the lower floors was being controlled. He had no idea of how miserable things had become on the roof. It would only

Fig. 9–16 Never make roof ventilation holes where they will cut off your means of escape. Always have a second way off a roof.

take a short while before the fire burned through more of the roof, forcing us into our last area of refuge. From there it would mean that somebody would have to put up a ladder to us; otherwise, we would be forced to slide down our lifesaving rope—not our favorite option! Fortunately for us, a chauffeur of one of the ladder trucks had monitored our transmissions and was able to reposition his apparatus to where we could use its ladder. While it may not have been the closest any of us had ever been to death, it was certainly the least excusable. From that point on, if I didn't have a sure escape route available, I would never place a hole where it might interfere with our intended route of egress, even if that meant it wouldn't be in the best location for ventilation.

The second concern when positioning the cut is the wind direction. This may be the first concern in some cases, since the wind might be blowing the fire or smoke in a way that would cut off your retreat. Locate the cut so that, if the fire vents out, it won't endanger nearby exposures. In case further expansion of the hole is necessary to relieve conditions below, make the initial cut so that it won't unduly hinder further cutting. The sequence of cuts needed to produce an 8×8 ft opening must be logically planned so that members making the later parts of the hole won't be downwind of the earlier segments where heat, smoke, or fire might prevent them from completing their tasks. Of course, the cut should never be laid out so that a member, in doing further work, must step on a portion of roof that has already been compromised.

The actual mechanics of making the proper sequence of cuts requires that the saw operators be capable of cutting both left- and right-handed so as to ensure that they are standing on solid roofing at all times. Locate the supporting roof joists by sounding them with an axe or other tool. Strike the surface of the roof with the blunt end of a tool held lightly in your hand. When the tool lands in the space between the joists, it will have a slight bounce to it. When the tool lands above the joists, it will strike solidly, with little or no bounce. The cut should be made close to the inside of the selected joists. This will provide as large an opening as possible. It will also mean that those who are pulling the roofing will have to pull against as few nails as possible. If you were to shift the cut only a few inches to the outside of the beams, you would encounter an additional set of nails that would require substantially more effort to conquer.

The saw operator needs to keep in mind the process of pulling the cut open as he makes the incisions. The hole should be subdivided into pieces of a manageable size (generally 4×4 ft maximum), since it is virtually impossible to pull up an 8×8 ft piece of roofing. The saw operator should also make sufficient knockout holes, triangular openings at the corners of each section designed either to fall or be pushed down, similar to inspection holes. These knockouts allow members to insert hooks below the roofline for pulling or prying. It is next to impossible to get a purchase on the roofing without them. It only takes a matter of seconds to make these cuts. Simply insert the blade so as to cross over the interior angles of the longer cuts. By making these holes at several locations, you can pull stubborn roof boards from several directions if necessary. This is extremely valuable when those inevitable wind shifts cause the members on the roof to change their position (Fig. 9–17).

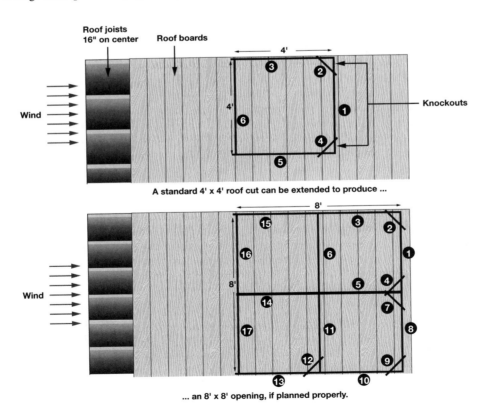

Fig. 9–17 Plan the sequence of roof cuts, taking wind direction, escape routes, and fire location into account.

The members can begin opening the hole once the 8×8 ft outline has been cut. It isn't normally a good idea to start pulling one cut section before the remainder of the cut has been completed, but doing so may be necessary under some circumstances. If you have only one saw, for example, the 55 linear ft of cutting required to make an 8×8 hole may take so long to complete that the interior forces are driven out of the building. If the first quarter is pulled after making the initial seven cuts, however, the attack team may be granted sufficient relief to remain in the area (Figs. 9–18, 9–19, and 9–20).

Fig. 9–18 Removing a cut section of roof covering requires teamwork.

If a second saw is available, it is usually best to complete all of the cuts before pulling the section of roof. Once again, the key factor is the direction of the wind. If it remains steady from the original direction, chances are that you will have no problem completing, or even extending, the original hole. If the wind is shifting, it's best to complete all of the cuts before you open up so that you won't be driven off by the smoke and heat.

Occasionally you will find that, even with a good-sized hole over the fire, conditions require further expansion of the vent. It is usually better to extend an existing cut than it is to start a new one. The first reason is that you are sure that you're in the right location. The second reason is that it will take less cutting, since the existing hole will act as one side of the extension. Cut to vent as many additional bays between the joists as possible rather than extending the channel between existing joists. Joists act as draft curtains, and the more bays you open, the faster things will clear up. Keep the wind at your back.

Fig. 9–19 Members pulling cut sections need to get a bite on the roof boards. Providing plenty of knockouts is the answer.

At times, it isn't just the smoke blowing out of the hole that drives you away, it might be the heat from venting fire. Even if the wind is blowing the fire away from you, the radiant heat can make close approach impossible. You can beat this temporarily by using the cut section of roof as a shield, buying you sufficient time to complete a few additional cuts or to start pulling the next section of the hole. Using a hoseline in any ventilation opening must be strictly forbidden. Remember why the hole is there in the first place—to let out all that bad stuff. A hose stream from the top simply blows it back in toward the troops below. It won't cool the burning material under the roof, so it can't put out the fire. Of course, working this closely over the seat of the fire demands continually evaluating the stability of the roof. If it is a standard roof, it is highly unlikely that the supports will weaken before the roof boards have failed. By the time failure occurs on this style of roof, the fire forces will have been driven back by fire venting through the boards. Beware, however, of roofs that have been weakened by previous fires or rot.

Fig. 9–20 Use a cut section of roofing as a shield if fire prevents close approach while extending the cut.

The inverted roof

The second style of roof is the inverted roof. It is similar to the standard flat roof in that it is constructed with a cockloft, with one surface separated from the main roof joists by a light wooden framework. In this case, though, instead of the ceiling being suspended below the joists by the framework, the roof boards are raised above the joists on a framework of 2×4s. This style of roof was very common on many large-area flat roofs built from the 1920s to the 1960s. Flat roofs aren't usually truly flat. Some pitch must be built into them to provide drainage. Particularly where the roof beams are supported on bearing walls of masonry, it becomes quite complex to arrange the 2×12 or 3×12 roof joists at precisely the right angle to create the desired runoff across a 150×200 ft H-shaped building (Figs. 9–21 and 9–22).

When these buildings were being built, a simple means was to set all of the large timbers at a flat, level height and create the pitch by means of upright posts of varying length. These would support a roof frame. Then 2×4 uprights can easily be cut to varying lengths. (Remember, this technique was most popular before power tools became common.) By starting at one edge of the roof and cutting each successive row of supports a ½-in. shorter than the last, you can easily provide the desired pitch to a roof. A plate is nailed at the top of each row of uprights, to which the roof boards are in turn nailed. Typically, each row of uprights consists of posts 2 ft on center, with the rows spaced 4 ft on center. As you may imagine, this lightweight construction doesn't feel as solid underfoot as would the standard roof, with its roof boards nailed directly to the 3×12 joists. In fact, the inverted roof has a very definite springy feel to it. This is normal and shouldn't be taken as a sign of impending failure.

Fig. 9–21 An inverted roof undergoing repair after a fire. Note the slightly charred 2x4 frame lying over in the foreground.

Fig. 9–22 An inverted roof from below. The main roof joists are at the level of the ceiling, and the roof itself is raised 4 ft above on 2x4s.

The inverted roof is actually relatively stable under fire conditions, for while it is supported on lightweight framework (2×4s), the frame is designed so that each portion rests directly on a support leading to the main joists at the ceiling level. Unlike modern lightweight wood trusses, which use 2×4s held together with pins or gusset plate, the inverted roof doesn't depend on a fastener to hold it in place. Gravity does the job of that, pulling each piece down onto the piece directly below. Even when the 2×4s burn through, the roof will only drop down a foot or two to the level of the main joists. These main joists, being at the bottom of the cockloft, aren't subject to the highest temperatures and will remain in place to support the framework above. This is in direct contrast with the newer lightweight trusses, which lose their entire strength once a portion of the truss burns through, often resulting in total roof failure.

Cutting an inverted roof should proceed the same as cutting a standard roof. Use extra caution to avoid making too deep a cut, thereby severing parts of the 2x4 framework. Early recognition of the presence of an inverted roof, plus an understanding of how one behaves under fire conditions, will allow members on the roof to evaluate more realistically just how bad the conditions are. Just because the roof shakes slightly when you strike it with an axe doesn't mean that you must abandon it. The preferred way to recognize an inverted roof is during prefire inspections. If that isn't possible, then inspect the supports when you make your first hole in the roof. If you find 2×4s, put up the caution flag. You must examine further to find out if the roof is of inverted or truss construction. If gusset plates or other types of truss connectors are visible, or if the 2×4s join the top plate at other than right angles, you have a truss roof and must take appropriate steps to evacuate. On the other hand, if the 2×4s are simply nailed directly atop the lower supports, you have an inverted roof and should be able to carry out required functions as safely as on a standard roof (Fig. 9–23).

Rain roof.

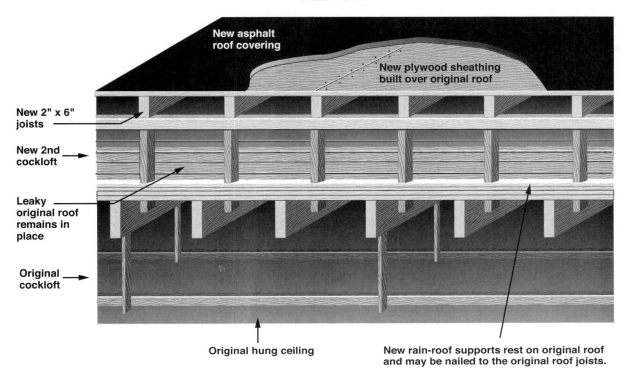

New asphalt
roof covering

New plywood sheathing
built over original roof

New 2" x 6"
joists

New 2nd
cockloft

Leaky
original roof
remains in
place

Original
cockloft

Original hung ceiling

New rain-roof supports rest on original roof
and may be nailed to the original roof joists.

Fig. 9–23 A rain roof is similar to an inverted roof, but has the added danger of extra weight and a second concealed void. It is far more dangerous than an inverted roof. To detect the rain roof, poke down with a hook or pike pole. If you hit solid roofing below, alert the IC.

One dangerous relative of the inverted roof is the rain roof. You can encounter this oddity on any type of roof—flat, curved, or peaked—of standard or truss construction. It is built over an existing roof that has become so porous that it no longer keeps the rain out. The cheap way out of this situation is often to leave the existing roof in place and to build a whole new one on a raised framework above the original. This poses several dangers. First, the added weight was never designed into the original roof supports, which may be approaching the point of failure. The second is that the two layers can substantially delay, or entirely prevent, venting. I operated on one of these roofs over a supermarket fire (flat standard roof) and it took us 45 min to open a hole approximately 6×6 ft. Finally, the multiple layers create multiple void spaces where fire may be hiding. This can result in conflicting estimates of a fire's intensity.

At the 1978 Waldbaum's supermarket fire in Brooklyn, which killed six firefighters, personnel operating on the roof initially reported little heat and smoke exiting from their ventilation openings. The main body of fire was in the cockloft below the original roof, which hadn't been penetrated by vent holes. The firefighters were unaware that they were operating for a substantial length of time atop very heavy fire in a truss roof. The result was that they paid the ultimate price for this lack of knowledge.

The Problem with Truss Roofs

Of all the various types of roofs on which firefighters may have to operate, the various types of trusses are the most dangerous. This is due to their lightweight construction (Fig. 9–24).

Both flat and pitched roofs may be constructed with either of the following two classes of supporting members: standard or lightweight construction. The primary difference between them lies in the mass of the beams that carry the dead and live loads. The standard roof, as described previously, has relatively large members, such as dimension lumber joists or heavy steel I-beams. This was the most common type of construction until the 1960s. As such, it was the type of roof on which firefighters gained their experience. A roof of 1×6-in. sheathing or ⅝-in. plywood placed on 2×10 joists spaced 16 in. apart is a substantial roof. It isn't normally prone to sudden collapse, unless a supporting column is destroyed, since the 1x6 sheathing or plywood would usually burn through before the beams supporting it lost their strength. This type of roof has several drawbacks from an architect's point of view, the first being the area that can be spanned. Using a standard wood

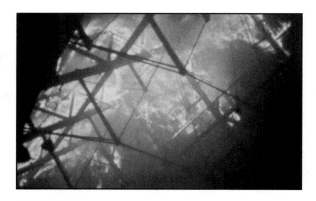

Fig. 9–24 The timber and steel rod trusses are clearly visible in this 120-year-old waterworks. The truss collapsed shortly after the photo was taken.

beam, the maximum length that can be spanned is about 25 ft. Past this point, there must be a support, either a bearing wall or a column. Neither one may be acceptable in buildings such as gymnasiums, which require large, open floor areas. The second problem is that they are expensive. It costs money to space these short beams this close together. As a result of these pressures, a new class of construction is being used more often—lightweight construction.

The Truss

Lightweight construction makes the best use of the truss design. Trusses aren't new inventions. In fact, they date back hundreds of years, supporting the roofs of churches and cathedrals. A truss allows large areas to be spanned with pieces of material that are much smaller than, and may be spaced farther apart than, a standard beam. This makes it attractive to architects but creates dangers for firefighters. The danger of a truss exists only under fire conditions. Normally, truss roofs are perfectly safe and strong, but their design results in several flaws from a firefighting standpoint. The first flaw is the lack of mass as compared to a standard beam. A truss is designed to be lighter and comprised of thinner members than a comparable standard beam. When exposed to fire, however, this means that a truss has less to give up before it fails. Also, the rate of heating of a smaller member is much faster than that of a larger one. A ⅛-in. thick piece of angle iron will heat to its failure point before a ⅝-in. thick piece will. That is the problem with lightweight steel trusses, also known as *bar joists*.

The second problem centers around the connectors. In a truss, short sections of material (less expensive) are joined together to span large areas. These connectors may be of different types, such as gusset plates, metal pins, or through bolts. These connectors can conduct heat to the inside of the material, hastening its destruction at the same time that it is being attacked on its exterior. If the connection fails, the truss may fail (Fig. 9–25).

A truss depends on every one of its pieces to hold it together. This is what poses its greatest danger to firefighters. The destruction of any one of its members, or the failure of its connectors, may cause the entire truss to collapse. Like a chain, it is only as strong as its weakest link.

Fig. 9–25 Gusset plates are the weak links in lightweight trusses, only penetrating the wood by ¼ in. and often pulling loose **prior to** installation.

Trusses consist of several elements joined together. They must all include a top chord, a bottom chord, and web members that connect them. These elements are arranged in a series of triangles creating a rigid shape. It is a geometric fact that, for any three pieces of a given length joined together, only one triangle may be formed. As long as they remain joined together and each piece stays intact, the triangle will retain its integrity. That is the key to understanding the dangers of truss roofs: they only hold their shape as long as the entire structure is intact. If it isn't, the entire truss may suddenly fail, causing the roof or floor to collapse. In examining the design of a truss, we find that the top chord bears a compressive load. The bottom chord, under tension, acts to *pull* the top chord into position. The web acts to transmit the load back to the other two at short intervals (Fig. 9–26).

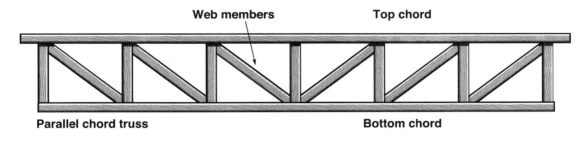

Fig. 9–26 The parallel chord truss is very popular in both residential and commercial buildings.

Picture a hunter's bow as a truss, with the bow as the top chord. If we push down on it, it will flex downward. Once the bowstring is attached, however, and has drawn the bow into its working shape, pushing down on the bow will have little effect. The string under tension keeps the bow pushing back up. Imagine the catastrophic effect that cutting the string would have on the loaded bow—this is what happens to a truss if either chord fails.

Trusses of various types are found in most types of lightweight roofs. Since there is a high probability of early roof collapse, their presence should immediately be made known to the IC. Members who are used to estimating the stability of a standard roof may not be aware that a lightweight roof is about to collapse under them, since there may be no obvious signs of impending failure. Dozens of firefighters have been killed by the failure of truss roofs over the last 35 years. The decision as to whether to commit members to roof or interior operations must be based on a well-informed size-up. Factors to consider include the severity of the fire, how long it has been burning, and whether or not fire has reached the truss. Once heavy fire has involved the truss area, it is too late to commit interior forces, especially if there is no civilian life hazard. Under fire conditions, the presence of a truss should at least be viewed as a red flag of danger (Fig. 9–27).

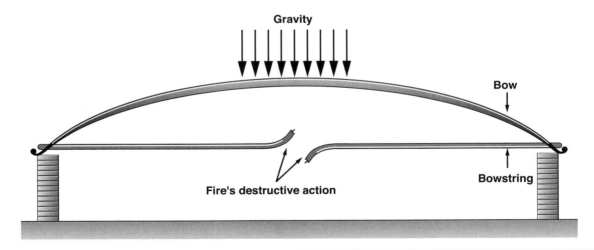

Fig. 9–27 A bowstring truss behaves much the same as a hunter's bow.

Types of Trusses

Of all the modern styles, bowstring truss roofs are the deadliest style of roof construction per incident. There is no excuse for these firefighter deaths, since the bowstring is by far the most easily recognized type of lightweight construction in use. The classic hump shape may not be visible from the street, being hidden by a high parapet wall, but any member reaching the roof should instantly recognize and report its type. The bowstring truss roof typically spans large floor areas, and it should be expected in occupancies such as bowling alleys, auto dealerships, and supermarkets. The web and chords consist of what appear to be massive timbers, but which are typically 2×10 in. boards bolted together with spaces between alternate boards. The usual spacing is 20 ft between trusses. Thus, the failure of a single truss can instantly open a 40-ft hole the full width of the building.

Failure of the end truss usually results in the hip rafters pushing out on, and collapsing, the end wall. At one incident in Los Angeles, several firefighters outside the building narrowly escaped death when this occurred. Another nearly identical incident in Brooklyn caught two FDNY firefighters setting up a defensive position well outside what was perceived to be the collapse zone. The collapse zone at these end walls that have rafters resting on a truss and bearing on a wall should be at least *twice* the height of the wall. If this isn't possible, members should be kept out of the collapse zone and operate instead from flanking positions, off to the sides of the wall in question. They might also operate above the wall in aerial platforms or from adjoining buildings.

Sometimes the failure of one truss pushes additional trusses over in a domino-like fashion. Noted fire service and construction author Francis Brannigan feels that bowstring trusses are so dangerous that all fire forces should immediately be placed in safe, defensive positions once heavy fire has involved the truss area. In fact, heavy fire isn't even a necessity, since many bowstring trusses fail simply from old age. Remember, there is little likelihood of a civilian life hazard in these occupancies. The occupants are normally all awake. As soon as they become aware of a serious fire, they will self-evacuate. Don't risk firefighters' lives simply for a structure that can be rebuilt (Figs. 9–28, 9–29, 9–30, and 9–31).

Fig. 9–28 Close-up of a wooden bowstring truss, showing the 2x4 web members. The failure of any one of the members can cause collapse of the entire truss.

Fig. 9–29 Bowstring trusses can bring down entire roof structures, even while entire sections remain intact.

Fig. 9–30 A particularly deadly type of bowstring truss has hip rafters that extend from the end truss down to rest on the end walls.

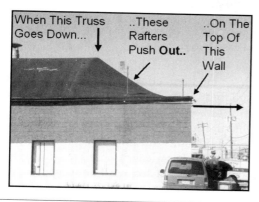

Fig. 9–31 Collapse of the first truss-supporting these hip rafters will push the rafters outward with explosive-like force against the wall they bear upon.

Fig. 9–32 Flat 2x4 trusses create a cockloft (more accurately, a **truss loft**) on each floor.

Lightweight trusses

As dangerous as bowstring trusses can be, at least they tip us off to their presence by their classic hump shape. Other styles of trusses offer no such warning. A new way of creating trusses, using very light pieces of lumber of short dimension joined together by questionable fastening methods, has led to truss designs of varying shapes. These lightweight trusses can be of a parallel chord type for flat use, such as in floors or flat roofs, or peaked, as in many private dwellings. There are a number of possible styles. The most common type has short scraps of 2×4s joined together by means of a stamped sheet-metal fastener called a gang nail or gusset plate. A second style consists of 2×4 top and bottom chords connected by tubular steel webs. These steel webs are pieces of tubing stamped flat at each end. The top and bottom chords are notched in the center to allow this flattened end to sit in the chord, then a hole is drilled through the side of the chord, into which a pin is inserted to keep the web connected to the chord. Both methods, while structurally sound under normal conditions, have serious implications for firefighters. Their biggest drawback is a lack of mass to resist fire. As stated earlier, a truss depends on all of its pieces remaining intact in order to maintain its shape and strength. All we need is for a single 2×4 to burn through, and failure of the entire truss becomes possible.

To make matters worse, however, both methods of joining the pieces together worsen the effects of fire. The gusset plate penetrates the wood only about ¼ in. When fire attacks the surface, this gusset plate readily pulls free, leaving the truss free to collapse. The pin drilled into the steel tube and wood-chord truss has two effects. First, drilling the hole (up to ¾ in. in diameter) further weakens the narrow dimension of the 2×4. Second, the steel pin and hole allow fire to attack from both the inside and outside of the chord. A concern common to all of the parallel chord 2x4 trusses is the risk of penetrating the top chord with a power saw blade during ventilation or overhaul. Since these trusses are often covered with only an inch of roofing and plywood, the 2×4 (1½ × 3½ in.) laid on the flat is vulnerable even to a shallow cut. This danger is quite real, and it could result in serious consequences. Tags attached to each individual truss by the manufacturers warn, "Anybody who cuts through any part of the truss shall be responsible for the cost of repair and engineering calculation needed to ensure stability of the truss." Pity the poor firefighter who cuts through one, causing a collapse that throws him into the inferno below. If he's lucky enough to survive, he may find the building owner suing him for damaging the roof (Fig. 9–32).

Another problem common to all open-web trusses is the large void spaces that they create. Professor Frank Brannigan refers to this void as the *truss loft*. Particularly with parallel-chord trusses, which are often used in place of floor joists, these spaces present a new challenge of uncontrolled fire spread. Structures using solid joists as floor supports often had the ceiling attached directly to the beams of the floor above. This acted as a horizontal fire stop, keeping the fire contained within that particular bay. Thus, if the fire remained within that bay, only two joists were subject to burn-through and possible collapse. The open-web truss of either wood, steel, or a combination of wood and steel doesn't have that fire-confining benefit. A fire that exposes one truss almost simultaneously exposes them all (Figs. 9–33, 9–34, and 9–35).

Fig. 9–33 Flat 2x4 trusses covered with plywood can easily be severed by power saws.

Fig. 9–34 The stamped metal fasteners that act as web members of these trusses are likely to weaken even before they get a chance to pull out of the chords.

Fig. 9–35 The thin tubular steel used as web members of these trusses is connected to the 2x3 top and bottom chords by drilling through the 2x3 and driving a metal pin through all the pieces. Early collapse is virtually guaranteed.

The firefighters of tomorrow are going to be paying with their lives for the practices of today's construction industry. Every effort must be made to identify those buildings where lightweight wood trusses are in use and to create or modify tactics to ensure safe operations at those buildings. Unfortunately, these buildings aren't as distinctive as bowstring roofs, so they must be identified before the fire and before they are covered over with plasterboard and other finishing. Today's firefighters, who have gained their knowledge of fire behavior from standard roof and floor systems, must be aware that the time frames and danger signs to which they are accustomed are no longer reliable. No longer will they have 10–20 min in which to search, rescue, and attack before they notice the floor or roof sagging. Many of the new systems fail in less than 10 min, dropping their entire load, often without any noticeable sag prior to snapping.

One particularly dangerous entry is the plywood I-beam. The marketers of this product are quick to assure owners that they will never be disturbed by squeaky floorboards again. This parallel chord beam, while not a truss, poses many of the same problems inherent to any lightweight system, mainly a lack of mass. It consists of a top and bottom chord of $1\frac{1}{2} \times 2$-in *plywood* with a web of ⅜-in. *plywood*. Does anybody in the lumber industry have any idea of how fast ⅜-in. plywood burns through completely? This joist has been shown to produce no noticeable sag until it suddenly fails. The owner of the nice, quiet floor should hear the screams of the firefighter as he plunges into the flames below (Figs. 9–36 and 9–37).

Fig. 9–36 Plywood I-beams are very quickly destroyed by fire and fail with little or no warning.

Operations on Lightweight Roofs

Until means are developed to vent fire-building roofs remotely, and until they become popular (as opposed to the unpopular reaction to the Philadelphia ventilation method used at the MOVE siege), firefighters are going to be held in check from ventilating lightweight truss roofs. If roof ventilation isn't possible due to the danger of collapse, then interior operations must also cease. Where fires exist in occupied structures built using these methods, the IC must weight an extremely heavy burden. When do you decide that the risk to firefighters outweighs possibly saving the lives of occupants?

As a guide, try to determine whether fire has penetrated the voids, thereby attacking the trusses. If so, firefighting should be done quite carefully and from a distance, using the reach of the stream to extinguish fire well before allowing members to operate directly over or below the damaged trusses. Opening up the building's sidewalls and directing a stream into the trussloft may be quite useful. Tower ladders are the tool of choice for this operation. Then place the primary emphasis on assessing the structure. Proceed slowly, providing total window ventilation and maximum lighting. Operate with the absolute minimum of personnel in the danger area until the trusses have been exposed and examined, and the safety of further operations has been well evaluated. Provide emergency shoring as needed. CADS-generated reminders of a building's trusses are invaluable pieces of information for today as well as for 20 or 30 years from now (Fig. 9–38).

Metal Deck Roofs

One very common type of truss roof, especially on commercial buildings, consists of open-web steel bar joists supporting a corrugated metal deck. Steel bar joists are formed by bending a steel bar in a repeating Z pattern. The top and bottom chords are formed by welding two sections of angle iron to the bends. These joists can be up to 60 ft long and may be spaced up to 6 ft apart. The roof decking—steel sheets approximately ⅛ in. thick, 6 ft wide, and 20 ft long—is laid on top of the bar joist and tack-welded in place. A layer of tar is hot mopped over the metal deck to help hold down the insulating material, either felt or rubber sheeting up to 3 in. thick. This insulation is then covered with another layer of tar as a sealant. This type of roof poses some unique hazards that the firefighter must understand.

Fig. 9–37 The top and bottom chords of this plywood I-beam are also of laminated construction.

Fig. 9–38 Fires in buildings with lightweight roofs may sometimes benefit from opening the sidewalls from a platform, as was done at this very large windowless, metal-walled furniture warehouse.

The primary hazard to the firefighter is the danger of falling into the ventilation hole while he or she is cutting it. This can occur due to the unique design of the metal deck and the spacing of the bar joist. The deck is given its corrugated shape to ensure rigidity down its 20 ft length. Otherwise, being so thin, it would bend quite easily. The decking is laid perpendicular to the bar joists, spanning several joists, giving it strength in the narrow direction. The deck sheets overlap at each edge. The danger to a firefighter cutting a hole is that he has no idea where the joists are below and no idea of where the ends of the steel sheets are. Where the joists are 6 ft apart, he may be making a cut that falls entirely within the space between the joists. If this is so, and if he steps on any part of the space between the joists, he may plunge through the roof, since he is in fact standing on metal deck that is supported only at one end. If he is at the end of the sheet, the remaining section can buckle under his weight, throwing him into the hole (Figs. 9–39 and 9–40).

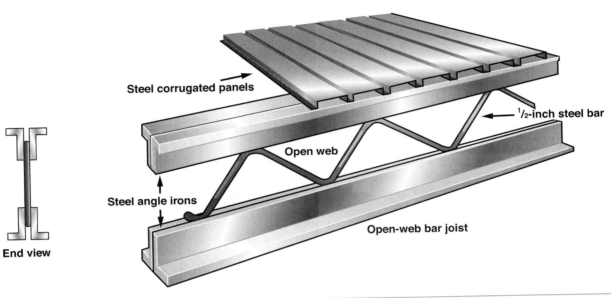

Fig. 9–39 A metal bar joist consists of four sections of angle iron welded to a round **bar** that is bent in a repeating **Z** pattern.

Fig. 9–40 Firefighters critique roof operations on a metal deck roof. Note the sagging roof decking along the left side of the hole.

The second problem with truss roofs—the danger of sudden collapse—isn't as likely in steel-bar joist-supported roofs as it is with other styles of trusses. Although the bar joists are very seriously weakened by exposure to fire, they sag initially rather than snap. Past experience has shown that while these roofs fail in less than 5 min of exposure to heavy fire, they don't fail without sufficient warning. In other words, the roof or floor will drop, but there should be ample warning to firefighters on the level of the deck over the fire. Of course, firefighters advancing below the roof may not be able to see this sag because of heavy smoke or fire, and could thus be endangered. For this reason, it is best to use large hoselines on straight stream to sweep the ceilings well ahead of the line's advance. Provided that it hasn't already pulled free of its support, the steel will regain its previous strength and load-carrying ability when it is cooled by the hose stream.

Many authorities recommend not committing firefighters to bar joist-supported metal deck roofs due to the danger of early collapse. In the spring of 2001, the FDNY made it official department policy not to cut metal deck roofs due to the hazard to firefighters. The effect of this decision is a cessation of roof ventilation efforts. If no ventilation takes place, the interior forces will likely be driven out of the building, resulting in surrender of the structure. If a large portion of the building is already heavily involved on arrival (50% or more of large buildings), this may be an inevitable event.

Still, to make a blanket policy of not operating on metal decks roofs is to give up many buildings that would otherwise be savable. I will not risk a firefighter's life just to save property, but I know that it is possible to operate safely on these roofs. What is needed is an evaluation of the extent of fire involvement, coupled with a coordinated fire attack. In these structures, the attack on the fire area should be conducted carefully, as described previously, using the reach of the stream. Operations in the areas adjoining the fire area (for example, three stores over from the fire store in a row of attached stores) require a hoseline for cooling the unprotected bar joists, preventing their failure. With the roof supports protected, the roof ventilation team may safely vent this exposed store, permitting the hoseline teams to stop the fire's advance. It is critical however that this roof cutting *not* take place over the fire, due to the danger. If it is not possible to operate on the roof due to a lack of cooling streams, it may be possible to provide some relief by opening walls from an elevating platform basket. For a further discussion of this matter, see chapter 14 (Figs. 9–41 and 9–42).

Sweeping the ceiling with large streams also helps reduce the third problem with metal deck roofs—the metal deck roof fire. This can occur without any major involvement of the items directly below and can result in extension of fire over the heads of firefighters who assume that all is under control with a steel roof and no visible fire below. In fact, this may be very untrue.

Once fire exposes the underside of the roof, the metal deck acts as a gigantic frying pan, heating up the layer of hot tar directly above it and distilling flammable gases from the tar. These gases are prevented from escaping upward by the layers of insulation above them. As their pressure increases, they are driven downward through the seams in the metal deck. There they are ignited by the open flame, and they distill yet more gases as they burn. The result is a self-accelerating reaction that can spread quite rapidly, raining little flaming balls of tar on everything below. The only way to stop this fire spread is to cool the bottom of the frying pan, thereby preventing the escape of flammable gases. The reach and cooling power of large 2½-in. hose streams allows the firefighters to catch up to a well-involved roof. It may otherwise outstrip their efforts and result in total destruction of the building. Of course, ventilation will be required to allow firefighters to remain in a tenable area in the event of either a metal deck roof fire or fire in the contents below (Figs. 9–43 and 9–44).

Fig. 9–41 This metal deck roof collapsed after exposure to fire.

Fig. 9–42 An aggressive attack from the doorway with two 2½-in. lines stopped this linen and bath store fire in its tracks. Roof ventilation was done after the steel was cooled to help vent the heavy smoke condition.

Fig. 9–43 If the steel roof joists ahead of the fire can be cooled with hose streams, the remaining stores in a strip mall can be saved.

Fig. 9–44 As dangerous as bar joists, their cousin the "C-joist" and it's variants, will also fail rapidly when exposed to fire.

Insulpan™ Panels

Another new type of roof construction is appearing that has similar potential to a metal deck roof fire for raining flaming droplets on materials below while impeding access for hose streams; the *Insulspan*™ *panel*. Insulspan™ panels are self-descriptive, in that they are panels made of Styrofoam insulation, sandwiched between two layers of oriented strand board (OSB) creating a very effective insulating material. The panels typically come in 8-ft widths and are up to 24 ft in length. The Styrofoam and OSB are bonded together in such a way that the pieces act as a single monolithic unit. The panels require very little other support, thus spanning an area as large as 24 ft without joists, creating large *cathedral* ceilings. This type of roof is being installed on new garden apartment and townhouse developments around the United States (Figs. 9–45 and 9–46).

Fig. 9–45 **Insulspan**™ panels extend from the ridge pole to the outer walls. There are no joists required. The panel is self supporting.

Fig. 9–46 An installer prepares to trim a 10-in. thick Insulspan panel using a chain saw. This view shows the OSB **sandwich** that surrounds the **Styrofoam** insulation.

Products of this sort create several difficulties for firefighters. The first is the nature of the product—Styrofoam insulation is highly combustible, producing tremendous quantities of toxic smoke and intense flame, OSB is also combustible, adding to the fire load. The product is supposed to be installed with a gypsum ceiling separating the panel from the occupancy, but there is certainly the potential for fire venting out a window to ignite the edges of the roofing and spreading into the interior of the panels. Stability of the roof under fire conditions is questionable at this time. I would recommend cutting these roofs (if you can) only from the basket of an elevating platform.

The second problem with these panels is their thickness. Units being installed in Brooklyn, New York, are as much as 8½ in. thick. The typical 12-in. circular saw has a depth of cut of only 4 in., rendering typical ventilation techniques impossible. The panels are readily cut with a chain saw that is at least 9 in. depth of cut, but many of the fire service chain saws have a chain guard that does not readily allow such a deep cut. The good news in this style development is that with the cathedral ceilings on the top floors, there is no common cockloft extending down the row to further spread the fire.

Trench Cuts

The trench cut is a specialized type of cut used to create a firebreak along the surface of the roof. During the 1960s and 1970s, there was a dramatic increase in fire activity in New York City involving large, multi-winged apartment buildings called *H*-type buildings, due to their general shape as seen from above. In fact, their shape often varies from *H* to *E* or *O* shaped, and other layouts. The primary factor found in all of these buildings is a common cockloft over all of the wings. The adjacent sections of large areas are joined by a fairly narrow connecting section called the throat. Many serious multiple alarms occur when fire is encountered in this cockloft space. Once there, it has full access to the rest of the building, often burning off the entire roof and top floor. In the past, firefighters faced with a large body of fire in one wing often try to stop the fire from spreading to the other wings.

The most logical place to do that is in the throat area, where the narrowness of the opening allows a limited number of personnel to fight a holding action without being outflanked by fire. The open courtyard on both sides of the throat assures that. Members cutting vent holes in the roof realized that it was possible to make two parallel cuts from wall to wall faster than it was to cut several smaller vent holes, if the cuts were made at the very narrow spots of the building. Thus was born the trench cut, sometimes referred to as strip ventilation. I prefer the term *trench cut* over *strip ventilation* for two reasons, the first being that the trench should normally be used only as a defensive tactic. To imply that it is part of an acceptable routine offensive ventilation method will result in using trenches at the wrong times and places. The second reason is strictly semantic. The word *trench* infers a deep opening that is crucial to its effectiveness. To be most successful, the ceiling below the roof opening must be pushed down for the entire length and width of the cut, removing as much fuel in the fire's path as possible. This exposes the fire to attack by hoselines on the top floor, and it also serves to improve the conditions under which the hoseline crews are working.

As it becomes better known, a major problem has begun to develop with the trench cut, in that it is often used improperly. As mentioned before, this is a defensive tactic, and it won't replace a properly located ventilation hole. Yet, time and time again, I have seen members rushing off to cut trenches even before the primary vent hole has been completed. The result is that fire is rapidly drawn across the cockloft toward (sometimes past) the trench and into the next wing before the trench can be completely pulled. At other times, I have seen the dramatic reversal that is possible when more experienced members have taken control of the roof operations and have completed a main vent hole over the fire. Conditions rapidly improve on the fire floor and in the cockloft, with fire extension reversing, to come out of the main vent hole, instead of spreading out through the rest of the cockloft. In these cases, fire never reaches the trench, extension is halted, and extinguishment operations are begun, instead of scurrying around trying to get ahead of the fire. Like any other tactic, a trench cut must be used properly to be effective (Fig. 9–47).

Fig. 9–47 This trench cut was located just behind a fire partition in the cockloft and stopped the fire from extending to a large wing of this apartment house. Note the snow on the safe side (foreground) and the steam and dry spots on the fire side.

Cutting the trench

Making a trench cut is actually a process that should be divided into several steps as follows: locating the trench, cutting the inspection holes, cutting the trench, and pulling the trench. I cannot overemphasize that this operation must not begin until the main ventilation hole has been completed. An exception might exist in larger departments where four or more saws are immediately available. In this case, two saws may begin cutting the trench while the remaining members are all committed to the other roof ventilation functions, venting over stairs and shafts, venting the top-floor windows from the roof, and cutting the 8x8 ft main vent hole. Any other sequence of cutting and venting risks pulling the fire faster than you can hope to control it.

Several considerations go into choosing the right location for the trench. First, it must be close enough to the fire to limit unnecessary extension. We don't want to give up too big an area without reason. At the same time, it must be far enough away that you will have time to complete the cutting and pulling, as well as for pushing the ceiling down, before the fire passes it (Fig. 9–48). (About 20–25 ft from the main vent hole should be sufficient, assuming that the main body of fire is below the main vent hole and not closer to the trench.)

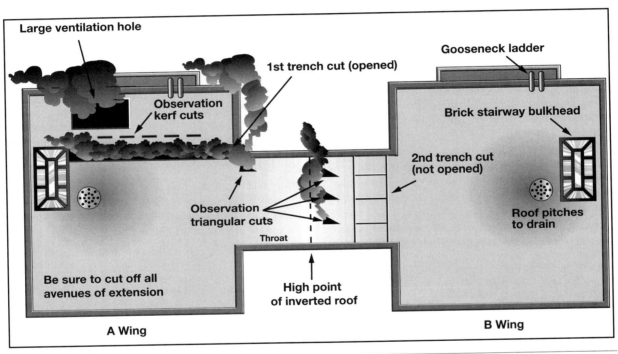

*Fig. 9–48 Top view of an **H** shaped building, showing the wing designation, and the proper location of vent openings and trench cuts.*

The second concern is to locate it by taking advantage of the building's construction to reduce as much as possible the amount of cutting required. This is usually done by making the cut at the narrowest point. Be sure to take advantage of any skylights or stair bulkheads that penetrate the roof and act as fire barriers. The less cutting and pulling required, the faster the cut can be made, and the greater the likelihood that it will successfully stop extension. A final consideration is the likely direction of fire spread. If you are faced with a situation where fire may spread in more than one direction, yet you only have sufficient personnel to make one trench, estimate the most likely path and use the trench and hoselines at that point. The other direction will have to make do with hoselines alone until reinforcements arrive. The most likely direction of fire travel is usually to the downwind side, but if there is little or no wind, the pitch of the roof may come into play. Fire will spread fastest from low points to high points, so position the trench to block its access to the high points.

The next step in making any roof cut is to make some examination holes. These should be positioned on the fire side, 5 or 6 ft from the intended edge of the trench. Their purpose is twofold—to ensure that fire hasn't yet passed the trench area, so that you don't waste your time, and also to serve as indicators of when to begin pulling the trench. In addition, more examination holes should be made on the safe side of the trench, just to be sure no fire gets past.

If no fire is found at the inspection holes, you can begin cutting the trench. Make two parallel cuts 3 ft apart from outer wall to outer wall of the building. (**Note:** The trench shouldn't be made any narrower than 3 ft or fire may jump past it. Making it wider than that also isn't recommended, since you may have to step over it to get back to the fire side once the fire has been brought under control. Also, wider cuts take longer to complete and pull, since more cutting is required and more nails will be holding the roof boards in place.)

With the two parallel cuts made, the next operation involves slicing the length of the trench into shorter sections and creating knockouts on the fire side where tools can grab the roofing. The cuts must be made parallel, as close to the outer walls as possible, and then toward the center so that the trench is divided into a series of 4-ft-long sections, with knockouts at each section. All personnel on the fire side of the trench must at this point have at least two ways off the roof not counting crossing the trench, since this area may suddenly become impassable. One or more hoselines should be positioned directly below the trench, and the ceilings in this area should be pulled or pushed down from the roof. (Usually pushing down is more effective, since the members on the roof are in exactly the right location, while the members on the top floor do not know exactly where the trench is located, how it runs, etc.)

Remember, this is a defensive strategy. These lines should be committed early and be separate from the attack lines. To expect attack lines to shift back to this position after they have been unsuccessful in putting out the seat of the fire may be too little too late. These members may be unable to reposition in time if they are outflanked by fire reaching other areas. Once the trench has been cut, the roof team must avoid the temptation to pull it open immediately, lest they unnecessarily draw fire toward the fresh air. They must have the patience to wait until fire shows at the examination holes on the fire side of the trench. At this point, the entire trench should be pulled. Take care to remove all material below that the fire could travel along. If the ceilings below haven't been completely pulled, the members on the roof should assist by pushing them down.

Keep a close eye on the inspection holes on the safe side to be sure that you aren't being outflanked. By this point, all interior forces on the fire side of the trench should have been withdrawn. That area has been surrendered. The objective now is to be certain that the remainder of the structure is saved. It is perfectly acceptable at this stage to use hoselines either from below or from the roof to drive fire away from the trench (Figs. 9–49 and 9–50). (**Note:** The use of hoselines into roof ventilation openings should be strictly forbidden, as described earlier in this chapter. A trench cut, however, isn't meant to be an offensive ventilation hole but rather a defensive, home-built fire stop. As such, using hoselines through this type of opening may be acceptable. This is one more reason for calling it a trench cut rather than strip ventilation, since we don't operate hoselines into ventilation openings.)

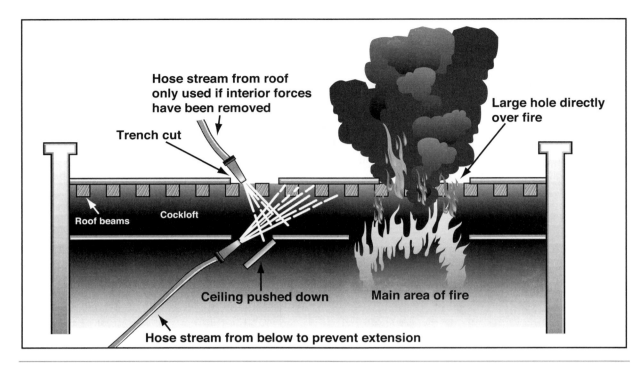

Fig. 9–49 The side view of a trench cut shows the need to push ceilings down below the trench to allow stream access.

Fig. 9–50 A handline should be positioned on the roof on the safe side of the trench as a last-ditch insurance policy to make certain fire does not get past the trench.

Trench cuts are particularly well suited for use in the H-type of building previously described, where narrow areas form choke points between larger sections. Other structures, such as garden-type apartment complexes, may also be suitable candidates for trench cuts. Large-area structures such as supermarkets and taxpayers, having a sizable body of fire already in the cockloft, don't generally benefit from trench cuts, since the length of cut needed to go from wall to wall demands so much cutting that it is impractical. (A 75-ft deep store would require two 75- ft-long parallel cuts, and 20 to 25 3-ft long crosscuts to subdivide the trench into manageable pieces, for a total of more than 225 linear ft of cutting.) By the time the cut was made and pulled, the fire would be far past it. These buildings are better fought putting all available saws to work cutting multiple large vent holes over the fire so that hoseline crews can put out the fire.

Some Additional Considerations when Venting

Never ventilate where you will create an exposure problem. Horizontal and vertical ventilation can both be problems if fire venting from the opening exposes nearby people or buildings. For example, if you will expose people on the fire escape above by breaking a fire escape window into a fire apartment, you must withhold ventilation until those people have been removed or protected (Fig. 9–51).

Fig. 9–51 When fire has gained a strong hold on a building and will likely vent through the roof, lines must be positioned to protect taller nearby structures. In this case fire entered the five-story apartment building to the left through upper floor windows when the roof burned through.

Serious fires that threaten the very existence of a structure demand *total* ventilation if interior forces are to do their job. I once responded to a fire on the ground floor of a large, three and a half story wood-frame restaurant. I was heavily involved in roof ventilation of a one-story extension that had a lot of fire below. Hoselines attempting to operate on the second and third floor and attic of the main building were repeatedly being driven back. When I left the extension and made my way around the rest of the building, I saw the reason. All of the windows on the remaining three sides of the building were still intact, except those that were in the stairway. The desire not to do any extra damage resulted in the total loss of the structure. *Glass is cheap!* If you don't let out the products of combustion, they will drive *you* out of the structure.

That brings up another question. When should you break a window as opposed to opening it from the inside? A good rule of thumb is to let the heat be the guide, not the smoke. If you can stand up and manipulate the locking devices, go ahead and open the window. If it seems stuck and you aren't being subjected to a lot of heat, and if smoke isn't impeding any other operations, you should probably look for another means of venting. However, if when you try to stand up you are driven back down by heat, or if the smoke is severely hindering visibility or mandating SCBA use, that glass should be broken when the hoseline is charged. A problem that the fire service is just beginning to recognize involves the construction of the newer types of windows on the market. The following two factors inherent in these windows make them different from most older models: their material and style of construction, and the so-called thermopane glass. There appears to be a trend developing, evidenced in injury statistics, indicating that firefighters are increasingly being caught in flashovers.

During the time that I was first writing this chapter, I myself was caught in a room that lit up, and I was forced to bail out of a window to escape. The occupant that I was searching for died. The building was fitted with vinyl, double-glazed replacement windows. A routine mattress fire turned an entire room into a wall of flame in a matter of seconds, with all the usual classic warning signs of flashover (open flame growing progressively to the ceiling, then rolling along the ceiling; high heat and smoke; rapid ignition of surrounding combustibles). This isn't your usual or routine mattress fire! I feel that part of the problem was due to the tightness of the building. Nothing was showing on arrival, yet plastic fixtures in an adjoining room had melted and dropped to the floor prior to flashover of the first room. There was an extreme heat condition present, and members were only able to stand for brief periods. Yet, when the mattress lit up due to a window having been vented, the surrounding combustibles ignited the entire room and an adjoining room in a matter of seconds. In my opinion, these windows and the accompanying insulation (6 in. of cellulose in the ceilings and walls) kept nearly every BTU from the smoldering fire within the structure. This raised the temperatures up to the 400–500°F range. Thus, once the open flame developed, the entire top 4 ft of the room ignited quite rapidly.

As I have stated, this had been a routine mattress fire with no unusual heat condition. I crawled right past the mattress, within 2 ft. I have done hundreds of them and have never experienced anything similar to this. I have experienced heat conditions similar to this in thermocouple-equipped test facilities, and that is what I base my temperature estimate on (plus the ignition temperature of ordinary combustibles, such as the books, clothing, and wood that ignited there).

It seems to me that today's common energy conservation practices turn occupancies into kilns, where a slow-burning fire can dry out all of the combustibles in the area. If the rate of burning reaches the open-burning stage, then the entire area is ready to flashover. This wasn't as common in older buildings. In the past, it was quite possible to work in a room directly adjoining a free-burning fire without severe danger. Once fire started to invade your room, you had time to retreat, since the fire's energy in that room still had to heat the surrounding combustibles. In the three months preceding the date that I first wrote this, two firefighters died in flashovers in the New York City area, and a number of others were badly burned. These were all people who had *been there before*, not inexperienced men. Something seems to be changing our frame of reference. (**Note:** For those who will jump on the *overprotected* bandwagon, let it be known that none of the members were wearing hoods.) Until more data is available on this subject, all firefighters should be cognizant of any double-glazing, or even triple-glazing. They may find themselves caught in a rapidly changing fire condition.

The other factors that these windows introduce are the difficulty in venting them with a hose stream, and the difficulty of clearing the entire sash area for entry or escape. FDNY has recognized these aspects of the problem and has taken steps to avoid difficulties. Their procedures call for a member to vent the windows in the fire area thoroughly as soon as they realize double-glazing is present. In the past, some ventilation was left to the hoseline, especially if the forces were shorthanded. Double-glazing doesn't blow out as readily when struck by a hose stream, however, and it requires manual removal to keep steam and hot gases from rolling back on the interior crews.

The inability to clear the horizontal sash in the middle of a double-hung window is possibly a more dangerous effect. Many of the new windows are of rolled vinyl construction that bends and bounces back into shape when struck by a tool. Considerably more effort is involved in removing it than is required for a wooden sash. Other styles of these windows are aluminum framed, and may defy even repeated blows with an axe. I was once forced to exit headfirst through the bottom of one of these windows after being unable to break the sash. It wasn't the type of experience that I would like to make a habit!

The immediate assignment of one or more members to vent the fire-area windows from the outside is an exceptionally important assignment to be made. After the placement of the initial attack hoseline and designation of an interior search team, quite possibly the next most crucial position to be covered is the outside ventilation (OV) team. The OV team is responsible for an immediate exterior survey of the structure, locating victims, locating the seat of the fire, and creating firefighter escape routes. The importance of this effort cannot be overstated. Three firefighters in Pittsburgh died when they were unable to locate the windows in the room they were in—the window in the room was covered with a plastic glazing that fit flush with the wall. An OV team assigned early in the incident should locate and clear such life-threatening obstructions.

Roof-venting Hazards

Just as interior operations pose special hazards, roof operations also present some unique hazards and difficulties. Some problems may be remedied by simple changes in actions or policies, while others may require massive engineering and construction activities to resolve. In all cases, awareness is the key to initiating protective action, since you must recognize that a problem exists before you can solve it.

Probably the greatest danger to firefighters engaged in roof ventilation is the fellow firefighter—particularly the member handling the power saw, but it could also be the member with the axe, hook, or any other moving object in conditions of poor visibility. Firefighters have had toes cut off and limbs slashed by close encounters with power saws. In almost all cases, the victim is someone other than the saw operator. In most cases, the saw operator must hold the greatest responsibility for the injury if he or she fails to perform the most important act on the roof—to stop the saw from spinning whenever it is raised from a cut.

This simple act can reduce the number of traumatic accidents on the fireground. Yet, every day, at almost any roof operation you might observe, you will see or hear power saws with their engines being gunned while the saw man awaits his next cut. This dangerous practice should be banned on the fireground. Gunning the engine engages an automatic clutch on all power saws and sets the blade or chain to spinning. It takes several seconds for this blade to stop. During this time, it can cut through clothing, flesh, bone, and just about anything else. The way to stop this danger is to train all firefighters never to allow a spinning blade or chain to get more than 6 in. above the roof. Don't begin to gun the engine until you are just about to penetrate the roof. After the hole has been cut and you are lifting the saw back out, lift it just clear of the hole and *then immediately set it back onto the roof to stop the blade*. If it is necessary to gun a balky saw to keep it running, you must clear all personnel out of the area. Cut the entire hole and then shut down the saw as soon as you make the last cut. A better move would be to get another saw up to the roof.

The danger from live saws spinning on the roof is greatest when visibility is poor. Under these conditions, however, the hazard to firefighters isn't just from the saw, but from their own movements as well. Firefighters moving around in smoke and darkness must be absolutely certain of what is in front of them before they commit any weight to an area. They can do this only by probing ahead of them with a tool. Unfortunately, this is often easier said than done. For one thing, the saw man may only have his saw. In this case, he should roll the saw along the roof in front of him and gingerly test each footing before stepping so as to avoid falling through. When visibility is completely obscured, the saw operator should completely stop all movement. Again, this may not be possible due to heat or fire being driven toward the firefighters. In this case, the members should drop and crawl toward safety. It is better to take some punishment moving cautiously than it is to walk into a hole or off the roof.

I once operated at a store fire where a member walked off the roof. He was walking toward the rear wall in smoke and saw solid material for several more yards in front of him, so he kept walking. Instead of the roof, what he has seen was the roof of a truck parked behind the building. He stepped into a 12 or 14 ft abyss and struck his head against the building on the way down. He walked out of the narrow space, collapsed, and was pronounced dead several hours later. His mistake was trusting his eyes and not probing his footing.

Sad as that tale is, not three years later, I was at another fire diagonally across the street from that building when another of that same department's firefighters walked off another roof, breaking his leg in several places. In still another instance, a firefighter with whom I worked walked into a shaft between two buildings and was severely injured, breaking his leg and ribs, receiving numerous bruises, and being nearly overcome with smoke. In all three cases, the members failed to use a tool to check their footing *before* taking each step.

Another concern for all members, not just the roof team, involves steel plating added to roofs as a security measure, an increasingly common situation as crime rates go up. It is typically encountered in high-value occupancies, such as jewelry and appliance stores, gun shops, and especially in high-crime neighborhoods. It consists of 4×8 ft sheets of steel, from ⅛ in. to ¼ in. thick, laid onto the entire surface of the roof and welded together. This obviously makes ventilation impossible using conventional methods. A department discovering such a condition must realize the potential dangers and take steps to counter them. Installation of a sprinkler system should be mandatory, but this takes time to complete. In the meantime, warning signs should be posted, visible to firefighters both on the roof and those entering at street level. Since roof ventilation won't be possible, firefighters will have to alter their usual tactics. In the event of a serious fire, early collapse due to the extra weight (8,000 lbs on a 20 × 50 ft roof) is likely. The presence of extra weight on the roof should also be highlighted in department surveys and CADS reports (Fig. 9–52).

Fig. 9–52 Dangerous occupancies and structural hazards should be posted to warn firefighters.

One dangerous weight that can't be planned or regulated is the accumulation of snow or water. I once operated on the roof of a burning tire store that was well involved. As I reached the roof, I found that snow had drifted right up to the top of the parapet wall. Checking the roof for stability with my hook, I found the snow to be 2 ft deep. Although we did operate on the roof, cutting it after shoveling an area clear, we did so only because the fire was largely confined to the first floor and was relatively light on the second floor. If we had had a heavy fire right under the roof, I seriously doubt that we would have operated there, since the added load was substantial.

Roof operations hold many dangers for firefighters. The risk, as well as the importance of the task, demands that the members assigned be among the best that a department has to offer. Failure to complete the assigned tasks can result in destruction of the building or even a loss of life. At the same time, since roof personnel are often operating without the direct supervision of chief officer, they must be the eyes and ears of the IC. It is incumbent on them to recognize and report any dangerous conditions immediately. Failure to do so may result in injury or death to civilians and fellow firefighters alike.

Critical items for the roof team to report to the IC

- Any visible life hazard

- Size and shape of the building, particularly depth, which may not be visible from the street

- Construction (trusses)

- Condition of roof—sagging, spongy

- Fire showing through the roof (report whether it is burning through the roof or if it is coming through man-made openings such as skylights, ventilators, etc.)

- Threatened exposures

- Weights resting on the roof (air conditioners, etc)

- Location of parapet walls

- Shafts

- Setbacks

10 Search and Rescue

Almost all firefighters remember their first really tough job, particularly if something goes wrong. This is true in my case. I had responded to a dinnertime fire in a large, frame, private house. As we pulled up to the block, a woman came running out of the first floor, her hair and clothes ablaze, screaming that her husband was trapped upstairs. Normally, with just two months in the department, I would have been one of the last people into the building, and probably just pulling hose more than anything else. But this situation was different. I was the only free pair of hands with a mask on his back. The officer leading the handline ordered me to "Get upstairs and get the husband!"

I found the interior stair easily enough and climbed rapidly to the second floor. I had a large handlight and a halligan tool. Quickly I thought of the training that I had received regarding search. I knew that a search has to be systematic. I knew that if I put one hand on a wall and just followed that wall around, I would eventually come right back to where I started. I knew that a search also has to be completed as quickly as possible if the victim is to stand any chance of surviving in a low-oxygen, highly toxic atmosphere. I began my search with a door on the right at the top of the stairs. This bedroom door had been closed, and even though visibility in the hall was poor, this room was relatively clear of smoke. It took only a matter of 30 or 40 sec to search it thoroughly. I made my way back out to the hall and, continuing my pattern, made a right turn. Just about 2 ft away, I found another door to another bedroom.

This door was open, however, and searching it took considerably longer due to the smoke. The results were still negative. Continuing to my right, I came to another doorway. I looked in below the smoke with my light and saw one wall about 2 ft ahead and walls to each side. A closet, I thought to myself. I moved on. The next room faced the front of the building and, since someone had vented its windows, it too was relatively clear. At this point, my air bottle's alarm began to sound, so I made my way back to the staircase, where I met two more members arriving to search. I told them to go to the left; that I had covered all but the one room in the front, just to the left of the stairs. When I got to the street, the situation had worsened. The fire had begun in the cellar, and little headway was being made. Gasoline stored there kept reigniting around the members, injuring several of the original hose crew. I was directed to get another bottle and act as relief on the line in the cellar. My thoughts soon turned away from the search that I had just made.

After a long, drawn-out battle, the cellar fire was finally extinguished. I was relieved in turn and was reassigned after a rest period to check the second floor for any tools that might have been left behind. Upstairs, I passed the two rooms on the right that I had searched earlier.

I was drawn to a doorway directly opposite the stairway. Instead of being a closet, however, it led to a short interior hall to two rooms that I hadn't seen during my initial search. The missing husband had been working in one of these rooms when the fire broke out.

What had gone wrong? I almost panicked as I realized that I had missed those rooms entirely. I was sure that I had looked into the doorway, but it had appeared to be a closet. My heart sank. I had failed in my first assignment. I had been caught in a trap that happens all too frequently to inexperienced searchers: I had relied on my eyesight alone to determine what the conditions were. As can happen, the smoke and the handlight beam met in such a way that they appeared to be a solid object. A brief probe with a tool or a free hand would have dispelled that illusion, but I didn't realize how disorienting the fireground can be. What had happened to the husband? He had jumped out the rear window just prior to our arrival and was making his way around to the front just as we made entry to the front door.

As anyone who has ever been inside a burning structure knows, conditions on the fireground aren't the most conducive to searching a building rapidly for sleeping, unconscious, or otherwise unresponsive victims. In many ways, it's comparable to finding a needle in a haystack. The smoke robs you of your visibility. If the victims are unconscious or otherwise unable to speak, we can't follow their voices. Even if they can make a sound, there's a chance that, with the crackling of the fire and the noise of the attack, you might not hear them anyway. Even the sense of touch is confounded. The gloves we wear aren't meant to allow us to perform delicate examinations. Even familiar objects may be so oddly distorted by heat as to be almost unrecognizable. How, then, do we go about searching the premises to maximize the likelihood of finding anybody? The answers to this question are many and complex, but they all begin with one thing—a plan.

To be efficient, all searches must be planned events. That is to say that there is no room for uncoordinated wanderings. Each member of a unit performing search functions must have a clear idea as to what to look for, where to look, and how to look. This function may be performed by the second man on the line as the attack is being made or it may be performed by a specially trained rescue company, but the objective is still the same and the methods are very similar (Fig. 10–1).

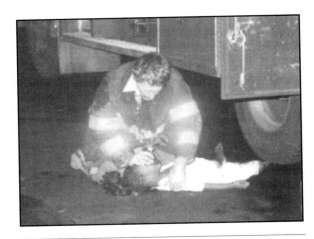

Fig. 10–1 The rescue of an unconscious victim is the ultimate challenge that all firefighters must be prepared for, both physically and mentally.

Primary and Secondary Search

Much as overhauling can be broken down into precontrol and postcontrol phases, so too can search be divided into primary and secondary searches. Overhaul is searching for fire, while search is for human life. The methods and degrees used in the two phases of search will probably be somewhat different, since the conditions under which each is performed vary greatly.

Primary search is a quick search for live victims before the fire has been brought under control. In some cases, it will be performed even before a hoseline has water. Due to the conditions under which it is performed (high heat, no visibility, often just ahead of the fire), it won't be so thorough that a determined person couldn't hide from you. Rather, it is a quick once-over of the entire accessible area, with an emphasis on checking the most likely locations to find victims.

On the other hand, a secondary search is performed after the fire is under control. At this time, the heat is off, literally and figuratively. The need for speed has lessened somewhat. Either the victim has survived or he hasn't. The secondary search must be an extremely thorough search to ensure that there is no possibility of a fire victim remaining undiscovered. Since the conditions of heat and visibility have been improved, it is a somewhat less dangerous, though no less vital, task than the primary search. Because the conditions vary so much between the two types of searches, it is only natural that there are some differences between them. Let's look at the key elements of both.

The primary search is often the first action taken on entry into a fire building. Although it may occasionally be part of a specific rescue attempt, prompted by reports that someone is definitely trapped, primary search must be a routine function to be performed to some extent at all structural fires. When manpower is very limited, the primary may consist of no more than the nozzle man looking along the floor under the smoke as he tries to locate the fire. If he spots a victim, a rescue attempt would surely be the next action. However, as soon as all critical hoselines are adequately manned (generally only two members are needed for midsized 1½- to 2-in. lines), any additional personnel available should be committed to the primary search.

Members assigned the task of making a primary search should be mentally and physically prepared for the task, as well as properly equipped. Preferably, they should be thoroughly experienced personnel. In the preceding pages, I have pointed out some of the reasons for this. Additionally, the trained, experienced member is more likely to be able to extricate himself from any dangerous situations he may encounter. At the same time, she will be better able to recognize her surroundings in the smoke and darkness. If a victim is discovered, the searching member must be able to drag the dead weight of the victim to a safe area. Physically, this is a very demanding task, and all members should realize the importance of maintaining their own optimum physical condition.

Search Safety

When making tool assignments, ensure that each interior search team is equipped with forcible entry tools, both to gain access to any locked areas as well as to facilitate egress, if necessary. A pike pole or hook is very useful for venting windows, probing under furniture (be sure to use the blunt end, not the hook end), and for closing doors to isolate the fire, allowing the search to proceed. Each search team, and preferably each member performing a search, should be equipped with a radio to be able to call for help. A good light is an absolute must, even though visibility may be severely restricted. You may have to put your face inches away from an object and shine the light directly on it to be able to determine whether a given object is a victim or a doll. Preferably, you should arrange the light in some type of quick-release sling so that you can carry it while still keeping both hands free for carrying tools and climbing ladders. The quick-release feature is needed in the event that the sling becomes caught on an entanglement. You must be able to drop the light to escape (Figs. 10–2 and 10–3).

Fig. 10–2 Search operations require a powerful light for each member. Shown here is a lantern on a belt, allowing both hands to remain free. A critically important feature is a quick-release device (a seat belt connector), which allows you to ditch the light if the strap becomes entangled on an obstruction.

In addition, all members performing interior operations should be taught two emergency maneuvers for use with their SCBA. These relatively simple operations can be lifesavers if a member gets hung up or caught in a tight space. They are the reduced profile maneuver and the quick release. Very simply put, wearing an SCBA on your back will give you an average front-to-back profile of about 18–20 in. This may be too wide to allow you to pass through some openings, between the building wall and the stair on a fire escape, for example, or between the studs of a partition wall, commonly 16 in. on center (see Emergency Escape Procedures). You can remove your right arm from the SCBA harness by fully extending your right shoulder strap. Keeping your face piece in place, spin the SCBA bottle around to a position in line with your body alongside your left shoulder. This substantially reduces your overall profile, allowing you to fit between the obstructions.

The quick-release or emergency escape maneuver is an extension of the reduced profile. It may be necessary if you become hung up on a wire or some other obstruction. Again, keep your facepiece intact on your face. Grasp the left shoulder strap firmly in your left hand, and keep it there until the SCBA is secured in the normal position on your back. Fully extend the right shoulder strap and remove your right shoulder from the harness strap. Also remove the waist and chest straps, if present, then pivot around to face the SCBA and locate the entanglement. During the entire process, your left hand *must* maintain a firm hold on the left shoulder strap. Once you have cleared the obstacle, swing the SCBA back into place and re-don it. Proficiency in both of these maneuvers should be a requirement of SCBA training before a member is permitted to enter a building.

1. Fully loosen the right shoulder strap and remove it.
2. Grasp the waist strap with the right hand.
3. Grasp the neck of the cylinder with the left hand.
4. Twist the mask assembly to the left far enough to allow you to pass the obstacle.
5. Pass the obstacle.
6. Return the harness to the normal position.
7. Slide your right hand straight back along your right side to secure the right shoulder strap.
8. Adjust the right shoulder strap for comfort.

Note: Face piece shall remain on if this maneuver is performed in a contaminated atmosphere.

Fig. 10–3 All firefighters should be able to perform the reduced-profile maneuver to help them escape a tight spot.

Before conducting the search, you should gather as much information as possible regarding the presence and whereabouts of any victims. Bystanders will often be clamoring for your attention, pointing to specific locations where victims were last seen. Occasionally, especially if it is a family member who is trapped, you may have to ask that person's specific location and get the bystanders to point out the correct room or window. At times, bystanders or escapees are too shocked by what is happening even to say a word to firefighters. I once responded on the first-due engine to a midmorning fire in a garden apartment. Another member and I were donning our SCBA, preparing to force the door, when I asked one of the five or six people watching us whether this was their apartment and, if so, whether they had the keys. They all said that it wasn't their apartment, then clammed up. I asked whether they knew who lived there, and they said, "An old lady."

I then had to ask whether they knew where she was, and they all said, "Yeah, *in* there!"

"*In* there?" I asked incredulously.

Again I got the same reply, "Yeah, *in* there."

"Why didn't somebody say something sooner?" my partner asked, but he only received mumbles in reply. Here it was, 11 AM with five or six people standing around and a victim still inside, and it was a major task to pry information out of them. Fortunately, the victim was behind a closed bedroom door, and although the living room had flashed over, we were able to get past it to remove her via the window to safety. She was transported to the hospital unconscious, but she soon recovered. If a primary search hadn't been instituted immediately, however, the extra minutes that it would have taken to get to the back bedroom following suppression activities may have made her survival less certain (Fig. 10–4).

When offering hints on searching, many manuals describe the technique of following the walls around the perimeter, as I described earlier in this chapter. This is a fine technique for maintaining a firefighter's sense of direction in heavy smoke, but it leaves a lot to be desired in making many primary searches. For one thing, while it is systematic, it is a very slow way to cross over several rooms, either to get to a victim or to get out of a dangerous situation. If you are entering an area that is so heavily charged that you have zero visibility, it is an excellent idea to have a hoseline with you, both for protection from fire and for a guideline along the escape path. If no hoseline is available, a light length of rope, say ⅜ in. diameter, is an alternative. This is a much faster way of finding your way back out than by following the walls, since you may have to travel 30 ft along three walls to cross over a 10 ft-wide room. Doing so is very time-consuming and could be dangerous.

Fig. 10–4 The evacuation of conscious victims shouldn't lead rescuers to assume that everyone is safe. A thorough search is always required.

When making a primary search of the fire area, it is usually best done by checking first as close to the fire as is safe, then working back toward the entrance. This assures that any victims in the immediate area are removed in the event that fire extends into the next room before it is darkened down. By moving to this area, you have located the seat of the fire and can now direct the hoseline so as to begin knockdown. The search team now moves back to safer areas, where any victims may have been less exposed to the elements of fire, giving them longer survival times. If you had started the search at the front door and progressively moved inward, you might not have reached the most endangered victims in time. Remember that someone must search the areas through which the hoseline is advancing as it darkens down the fire.

When you are beginning a search of areas above the fire, it is often better to begin the search as soon as you enter the floor area and progress in the direction of the fire. This differs from the fire-floor practice because of the differing conditions on the two floors. Whereas it is often possible to move in below the heat and smoke on the fire floor, the floor above the fire is often banked down to the floor, even with just a moderate fire condition. In addition, the likely spread of fire from room to room within an occupancy will usually be much faster than from floor to floor. All of the occupants above the fire are directly threatened by gases, while those in the fire area are threatened by both gases and fire.

Probable Locations of Victims

Human behavior during fires, and actual statistics, tell us that certain locations are prime places for finding victims. Almost all living creatures (firefighters being the notable exception) will flee a threatening fire as soon as they become aware of it. What may happen, however, is that the fire or its by-products will overwhelm those victims before they can complete their escape. For this reason, place great emphasis on searching the routes that people normally use to enter and exit the structure. Generally, that means through the main door and up the main interior stairway, if present. This, of course, is one of the key locations for placing the hoseline. Partly for this reason, those exit ways and stairways must be searched as soon as possible. Pay special attention to the area directly behind the door, particularly if it seems difficult to open fully. A victim in the process of manipulating the locks may have been overcome and fallen before being able to escape. The areas immediately below windows deserve special attention for the same reasons.

People who are unaware of the presence of fire cannot flee from it, nor can those who are physically unable—infants, invalids, and the like. For this reason, place a high priority on searching the bedrooms. As discussed in chapter 7, this may mean doing an end run around the fire instead of trying to charge right through it. The best way to do this is by the VES approach (vent, enter, search). A six-member crew can be split up into three teams of two, each taking a separate avenue of approach to the bedrooms. The team that enters through the front door with a hoseline is responsible for searching the immediate fire area and the exit paths, and then proceeding up the stairway to search the interior hall, making their way into un-searched rooms. The other two teams should work on opposite sides of the building in homes and smaller commercial buildings. In larger buildings, only one VES team may be available, due to the need to deploy the other team for roof ventilation, victim removal, or other tasks.

Coordination is the key. You are trying to get to the most severely endangered victims in the shortest period of time. You cannot afford to have teams duplicating primary searches of areas that have already been searched. The most certain way to do that is by assigning the different locations that must be covered as part of your SOPs. Make sure that you aren't duplicating assignments until you have enough personnel available so that this would no longer be a problem. Some manuals recommend flipping mattresses into a *U* position to indicate that a given room has been searched. I don't care for that idea for two reasons. First, I haven't yet seen a mattress that will stay in that position once you let it go. For another, by flipping the mattresses around, you can end up covering the victims with bedding. This has been the case in at least one fire I know of. A better way of preventing duplicate searches is to issue latch straps that not only indicate that a room has been searched, but that also prevent the door from closing and locking again. If at all possible during the primary search, don't disturb the furnishings too much. Instead, you should use pieces of furniture as landmarks. This can be useful in finding your way out again under severe conditions. If you carelessly toss a chair or other piece of furniture aside during your search, you not only risk having it cover a victim, you might also have removed a valuable reference point for yourself or another firefighter. Similarly, if you pull curtains off of the windows while venting, make sure that the area where you throw them has been examined first.

When actually performing the primary search, make the conditions as bearable as possible. Vent as you move along, provided that it won't cause the fire to light up. When you pass a window, keep its location in mind, since it may be your nearest point of escape in the event of an emergency. Closing doors between you and the fire is usually a good idea, provided you are certain of another way out. It is often possible, while making a determined search, to get so preoccupied that you forget about the fire, especially if you close the door to the room you are searching. When you make your way back and open the door, you may find that the original way in is blocked by fire. If this is a danger, leave one member of the team at this doorway to monitor conditions and call the other searchers back before conditions deteriorate too severely.

In chapter 9, I described the effects of increased ventilation on a mattress fire in a building equipped with thermopane windows. Another firefighter and I had gone past the smoldering mattress after darkening it down with a water extinguisher. I had gone all the way to the back of the apartment while my partner searched the front. Like a deja vu experience, I found myself looking into a closet off the kitchen of this apartment. Thinking back 19 years to my first fire, I thought to myself, "I can't be fooled like that again." I probed with my light, and the smoke opened up to reveal a toilet. The closet was actually a bathroom. I started to search the small room, looking into the tub, when the room brightened up all around. I no longer needed the hand light—the fire had lit up and was coming straight for me. I immediately retreated to the kitchen, calling for the hoseline to open up. (Never retreat into a bathroom if there is any alternative. Bathroom windows tend to be too small for escape.) The hoseline crew replied that they still didn't have any water. Unable to close any doors between the fire, and myself I was forced to bail out through the kitchen window onto a ladder that was about to be raised to an adjoining window. The primary search was completed as far as was possible, and the results were negative.

The Secondary Search

Once the visible fire has been extinguished and conditions start to improve, it is time to begin the secondary search. This is often combined with the overhaul phase. As areas are examined for hidden fire, they are also inspected for victims. Take care not to bury either smoldering material or casualties under debris when you open the ceilings and walls. Charred bodies can be very difficult to distinguish from debris if ceilings are pulled on top of them. Since the fire has been darkened down, the pace should be somewhat slow. Before moving items, make a thorough examination of the area where they will land. The secondary search must include the perimeter of the building, including any rooftops or setbacks to which people may have jumped, as well as the areas beneath the windows. Be sure to examine any shrubbery in these areas that might be concealing an unconscious victim. You must examine all of these areas before allowing any debris to be thrown out of the windows.

An excellent policy is to have different people do the primary and secondary searches of a given area. This is necessary because we all tend to look at things differently. For instance, the person who made the primary of a room may only have had time to make a cursory examination of the area beneath the bed. In being sent back to the room, he or she might have a tendency to say, "Oh, I already looked under there, I'll look around someplace else." By switching crews, or even just trading off rooms between two searchers, each area becomes a new, uncharted area that must be examined thoroughly.

My company once did a primary search in a very cluttered apartment from which one severely burned victim had just been removed. As my officer and I searched the bedroom, I swept a bed, felt it, and looked under it. The officer had to pull apart a large pile of clothing from a narrow space between the bed and the wall. As he did, he looked at the space, and then we moved into an adjoining bedroom. Both rooms were negative except for the one victim who had been removed to the street. Once we had completed all three rooms, the officer radioed the report "Primary search negative on the top floor" to the chief out in the street. The chief promptly radioed back that we had better take another look, because an occupant of the apartment was positive that there was still another person in there. Before we could complete the secondary search, another company that had been specially called to the scene for relief purposes was sent in, and the chief ordered us out. A member had been seriously injured, and we were to treat and transport him to a nearby hospital. As we were leaving the hospital, having dropped off the injured member, we heard the report of the discovery of the second casualty.

We headed back to the fire scene, expecting to be chewed out for missing the victim and curious as to where he had been found. Reporting back to the chief, we found out some of the circumstances. The fire had been a revenge fire, set with accelerants in the stairway. The front apartments had no fire escapes, an illegal alteration. One survivor of the top-floor front apartment, an 85-year-old man, had been sharing his rooms with two other elderly men when the fire broke out, blocking the stairs. This gentleman escaped by climbing down four stories along a clothesline that he kept in his room for just such an event. When the other two occupants didn't follow him, he was certain that they had remained inside. Faced with this story, the chief in charge knew right away that if we had missed the one victim on the first primary search, he must have been well concealed. From his many years of experience, he knew that, while our unit certainly would have found him eventually, it would be faster to bring in a fresh point of view. As it turned out, the victim had been in the space between the wall and the bed. Once he realized that the stairway was impassable, he had apparently tried to hide from the fire. He must have thought that the piles of clothing would insulate him. He succeeded in not getting burned but was overcome by the gases. As for the elderly rappelling artist, when questioned about what made him think of the rope as a means of escape, he replied, "Son, look at me. You don't get to be 85 years old living in Harlem unless you know how to get away from fire."

You must be sure when conducting the secondary search that any space that can possibly hold a human being, including an infant, has been examined. You must be exceedingly thorough. I have found infants left sleeping in a dresser drawer near the parents' bed. Personnel overturning furniture in a hasty search would almost surely close the drawer or cover the child. Children are often found hiding from fire in toy boxes, in closets, and under beds. An excellent training technique relevant to searching for children is to play hide and seek with young relatives. Pay attention to the patterns in their hiding places. They can be quite ingenious and, if faced with danger, will almost invariably pick ingenious places to hide. Even refrigerators, freezer chests, and kitchen cabinets can be a child's refuge (Fig. 10–5).

Fig. 10–5 Any space that can hold an infant or a small child must be examined as part of the secondary search.

During training, one method for developing good search techniques is to cover the SCBA facepiece with an elastic shower cap, then have the members search an area. Have them describe the objects that they find and determine what room they are in based on this deduction. Members should easily be able to recognize bathrooms and kitchens. Extra practice will enable them to recognize children's cribs by their barred sides and high legs, and bunk beds by their low mattresses, usually confined at one or both ends. Cribs and the top level of bunk beds are two areas occasionally overlooked during primary search.

Another difficulty in performing the secondary search is simply recognizing the remains of severely burned persons as being human, particularly when debris has landed on the body. Completely examine all items before tossing them out of a window or dragging them to the street. This procedure should be thoroughly ingrained into all personnel, and it should be routinely performed, even if no one is reported missing. More than once, parents have assured firefighters that everyone is out of the house, having forgotten about a guest or relative sleeping over. In past instances when a body was overlooked, the fire service has received not only bad press coverage, but also the threat of negligence suits.

Even at seemingly routine fires, you must search *all* of the areas above the fire and *all* of the accessible areas below it. If a room or area on the fire floor or the floors above is locked, forcible entry may be required. Realize that shafts, pipe chases, and the like can channel deadly concentrations of smoke to areas remote from the fire without visible indications on the stairways. Generally, areas two or more floors below the fire don't warrant forcible entry unless they show signs of specific problems. An exception may be below-grade areas, where carbon monoxide from smoldering fires may settle. Members should wear masks whenever they descend below grade in a fire building, even if the fire was confined to the first floor and there appears to be little or no smoke in the cellar.

Performing an adequate secondary search requires a sufficient commitment of personnel to complete it in a reasonable time. A good rule of thumb would be to send a two-man team to each moderate-size area of an open-floor area such as an office or a store. A similar two-member team should be available for each apartment in a dwelling. This number can go up or down, depending on smoke and fire conditions. In light smoke conditions, the two-member team will be able to move rapidly through an area and won't require relief. If considerable smoke remains, or if the area is cluttered, then additional personnel may be required.

Guide Ropes

It may be too dangerous at times to enter a fire building without having a way to guide you rapidly back to a safe exit point. As mentioned earlier, if there is a danger of fire overrunning the searchers, a hoseline must be in place. There are times, however, when the immediate danger isn't from fire—it is more a problem of losing one's way due to disorientation, a mazelike arrangement, large open areas, or extremely heavy smoke that cannot be readily dispelled. In such circumstances, you should use a guide rope, also called a tag line.

Using a small-diameter rope for directional purposes is a relatively simple idea, yet it is an idea that is probably misapplied almost as many times as it is attempted. To be successful, you need more than just a rope. You need a plan that is understood by all of the members, as well as some additional support equipment and personnel who are properly trained in carrying out each step of the plan. To use a guide rope for members to follow in or out of the fire building, you first need a rope of sufficient length to reach the desired area. Since the purpose of this rope isn't to carry any weight, but merely to be used as a guide, ¼- or ⅜-in. nylon works quite well, allowing 200–300 ft of rope to be stored very compactly. You should be able to connect several of these ropes together for use in large-area buildings if the need arises. The next requirement is radio communications to the outside, preferably on a channel that is separate from any other activities. Members entering a structure that is unusual enough to require a guide rope need immediate contact with the outside. A good working flashlight is a requirement for each member, while working Personal Alert Safety System (PASS) devices are also highly recommended. In addition, each team should be equipped with a set of forcible entry tools. Position a large, high-intensity floodlight at the entranceway to act as a beacon. Once all of these items have been assembled, plus SCBA and any other needed equipment, the members may proceed with their task (Figs. 10–6 and 10–7).

Before entering the building, secure one end of the guide rope to a substantial object on the outside. The remaining rope is then carried by one member, designated by the officer. This member's job is to ensure that the rope remains taut at all times. If it becomes necessary to backtrack at any time, he must retrieve the rope that has been paid out and re-deploy it along the new route. The intent is to ensure that the most direct means of safe egress is maintained at all times. As mentioned previously, heavy smoke on arrival is only one of the reasons for using this technique. You should deploy a guide rope in any building where disorientation is possible. Large-area office buildings, hospitals, and schools are excellent candidates for the guide rope, as are tunnels, ships, and subways. It isn't necessary when using a guide rope in this manner that each member remains attached to it. Instead, as long as they maintain contact with the one member who has the rope, the others are free to perform their tasks, whether in good visibility or not. If it becomes necessary for a member to exit, it is a relatively simple matter to follow the rope back to safety. (Note: Anytime one member of a team must leave an area, he must be accompanied to safety by another member. In effect, if one member of a three-person team has to leave, the entire team should leave.)

Fig. 10–6 A search guide rope should accompany each team entering every commercial building. The bag worn over the shoulder here carries 200 ft of light line.

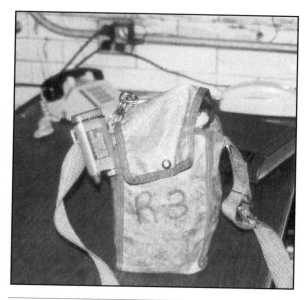

Fig. 10–7 A strobe light attached to the bag aids locating it in heavy smoke.

Team Search

Another use for the guide rope is as a tether when performing team searches of areas. This differs from the simple use of a guide rope in that all of the members must remain in contact either with the main search rope or a short length of personal rope attached to it. This practice is necessitated by poor conditions. Generally, a search team is only used under such extreme conditions if there is a definite report of a person missing. In this case, provisions will have to be made for the prompt rescue and removal of the victims to safety once they have been discovered. Preparations for a team search are similar to the general use of a guide rope, but additional people and equipment are required. At minimum, six people (preferably seven) must be trained and ready to go.

One person acts as the control man, remaining outside the area at the point where the search rope is secured to a substantial object. He must be equipped with a portable radio, as must each of the other members participating in the team search. Again, these people should be on a channel that is separate from all other radio traffic. Again, position a large flood or spotlight at the entranceway to indicate its proximity to the searchers. A device that has proved exceptionally useful for marking the entrance during either guideline operations or team search is the Target Exit Device™ or TED. This strobe device also emits a series of high-decibel chirps, 4 times in 2 sec, followed by 10 sec of silence. I have used this device under very heavy smoke conditions while searching an apartment on the floor above a serious store fire. The sound provided an additional sense of depth and direction that was very reassuring. Some teams tie a series of small knots on the search rope to indicate the distance that has been traveled and thus how far away they are from the exit. I find that anything on a rope that changes its contour tends to hinder it from paying out properly.

In addition to maintaining radio contact, the control man must have a prearranged communications system using a series of tugs on the line in the event of poor reception; perhaps one tug for *stop*, and, if given from the inside, *we are stopping*; two tugs for *okay to go* or *we are proceeding*; three tugs for back up or *we are backing up*; four tugs for *send help* or *we are sending help*.

The control man's job is to monitor how many people enter, who they are, and their time of entry and exit. When performing searches in extremely large or complex areas where smoke or toxic fumes are present, the low-pressure alarm on the breathing apparatus may not provide sufficient warning time to escape before completely running out of air. By assigning a maximum 20 min of submergence time for nominal 30 min SCBA, the control man can indicate at the 10 min interval that the members inside have reached the halfway point and now must return to fresh air. (Note: Although using a guide rope will speed exit time, safety considerations should indicate whether or not the halfway rule can be extended slightly or not at all. The temptation is to say, "I'll only go a little bit further before I turn around." The result may be that one of the searchers becomes an additional victim.)

If the operation will take longer than the 10-min work time allows, it must be performed in relays, with each team extending the area further. Another option is to use longer-duration breathing apparatus. One-hour bottles can provide 20 min of work time and 20 min of exit time. Whatever breathing devices are being used, however, they should all be of the same type and duration to allow for emergency buddy breathing, if needed. The control man must be aware of the working time permitted and be prepared to send in the rescue team if the searchers aren't out at the expiration time.

The search team should consist of a leader and a minimum of two members (maximum four—*always* even numbers of personnel). The team leader should be secured to the main search rope and pay it out as he goes. In addition, he should take another large floodlight to serve both as illumination of the search area and to act as a beacon for returning personnel. A strobe light at this end of the rope can also be helpful. Each additional searcher must be attached to the main search guide rope by a 25- or 50-ft length of rope. Their ropes should be attached to the main guide rope just behind the team leader. Snap hooks attached to each end of the ropes simplify these tasks and facilitate quick escape.

As the search begins, the team leader enters the area and moves a desired distance in the proper direction. Where the search commences can vary with the situation. If information points toward a certain area as the last known location of the victim, the team may move to that vicinity before starting the search. If no information is available, the search may have to start at the entrance to the area, in which case the two searchers remain on opposite sides of the guide rope just inside the door.

The team leader then advances until the two short tethers draw taut. He stops there and signals the members to begin sweeping forward. If they cover the entire area without finding the victim, they will end up at the forward end of the rope. The team leader then moves up, and the process begins again, repeating until either they find the victim, they reach the end of the building, or the team's working time expires. If the work time runs out, the search team will be replaced by the rescue team (at least two additional members, who are standing by in the event of an emergency or for when the victim is found). To speed up and simplify this process of changing places, the end of the search line inside the building must be secured to a substantial object and the floodlight directed back along the search rope. This will enable the new members to move rapidly to the point where they are to commence searching. Of course, a new rescue team must be put in place at this point. The rescue team should be equipped with a spare SCBA to be placed on each victim, a stokes stretcher to facilitate removal, and any hardware that might be needed, such as a lifesaving rope for hoisting the victim (Figs. 10–8, 10–9, and 10–10).

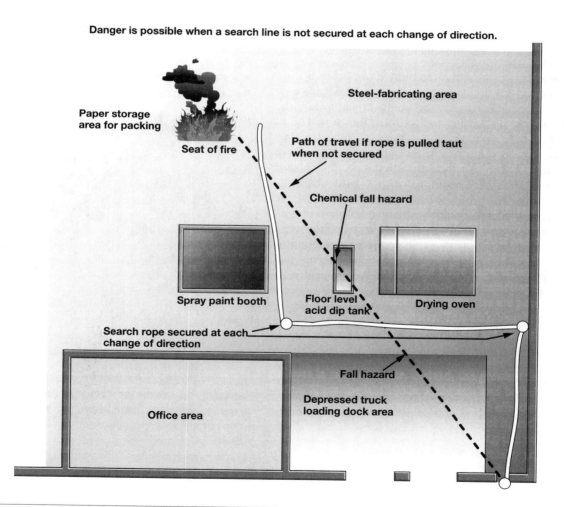

Fig. 10–8 Search ropes are intended to improve firefighter safety, by allowing members to make a speedy retreat without encountering unexpected hazards while withdrawing.

When deploying the search rope, it is important to secure the end of it outside the structure, or inside a stairwell on an upper floor, regardless of the conditions when you start the operation. Conditions can change substantially during the course of what is likely to be a rather lengthy incident. At one such incident, a rescue company located several members of a truck company who had become totally disoriented in a large, windowless building. The members had been investigating what they'd thought to be a relatively small fire, but it had escalated rapidly when it reached some flammable materials. While the fire was being held in check by the sprinkler system, the smoke that was being produced had enveloped the truck crew almost instantly. Although their SCBAs were protecting them for the moment, they were totally lost several hundred feet into the building. Their air supply couldn't last against the lengthy process it would take to clear out all of the cold, moist smoke that remained. The rescue company was directed to implement the team-search technique. Aided in part by some other newfangled gadgetry, they were quickly able to locate two of the missing members.

One member was still unaccounted for, however, and the searchers had to continue on. They placed the guide rope into the two members' hands and pointed them toward the exit. Since both still had air, it was felt that they would be all right. When they got to the end of the rope, however, they found themselves some 50 ft from the outside and still in heavy smoke, unsure of which direction to go. They placed a call to the outside for additional help to guide them the rest of the way. The problem had come about because the rescue company had tied off the end of the search rope in an area that had previously been perfectly clear. Conditions had worsened so dramatically that it was now at zero visibility. Although this operation was a great success—saving the lives of three firefighters—the lesson of where to tie off the rope was a hard one to learn.

An additional lesson is the need to secure the rope to a substantial object whenever the rope makes a change in direction. This is necessary to ensure that the exit path follows the entry path, and that it has been searched for any hazards such as floor openings and obstacles. If the line isn't secured at each change of direction, the rope will be pulled on a diagonal, which may lead personnel into unknown hazards.

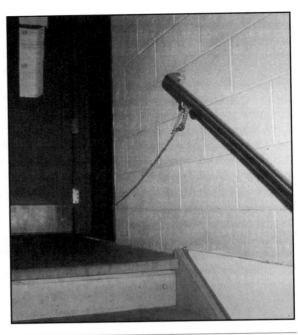

Fig. 10–9 Always secure the safe end of a search rope in a stairwell (in this case to the handrail) or outside the building. Conditions can change rapidly.

Figs. 10–10 TED uses a strobe light and an audible signal to indicate the location of an exit. It should be used in conjunction with a search rope.

Applying Technology to the Search Process

The conditions under which firefighters are forced to operate are obviously life threatening. Explosions, backdrafts, and the collapse of walls, roofs, and floors can all kill instantly. There are many situations, however, where if a firefighter in distress can be located and removed, his chances of survival will dramatically increase. Fortunately, modern technology can be applied to many of these situations to speed up our search and rescue efforts. One of the obvious means is the two-way portable radio. If the trapped member is conscious and radio equipped, he may be able to call for assistance and direct help to his location. Some models of portable radios can be equipped with a distress setting that reports the identity of the member in distress to other radios by a coded identifier: it is a manual operation however. What can be done for the unconscious victim? Having a working PASS device, set in the arm position (or having an automatic, integrated PASS) can help by providing a sound source to follow. Lacking these devices, the searchers may be able to use the trapped member's radio to home in on him (Figs. 10–11 and 10–12).

This process, called feedback-assisted rescue (FAR), is the result of an otherwise annoying phenomenon of radio behavior—feedback. When two or more walkie-talkies are in close proximity to each other and the mike is keyed on one of them, a high-pitched squealing noise is broadcast over all of the radios tuned to that channel. Firefighters can intentionally create this squeal, then listen for it coming from the missing member's radio. For this to work, the missing member must be wearing a walkie-talkie; it must be on, and the channel must be known. All of the other radios must either be off or tuned to a different channel, otherwise the searchers won't be able to pinpoint the direction of the lost member. You must take the two radios that will be used to create the feedback well away from the scene. Place the microphones of these two radios 1–2 in. apart, then key the mike on one of them. You must continue this until the member is located, which could take some time. Obviously, another radio channel or some other means of rapid communication must be available for other messages (Fig. 10–13).

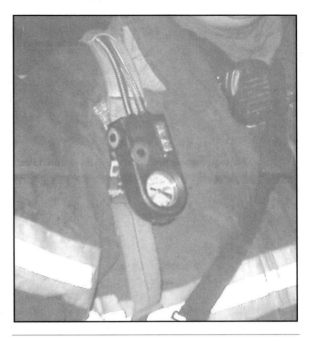

Fig. 10–11 An integrated PASS device on the SCBA ensures that the wearer will activate the system when the mask is donned.

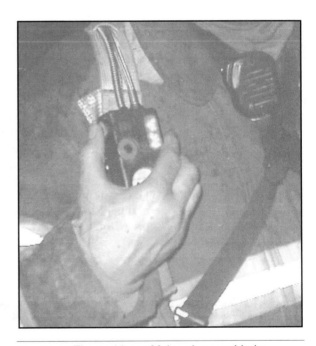

Fig. 10–12 The problem of false alarms with the integrated PASS requires all members to make a conscious effort to check and reset their alarms whenever they hear one sounding. Otherwise complacency may lead to ignoring the device when it's a real emergency.

A similar procedure is possible for members who aren't wearing walkie-talkies but who are equipped with pagers. In this case, the base station must first activate the alert tone (again, make sure everyone else's pagers are turned off), then use the two radios to produce the feedback effect on the dispatch frequency. It isn't advisable to use a base station or a mobile radio as the feedback transmitters in either of these instances because of the range over which they transmit and the amount of havoc that they will wreak on other communications, some of which might be just as urgent. Two portable radios are sufficient to produce the feedback effect for up to a half mile.

Although both FAR and pager-assisted rescue are assets in locating downed firefighters, they both have some big drawbacks. The biggest is that they are manual operations, requiring first that someone recognize that a firefighter is missing. In addition, it can be a time-consuming operation to get all of the other radios on the fireground turned off or switched over. A far better method of accomplishing basically the same task without the previous drawbacks is to equip each interior firefighter with an automatic distress sensor and the means of sounding an alarm. PASS devices have rapidly proven their worth. They are compact, lightweight motion sensors that, if allowed to remain motionless for approximately 30 sec, emit a low-level pre-alarm, indicating to the wearer that the alarm

Fig. 10–13 To conduct FAR, dispatch a member with two radios to a remote area. Have that member press one transmit button.

is about to sound. If the situation is normal, all the wearer has to do is shake that part of his body to which the PASS is attached, causing the device to recycle itself. If the wearer is unconscious, pinned, or otherwise immobilized, however, the unit will automatically go into full alarm approximately 5 to 10 sec after the pre-alarm sounds.

The full alarm is a piercing siren (95–100 decibels) that will last up to 10 hrs. The units have the following three-position switches: off, on (which instantly turns the siren on manually in case the wearer wants to call for help), and arm, which puts the device into the automatic mode. Once the device has gone to full-alarm status, it can only be silenced by manually turning the switch to off. The usual practice is to attach the devices directly to the SCBA so that, anytime a member enters a building, he will be equipped with a PASS. The trick is to get members to use the device. Although the benefits of these devices should be obvious, there have been instances where trapped firefighters have died with their PASS devices in the off position. Obviously, the best option is the automatic integrated PASS, which is built into the SCBA and which automatically arms itself as soon as the cylinder is turned on. They also have a manual override in case the wearer isn't immediately going on air. It should be a strictly enforced policy that these devices be armed as soon as the SCBA is turned on. If you make this a habit, your firefighters stand a better chance of survival if they ever find themselves in need of assistance.

Thermal Imaging Cameras (TICs)

One of the latest advances in the field of search and rescue is the TIC. This device resembles a video camera that sees objects in a black and white negative screen or in a temperature variable color display. Instead of seeing the object's actual color, it reads temperature differentials. Unlike the various infrared thermometers that are used to find hot spots and hidden fire, these devices actually show you a picture of the area at which you are looking. The surrounding room is gray, while warmer objects are plainly visible, since they appear to be all white. Similarly, cooler objects appear black. The hotter or colder the object, the greater the contrast with the surroundings. The camera can see through smoke and mist, and the human form can be clearly visible if the temperature difference between the body and the background is significant. Some of these devices can detect as little as .5°F difference between objects. In fact, the device will even briefly pick up the radiant energy that a person transfers to an object after that person moves away (Figs. 10–14 and 10–15).

Many of these cameras come equipped for connection to a remote monitor outside the immediate fire area, either by means of a coaxial cable or, in some cases, by wireless radio frequency (RF) transmitter. In the future, it may become practical to send a camera-equipped remote-controlled robot into an extremely dangerous area to perform reconnaissance. There are a number of such cameras currently on the market, both handheld and helmet-mounted, with prices ranging $8,000–18,000 per unit. This decrease in cost from early models has made it possible to equip every engine and ladder company, as well as special units with the devices.

FDNY has been using various types of TICs since 1985, and with great success, locating victims and hidden fire, as well as the source of nuisance odors that might otherwise have taken many man-hours to locate. When I was the captain of Rescue 1 in Midtown Manhattan, we were equipped with two handheld units and, for almost a year, an additional helmet-mounted unit. It wasn't uncommon to use all of these devices simultaneously, several times a day, chasing fire traveling within ductwork running throughout some very large buildings. Both the handheld and helmet-mounted units have their pros and cons. Although they are designed to be rather rugged, they must be treated carefully. After all, they are expensive pieces of electronic equipment. As they receive more field use, it is likely that newer generations of these devices will incorporate features to make them sturdier against fireground handling. The U.S. and British Royal Navies have already equipped all of their warships with these devices for use in shipboard emergencies. The military and the fire service are probably the two roughest environments in which such a device will see service.

Firefighters who use these devices must still perform all of the other basic procedures of a fireground search. These devices, especially the helmet-mounted models, can produce a kind of *Superman syndrome* if the user isn't wary. Being able to see a victim clearly in a smoke-filled room is a great benefit, and members tend to move to that area immediately to perform rescue, and rightfully so. Yet the firefighter must remember that he is using an electronic device that, like all others, is subject to breakdown. We've had cameras malfunction without warning. If the firefighter hasn't done the basics (maintained contact with another member, a search line, a hoseline, or a wall, and ensured that his or her

Fig. 10–14 TICs have many uses on the fireground, from primary search for victims to overhauling to hazmat monitoring.

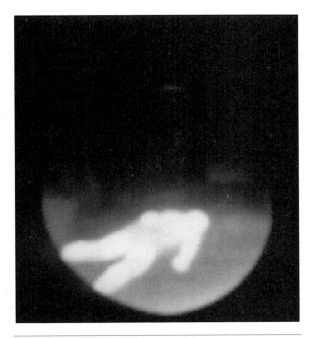

Fig. 10–15 This photo of a victim was made with a TIC. The form of the victim is clearly visible through dense smoke. Note: This is a civilian victim in an area remote from the fire. A downed firefighter in bunker gear may not show as clearly white, and could have large areas of black.

landmarks could lead back toward safety), he might find himself worse off than if he didn't have the camera in the first place. The handheld units don't seem to be as incongruent with safety procedures as the helmet-mounted models, since the members can use the camera for several seconds to scan an area, but they must put it down again as they move on, following a wall, search rope, or hoseline. The helmet-mounted units offer hands-free operation, to a point! The stereovision screen tends to require adjustment to maintain the quality of the picture. This tends to cause tunnel vision, capturing all of the wearer's attention. Persons who have weaker vision in one eye tend to have some difficulty with the dual screen. In addition, the unit is heavy on the wearer's head and neck, a factor that is compounded by the cable that connects the helmet to the belt-mounted battery pack. Additionally, there is a sanitary consideration if many different firefighters wear the same helmet on different shifts or alarms.

There is no guarantee of a successful search and rescue, even with all of the modern hardware that is available. At times, all that we can do is to try our best. Whenever circumstances result in someone becoming trapped, the firefighter must be willing and able to do his absolute best. In such a case, a determined firefighter struggling to reach a victim before the fire does will ignore injuries that would stop him or her if only property were being destroyed. When firefighters feel that they cannot possibly bear any more, they lower their head and take yet another step forward. And if they are fortunate enough to reach that person who needs them, they are the most fortunate of all mankind, for they have performed a task that few others on earth can or would do. They have saved a life.

Specific Fire Situations

11 Firefighter Safety and Survival

Aconcerted effort to reduce firefighter deaths over the last 20 years has seen some successes. Firefighter deaths have dropped from a high of 140 per year during the 1970s to around 100 per year in the 1990s. Still, there is a disturbing trend in the statistics—a trend toward firefighters dying from exposure to fire and products of combustion while trapped within the fire building. According to one survey by Chief Vincent Dunn, out of 173 firefighters who died on the fireground during a recent 10-year period, 113 were caught or trapped and subsequently died from exposure to products of combustion. The following are many reasons for these deaths: reduced staffing, increasing amounts of flammable fuels in buildings, modern building construction and sealing techniques, and possibly the overconfidence of firefighters themselves, having too much faith in their state-of-the-art protective equipment. All of these problems need to be addressed to resolve their long-term implications. The immediate problem is how to keep our current generation of firefighters alive while the proper solutions are being developed.

Your department can take the following three steps immediately to help reduce the rates of firefighter injury and mortality:

1. Improve hazard awareness and recognition.

2. Provide emergency escape or self-rescue capability.

3. Provide rescue capability by deploying RITs (Rapid Intervention Teams).

The Survival Syllabus

In this day and age, when we are forced to sit through all sorts of mandated training, much of which is superfluous to the line firefighter, it is past the time that each and every one of us is trained in survival. The following are among the topics that should be included in such a firefighter survival course—a sample course outline, if you will (Fig. 11–1).

Hazard awareness

Firefighters must be trained to recognize the dangers that they face and the actions that they need to take to protect themselves before they get into trouble. They also need to be taught the proper *attitude* to avoid trouble. Hazard awareness should be developed as part of each firefighter's size-up. The six-question Firefighters Survival Survey is an information-gathering thought process designed to focus a firefighter's attention on doing his job as efficiently as possible while maximizing the chances of going home in one piece. As soon as possible, each firefighter who arrives on the fireground should determine the answers to each of the following questions:

Fig. 11–1 If time is crucial and conditions severe, you may have to escape in a fashion you did not anticipate.

(1) *What is the occupancy?* Fires in certain occupancies pose certain risks to firefighters. We should expect to encounter similar dangers in occupancies that have bowstring-truss roofs, for example. Firefighters responding to such alarms should expect early collapse and take defensive positions. The occupancy itself can also tell us about other dangers, such as hazardous chemicals in an exterminator's shop or a garden supply business. More importantly, the occupancy should tell us what our attitude should be at that particular alarm. Each of the previously mentioned occupancies has the potential of producing a large, spectacular fire. Although the civilian life hazard in each might be small (more people die in car fires each year than in store fires), each also poses a severe danger to the firefighters who conduct an aggressive interior attack.

(2) *Where are the occupants?* The greatest fire-related loss of civilian life each year occurs in one- and two-family dwellings. Firefighters might be required to take some actions at these incidents that they wouldn't take at a dry cleaning establishment or other small store. In an attempt to rescue trapped occupants, we may have to pass or go above the fire. Once the danger to occupants has been removed, however, we should slow down the pace of our activities and again weigh the consequences of our actions. Remember, you swore to protect life and property, but its life first (including your own), then property. You should always do a thorough search of all of the premises, but if the homeowner meets you in the front yard and tells you that no one else is inside, you shouldn't get killed doing a primary search. Similarly, if your primary search is complete and the fire conditions are worsening, you should get to a safe area and await fire control before conducting a secondary search.

(3) *Where is the fire?* Fires in cellars, attics, and windowless areas create great difficulties and dangers for firefighters. Cellars and attics pose similar difficulties in terms of limited access, escape, and ventilation. Also, many of these areas are unfinished. Structural members are thus exposed to fire, leading to early collapse. In cellars, there is an additional danger of fire extending into wall voids, possibly surrounding and cutting off the escape routes of unsuspecting firefighters. Pay attention!

(4) *How do we get in?* Firefighters should survey the building for the entry route that will take them to the particular part of the structure that they have to reach. Not everyone should be going through the front door. If your job is to get above the fire floor, the fastest, safest method may be to use a ladder to reach a window.

(5) *How do we get out when things go wrong?* More important than getting into a building is getting out when you're in trouble. Most trapped firefighters become disoriented or lost prior to getting trapped. Maintaining contact with a partner, a wall, a hoseline, or a search rope is a must. Every firefighter must be taught to include a survey of the windows in his pre-entry size-up. Recognize casement windows, security gates, and bars before you enter the building, not when the fire is chasing you out.

Norman's three rules of survival are as follows

- Never put yourself in a position where you are depending on someone else to come and get you.

- Always know where your escape route is.

- Always know where your second escape route is.

You might violate one of these rules and survive, but if you break all three, your chances of survival will plummet!

(6) *What is happening to the building?* The survival survey must include an up-to-the-minute evaluation of conditions within the building. Is there a potential for backdraft, either upon entry or from within a concealed space when you open it to check for extension? Are the heat conditions escalating to the point where flashover may occur? In this regard, you must consider the height of the ceilings in commercial buildings. High ceilings may hinder firefighters from recognizing potential flashover. Is the fire spreading out around you in concealed spaces, threatening to cut off your escape route? Also, what effect is the fire having on structural stability? Will the building fall down around you if you stay where you are? Is the hoseline making any progress? If the answers to any of these questions indicate potential trouble, you'd better make sure your escape routes are clear and that you will be able to reach a safe area in time to avoid trouble.

Escape training

After a firefighter has been trained in the use of the SCBA, he or she must be taught how to operate with it under emergency conditions. Among the items that should be taught are the reduced-profile maneuver, the emergency escape (quick-release) maneuver, emergency door opening, and mask sharing (buddy breathing). An optional skill is changing cylinders in a smoke-filled atmosphere.

Entanglement is a serious fireground hazard. Everyone who puts on an SCBA should know how to take it off blindfolded and with gloves on while maintaining the face piece and a constant air supply. At times, you may need to perform an emergency escape maneuver to disentangle your SCBA from dangling electric or cable TV wires, but such situations aren't all that common. Usually the best thing to do is simply to back up, get down lower, and then proceed, since the entanglement is usually between the cylinder and the wearer's back. Removing the SCBA from your back will most often be needed when breaching a wall, since this will allow you to pass between the studs. All such evolutions are difficult, last-ditch efforts that must be practiced beforehand if they are to be successful in a moment of crisis.

The reduced-profile maneuver is a simple task that allows firefighters wearing SCBA to squeeze through narrow spaces. The most common use for this technique is to get through the space between a building and a fire escape stairway. The most critical use might be when a firefighter has to squeeze through the 14-in. space between two wall studs to escape to another room if the doorway becomes blocked. To perform the reduced-profile maneuver, simply extend your right shoulder strap fully and slip your right arm out. Then, grasp the cylinder with your left hand, the waist belt with your right hand, and twist the cylinder to the left, in line with your body.

Carrying the reduced-profile maneuver a bit further, firefighters should be taught the quick-release or emergency escape maneuver. This operation may be required when a firefighter finds himself caught on hanging wires, bedsprings, or other entanglements. Both the emergency escape and the reduced-profile maneuvers should be performed with the face piece connected and the member breathing via SCBA. The emergency escape maneuver may also be required for a member who has to fit through such a small opening that the SCBA has to be taken off entirely and pushed through ahead. To perform the emergency escape with your face piece in place, extend your right shoulder strap fully and grasp the left shoulder strap firmly in your left hand. Never let go of this strap until you re-don the SCBA at the completion of the maneuver. Next, release the waist buckle with your right hand and spin to the left to face the SCBA. You can then free any entanglements with your right hand (Fig. 11–2).

Fig. 11–2 Knowledge of all the emergency procedures for your SCBA is crucial to survival. Practice them with gloves on and progress to blacked-out face pieces.

Another emergency action is replacing cylinders in contaminated atmospheres—a skill that isn't for the novice. Only after a firefighter has demonstrated competence in the routine activities should she be introduced to this complex task.

Still, changing cylinders may be useful in certain situations. Fires in the holds of ships, tunnels, sub-cellars, and windowless buildings may result in such long escape times that a firefighter risks running out of air while still in a dangerous atmosphere. One alternative is to bring in a spare air cylinder to change to as the first bottle runs out. You should only consider such emergency measures when longer-duration cylinders or hose-supplied air sources aren't available or practical. In this maneuver, remove the SCBA as in the emergency escape and set it down directly in front of you, cylinder up. Place the fresh cylinder right alongside the unit. At this point, crank down the valve on the nearly depleted cylinder until it is nearly closed, then draw in a deep breath. After closing the valve on the depleted cylinder, disconnect the high-pressure hose from the cylinder and reinstall it on the new cylinder, then turn its valve on. After securing the fresh cylinder in the harness, re-don the SCBA and exit to safety. This is an advanced skill, and it should only be attempted by those who are thoroughly familiar with the maneuver and only as last resort.

Even the most careful of us have found ourselves caught up by rapidly changing conditions. Let just one single aerosol can of hair spray or insect killer explode as you pass a room, and you can find yourself cut off by fire. Whatever actions you take next are likely to determine your chances of survival, injury, or demise.

If you are cut off by extending fire, you will have limited options, but some of the simplest choices are sometimes overlooked. First, try to find an area of refuge, even if it is only a temporary one, and *close as many doors between you and the fire as possible*. Second, call for help by whatever means possible—by radio, PASS alarm, or voice alone. Third, begin seeking another escape route or area of refuge. Locate any doors or windows that will get you someplace that is better than where you are. If a ladder can be placed to a window, you are halfway home, but climbing out onto a ladder with fire rolling over your head is no easy task. An emergency descent may be your only option. This is a fairly difficult maneuver that requires practice if it is to be accomplished safely. Similarly, an emergency bailout on a length of personal rope may be advisable if no ladder is available. If neither rope nor ladder is available, you are in trouble. If you are on the second floor of a dwelling, you may survive a leap from the window, but many have not, and others have survived but sustained severe injuries. A better way would be to get out over the window ledge and hang on as long as possible with only one arm and one leg exposed on the sill. If help doesn't arrive in time, unhook your leg first and hang feet-down. The length of the drop will then at least be shortened by your height.

As this third edition went to press, six firefighters, including two from my old unit, Rescue 3, were trapped by rapidly extending fire on the top floor of a four-story Bronx building. They were cut off from their primary escape route by the changing fire condition, and their secondary escape route was blocked by an illegal alteration. All six jumped nearly five stories into a rear courtyard. Two died and the other four were injured, three very seriously. Two members who survived used a short personal rope to reduce the height of their fall. The fourth firefighter miraculously escaped with relatively minor injuries. He was able to hang feet first from the sill as described previously.

Another avenue of escape may be to kick your way out through an interior partition or locked door. Using forcible entry tools to breach a partition wall is far faster than kicking. (It is virtually impossible to kick your way through an exterior wall in time to escape. Don't even think about it as an option.)

Under less severe conditions, where the firefighter isn't in danger of being overrun by fire but is merely lost, it may well be better initially to avoid unnecessary exertion so as to conserve air. The first action should be to call for help by radio, using all channels. A PASS alarm or a call by voice may be heard by other firefighters who are nearby and who may not have heard the radio request. Also, they may not have realized how close they were to your position. Stay calm and stay put to avoid wandering further away from help. While sitting still, conserve air. Hold your breath every few seconds and listen. You may be able to regain your bearings. Lay with your face down near the floor and turn off all of your lights temporarily so that you might be able to see other light sources. Be sure to turn them back on after your look around, to make yourself more visible to searchers.

If your air situation is critical, or if you are unable to contact assistance, you must go through the same actions listed previously for those cut off by fire. Bail out of a window, breach a wall, or move to a less hostile area. Whatever you do, don't give up. Use any and all means at your disposal to attract attention to your plight. Call continuously on the radio, and call the dispatcher on any working telephone that you find. The O button is bottom center on most push-button phones; the 9 is one row up on the right; the 1 is top left. If you are entangled and unable to jump, as a last-ditch effort, go ahead and throw your helmet out the window. Aim near someone who can help you, if you can still aim. Hopefully your helmet has your name on it so somebody outside will know who to look for. If there is any possibility that you may have to jump, keep your helmet on and fasten your chin strap tightly. The four survivors of the Bronx tragedy each kept their helmets on, as did the two who perished. It increases your odds of survival.

Other critical skills must be mastered before a firefighter goes above a fire for search. Forcible exit, unlike forcible entry, is performed to save a firefighter's life. When a firefighter finds that he is about to be overrun by fire or is out of air, he must take immediate steps to escape. The most common method of forcible exit is to break a window, but if no window is available, he may have to force a door or even breach a wall to reach an area of refuge.

Forcing a door is typically a two-person job for members using the conventional method. At times, though, a number of factors may prevent using these standard techniques. A member may be alone or with an injured partner, or they may not have the necessary forcible exit tools. At other times, severe heat may prevent the members from rising up off the floor far enough to force the door in the conventional manner. In these cases, it may still be possible for the members to kick their way through the door or even a wall. Forget the Hollywood stuff of standing on one leg and kicking with the other. Lay down flat on your back or side as your SCBA permits, draw both knees up, and then kick out violently at the door with the bottoms of both feet. Naturally, this technique works best on doors that open away from you, but it can be of value on doors that open inward if there is a panel on the lower section that can be kicked out. It's worth a try if all of your other options are gone.

A gypsum-board wall can easily be breached in a similar fashion to provide access to a safer area. Walls built of wood lath and plaster are more difficult to force in this manner, but given enough time, they may also be breached. A better way to breach wood lath and plaster, however, is to use a tool, like a halligan or a pike pole, to create the initial opening. First drive the tool all the way through the wall and out the other side, about 3 ft above floor level if possible. This is necessary to check for any impassable obstructions like toilets, refrigerators, or cabinets, or fire already present in the next room, before it is too late to find a more suitable location for escape. Then, drive a halligan or pike pole into the wall about 3 ft off the floor. Stick the tool down into the hole as far as you can, then pull sharply outward, pulling off the lath on your side of the wall. After this side has been sufficiently cleared, begin work on the lath and plaster on the opposing side. Either hammer it out or, better yet, kick it off as described previously for forcing a door. This will create an opening about 13½ in. wide. The trapped member may have to resort to the reduced-profile maneuver to fit through this space. If tools and time permit, it is best to widen this space by knocking one of the studs loose at the bottom

Even the simple act of breaking a window and climbing out onto a waiting ladder can be difficult when there is fire rolling out of the top half of the window over your head. In this case, the methods of mounting a ladder taught in Basic Firefighting 101 no longer apply, and you must execute an emergency bailout. You must get out now while staying low to avoid being toasted. This means a headfirst exit, sliding over the windowsill. That's why the tip of a ladder must be placed just at or under the sill when it is placed for routine entry. Insert one arm between the first and third rungs and grab the second rung palm up with one hand. With the other hand, grab the third rung from the top, palm down, then roll out of the window. Once your legs clear the sill, you can make a controlled rotation until your feet are toward the ground, then descend normally. (Fig. 11–3).

In a variation of this technique, grab the beams of both sides of the ladder, palms down, and slide all the way down the ladder headfirst, dragging your feet along the rungs to control your rate of descent. This is a more difficult maneuver, particularly when you encounter different sections of an extension ladder. It may be necessary to use this method if more than one member is trapped at the same exit and time is crucial. You may not be able to wait in a room full of fire while your partner is rotating his feet downward at the top of the ladder. Both methods require practice to develop the skill and confidence necessary to perform them under severe conditions. Start easy during practice, using a small ladder at a low angle. As you develop your skill, increase the angle and the height involved. Always have the trainees belayed with a rope in case they lose their grip.

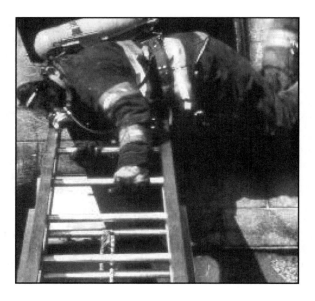

Fig. 11–3 The first time you exit a window headfirst should not be on the fireground, when visibility and control are likely to be less than ideal. Practice emergency maneuvers beforehand. Do them safely and have a safety line on members working at height.

The last emergency escape skill involves using a rope to exit a window when no ladder is immediately available. In the past, all FDNY members were issued a 40-ft personal escape rope and storage pouch that fits within the pocket of a turnout coat, as well as a Class 2 personal harness. The harness is attached to the wearer's bunker pants. All he has to do to attach the harness is to step into and pull up the bunker pants, then snap one hook. The rope is fitted with a locking hook on one end, allowing it to be quickly passed around a substantial object, such as a steam pipe, then snapped back on itself to anchor it. The wearer would then proceed to the exit window, attach the hook of the harness to the rope with four turns around the hook, and exit the window, rappelling down to safety (Figs. 11–4 and 11–5).

Fig. 11–4 When making an emergency rope escape from a window, roll low over the windowsill to avoid the heat and flame.

Fig. 11–5 Control your rate of descent with the friction of the rope as it passes around your body.

If you aren't equipped with a harness but do have a rope, it is possible to perform an emergency body wrap and perform the same feat of escape. After the rope has been anchored and deployed out the window, grab the rope just outside the window and pull in about 3 ft of slack. Pass this rope under your two armpits and around your back, over the SCBA cylinder if possible. Next, bring the two sides of the rope (the side leading to the anchor as well as the side leading to the ground) together in the firm grasp of one hand in front of you, then roll out of the window, low over the sill to avoid heat and smoke. The two sections of rope that you are grasping firmly prevent you from moving down the rope. At this point, gravity takes over and your feet will pendulum downward. Once you are vertical, bring up your free hand and grasp the end of the rope that passes under that armpit. Hold both sides of the rope firmly, then slowly separate your two hands horizontally until you begin to slide along the rope. Use your body to create friction to control your rate of descent. Understand that this is a life-and-death measure to be undertaken only under very extreme circumstances, since it is possible to free-fall right off the rope if you don't perform the maneuver properly. The critical points to remember are as follows:

1. Don't allow your hands to spread further than about shoulder width apart. Otherwise, you will experience a rapid loss of friction and may fall free of the rope.

2. Take all slack out of the rope just prior to exiting the window to avoid a sudden impact load that could break your ribs, pull the anchor free, or destroy the rope.

3. Although this is a relatively simple, easily learned skill, training in this technique is vital to success if you are actually forced to employ it during an emergency. During training, securely belay the members on a separate rope to avoid a potential free fall if someone misjudges his hand position.

4. Performing this technique imposes a great deal of friction on the body, particularly under the armpits. Never attempt it without a turnout coat to protect your body, and always wear gloves on both hands.

5. As with all rope rescue techniques, you should use an NFPA-rated rescue rope—in this case, one suitable for one-person loads for personal use.

Firefighter accountability

One of the key items when a firefighter is trapped or missing is to know who and how many people you are looking for. This will often help focus your rescue efforts in the proper location. This could be a very difficult criterion to deal with without some sort of hard information. Fortunately, it is easily addressed. In career departments with on-duty crews, this is most easily accomplished with a riding list—simply a small sheet of paper listing the name of each on-duty member. Alongside each name should be his assigned duty, such as nozzle man, hydrant man, roof man, and the like. Because each member's riding position on the apparatus is noted, such a list may be useful in identifying members in the event of a traffic accident. To be effective, the riding list must be filled out promptly in duplicate or, better still, in triplicate at the start of each tour and updated constantly to reflect any changes during the shift, such as a member who goes home sick or is relieved. One copy of the riding list should be kept by the officer on duty, who can use it to conduct a roll call if conditions demand. The other copy should be posted in a conspicuous place on the dashboard of the apparatus, where it can be retrieved by the officer in command if the officer or the entire company is among the missing. At large-scale operations, it is desirable to leave a third copy at the command post.

In noncareer departments, the names and number of members boarding the apparatus won't be known until the unit leaves quarters. Such departments won't be able to use the written riding list as such, but they should still have some other method of recording the names of those who have responded. One system that works well consists of issuing each member a small brass or plastic tag imprinted with his name. The tag is attached the member's turnout coat by a key ring. As each member boards the apparatus, he detaches the tag and either places it on a snap at his riding position or into a central collection container. By the latter method, the member's particular assignment goes unrecorded, but a central collection container can more easily be brought to the command post if it becomes necessary to take a roll call. In a similar version, each member is issued an adhesive name strip, which he then sticks onto a board on the apparatus.

Such a board can indicate the member's assignment, and it is also portable so that it can be brought to the command post. No matter how the system is laid out, however, it is useless if the members don't use it. There should be strict standing orders that before any member operates on the fireground, he or she must first deposit his tag at an apparatus or at the command post (Figs. 11–6 and 11–7).

At the command post, there must be provisions for recording the locations where units are operating. The simplest is a command chart, which is simply a preprinted form listing each unit, plus the floor and area of the building to which it has been assigned. This sheet is easily filled in by the chief or an aide as each company is put to work, indicating the floor and area next to the unit's number.

Now that you have a record of who is on the fireground and where each unit is operating, the control of firefighters is simplified. In addition, if it becomes necessary to account for a firefighter, you have a means of doing so without overlooking anyone. The size of many fire operations and the time it would take to cover all of the areas to conduct a head count, however, require some additional measures to enable us to identify a missing member rapidly. The most important item is a portable radio. This permits instantaneous reporting from remote areas, allowing search and rescue efforts to begin immediately. As an absolute minimum, at least one radio should accompany each hose team; so one radio may serve two people. Occasionally, however, it may be necessary for a member to operate alone for brief periods on the exterior of the building. Any member who may have to operate alone should definitely be radio-equipped so that he can call for help as well as respond to a head count. Another item that should become mandatory—as mandatory as SCBA—is the PASS device. These devices sound an alarm whenever a firefighter remains motionless for approximately half a minute.

Fig. 11–6 A detachable nametag does no good if it is still attached to the wearer. All members are responsible for establishing accountability.

Fig. 11–7 Accountability requires more than just knowing who's on the fireground. A system of tracking their location throughout an operation is vital.

Roll Calls

A roll call on the fireground is a sign of one of two things happening: either the situation is spiraling out of control with rapidly changing conditions, or the situation is well in hand and being managed very effectively. Right now a roll call at a serious fire requires contact with a large number of people who are likely to be operating in widely scattered locations. That means the command post must establish contact with each unit or member individually, which can take a considerable length of time in a crisis. For this reason, most ICs do not conduct a roll call until after some precipitating event.

Consider how long it takes to contact each unit at even a routine operation involving only three or four units, including having each unit verify the location and condition of all its members. (Since engine companies primarily work together as a team they are relatively simple to account for normally.) It is not unreasonable for this process to consume a full minute before a roll call is completed and a missing member is recognized. Try it at a training session sometime and you might be surprised at how long it takes, especially if someone is actually missing. I used to make it a practice when I was teaching live fire evolutions to remove one member of a company from an operation and *stash* him or her out in an apparatus, with instructions not to say a word until someone recognized that firefighter was missing. Sometimes companies went the entire evolution and never missed their member until they were directed to account for all their people. Sometimes even then they did not know they were missing someone. Technology is currently being developed that should make it much faster and simpler to conduct a roll call.

What we need is a wireless system, built into each member's portable radio, or better yet, their SCBA, that allows all members on the scene to check in simultaneously when the order for a roll call is given. A small laptop computer, integrated into a command and control chart could track each member on the scene and record their status. The system should include perhaps three status conditions as follows: Safe, need assistance, and Mayday. By building the device into the SCBAs integrated PASS device, the Mayday could automatically be sent to the command post under conditions where the member stops moving for a predetermined period. The technology currently exists to make this concept available on the fireground right now; all that is needed is the demand from the fire service for its implementation. The cost factor is not as high as you might think, this is *off the shelf* stuff.

A roll call should be ordered and recorded by the command post whenever any of the following events occur: PASS alarm activation, sudden fire extension, structural collapse, or the issuance of orders to withdraw from a structure. In addition, both company and command officers must be aware whenever contact has been lost with any of their team for an unreasonable length of time. For example, if an officer sends a team to vent the roof overhead and doesn't hear any activity or progress reports for several minutes, he should initiate radio contact with those members. If he is unable to make contact with them, he should request that a unit operating nearby either make contact or report their welfare. If they still can't be contacted or located, the officer must inform the command post immediately that he has a missing member. I take an informal roll call of companies continuously during a fire, by contacting them and their remotely operating members and asking for reports periodically or by personal observation. At an incident where only four or five companies are operating according to SOP, this is a manageable task, especially for a chief that is working solely with units he or she deals with every day. If the incident progresses beyond this scope or if units you do not see on a regular basis are involved, the process is much more difficult and will require the assistance of a tracking officer or aide.

Rapid Intervention Teams

When the Mayday call originates over the radio, or when an officer reports that a member is missing, somebody has to go get the person who's in trouble. Unfortunately, firefighters still end up in life-threatening circumstances, and the only thing that can save a trapped member is the response of others. Firefighters usually get into trouble when situations deteriorate suddenly, often when all of the available personnel are committed elsewhere. Hence the need for a RIT.

In its most basic form, the need for a RIT goes right back to the minimum staffing standards that have been fought over for many years. NFPA 1500 simply formalizes an inescapable fact: to rescue people or accomplish any other task, you must have able-bodied personnel available to do the job! The fireground will never in my lifetime be a place where a trapped firefighter is going to get out of trouble simply by saying, "Beam me up, Scotty!" Somebody's going to have to go in and get him, hopefully alive. If you don't have the personnel to effect the rescue, the trapped firefighter will die. It's as simple as that.

The makeup of RIT

Not all RITs are created equal! Nor should they be. NFPA 1500 and OSHA's respiratory protection standard, (also known as the 2 In/2 Out rule) each requires that at least a two-person RIT be available during the initial stages of serious fires. As an incident escalates, this number may easily be inadequate for the intended purpose. In fact, a two-person crew will have a great deal of difficulty in physically removing an unconscious firefighter from anywhere but the first floor or other places from which he can easily be dragged. Designating a two-person ambulance crew as your RIT may look okay on paper, but it will be woefully inadequate in case a firefighter needs rescue. Get a fully manned fire company (or two two-man or three-man companies) as your RIT and make sure they know their job. "Easier said than done," you say. "We only have four companies in our department." My answer to that is simple: Call more. Whether it's through mutual aid, automatic assignments, or a recall of off-duty personnel, you must provide this capability. You can do it early, in time to rescue anyone who needs assistance, or you can play the odds and wait until things have gone wrong. Either way, you'll need help if the incident escalates. The question is whether you call them in time to avert a tragedy or to recover a body.

In FDNY, we have had instances in which the lives of firefighters have been saved by the RIT and/or an assigned rescue company. RITs have proved their worth many times in the nearly 15 years that they have been in place. The same is true in many cities across America. So how do you make this concept work for you? There's more to it than just adding another unit to the assignment. The four prerequisites to a successful rapid intervention operation are people, policies, tools, and techniques (Fig. 11–8).

People are the most important resource. You must arrange for a team to be immediately available at every working fire or special operation. I believe four people are the absolute minimum to carry out a successful rescue of a downed firefighter in simple circumstances. In the past, FDNY used to designate an engine company for this task. They currently assign a ladder company, which is a far better choice. Ladder company members often have the experience of removing civilians and are more often equipped with the basic tools needed for this critical task, *i.e.*, radios, hands-free lighting, forcible entry tools, rope, ladders, and the like.

One key personnel issue that must be addressed when implementing a RIT policy is the attitude of the assigned RIT members. Every firefighter worth the title wants to get in and slug it out with the devil, but that is not the job of the RIT. Those members must understand that they are being entrusted with the lives of every firefighter in the building. They must show up ready for this task, properly equipped, and then stand fast, hopefully never to be needed. This can create some frustration, since we hope that they won't be able to actively perform their jobs, but it is critical that they realize that they are their brothers' keepers. Besides, they'll get their turn when someone else is standing by outside for them. Such adherence to duty must be stressed in the RIT policy. There is no room for freelancing. The team must stand by and wait for orders.

Fig. 11–8 This Firefighter Assist and Search Team (FAST) truck stands by with their rapid intervention gear at a multiple alarm. Note: The gear in the foreground from a previous FAST unit that has been deployed. Each unit should bring its own.

Another item that is essential to the success of such an operation is having a standard policy that defines the basic duties, assignment, tools, and position of a RIT. In larger departments, this is fairly easily to do. In many cases, however, the RIT will be coming from a neighboring department or through some type of mutual-aid agreement. Each participating organization must agree to provide the standard minimum crew with appropriate tools and training. The dispatch of the RIT should be automatic, usually upon the receipt of an alarm for a working fire. As soon as members are placed in harm's way, there should be some means of getting them out. The agreement must be reciprocal in that each organization should know that they are just as likely to be the RIT as they are to require it. This makes it to everyone's benefit to perform equally well.

An important part of this agreement is a stipulation as to when the RIT will be employed, what the chain of command is, and when the RIT will be released from its duties. Naturally, firefighter safety is the primary goal of the RIT. This doesn't necessarily mean that the RIT should never budge from the command post until a firefighter is down. Particularly in smaller buildings, the RIT could be employed to open up additional escape routes so that firefighters can avoid getting into trouble in the first place. If these members are so employed, the team must still remain intact, ready to respond to any emergency. In larger buildings, however, such actions may take the RIT away from their area of responsibility and could delay them from deploying rapidly in a crisis. Since the main task of the RIT is firefighter safety, the IC must avoid treating the team as just another unit that has been staged to handle fire extension. This is a fine line to walk in some cases. The IC must decide whether firefighters will be endangered by extension while waiting for additional units to be employed. For example, the RIT may have to be deployed to get a vitally needed hoseline in place to protect firefighters as they escape. It is better to keep firefighters out of trouble in the first place than it is to get them out after they're down. Put out the fire and everything else gets better!

One critical task the RIT can perform without reducing their readiness is to assist the IC with tracking the location of operating units and personnel. They will need this information in order to respond to a Mayday for a missing or trapped member. In the FDNY, the firefighter assist and search team (FAST) truck is issued a dedicated handie-talkie that provides a visual display of any unit that transmits a Mayday signal via the emergency button on a trapped member's radio. That is a key task for the FAST truck: monitor radio communications for Mayday or other messages that indicate members may be in distress.

The RIT should report to and be under the direct control of the IC or the operations officer if they are forward-deployed, as in the case of a fire in a high-rise. The RIT must be located close enough to the scene to accomplish its mission within a reasonable time frame. You may need more than one RIT in large buildings. It must be understood that they are committed to their assignment and must not be picked up by any other officers for other tasks. This can be a problem, particularly since they often arrive at a stage when the incident isn't going well and reinforcements are being sought. In FDNY, department policy requires that the dispatcher notify both the IC and the assigned unit of their designation as the FAST truck or RIT. This serves to remind all of the importance of the task. Perhaps more importantly, it establishes responsibility and accountability for these duties so that a crew could not get involved in firefighting and later claim that they didn't know that they were supposed to be standing fast.

One way to minimize the standing-fast syndrome, in which a company watches standoffishly and does nothing tangible, is to use the RIT as an attack company if the incident escalates. By this method, a new RIT is designated as each additional alarm is transmitted, and the previous RIT is then permitted to operate. Although this is probably the best practical alternative, it is certainly an imperfect solution from several standpoints. The members of a RIT that has been on the scene for some time have had a chance to size up the building, determine the location of operating units, and see how the situation is progressing. To replace them with another unit means that the new unit, before being able to operate efficiently, must spend valuable time gathering the information that the first unit has already learned. Moreover, those in a new unit will never have an accurate picture of how the incident evolved prior to their arrival, yet this is vital information if they are to locate missing or trapped members.

If you choose to call a new RIT and employ the prior RIT as an attack unit, you must wait for the new unit to arrive on the scene, stage their equipment, and receive a briefing on conditions before the first unit leaves the command post. Otherwise, the later-arriving unit is at a real disadvantage if something goes wrong. If the first unit is deployed immediately, there will be a gap of several minutes in RIT coverage right at the time that it is probably needed most. Yes, you may need a truck company upstairs to pull ceilings, but be very, very careful about using your RIT for this purpose before they have been properly relieved. If it looks as if the fire is starting to get away, that's when the RIT is most likely to be needed. Firefighters don't usually end up trapped at windows when everything has gone right! This isn't to say that the RIT must stay in the street until the last length of hose has been packed away, but you should make very sure that the fire is under control and that nothing is going to put firefighters at risk before you send in the RIT to open up walls for overhaul.

RIT tools

There are a number of tools that we know will be needed at any working fire. Think of the first group as those that every member stepping off the rig should have in his possession at all times. They include the following:

1. SCBA with activated PASS alarm.

2. A large hand light, preferably on a sling, for hands-free use.

3. A good, sharp knife, or better yet, a combination knife and wire cutter such as a *Leatherman* or other multipurpose tool.

4. A spare cylinder immediately available on the apparatus for each member.

5. An NFPA-rated personal rope—40 ft of $\frac{3}{8}$-in. nylon.

Additionally, each RIT officer should ensure that his team is equipped with the following tools when they report to the IC:

6. Portable radios for each two-member team within the RIT.

7. A 200-ft search/guide rope for each two-member team.

8. Forcible entry tools (halligan, flathead axe) for each two-member team.

9. At least one hydraulic forcible entry tool (rabbit tool).

10. One lifesaving rope and harness belt.

11. A spare mask for a trapped firefighter.

12. A stokes stretcher and resuscitator.

13. A power saw, either wood- or metal-cutting, depending on the size-up of the building.

14. Ladders that are suitable for the building—tower, aerial, or portable.

At the scene, the unit should also attempt to obtain a copy of the building's floor plan and a copy of the command chart indicating which units are operating and where they are located. Also, if possible, they should locate a hoseline that can be committed for rapid intervention, if necessary, and know what unit is supplying the line.

RIT size-up

As with every other fireground task, the RIT must perform a size-up. The specific duties of the RIT make its size-up somewhat unique. While encompassing all of the aspects of the standard 13-point approach (see chapter 2) and the survival survey, the RIT's size-up includes extra considerations. The RIT officer should:

1. Request a size-up of the building by radio while still en route, determining the size of the building as well as its construction, occupancy, and fire location.

2. Monitor tactical and command channels for urgent or Mayday messages.

3. Once at the scene, perform the RIT size-up of the building, occupancy, and the location and extent of the fire.

4. Know the location of units and members and their routes of access.

5. Monitor the progress of the operations.

When firefighters end up trapped or missing, the RIT must have prepared for the occasion with some techniques that have been practiced and honed. They must be able to perform them blindfolded, because that's exactly the way they're going to have to be done. The fireground is extremely confusing when a Mayday has occurred. Of course, no two situations are exactly alike, but some problems repeat themselves so often that we should expect them and have a game plan all worked out in advance.

The first problem is a lack of information about what has happened. The IC out in the street (or worse yet, down the street in his command van) may have no indication as to the severity of the problem, other than a garbled radio transmission. His first priority must be to gain control of the situation, clear radio channels of any unnecessary traffic, and establish clear lines of communication with the parties that are reporting the problem. This may require shifting all other ongoing operations to another channel, arranging for runners, or transferring command of firefighting or rescue operations to other officers. At an ongoing operation, one person cannot run a large-scale firefight while simultaneously directing an involved rescue effort. While trapped firefighters obviously demand a total commitment of force to remove them, we must still put out the fire, for an ongoing conflagration will continue to worsen the situation.

Once communication with the trapped members has been established, the IC must immediately be advised of the number of trapped members and, if known, their condition. Then, necessary resources such as multiple advanced life support (ALS) paramedic ambulances or additional ladders can be special-called. Ideally, he should ascertain the exact location of the members. With conscious victims, this can be accomplished by anything from direct voice contact to radio contact to tapping on metal objects. You may be able to locate unconscious firefighters by a PASS alarm, feedback assisted rescue (FAR), pager-assisted rescue, or by using a thermal imaging camera to see through the smoke.

If you can make contact with the victim or the rescuer who has reached him, determine the nature of the problem. Is the victim simply hung up on an obstruction, or is he cut off by fire and in need of a separate escape route, either by ladder, extinguishment, or a breaching operation? Is the member out of air? Is he pinned? You should immediately make provisions to handle a very prolonged, difficult operation.

Removing Unconscious Firefighters

As soon as the first report is received that a firefighter is down, three things should happen. First, if not already at the scene, an ALS ambulance should be called. Second, additional firefighting personnel must be special called. This might require transmission of multiple alarms or calls for mutual aid. In New York City, the report of a trapped firefighter routinely prompts the response of a second heavy rescue company to the scene. Third, a protective hoseline, a spare mask, and a resuscitator should all be brought to the vicinity, even if there is no obvious immediate need for them.

These incidents have a tendency to go from bad to worse. What may look at first glance like a simple removal can often disintegrate into a nightmare of failed efforts. Be prepared to protect the victim in place while additional resources are being deployed. Firefighter rescue is a lot different from the rescue of most civilians. For one thing, all firefighters are full-grown adults. There are no 55-lb children being thrown over one arm in these scenarios. Second, firefighters are further weighed down by turnout clothing and SCBA, which can get hung up on obstacles. Wet bunker gear can add 30–35 lbs to a firefighter's weight. The bunker gear also expands a firefighter's profile, making movement through confined spaces more difficult. Finally, there is the physical and psychological stress of knowing that one of your own will die if the rescue is unsuccessful. When a firefighter is overcome or trapped in a fire, we must remember that it is a trained firefighter who is trapped, not an untrained civilian. For a firefighter to become trapped, conditions are likely to be very severe. A successful rescue of a trapped firefighter depends on several elements; sufficient manpower available to perform the required tasks, training in a variety of potential situations likely to be faced, and utilizing the appropriate tools and techniques for the situation.

When a *Mayday* is transmitted for a firefighter in distress, there is a natural inclination for members on the fireground to drop what they are doing and respond to the *Mayday*. This well-intentioned response to this emergency may actually hamper the rescue effort of the distressed firefighter. If possible, the IC should assign a chief officer to supervise the rescue effort of the distressed firefighter. Some of the considerations that the chief in command of the rescue effort should be the following:

1. Are operating members still fighting the fire?

2. Is the area of access or egress being blocked by members that are not an active part of the rescue effort? Clear doorways and staircases.

3. Does the rescue effort need the support of a hoseline?

4. Is specialized equipment necessary to complete this rescue?

5. Is there a tactic that could help mitigate conditions in the area where the unconscious firefighter has been found?

6. Is a certified first responder-defibrillator trained (CFR-D) or paramedic engine company available to provide medical support close to the extrication point?

7. Are emergency medical services (EMS) standing by to administer *advanced life support* upon the removal of the trapped firefighter from the fire building?

8. It is important to emphasize that the operating units continue to fight the fire when a *Mayday* is transmitted for a distressed firefighter. The abandonment of engine or ladder company operations to assist in a rescue where resources have already been deployed to handle the situation places the trapped member and the rescuing firefighters in severe danger.

Incident Commander's Duties

When the IC is informed of a trapped or unconscious member he or she should take the following steps, as appropriate.

1. Take control of handie-talkie channel(s) and direct all non-essential handie-talkie (H-T) traffic to stop.

2. Gather information as to the identity, location, and nature of the situation.

3. Assign appropriate resources to remove the unconscious or trapped member. Assign a chief officer to direct and control the removal. The identity of units assigned to this task shall be announced over all handie-talkie channels in use.

4. Notify the dispatcher to have an ALS ambulance sent to the scene on a top priority basis if one is not already present at the scene.

5. Reevaluate—and modify as necessary—the firefighting strategy.

6. Call additional resources/ transmit additional alarms to replace units that are committed to the rescue effort. Designate a staging area at least a block away from the scene and have the dispatcher direct a chief officer to that location to gather the incoming units and maintain control over their assignment.

7. Consider the need for the following:

A. Additional chief officers to relieve the IC of firefighting responsibility, allowing him to control and coordinate the rescue effort

B. Additional chiefs for assignment as follows:

- Sector or group commanders

- Additional safety chiefs

- Victim tracking officers

- Communications coordinator if not already assigned

- Coordinator for support agencies

- Chiefs for relief, as necessary

C. Additional heavy rescue companies with specialized extrication equipment

D. Units for lighting, mask service

E. Support agencies, *i.e.* Red Cross, police, utilities, etc., for cranes or other heavy equipment, as well as control of utilities

F. Fire department chaplain

G. Press officer

H. Department photographer and /or video unit to document the scene

If resources and fire conditions permit, simultaneously begin multiple avenues of approach. Whichever team gets the victims to safety first will have proved effective, but if the first attempt fails, a backup plan will already be underway. At most incidents where a firefighter is trapped, the best resource you can have is additional firefighters. Much of the work, such as removing debris and transporting the victims, will have to be done by hand. A clear chain of command is critical in these instances, though, with a single person designated as the victim removal officer. This officer must make the decisions as to what steps will be taken and in what order.

Because a firefighter's natural instinct is to want to help in these situations, a clear plan of action is needed to prevent freelancing, which can endanger the victims as well as the would-be rescuers. Individual firefighters can prepare themselves for operations at these incidents by equipping themselves as described previously. The tools of a RIT will also serve them well in their daily routine.

Although the number of rescue scenarios is nearly endless, several situations repeat themselves often enough that we can prepare for them with some degree of certainty. The following actions could save a life:

1. Call for assistance immediately. Give your location, the number of victims, and details of any assistance you need.

2. Do not share your mask! Your mission is to get both of you out alive, not to double the number of victims. A rare exception might be when a firefighter is trapped, unable to move, and out of air. Be sure that reinforcements bring in additional masks.

3. Be totally familiar with all of the types of SCBA that your department uses, as well as those of responding mutual-aid departments. Your survival, or that of a trapped firefighter, can depend on your knowledge of how to quickly don, doff, and operate a mask in zero visibility while wearing gloves.

4. Have a sharp knife handy to free hung-up members. You may have to hack through entanglements or quickly cut mask, shoulder, or waist straps.

5. Once the victim is out of immediate danger, begin basic life support and await assistance and medical aid.

6. Keep it simple, stupid! Avoid setting up fancy systems when a simpler method will suffice. The more complicated things are, the more likely there will be a problem under high-stress conditions. The safety and survival of victims and rescuers is at stake. Do it right!

7. If fire conditions and logistics permit, positive-pressure blowers or fans to inject fresh air may prove useful around the rescue effort, even if they worsen fire conditions in remote areas.

Depending on the fire conditions and the nature of the injury, a variety of drags or carries may be appropriate. The techniques described below are intended for use under difficult fireground conditions. They depend on only minimal equipment and set up time. They are intended for use as lifesaving steps under extreme circumstances. They are not intended for use at routine removal situations, where time and equipment concerns allow other more suitable, sophisticated hauling and patient handling systems to be set up.

Once the rescuing firefighters arrive at the location of the unconscious firefighter, generally the first action taken should be to assess the victim's breathing status. If the victim is breathing, check the supply of air remaining in their SCBA. Having the RIT bring a fresh mask with them when entering for rescue solves one of the more serious problems quickly; if we can get a new mask on the breathing victim we have bought ourselves some time to accomplish other tasks. Once the victim has a mask on that is supplying air, that mask should be converted to a harness. This action will provide the rescuers with a secure grip of the unconscious firefighter. This will greatly assist the rescuers in dragging the unconscious firefighter by the shoulder straps of the SCBA. When this technique is completed, the SCBA will not come off the unconscious member being pulled by the shoulder straps. Horizontal movement is possible by a single member under most conditions.

To convert the SCBA into a harness, do the following:

1. Unbuckle the waist belt of the SCBA.

2. Fully loosen both halves of the waist belt.

3. Take half of the waist belt and put it behind the unconscious firefighter and bring it up between his/her legs.

4. Take the other half of the waist belt and bring it in front of the unconscious firefighter, and connect it to the other half of the waist belt.

5. The waist belt is now connected between the trapped firefighter's legs. Tighten these straps as necessary (Fig. 11–9).

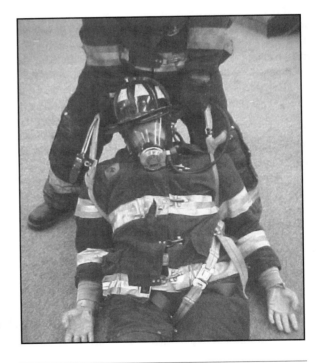

Once the breathing member is supplied with air, our priority is to remove him to a tenable atmosphere without inflicting spinal injury. If the member is not breathing, we must still remove him to a tenable area, where CPR and other resuscitation efforts can begin. Although the emphasis on preventing further injury takes a back seat to the need for speed, a nonbreathing member has to come out immediately. At times, fire conditions may be so severe that immediate removal of the distressed firefighter is critical, even in the face of a spinal injury. Removal of the distressed firefighter to a tenable atmosphere usually involves little danger from spinal injury if there was not a fall or other injury involved. Depending on the victim's location, fire conditions and the

Fig. 11–9 An emergency harness can be quickly created on a downed firefighter simply by passing one half of the SCBA waist strap behind one leg and in front of the other. Use the shoulder straps to drag the downed member.

mechanism of injury, various drags or carries may be appropriate. Grasping the SCBA straps will allow you to drag the member headfirst. This is the preferred position, if it is possible, since dragging a man headfirst protects his head from getting caught on most obstacles, and it also prevents his arms and legs from getting wedged in crevices.

If the victim does get hung up or snagged on an obstacle, the rescuer will have to back up and feel for the obstruction. The rescuer must be able to recognize by touch all of the components of the SCBA and differentiate between the mask and foreign objects. Avoid trying to pull the victim forcibly past any obstructions that could wedge him deeper or produce injury. Instead, try to maneuver the victim backward until he is clear of the obstacle. If this fails, it may be necessary to cut the victim free with a knife.

Any of the methods described here might be used at one time or another in order to get the victim to a place of safety. Once the victim is out of immediate danger, we should immediately begin to assess her airway, breathing, circulation (ABC) and treat as necessary. At this point, it might be wise to wait until the victim is secured to a backboard or stokes basket prior to moving her any further, especially if there is any indication that a spinal injury is present.

If two rescuers are available and space permits, they should work together side by side at the victim's head. If this is not possible, due to space restrictions in a narrow hallway, one member will have to work at the head and the other at the victim's crotch. This method will enable firefighters to move an unconscious member in a confined, cluttered or restricted area. It may also be appropriate where heat conditions require that the rescuing firefighters remain as close to the floor as possible (Fig. 11–10).

Fig. 11–10 The SCBA harness is critical to pulling a victim in the crawling position. The second member pushes with the victim's leg over his shoulder.

To work in this fashion, first convert the distressed firefighter's SCBA into a harness. Firefighter #1 takes a position on his hands and knees at the feet of the unconscious firefighter and places one of the unconscious firefighter's legs over one of his shoulders. This firefighter then wraps his arm around the victim's thigh as close to the top of the leg as possible. With the victim's knee placed over this member's shoulder, it will enable him to push the unconscious firefighter's weight with his legs. This firefighter will greatly assist the effort rather than having the firefighter at the head pulling most of the weight with his arms. The second rescuer positions herself on his hands and knees at the head of the unconscious firefighter. He loosens one shoulder strap on the victim's SCBA to get a handle to pull on. This member pulls as the other member pushes to remove the unconscious firefighter from the area. If conditions permit, a cadence count of one, two, three, pull will help to coordinate the effort.

All of the previous methods might be used at one time or another to get the victim to safety. Once the victim is out of immediate danger, we should assess his ABC status and treat as necessary. At this point, it might be wise to wait until the victim has been secured to a backboard or stokes basket prior to moving him any farther, especially if there is any indication of spinal injury.

Moving an Unconscious Firefighter Down a Stairway

If the victim must be brought downstairs before reaching a place of safety, the technique used will vary with the resources available and the weight of the victim. A light- to moderate-weight victim can be brought down by one person using the same drag used to reach the stairs. Use extra care to protect his head and neck.

A very large firefighter in turnout gear and SCBA can be difficult to move down a tight staircase. In this case, two firefighters and one length of rope can readily solve the problem. Again, using the drag that was used to reach the stairs, one firefighter starts down the stairs with the victim's head and shoulders, shielding these areas from harm. At the same time, the second firefighter, rather than trying to carry the lower body and risk being pulled down the stairs on top of the others, simply ties a clove hitch and binder around both of the victim's legs at the ankles. This rescuer can control the victim's descent by feeding the rope hand over hand from the landing. Since gravity is assisting the first rescuer, even a small-framed person should be able to move a large victim on the stairs. The rescuer at the top can provide braking action to control the rate of descent either by taking a half turn around an object or by laying the rope flat along the landing and passing the rope under the instep of the rescuer's foot, making the rope bend over the lip of the top stair.

This technique is a quick and dirty method of improvising a rope-lowering system using the barest of essentials. A 40-ft length of ³⁄₈-in. personal rope is more than sufficient. The rope won't be carrying the person's full live load, the staircase will. This method is designed to allow you to remove a single large victim when conditions are still threatening. This isn't the time or the place for fancy mechanical advantage systems, which are probably out on the apparatus anyway. This method can be put in place in a matter of 30–40 sec using material that should be readily available on the person of any firefighter operating above a fire for search and rescue. This method isn't meant to help move a victim upstairs or up an incline, although if sufficient personnel are available, it will work. Still, in the understaffed, highly confined conditions indigenous to firefighting, other simple, practical methods should be used instead.

Moving an Unconscious Firefighter up a Stairway

Removing a victim up the stairs is another challenge, usually necessary in below-grade situations such as cellar fires. The usually high concentrations of dangerous gases in below-grade fires severely restrict rescuers because of their need for SCBA. This can hamper attempts to crawl under the victim to carry him to safety.

As in most firefighter removals, the first step is to convert the downed firefighter's SCBA into a harness. Once that is done, rescuers drag the unconscious firefighter to the base of the stairway so that he is lying face up, and about three-fourths of a body length on the stairway. One rescuer positions himself on his hands and knees at the feet of the unconscious member. This rescuer places both the victims' legs over both of his shoulders, and wraps both his arms around the top of the victims legs, placing his chin as close to the victim's groin as possible to maintain a good grip on the victim. This rescuer must keep his chin in the victim's groin throughout this carry or else he will lose his grip and leverage (Fig. 11–11).

The other rescuer positions himself on the stairs at the head of the victim, facing down the stair. He must loosen one or both of the victims SCBA shoulder straps to provide a handle to lift the victim's head and shoulders up the stairs. Once both rescuing firefighters are in position, the member

Fig. 11–11 Moving an unconscious firefighter up a stairway is among the most difficult tasks on the fireground. Teamwork is essential. The top member should use the SCBA straps as a harness, while the lower member gets the victim's legs up on his or her shoulders to carry.

at the bottom raises himself to a standing position. At the same time, the member on the stairs lifts on the shoulder straps of the SCBA to lift the head and shoulders of the victim off the stairs. Both rescuing firefighters now walk slowly up the stairs in unison. They should lean one shoulder against the wall to assist them in maintaining their balance. The rescuing firefighter at the feet must be careful to move at the pace of the rescuing firefighter at the head, since that member will be walking backwards.

If conditions in the area are so severe as to prevent the members from standing, or if the weight of the unconscious member prevents the members from carrying the victim, it may be necessary to utilize another method. Rescuers should consider the following: Convert the victims mask into a harness as previously described. Position the victim on his or her back with their head near the base of the stair. Bring one end of a lifesaving rope or a personal rope down to the victim's location and pass it through the shoulder straps of the SCBA. Then tie or clip the rope back on itself, or run the rope back up to the members at the top of the stairs, providing two ropes that can be used by hauling firefighters at the top of the stairway. Rescuer(s) with the unconscious victim guide the victim up the stair, lifting the head and shoulders, while others pull the victim up by the rope. The member(s) with the victim must closely control the hauling, signaling the others at the top of the stairs (these members might be located outside the danger area) when to haul. Again a cadence count of one, two, three, pull helps to coordinate the separate efforts.

Another approach that may work involves a backboard. Lay a backboard along the stairs, and then drag the member to the base of the staircase. Secure the victim at her ankles with a clove hitch and binder around both legs, then run the rope up to the surface, where it is grasped by two to four rescuers. Run the free end of the same rope back down the stairs and tie it to the top of the backboard. Once the victim has been secured to the rope, give the signal to haul, and have the other members pull the victim feet first onto the backboard. One rescuer stabilizes the head and neck and maintains an open airway. Once the victim is fully on the backboard, the surface rescuers pull both ends of the rope at the same time, bringing up the board and victim together. Using the board is very important, since it helps prevent the victim from getting hung up on stair treads during the ascent. The operation can be made easier if you can place a ladder on the stairs for the board to slide on.

Moving an Unconscious Firefighter out a Window

Fig. 11–12 One means of raising an unconscious firefighter to a window sill involves rolling the victim onto a rescuer's back. The rescuer then uses his leg and arm muscles to rise up to windowsill height.

Removing a heavy firefighter out of a window and onto a ladder is a great challenge. If the windowsill is high or if the working room is constricted, firefighters on the inside will have difficulty removing him. One of the first steps is to get a member underneath the victim by rolling the unconscious member onto the crouched rescuer's back. The rescuer can then use his powerful leg and back muscles, not merely his arms, to push the victim up to the windowsill. A second rescuer inside can help keep the victim on the first rescuer's back as the victim is transferred to a third rescuer on the ladder. When possible, removal by ladder can be made safer and smoother by placing two or even three ladders side by side with a rescuer and a backup (this could be a civilian) on each (Figs. 11–12, 11–13, 11–14, and 11–15).

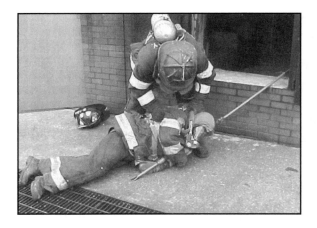

Fig. 11–13 A strong hook or a folding attic ladder can be used as an improvised ramp.

Fig. 11–14 The victim can be dragged onto it in steps…

Fig. 11–15…until the low end is raised to lever the victim outward.

If conditions prohibit the rescuer from getting under the victim, and if the members inside cannot lift the victim to the sill, then we have to arrange for the outside members to help. This requires using a rope and a ladder as a high-point anchor. The rope must be secured around the victim and then passed up, out of the window, over a rung of the ladder, down along the inside of the ladder, and then out along the ground. The end of the rope must be long enough to allow three or four rescuers to grab it and pull. The rescuers inside assist the victim up to the height of the windowsill, while the members outside perform most of the lifting effort. As much of the work as possible, including setup, should be done outside, where the visibility is better and manpower is more readily available.

Convert the victim's SCBA into a harness and drag the victim to a sitting position just under the windowsill. Bring the lifesaving rope up the ladder and pass the rope over an upper rung that is just slightly above the top of the window if possible. Pass the end of the rope to the inside members, who pass the end of the rope through the shoulder straps of the SCBA/harness. The rope should then be passed back out to the member on the ladder if time and conditions permit. The rope is fed back down the ladder to the ground, where now two groups of rescuers can pull on two separate lines of rope. The signal to haul is then given to the rescuers below, while the inside member(s) assists the victim out the window.

Once the victim's torso is clear of the windowsill he can be positioned out of the window and lowered to the ground by the members below. This is an exceptionally reliable (if not easy) method of removing an unconscious member. I have used it to remove a member who weighed well in excess of 275 lbs out a window where the windowsill was located 6 ft above the floor. No other method works under these circumstances. The addition of a single-sheave pulley on the top of the ladder, instead of using a rung, adds to the ease of the operation, but the lack of such a pulley or of the time to set it up shouldn't prevent you from using this system in a life-or-death situation. Similarly, if available, you can use a backboard as a ramp from the floor to the windowsill, and the victim can be slid along it. A word of caution: SCBA manufacturers are not in favor of this technique, and with good reason. Many of the SCBA straps that are found in the field are in terrible condition and could snap if subjected to these stresses. Examine your straps carefully during your mask checks. If they are defective or worn, replace them. It is your mask they will have to use to carry you out of that building if you go down (Figs. 11–16, 11–17, and 11–18).

Fig. 11–16 Removing an unconscious firefighter over a high windowsill is best accomplished using a rope and a ladder as a high point. This 250+-lb firefighter was successfully removed through a window where the sill was nearly 7 ft above the floor!

Fig. 11–17 To hoist an unconscious firefighter out of a room using a ladder as a highpoint, one end of a rope is passed over a rung that is at least as high as the top of the window opening. The rope is then secured to the downed firefighter.

Fig. 11–18 The other end of the rope is pulled by firefighters on the ground, hoisting the victim above the windowsill. These same members then lower the member to the ground for treatment.

Another alternate method of attaching the rope to the unconscious victim must be available for those situations where the member may not have an SCBA on or where the straps appear to be unable to sustain the load involved. This involves using the *handcuff knot* on the victim's arms or legs. The *middle* of the rope is passed up the ladder and over a rung slightly above the top of the window. The middle of the rope is then passed in to members inside who secure the handcuff knot around the victim's both forearms. The signal to haul is given to members below, while the inside member(s) guide the victim out the window. The member is then lowered to the ground. A word of caution about all efforts to use a ladder as a high point- one person must constantly remain footing the ladder. The members on the ground should be directly in line with the ladder, since the hoisting and lowering operation tends to pull the ladder sideways otherwise (Fig. 11–19).

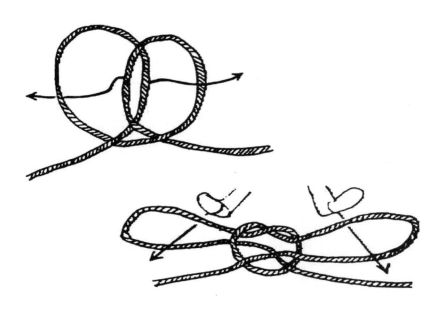

Fig. 11–19 Tying a handcuff knot is a simple three-step procedure.

Step 1—Make 2 loops in the middle of the rope, as if to make a slip-over clove hitch; place the right loop over the left.

Step 2—Pull the left side of the right loop down through the left loop, while pulling the right side of the left loop up through the right loop.

Step 3—Place the two loops thus formed over the victims feet up to the ankles. Pull the loops taut, then signal the other rescuers to haul on both ropes.

Vertical Removal through a Hole in a Floor or Roof

Another situation in which we find great difficulties in removing an unconscious firefighter is when the victim has fallen through a floor or roof to a lower floor. While the rescuers should attempt as many routes to the victim as possible, the most direct route is often through the same opening through which the victim fell. This could mean lowering a ladder through the opening if space permits. Otherwise, a member might need to rappel down or be lowered. Once rescuers have reached the victim, they should take the most appropriate removal route, perhaps up or down stairs or out a window. In some cases, though, the only available removal route will be back up through the opening. If a vertical removal is considered because a firefighter has fallen through a fire-weakened floor or roof, then the area around the perimeter of the hole must be surveyed by the rescuing firefighters. If there is a possibility that a secondary collapse may occur during the rescue removal, then the area around the hole must be reinforced. This may include removing doors from the fire building and placing them near the hole to spread out the weight of the rescuing firefighters. Portable ladders placed near the hole may also spread out the weight of the rescuing firefighters.

Since there is a fall involved, spinal injury is a major concern. If it is at all possible, the member should be secured in a spinal immobilization device. The victim might be placed on a backboard if conditions permit, and the board hauled up along the ladder as described previously. Alternately, the board and victim can be rapidly lashed to the ladder, and the whole ladder can be withdrawn from the hole. Of course, this means that the rescuers must have another way out, or they must wait for the ladder to be returned.

Another device that could be useful is a specially designed extrication harness, which provides spinal immobilization and doubles as a rated hoisting harness. This can be applied to the victim and then attached to any type of hauling device for the vertical lift. Of course, all of these methods are somewhat involved and time-consuming. Moreover, special equipment must be brought to the area.

A last-ditch method that requires only a rope may be the only thing that works. This method involves lowering the middle of the rope down through the hole to the rescuing firefighters who are with the fallen firefighter. A slip-on handcuff knot is placed around the victim's wrists or ankles. To save time, the members at the upper level may pre-tie the handcuff knot to the rope prior to lowering it down to the rescuers who are with the fallen firefighter. Either the life saving rope or a member's personal rope may be used for this purpose (Figs. 11–20 and 11–21).

If the victim is to pass through any tight spaces, it will be necessary to remove the victim's SCBA prior to attaching the handcuff knot. If surrounding conditions demand that the face piece of the SCBA remain on the victim during the extrication, then keep the face piece on the victim and remove the rest of the SCBA from the body of the victim. Attach the shoulder straps of the now-removed SCBA harness to the lowest snap hook of the victim's bunker coat by opening the coat snap, passing the straps inside and buckling the snap. This will reduce the profile of the victim as he is being hoisted through the hole.

If time permits, lower a second personal rope and place another handcuff knot right alongside the first one, on the same wrists or ankles. This option is desirable when an exceptionally heavy firefighter is being extricated. Once secured, the signal to haul is given and two or more rescuers on each end of each rope can begin to hoist. This allows at least four rescuers to hoist vertically, which is the minimum that should be used. One or more rescuers should be ready to pull the victim the rest of the way out of the hole, once he reaches floor level. Roll him onto his stomach if possible to make removal easier.

Fig. 11–20 The handcuff knot placed on the member's forearms permit him to be hauled vertically. Note: The SCBA is secured to the victim's coat, allowing it to be hoisted with the member in distress while minimizing the profile.

Fig. 11–21 By lowering the middle of a length of rope into the hole, the rope can be hauled by more members. In addition, the two sides tend to center the victim in the opening, preventing him or her from getting hung up on the edge.

This method has several advantages under severe conditions. By hoisting on the extremities, the arms are always extended overhead, reducing the body's profile at the shoulder area, which is the widest part of the body when the arms are down. The knot is exceptionally easy to apply. It can be tied and placed in less than 30 sec. Using the middle of the rope allows two teams to hoist. Placing the knot around the victim's ankles and hoisting head-down can also be done (this serves to anatomically align the head and spine). This way, the weight of the head places traction on the spine. This also ensures an open airway for the victim. If conditions permit, place a backboard at the mouth of the opening so that the member will slide onto it. He can then be strapped and transported from the area.

If the rescuer that reaches the victim is equipped with an extra large carabiner, it is possible to use the rope and the SCBA harness to create a 2:1 mechanical advantage system for hauling. Pass the SCBA waist strap through the legs to improvise the harness, then clip the carabiner around the shoulder straps. Pass the life rope through the carabiner and secure it to an anchor. By hauling on the free end of the rope, the carabiner and victim will be pulled toward the anchored end. In a vertical removal, anchor the end of the rope to a hook or attic ladder that spans the opening, resting on two rescuers' shoulders as illustrated in the photos. This is an effective way to hoist heavier weights (Figs. 11–22 and 11–23).

Two factors that have repeatedly shown themselves to be problematic are the lack of planning to meet emergencies and allowing a panic mode to set in because it is a firefighter who is trapped. Training and drills in firefighter removal procedures help to deal with both of these problems.

Fig. 11–22 A halligan hook or an attic ladder can be used as an anchor point for a 2:1 hauling system. The rope is secured near one member's shoulder. The two members remain squatting or kneeling on one knee.

Fig. 11–23 Other members haul on the free end of the rope, creating a 2:1 mechanical advantage. When the victim's head approaches the floor level, the two kneeling members stand, quickly raising the victim several additional feet.

12 Private Dwellings

Fires in one- and two-family homes make up the vast majority of structural fires in the United States. More than 425,000 homes burn every year. Private dwellings (PDs) are responsible for more than 70% of civilian fire deaths annually. Much of the blame for these statistics rests with the way our homes are built and the myriad devices we use within them. We know from past experience that the majority of fires in PDs begin below the second story. This shouldn't be a surprise. Most of the activities and devices likely to produce a fire are located in either the cellar or the first floor, including the heating plant, the electric service, the water heater, and cooking equipment. In addition, quite a lot of any cigarette smoking occurs in the living room. As mentioned earlier, the majority of victims who are still inside will be found in the bedrooms. Firefighting activities, then, should use the following scenario:

1. Secure the water supply

2. Stretch a charged hoseline into position to protect the bedroom exits

3. Perform VES of all areas

4. Extinguish the fire

PDs suffer from several defects in design and construction that make them uniquely susceptible to the rapid spread of fire. In multistory homes, the most serious of these is the lack of an enclosed stairway. An open staircase is nearly unheard of in any other occupancy due to building code restrictions, yet it is almost universal in two-story homes. This open stairway acts just like a chimney, funneling fire and gases upward toward occupants and exposed areas. The second major problem is the lack of code requirements for interior finishes in these structures. Often, flimsy sheets of wood paneling are nailed right over wood studs to create a partition. In the event of fire, this creates an extremely fast, hot conflagration. These fast-spreading fires require a line that can put water on the fire immediately. Of course, the line must flow enough water to extinguish the fire or it might as well not even be stretched.

The design of PDs does have one thing in its favor—relatively small room size. This means that, if the fire is confined to one room, it is easily extinguished with a midsize handline (1½–2 in.). Unfortunately, people tend to have a desire for open spaces. Most often, doors within a given home are left open, thus exposing greater floor space. For this reason, when attacking a working fire, it is a good idea to stretch another line to back up the first one. This should be in addition to any line that is required to cover the floor above the fire. It is a rare house fire that will require three lines, but the ones that do give us a bad time. If you are only prepared to put two lines on the fire and you suddenly find fire in an unsuspected area, or if you find fire of unusual intensity, you may be overwhelmed before you get a chance to react. Plan for three lines in advance. Hopefully you won't need them, but if you do, you'll be ready (Figs. 12–1 and 12–2).

When establishing a water supply to feed these attack lines, you must plan to be able to supply at least these three lines, a flow of about 400–600 gpm. An additional 250 gpm may be required for protecting exposures when wooden structures are closely spaced. The hose bed and the pumper-hydrant system must be capable of rapidly supplying this needed flow. The best method of doing so involves using an in-line stretch from the hydrant into the fire. This is by far the fastest available method of applying water, since the use of preconnected lines is expedited by having the apparatus in front of the building. These lines, rapidly stretched by two members, can be charged with water from the unit's booster tank as soon as the chauffeur gets out of the cab. With one 1¾-in. line flowing 180 gpm out of a 500-gal booster tank, the chauffeur will have more than 2½ min to get the supply line broken and tied into the intake of the truck before the booster tank empties. Remember, 2½ min of actual full-flow firefighting is a long time in a dwelling. The odds are that the fire will be all but fully extinguished long before a 500-gal tank runs out. Nonetheless, you must be prepared for the 1% of fires that don't go out with 500 gals or even with two lines operating simultaneously.

The proper position for the first engine (using an in-line supply) is just past the fire building, unless the fire condition dictates spotting it elsewhere, to use the preconnected master stream (most often for exposure protection). By pulling just past the building, you leave the front of the structure clear for an aerial device. Although aerials aren't normally required at lower floor fires in these buildings, the spot should be left available anyway (Figs. 12–3 and 12–4).

A slightly different deployment of hoselines from that described previously may be required if the fire originates in the basement or cellar of a balloon-frame home. Balloon framing was very common in homes built in the late 1800s and early 1900s, particularly in the East and the Midwest. Balloon framing consists of long, one-piece 2×4 studs rising from a foundation up to the eaves without any fire stopping along the way. The second-floor joists rest on a small board (1×4 in.) that is notched into the upright 2×4s, called a *ledger board*. This sort of construction allows any fire that penetrates from the cellar into the upright space

Fig. 12–1 Upper floors normally contain sleeping areas, and so must be the prime target of VES attempts.

Fig. 12–2 Open stairways are chimneys that rapidly cut off escape. The first hoseline must be stretched to get water between the fire and this critical area. Usually that means going through the front door.

Fig. 12–3 In-line pumping is the ideal tactic for fires in PDs, putting the first line in operation off the booster tank as soon as the pumper stops near the front of the house.

Fig. 12–4 In-line pumping with LDH should prove adequate even when faced with a fully involved dwelling. The advantage of having the deck gun in front of this building should be obvious.

Fig. 12–5 Balloon framing allows fire to have a ready means of traveling from cellars right to the attic.

Fig. 12–6 Platform framing has a solid 2x4 bottom plate at each level, which acts as a fire-stop to reduce extension within walls.

to travel unimpeded to the attic. (The fire can also spread horizontally out into the space above the first-floor ceiling, but this is usually a much slower process).

If you arrive to find a serious cellar fire in an older home, you should suspect the possibility of balloon framing and take steps to cut off the upward spread of the fire. Have a team on the first floor get to an outside wall (that you hope is over the main body of fire), tear off the baseboard molding, and then punch a hole through the lath and plaster. (These homes were originally built with wood lath and plaster, making for extremely rapid fire spread.) If the inspection reveals a 2×4 lying flat along the floor, this is a plate, and balloon framing is not present. It would be wise to check all of the exterior walls in the vicinity of the fire, however, since additions or extensions made to the house using more modern methods could conceal the fact that an adjoining balloon-frame wall remains from the original construction (Figs. 12–5 and 12–6).

If fire is found traveling up a balloon-frame wall, the race is on. You will need a line at the top floor or the attic as well as one on the first floor and one in the cellar. You must commit a large number of members with hooks to open the first-floor walls, as well as the top-floor ceiling if the attic isn't readily accessible. Speed is of the essence here. Make a small, fast examination hole into each bay between the studs. Where fire or high heat is present, quickly drive the hose stream upward. Don't bother opening bays directly over or under a window. Move on to other continuous bays. The windows and doors act as fire stops, halting vertical extension. Once you have made holes in all of the exposed bays, begin expanding the wall openings further to expose any hidden fire. Be sure to let the team on the top floor know which bays had fire in them or at least which walls to open first. This will speed their efforts in locating hidden fire.

A fire in the cellar or basement of a balloon-frame house is one of the few cases where roof ventilation in a PD will be justified. Another notable exception to a general ban on roof venting at lower-floor fires is the lower-floor fire that vents out a window and ignites combustible exterior siding, extending up to the eaves. The most serious case of this involves asphalt sheathing. This material isn't commonly installed anymore, but many homes are still covered with it, often covered over by newer, but just as flammable, vinyl siding. Also known to firefighters as gasoline siding, asphalt sheathing burns rapidly once ignited, producing tremendous radiant heat and great volumes of black smoke. Fire will readily burn through the eaves into the attic, often bypassing the floor above the fire.

Tactics require that the fire on the outside of the structure be darkened down from the outside and that the stream sweep the eaves. The first line should generally be placed to the interior fire to protect the stairway as usual. However, in densely crowded areas with many homes sheathed in asphalt, the first line may have to be an exposure line. After the interior line has been placed, another line must be brought to the attic to check extension into the structure. Roof ventilation may be required in this case, even though the main body of fire was in the cellar or on the first floor. Depending on the proximity of the fire to the front door, it may be best to hit the fire on the exterior siding and to sweep the eaves as soon as the line has been charged, then advance the line through the door to extinguish the main body of fire. This is particularly true when manpower is inadequate to position three lines rapidly and to provide the necessary ventilation (Figs. 12–7 and 12–8). A fast exterior hit buys time to get to the attic. (See chapter 4 for additional discussion of asphalt-siding fires.)

As you can see, there is little variation of the first hoseline's position, regardless of the fire situation. Certain basic tasks, such as protecting the stairway or protecting occupants, are so important that the fire situation doesn't actually affect the general attack plan. The first hoseline must go to the seat of the fire via the interior and cut off any threat to the interior stair or sleeping areas in the process. The same is true of the various truck duties. In a house fire, the emphasis must be on searching for and removing all occupants. The time span available for this to be accomplished is so short that, if the occupants are to survive, search must begin as soon as personnel arrive. If no separate unit arrives to perform the search and other ladder functions, then the first unit to arrive will have to ensure that these tasks are performed as part of their operation. Assuming that six people initially arrive on the fireground, the critical tasks can be started, if not necessarily completed. One member acts as the hydrant man, stopping at the best hydrant on the way in, and drops a supply line. This ensures the vitally important water supply. He or she connects the supply line to the hydrant and awaits orders to charge the line.

The pumper then proceeds to the fire, where two members stretch an attack line to the seat of the fire. These two will be responsible for the primary search of the fire floor in addition to their attack duties. Although this dual responsibility isn't the most desirable situation, it may be necessary given the manpower constraints. The chauffeur will at this point be heavily engaged supplying tank water to the line, breaking the supply line, and connecting it to the intake of the truck. He may be able to assist in other functions, such as operating a prepiped deck gun to provide exposure protection, venting

Fig. 12–7 Asphalt siding is extremely combustible. In this photo, flame has almost completely involved the exterior, except where there is asbestos siding (white material on the lower half).

Fig. 12–8 Older asphalt siding may be covered with newer vinyl, which is just as combustible, and adds to the difficulties. This fire spread from the cellar to the attic (bypassing the second floor) largely due to the vinyl covered asphalt siding.

ground-floor windows (especially those on the front of the building), or placing a portable ladder. The two remaining members should primarily be responsible for search of the upper floor. The hydrant man should immediately return to the apparatus after he has charged the supply line. He or she may be used as necessary to augment either attack or search. When less manpower than this arrives initially, some tasks will be delayed or will simply not be accomplished in time. When assigning members to tasks, keep the priorities of search and attack in mind, as well as the need to operate in pairs. If more manpower is available, arrange the assignments to ensure the most efficient use of all members without duplication of effort.

Venting Peaked Roofs

In chapter 9, we discussed the critical nature of ventilation and techniques for implementing the principles in larger flat roof structures. PDs are primarily constructed with peaked roofs. In this chapter, we will discuss the unique features of PDs and how ventilation should be conducted in these structures.

Few depictions of firefighters at work are more traumatic to the homeowner than that of a firefighter chopping away at a roof. This and the resulting damage often appear unnecessary to the layperson; however, when roof ventilation is performed as needed, such action is absolutely essential. Proper roof ventilation will actually limit the overall fire damage by reducing the lateral spread of the fire and by speeding up the advance of the hoseline and extinguishment efforts. By allowing hot gases to escape and disperse, or even to burn above the roof hole, you ensure that those gases aren't burning on the inside. Search, rescue, attack, salvage, and overhaul are all facilitated by being able to enter and operate in better conditions.

On the other hand, unnecessary roof ventilation not only increases damage, it may also delay vitally important operations in other areas. Roof venting takes manpower—manpower that, in the early stages of the fire, may have to be committed to search and hoseline operations.

Generally, if the fire hasn't extended into the area directly under the roof, then horizontal window ventilation will be sufficient. A kitchen fire on the first floor won't usually benefit from venting the roof in a two or two-and-a-half story home. It would be much faster to vent the kitchen windows, either by the use of hose streams (straight streams will break glass), by a member on the inside with a tool (this member is searching for life and simultaneously checking on extension), or by a member on the outside, if conditions dictate. It would take at least two firefighters to open the roof quickly and, since the fire is remote from the roof, little would be gained.

In well-advanced fires in the attic or loft areas, however, roof ventilation may not only be beneficial, it may be a necessity. Generally, attic areas don't have many windows. Many times, the only structural attic ventilation consists of a small louver at each end of the house or, worse yet, only vents in the soffit (underside) at the level of the eaves. Heavy fire in this area would prevent the hoseline from advancing and could result in steam burns to members when water is applied (Figs. 12–9 and 12–10).

Fig. 12–9 Attic louvers don't provide sufficient ventilation at working attic fires.

Fig. 12–10 Fires in the attic itself demand a ventilation opening be made quickly and in the correct location. Here, a single quick-cut proved adequate.

Early recognition of the presence of fire in an attic or loft is crucial. If fire enters these spaces, the roof must be opened, even if this means calling additional resources to get the job done. Remember, the stability of the roof is quickly being destroyed. Cut early and make your openings properly.

Roof Design

Problems encountered in roof ventilation are often due to the peaked-roof style, the pitch of which makes them difficult to walk on under the best of conditions and outright dangerous during a fire. Additionally, peaked roofs are designed to shed snow better than flat roofs, meaning that the roof joists can be of a smaller size. So, 2×6-in. lumber, 16–24 in. on center is common. You may encounter 2×4s in many older homes. Lightweight wood (2×4s) and gusset-plate trusses can be found in many new buildings, which is dangerous. Remember, these roofs won't carry as great a live load as a flat roof, so use extreme caution.

Fig. 12–11 Older roofs had no solid decking but consisted of wood shingles nailed to 1x2-in. furring strips. These strips are rapidly weakened by fire.

The roof sheathing rests above the joists that support the roof. On older homes, this sheathing is commonly only 12-in. furring strips nailed to the joists, 6 in. on center. These serve as fasteners for the shingles. This sort of roof is very easy to cut and step through. Firefighters must use extreme caution when putting any weight on the roof, lest they find themselves penetrating a well-involved attic. At one such attic fire, my brother, Joseph, fell through the roof decking up to his armpits. He was saved from becoming a casualty by his protective clothing and a nearby firefighter who was able to haul him free of the hole. When in doubt, caution dictates working on a roof ladder or out of a platform bucket (Figs. 12–11, 12–12, and 12–13).

Fig. 12–12 A tongue-and-groove roof is generally more stable than one of either furring strips or plywood.

In the next phase in roof construction, the furring strips were replaced with 1×6-in. tongue-and-groove boards for sheathing. These allowed roofers to be less precise when nailing the shingles, since there was always wood under the shingle instead of a 6-in. space. Tongue-and-groove boards offer a little more support for the firefighter, but they also require more work to open.

The latest method of covering roofs involves sheet roofing with plywood or, in some cases, particleboard for the surface. This construction offers the same or less stability as tongue-and-groove, but it can be much more difficult to open. Plywood resists chopping with an axe and, because it must be pulled up in one large piece, there are many more nails holding it to the joists.

Fig. 12–13 A plywood roof is difficult to open using the standard roof cut. You should use the quick-cut method instead.

The layer above the sheathing, made of overlapping shingles, performs the main function of the roof—to keep the weather out. Many times, a layer of felt roofing material or tarpaper is applied between the sheathing and shingles as a barrier to moisture.

Shingles are made of a variety of materials. Older wood shingles are easily cut, but they pose a severe danger of fire spread. Many building codes across the country reflect the conflagration potential of wood shingles, mandating their replacement with asphalt shingles or some other slow-burning material. Asphalt shingles are similar to wood shingles in that they can be cut easily, often permitting the shingles and sheathing to be breached in a one-step operation.

Hard roof coverings like slate and tile are more difficult. The best approach to venting this kind of roof is to remove the covering before attacking the sheathing. Slate may be scraped off using a shovel or the blade of an axe on the flat. Tile roofs may be pried off with a pointed tool. Both slate and tile can be broken loose by striking them with a blunt object. Use caution with all methods: these roofs produce heavy, sharp-edged debris that, after falling from the roof, can inflict great injury to anyone caught in the way. Both types are extremely slick when wet.

Regardless of the type of roof covering, firefighters should operate safely. Remember, roofs are built to keep water out, not to support a fire attack (Fig. 12–14).

Making the Cut

Although power saws are preferred over axes for ventilation, the principles of making a roof opening with either tool vary only slightly. The safety of the members performing the task is always the first concern. The stability of the roof plays a key role, determining the need for roof ladders, platform baskets, or aerial ladders. In all cases, personnel shouldn't be committed if the roof isn't structurally sound or if it is badly damaged. In such cases, it usually takes only a few minutes for the fire to blow through anyway! If in doubt, provide the safer means of accomplishing the goal. Additionally, members performing roof ventilation should be in full protective clothing, regardless of how they will perform the task. Their bunker gear should be fully closed and they should be wearing SCBA. Fire suddenly erupting from a vent hole can do a lot of damage to the human body.

Once the method for venting has been determined, members and tools must be assigned. Safety again dictates that at least two members (preferably experienced) be assigned to the task. On low-pitch roofs (15–20°), a power saw may be used to make the cut. On roofs with higher pitches, both circular and chain saws can be extremely dangerous. Any sudden push or pull from kickback could throw the member off balance, possibly from the roof. On these high-pitch roofs, if you can't operate the saw from the basket of a platform, use an axe (Fig. 12–15).

Fig. 12–14 Tile roofs present several problems and hazards to firefighters.

Fig. 12–15 Pushing down a ceiling with a pike pole. Note that the hook end is up. The blunt end is more efficient at pushing down ceiling material and is less apt to get snagged on wiring and lath.

Once the cut has been completed, other tools are required. A 6- or 8-ft pike pole can be used to pull up roof boards. Pry them loose if necessary and push down ceilings, while giving the member enough reach to stay out of the heat and smoke. You should always bring a pick head axe to the roof in case the power saw fails. The pick end is useful for prying up roof boards, shingles, and the like. A variety of other tools might prove useful or necessary as follows: lights, rope, halligan tools, and so on. However, resist the desire to send too many people to the roof. They can overload it and are often in each other's way.

Determining where to cut a peaked roof is usually less complicated than on larger, flat-roof commercial buildings. Because the roof is proportionately smaller, you can't be that far off. The basic guidelines still apply. First, the members should size up the building as they make their way to the roof. Fire showing out of an opening is a good indication, as are the hot spots evidenced by bubbling tar, steam on a rainy night, or melting snow.

These signs will usually place members at the right end of the roof, as close over the seat of the fire as is safely possible. Heat rises, so the hole is made as close to the peak as possible to release the greatest amount of hot gases. In most cases, the side of the peak in which the hole should be made is determined by the direction of the wind. Making the hole on the downwind side of the peak helps the gases escape. A hole on the upwind side causes heat to be blown back toward the hoseline crew (Figs. 12–16, 12–17, and 12–18).

One problem recently encountered is the presence of solar heating panels on roofs. If the panels are located directly over the best site for ventilation, the venting team may have to decide whether to move the hole farther away from the seat of the fire or to place it on the other side of the peak. The conditions at the scene must dictate this decision. If a strong breeze is blowing onto the other side of the peak, the hole will have to be made farther away, even at the risk of pulling the fire.

Satellite dishes mounted on roofs present a similar dilemma. In addition, they are often installed on existing roofs, with little or no consideration given to the load. They also pose the added risk of extra weight—weight that under fire conditions could be the straw that breaks the camel's back.

There are several ways to cut a roof, depending on the situation. Most authorities recommend a hole of approximately 4×4 ft, which makes a pretty fair-sized piece of roof to pull after the cut has been made.

Fig. 12–16 Cutting a long, narrow strip along the peak of a roof vents a large number of bays and, because it is located at the high point, allows the greatest amount of heat to escape.

Fig. 12–17 Solar panels could prevent you from making a vent hole at the most advantageous position.

Fig. 12–18 Satellite dishes pose a danger of overloading the roof as well as obstructing part of it.

If working under poor conditions—limited manpower, working off a roof ladder or platform, or straddling the peak—the size can be even more daunting. The quick cut works well under these conditions. It has a big advantage over the standard 4-ft-wide hole because it only spans one joist. This allows the roofing to hinge on the joist by pushing down one side of the cut material (Fig. 12–19).

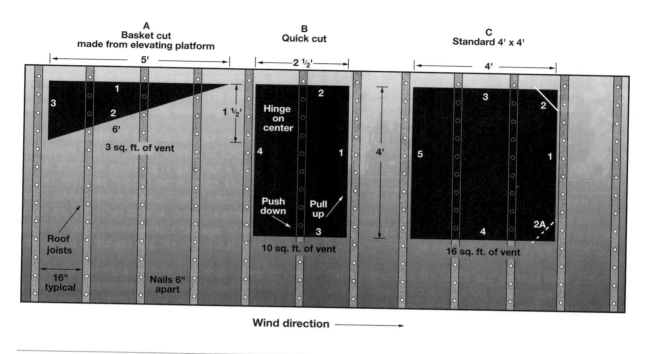

Fig. 12–19 Members venting a roof must know a number of types of cuts, depending on the conditions and type of roof encountered.

The quick cut is especially effective on plywood sheathing, where the standard 4×4-ft hole will span at least two joists of a 16-in.-on-center joisted roof. That means approximately 14 nails are holding the roof to the joists after the cut has been made, providing great resistance to being pulled. Hinging the cut section on one joist will speed up opening the hole and ensuring ventilation. Larger holes can hamper or even prevent getting the hole pulled. If sufficient relief isn't accomplished with the hinge method (which is rare), the cut may be extended or another hole begun.

Another cut that is often possible from the basket of a platform involves cutting a triangular opening. Again, this hole usually spans several joists and is difficult to pull. It also provides only limited ventilation. Its main advantage is that the member never has to set foot on the roof. The size of the cut is limited only by the member's reach and his or her ability to position the basket.

Regardless of the style of cut to be made, use the following basic rules:

1. Ensure the stability of the area before entering.

2. Always have two means of escape.

3. Plan the cut and inform fellow members of its layout.

4. Arrange the sequence of cuts to keep the wind at your back.

5. Cut adjacent to the joists for maximum support and the least bounce.

6. Never step on the cut.

7. Don't cut the roof supports.

Roof ventilation demands that firefighters be able to perform with both their left and right hands, because when the outline of the opening is made, the member must always be on the supported side of the cut. On one side it will be to his or her left; on the other, to the right.

When the need for roof ventilation exists, a fire officer must take action quickly to see that it is accomplished. Fortunately, an aid to our venting efforts is becoming more prevalent as architectural designs evolve, even if not for the reasons we want. Skylights in homes are becoming more and more popular. In some cases, on your arrival, fire will have already melted them, simplifying your problem. If you must vent them manually, make sure they're in the right location and that opening them will have the proper effect. Beware of the bounce of the tool—it could strike you or cause you to lose your balance. In some cases, it might be easier and less damaging to pry the entire unit loose (Fig. 12–20).

Fig. 12–20 Skylights are increasingly common in residences. Vent them using the same criteria as for cutting a hole—open the leeward side.

The average house fire, where there is no major involvement of such blind spaces as attics or cocklofts, will normally be handled with 6–10 well-trained members who are there to work. More manpower at the relatively minor fires (room and contents) will allow quicker attainment of goals. Fires that extend or originate in such problem areas as cellars and attics; fires where people are trapped or firefighters are injured; and serious fires in very large homes are all situations that demand immediate assistance. A 6- or even a 12-member crew won't be able to accomplish all of the normal tasks in addition to the added demands placed on them by extenuating circumstances. The initial-arriving units shouldn't hesitate to call for assistance whenever they are faced with any of these conditions.

13 Multiple Dwellings

Multiple dwellings are those buildings that house three or more families. In some areas, they are called apartment houses or tenements. These occupancies pose a high life hazards regardless of the time of day due to the varying sleep patterns of the residents (Fig 13–1). Place an extremely high priority on searching all of the apartments on and above the fire floor, since the occupants of an apartment above may not become aware of the fire until after the means of egress have been blocked. Multiple dwellings often have a number of construction features that create difficult fire-control problems, such as the practice of stacking kitchens and bathrooms over the ones below, thereby creating pipe chases that run the height of the building—ready avenues for vertical extension.

Only an early recognition of the common problems will enable us to deploy units to solve them. In many of the larger apartment buildings, the officer in command must take into account the time factor necessary to implement a tactic. This requires the ability to recognize the situation as it currently exists and to project what that situation will become during the time it takes to set up. For example, if the fire is in a couch and can easily be controlled by a water extinguisher, but if the extinguisher is on the apparatus and the couch is on the sixth floor, it wouldn't be wise to order only an extinguisher to be brought up. By the time it arrives, the fire will probably require a handline, having ignited the rest of the room.

One of the first difficulties that may be encountered is simply to recognize the presence of a multiple dwelling. Many, many older homes have been converted, legally or illegally, into multiple dwellings. The illegal conversions are usually a much greater life hazard, since many fire safety features required of legal multiple dwellings may

Fig. 13–1 Fires in multiple dwellings place the lives and property of dozens of residents at risk and demand swift aggressive action by firefighters.

be absent (such as an enclosed stairway, a fire escape, or a sprinkler system). In many areas of the country, response assignments to PD fires bring only two engines and one ladder, whereas an alarm for a large hotel or apartment house will bring three or four engines and two or more ladders. The fire administrations in these cities have recognized the potential loss of life in the multiple dwellings and have assigned a greater number of personnel to deal with it. When you arrive to find a converted multiple dwelling, you must make the same adjustment and request whatever additional assistance you need, usually personnel for search and aerial equipment for removal (Figs. 13–2 and 13–3).

There are a number of ways to recognize multiple dwellings apart from their size and obvious indications. Counting gas or electric meters on the exterior should be a tip-off, since there is usually one per apartment. The same is true of doorbells and mailboxes. A fire escape is another indicator, although you may find one on a larger one- or two-family home as well. Even information received as part of the alarm can be useful, such as "smoke from apartment four." Once inside the building, the presence of padlocks on interior rooms usually indicates the presence of an SRO, where a person or even a whole family rents a room and shares a toilet and other facilities with the other occupants. Due to the potential number of occupants, SROs demand fast reinforcement for the search effort. Normally, each and every room must be forced and searched, since even those found padlocked from the outside have been found to contain children locked in by their parents. Unfortunately, the low-income areas where these types of occupancies are found are also the areas with the highest incidence of fires. Heavy losses of life are sometimes unavoidable.

The loss of life in multiple dwellings is often caused or compounded by building features that promote a fast-spreading fire. There is an open interior stairway in many older multiple dwellings that runs from ground level to the roof, leading to rapid mushrooming on upper floors. A series of catastrophic fires in tenements in New York City in the late 1890s showed the need for separating the apartments from this veritable chimney by some fire-resistant material. As a result, most apartments are now equipped with fire-rated doors, while the stairwell is usually a fire-rated enclosure as well. This will serve to keep fire within an apartment or keep fire in the hall from extending into an apartment, but only as long as the doors remain closed. If a fleeing occupant leaves an apartment door open, the fire has ready access to the stairs. For this reason, many codes require these doors to be equipped with approved self-closing devices. Newer fireproof buildings will be discussed at the end of this chapter.

Fig. 13–2 The four electric meters on the Exposure 2 side identify this structure as a converted multiple dwelling.

Fig. 13–3 This fire escape is evidence of a large number of occupants. The top floor of this residence is a single-room occupancy (SRO).

Larger multiple dwellings require a means of bringing ventilation and daylight to rooms in the middle of the building. These light and airshafts pose a danger of fire extending horizontally across the shaft (often bypassing a fire wall), as well as vertically, due to auto exposure or lapping flames. This is common because of the window layout, often directly above or opposite the window of the fire apartment. Since there is no roof to cause mushrooming, this means of fire spread isn't as rapid as it is up a staircase, but you must check for it and, if necessary, position lines to cut it off. While waiting for the arrival of a hoseline, search crews can greatly slow the entry of fire by removing all combustible materials, such as curtains, from around the windows, and by judicious use of portable extinguishers or even pots of water from the sink. Another method that may work is to open a window or a door opposite the shaft in the exposed apartment. Fire venting into a fairly narrow shaft creates a rather strong updraft. I have watched as fire was pulled out of an exposed apartment into the shaft when a door was opened, allowing a flow of air. The fire then blew back into the apartment when the door was closed. Naturally, we immediately blocked the door open while we awaited the arrival of an exposure protection line (Figs. 13–4 and 13–5).

Fig. 13–4 Light and air shafts between multiple dwellings present serious exposure problems to the adjoining building.

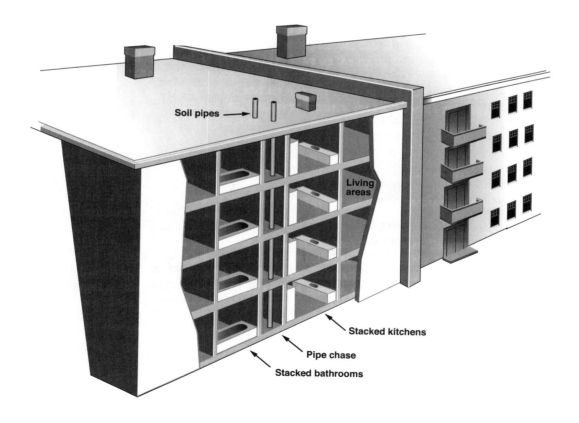

Fig. 13–5 Multiple dwellings share some common features, including stacked apartments and pipe chases.

Multiple dwellings are often found to contain many other types of shafts, elevators, compactor chutes, pipe chases, and channel rails. Each of these has the potential for spreading fire, heat, and smoke to the upper stories. The two greatest threats to the structure in the event of fire are from pipe chases and channel rails, the furred-out spaces around vertical steel I-beams. Elevators are generally required to be built with at least 2-hr fire-resistance ratings, and they don't generally account for serious extension hazards, although they may transmit smoke and heat rapidly. Compactors, on the other hand, have been the cause of several serious fires when rubbish burning within the vertical chute extended to combustibles on several floors at once. (For more on compactors and incinerators, see chapter 17.)

Economy of construction often results in bathrooms and kitchens being back-to-back in adjacent apartments. This way, one riser can feed all of the faucets. Carrying this procedure a step further, builders usually align the kitchens and baths from floor to floor. When fire has involved either a kitchen or a bathroom, it must then be a high priority to get to the same location on the floor directly above, and on the top floor, to see whether there is fire traveling within the pipe chase. Whenever you are searching for fire or taking a line above the fire, and you are working in very heavy smoke conditions, try to find the wall directly behind the toilet and open it up. The toilet requires the largest diameter pipe, so it is usually in this wall that the biggest pipe chase is located. If you open it up and get water going up and down that shaft, you might cut off extending fire. The members assigned to the roof position, after performing their initial duties, should feel each soil pipe that comes through the roof. If one is hot to the touch, it indicates that fire is traveling up that pipe chase. Inform the members below and begin ventilation of the roof adjoining the pipe. By cutting and pulling a hole over this chase, you will let the fire go straight up and slow the horizontal extension into the cockloft.

The channel rail void is much more difficult to locate than the plumbing chase, since we don't have any outward sign to tip us off where it might be. These voids may be found in any building that is built with steel columns or I-beams supporting the floors. Since they are a structural requirement rather than a lifestyle feature, they can appear anywhere in the floor layout. Usually the architect tries to hide them within walls where their extra thickness won't be noticed, such as in a closet. If you are lucky enough to see an actual projection from a wall or a ceiling where the beam is boxed out, open up around it first. If you have no indication of where the beams are located, you must begin a time-consuming hunt. It helps to know in which buildings you will find them and where they will be located so as to speed your efforts. Generally, steel framing is used in Class 3 buildings (ordinary or non-fireproof construction) that are more than 25 ft wide (Fig. 13–6).

Although there are buildings that meet this description that don't have steel framing, they are the exception. If you start looking for the channel rails and find out that the joists rest on another masonry bearing wall, you will be pleasantly surprised. In the other hand, if you don't look for the channel rails and fire has gotten to them, you will be very unpleasantly surprised. When you find fire in any vertical void, you must be certain to check the top of the shaft where it enters the cockloft and the bottom as well,

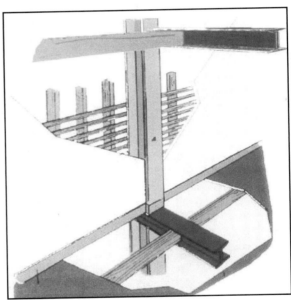

Fig. 13–6 A channel rail is a steel column that runs the height of the building supporting horizontal girders. They are often boxed out with wood lath and offer fire a highway for extension to the cockloft.

since falling embers could result in fire breaking out below the operating forces. Such a situation resulted in the total destruction of at least one large hotel that I know of. It is usually best to send someone who has actually seen the location of the voids that the fire is in to show people where they are, rather than try to explain it over a radio. I've seen cases where companies following such instructions have found other voids that they pronounced clean while fire swept right past them in the original shaft.

Operations at multiple dwellings where so many people may be trapped should focus on removing those in greatest danger first, while protecting the greatest number in place. This usually means getting the first hoseline in place to protect the interior stairway, regardless of whether or not people are showing at a window. The only deviation from this plan should be where victims are being directly threatened by fire and a hoseline is needed to keep the fire away from them, or where a ladder must be raised to remove them immediately and no other manpower is available, including civilian. The person who is screaming and waving for help can at least scream and wave. The people who are already overcome cannot. They need your assistance more than those who are clamoring at a window. In this regard, all available personnel should be committed to getting the first hoseline stretched and operating before beginning to stretch the second or third lines. The importance of quickly positioning a line to protect the main vertical artery, the stairway, cannot be overemphasized. Lose it, and there is a good chance of losing at least the top of the building, not to mention any occupants who are still above the fire. Put out the fire and you will have improved conditions greatly for all of those exposed people (Fig. 13–7).

Maintaining control over the interior stairway can in itself cause delays in the actual fire attack. At times, occupants of upper floors may still be descending the open staircase while the line is ready to advance. The door must be kept closed in this case, keeping the smoke and fire within the apartment until the stairway is cleared of occupants. During this time, it is very helpful to send a member with an axe or other heavy tool to break any windows that are found on the floor landing or on the stair between the fire floor and the floor above. These windows on the stair half-landing are often glazed with wired glass and require a lot of work to clear. A heavy tool is mandatory. The ventilation that this effort provides is very beneficial, particularly when ventilation of the top of the staircase is delayed (Figs. 13–8 and 13–9).

Forcible entry techniques used at multiple-dwelling fires should reflect this desire to maintain control of the door to the fire area. Knocking out part of the door panel isn't recommended, nor is removing the door entirely from its hinges, since if there is any problem with fire control, such as a blown length of hose, the fire may extend out into the staircase area. For this reason and others as well, multiple dwellings are the made-to-order situation for Hydraulic Forcible Entry Tools (HFTs). The HFT will ensure entry to these doors via the side of the lock in nearly all cases, leaving the door intact on its hinges. Also, the number of exposed apartments on a floor requires a tremendous forcible entry effort, which can be greatly hastened with an HFT.

Fig. 13–7 Even a two-room fire in a typical apartment building is a serious problem, since the potential for fire spread and life hazard is enormous.

Fig. 13–8 The window on the stairway half landing above the fire is an excellent portal to vent.

Fig. 13–9 Stairway windows are often apparent—as are these on the half-landings—midway between the floors.

With the HFT, one member can force all of the doors on a floor within minutes, while other personnel begin the search of each of these apartments. This is particularly critical when the forcible entry is being performed in the open stairway landing directly above the fire apartment. The members above can rapidly gain entry to one or more areas of refuge should fire extend into the public hall and stairs. When going above a fire to search, be sure to force entry first into an apartment *other than* the one directly over the fire. If you then force the apartment directly over the fire and find that it is already full of extending fire, you have an escape route, even if the stairs were also to become blocked (Figs. 13–10 and 13–11).

A serious fire in a multiple dwelling is a fast-spreading affair when a large body of fire is present. While intermediate-size handlines are sufficient for fires that are confined to a single apartment or even those that have extended through to the floor above, if heavy fire is showing in two or more apartments on one floor, it is time to bring out the 2½-in. line. If you are inside with a lot of fire and a possibility of many occupants, this isn't the time for a knockdown, drag-out battle. You want to walk squarely up to the fire and drop it with one punch. Get that big hose moving!

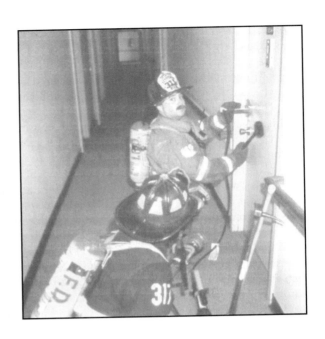

Fig. 13–10 The HFT is essential in multiple dwellings (MD) where large numbers of doors may have to be forced.

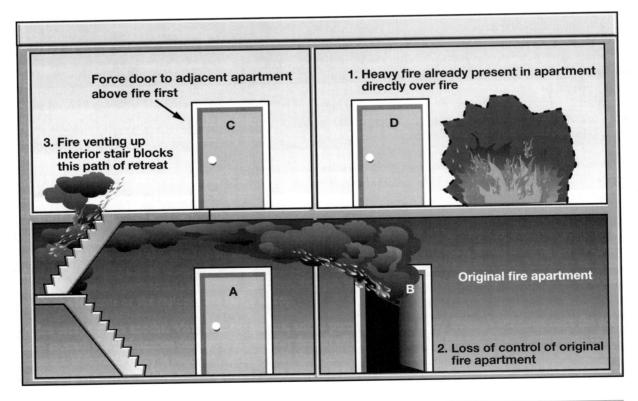

Fig. 13–11 When searching the floor above the fire, force the door to an apartment other than the one over the fire apartment first, to create a safe refuge if fire vents up the stair.

I once witnessed two 1¾-in. lines operating into a fully involved top-floor apartment and cockloft and having fire simply roll right around the nozzles. The 300 gpm they were throwing just wasn't any match for the volume of fire and the wind pushing it. Clearly, this required a change in tactics. In this case, we were able to close the fire-rated door to the apartment, while cutting off extension in each direction until the fire lessened in intensity. Other circumstances might require attacking from the opposite side, with the wind at your back. The decision as to what size line is required, however, should be based on a realistic size-up. If need be, two 2½-in. handlines or a 3-in. line fed into one of the modern, lightweight portable deluge sets might be an answer. Don't underestimate the amount of fire possible in these Class 3 buildings, however, just because they are residential occupancies (Fig. 13–12).

Stretching lines to the correct area at a multiple-dwelling fire is obviously crucial. At many larger buildings, you will have a choice of more than one staircase leading from the ground floor. You must be sure to select the proper stair before committing the hoseline, or you may find yourself on the fifth floor with no access to the seat of the fire. Apart from holding the hoseline in the lobby until someone climbs all the way up to the fire floor to verify the proper stairway (which may have to be done at large-area complexes), the fastest method of selecting stairways is to have a member climb to the second floor and locate the proper apartment line. Apartments in most of these buildings are designated with alphanumeric systems, such as apartment 5B. Usually the number indicates the floor and the letter is the apartment, with all of the similar-lettered apartments stacked one on top of the other. Thus, 5B would be on the fifth floor, over apartments 4B, 3B, and 2B. The layout of the ground floor shouldn't be taken as indicative of the upper floors. The ground floor is often different because of lobbies, machinery rooms, offices, stores, and the like.

If fire is reported in apartment 5B, and if the member at the second floor finds apartment 2B just off that staircase, he should check at least one more door to find a 2A or 2C, then call down to have the line stretched. That way, if he doesn't find the B line of apartments in that area, it is quick enough to drop down to the lobby and check the other stairways. This type of stairway layout is called an isolated staircase, since the members will be isolated from the fire area if they choose the improper stair. A similar type of stairway is the wing stair, where a stairway provides access only to those apartments in its particular wing. Members discovering either of these types of stairways should relay this information to the officer in command and all members on the fireground.

Alphanumeric systems usually indicate the floor by means of the number and the apartment line by the letter, but this isn't always the case. The managers of some buildings choose to let the letter designate the floor. In the example cited above, the member finding apartment 2B right opposite the stairway door should refer to at least one more door. If he finds 1B, 3B, or any other number with a B suffix, it should tell him that this is the B floor and that 5B is to be found here (Figs. 13–13, 13–14, 13–15, and 13–16).

Fig. 13–12 Larger MDs can have several wings with multiple staircases. (Note the bulkheads.) You must select the correct one to stretch the attack line. (How would you describe the shape of this building to the IC? Maybe this way: "A big U, with wings off the tops of the U.")

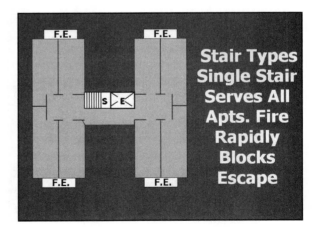

Fig. 13–13 Single stairs serve all apartments, but fire can rapidly block this single escape route.

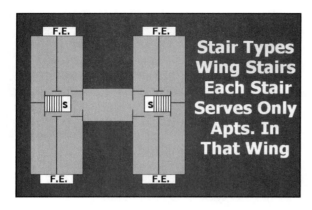

Fig. 13–14 Transverse stairs are an aid to FD operations; allow several approaches.

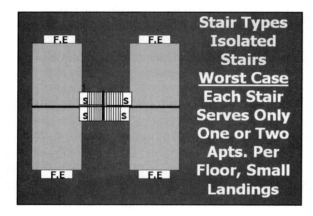

Fig. 13–15 Wing stairs are a problem during a fire; they only allow access to one wing.

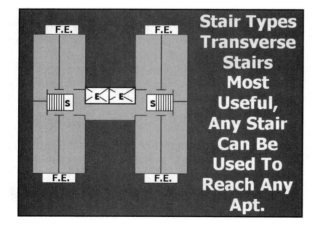

Fig. 13–16 Isolated stairs are the worst-case scenario; each stair serves only one or two apartments per floor and they have very small landings.

Knowing how to find the number layout can pay off at larger buildings, even when all of the staircases connect to a common hallway on each floor. This arrangement is often called transverse stairs, since firefighters can cross over from one stairway to another on all of the floors. This can greatly assist firefighting, since lines may be stretched up several stairways. You must take care, however, to avoid using opposing streams if a serious fire engulfs the hallway.

The stairway nearest to the fire area may in many cases be remote from the main entrance. In this case, it is better to stretch across the building on a floor below the fire, out of the heat and smoke, and then come out of the nearest stairway and attack the fire. Particularly when the door to the fire apartment has been left open, the hallway may be pitch black, making it difficult to locate the fire apartment. In this instance, by dropping down to the floor below and locating the appropriate apartment line, you can get your bearings as to the location of the fire.

Where the staircase maintains a constant location on each floor (called return stairs), it is a simple task to count the number of doors on the floor below from the stairway to the proper apartment, then go back up and follow the same steps. This only takes seconds to do, but it can save many minutes in locating the fire area (Fig. 13–17).

As mentioned previously, the first line is committed to the fire floor via the interior stairway. The second line should be stretched the same way to the same location to cover any additional involved apartments or to assist the first line with an extreme fire condition. Otherwise, it may be stretched up one more flight to cover the floor above the fire. Whenever a line goes up above, the members must

Located at points remote from each other, but a person can go from one stairway to another via public hall in all floors of the building. An asset to fire operations.

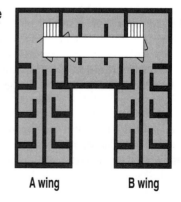

A wing   B wing

Transverse stairs

Fig. 13–17 Return stairs allow firefighters to maintain their bearings in heavy smoke.

let the first crew know that they are doing this, lest the first crew be relying on the second line as a backup. This also allows the second crew to be warned in case the first line has to back out, leaving the second crew exposed above the fire. After these first two lines have been stretched and are protecting the stairway, any additional lines that are stretched should take an alternate route to avoid overcrowding the stairway. (Stretching three or four lines up one staircase can result in a knot of hose, leaving no line able to advance.) In practice, this is often a more desirable procedure for other reasons, such as the need to protect the secondary exits, to get a line to a remote area quickly to cut off extension, or to get lines above the fire before the interior stairway is fully under control.

Remember, avoid tying up aerial devices as temporary standpipes. The temptation is often great with a gated outlet at the platform basket, but it will come back to haunt you. These are your defensive weapons and, if the situation demands, they must be free to maneuver to cover the interior forces' retreat or evacuation. It is far superior to ascend via the aerial device (or the interior stairway) and haul the supply line up the outside. When doing so, give consideration to the possible need to supply multiple lines. It takes almost the same amount of time to pull a 3-in. line up to the fifth floor of an apartment house as it does to pull up a 1¾-in. Still, if you pull up the 3-in. line first, equipped with a water thief, supplying a second, third, or even a fourth line is greatly simplified (Fig. 13–18). Remember, this is not a small home. There is a lot at stake here, and genuine team effort is required.

Fig. 13–18 A 3-in. line stretched up the outside of the building can supply multiple handlines on the upper floors. Note: The line is entering a floor below the fire.

Whenever you are faced with a situation that is so serious as to require three or more handlines, coordination is bound to become a problem unless all of the forces know exactly where they are, as well as where other units are operating. A system of communicating one's position within the structure is essential in large-area multiple dwellings, as well as in schools, hospitals, and the like. In many cases, these types of structures are segmented into wings. Since incoming units may not know the internal system of a given building, a standardized fireground terminology for use in all similar buildings is useful.

The simplest system that I have seen so far is to designate the far left wing (as viewed from the CP) as the A wing, the next to the right as the B wing, and so on. (Avoid calling the left wing the exposure 2 wing or the far right wing the exposure 4 wing, since this is bound to lead to confusion if there is any threat from the fire building to exposure 2 or 4. You must remember that these are separate buildings.) The wings of these large structures are usually joined together by a relatively smaller section called the throat. It is in the throat that items such as the stairways, elevator shafts, compactors, incinerators, and other building features are often found. The throat is usually the place where you will want to make a defensive stand if severe fire conditions prevent your advance down the hall toward the fire area. This can happen quite easily at top-floor fires, where heavy fire is present in two or more apartments and has entered the cockloft area (Figs. 13–19 and 13–20).

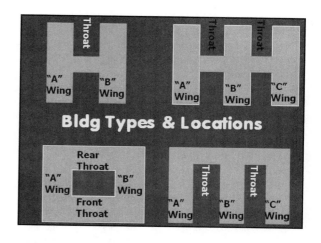

Fig. 13–19 Overhead views of various shape buildings and their reference locations.

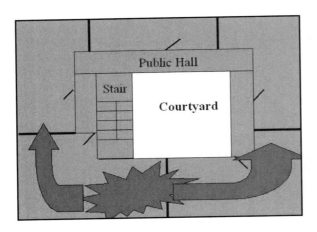

Fig. 13–20 Fires in O shaped buildings wreak havoc as fire spreads in two directions at once.

Fig. 13–21 Heavy fire in the B wing threatening to extend across the throat to the A wing.

Fig. 13–22 If a quick knockdown by interior forces isn't possible, the elevating platform is the tool of choice.

In this case, interior handlines may be driven off the top floor by the volume of fire and by wind pushing toward them. Here, the aerial platform and the trench cut form the best defense. To be effective, position the platform with the turntable in line with the throat. Keep that position clear in the street in the event that a platform responds later in the incident. A telescoping boom platform enables members to operate at the top-floor window with a master stream that will penetrate ceilings and extinguish fire in the cockloft, as well as within the apartments themselves. In some cases, the front of the building is enclosed in a decorative sheet-metal cornice that can be opened up from the basket, allowing stream access directly into the cockloft. At newer buildings, this same effect may be possible through screened cockloft vents. The elevating platform, however, is a necessity, since it allows close observation of the stream, a stable working platform, and an almost surgical operation of the stream in all directions through the opening. Aerial ladder pipes don't offer these options (Figs. 13–21, 13–22, and 13–23).

Fig. 13–23 Smoke emanating from the cockloft vent above the left window indicates heavy fire in the cockloft. Many lines operating on the top floor, with truck support, will be needed. An elevating platform should be positioned as a defensive measure.

Roof Operations

Several factors that we have discussed in this chapter require the *immediate* commitment of personnel to the roof in multiple dwellings. Unlike PDs, there is usually a staircase that goes from the ground floor through the roof. The top of this stairway is enclosed in a bulkhead on the roof. Many times, there is a skylight atop the bulkhead to allow in natural light. Being the high point of the *chimney*, the stairway that leads to the bulkhead is likely to be one of the hottest places in the building other than the fire apartment. People have tried to escape many times by running up the stairs to get to the roof. Often they don't quite make it and collapse on the stairs or just inside the door on the landing. This means that two members must proceed immediately to the roof at all multiple-dwelling fires.

They should immediately force open the door to the stairway and search quickly for any overcome occupants. If there is no need for a rescue, they should then vent the skylight atop the bulkhead. If the stairway ends at the top floor, it is very common to find a skylight over the stairwell. Again, it should be removed or broken to *take the lid off the chimney*, thus preventing mushrooming on the upper floors and allowing interior search and attack teams to operate with improved efficiency. When breaking glass in skylights, remember that people are likely moving up the stairs. Break a small pane first, pause a few moments as a warning to those below, then continue to clear the entire skylight, trying to pull as much back onto the roof as possible. Members ascending the stairs below should move close to the walls when skylights are being taken out. Keep your head erect and your hands off the banister, lest a piece of glass impale your hand, glove and all, to the rail (Fig. 13–24).

A folding attic ladder secured to the top section of the aerial ladder or elevating platform boom allows quick access to skylights located on top of high bulkheads. After vertical ventilation has been accomplished over the stairs, the members on the roof should quickly examine the rear and sides of the building from the roof, noting any trapped occupants or fire extension that may not be visible to the IC. Good communication from this level allows the officer in command to form a picture of the conditions without having to make the trip personally.

Fig. 13–24 A serious top floor fire will require extensive roof operations. Venting top floor windows from the roof helps speed the operations on the top floor.

The benefit gained by sending members to the roof for vertical ventilation far outweighs any other possible duties that they might be assigned. Even if there were people jumping out of windows on arrival, at least one member should be sent immediately to the roof, since by channeling the products of combustion to an area where they can do no harm, that member may well prevent others from jumping or, more importantly, may channel fire away from overcome victims.

"Great!" you may say, "but how does that member get to the roof if there are people jumping out of windows? Obviously I need the aerial device for rescue." The answer will depend on the construction and features of the building. Since it is totally independent of the fire structure, the aerial device is *the* preferred way of reaching most roofs. We can see that the members have reached their objective, and we know how they can get off the roof if they have to. At times, however, using the aerial just isn't possible because of setbacks, trees, wires, the height of the roof, or because the unit is needed elsewhere. In such a case, the fire escape *may* be the solution.

A word of caution on using fire escapes—most fire escapes don't go to the roof, only to the top floor. (Fire escapes in the rear of buildings usually go to the roof via a narrow, vertical ladder called a gooseneck ladder, which is difficult to climb while laden with equipment.) Also, to me, fire escapes resemble nothing so much as barbecue grills waiting for some meat to land on them while fire vents out of windows below. You must be extremely careful to select a fire escape that isn't likely to be threatened by flames venting below you. The third point about fire escapes is that they are simply dangerous. Most have been hanging outside of the building since the day they were erected, with nary a fresh coat of paint to protect them. I've fallen through too many rotted treads to have any respect for these firefighter traps. That doesn't leave too many options, though.

The interior stairway of the fire building is definitely out unless there is a separate wing remote from the fire area with its own stairway to the roof. Preferably, this should be separated from the fire area by a firewall or an enclosed stairway. Trying to get to the roof by way of the stairs in the fire area will likely result in the firefighter becoming a victim rather than saving one. There is one other method available in some areas where older multiple dwellings are common. Many times, the buildings were built in rows, abutting each other. If this is the case, the adjoining building should be the first choice for reaching the roof, even over the aerial device, since it is usually much faster and safer (Fig. 13–25).

Fig. 13–25 Many times, you can reach the roof of the fire building via the adjoining buildings roof. Note the large shafts between buildings. Check the roof in front of you before stepping over any parapet walls!

Once the members have vented over the stairways and other vertical shafts, the duties that they will perform next depend on the location of the fire. If the fire is on a lower floor, the roof team may descend to help in the search. They shouldn't attempt to descend via the interior stair, however, since it is possible that they may be caught by erupting fire. If there is a rear fire escape, they should use it. If not, they will have to descend the way they got up and re-ascend via the interior stairs. If there is a fire escape, then they have a decision to make: What floor should they search?

Generally, the three greatest areas of danger to the occupants, in order, are

- the fire floor
- the floor above
- the top floor

Any intermediate floors come later. The interior search team will be handling the fire floor. That leaves either the top floor or the floor above the fire. In many cases, the decision will depend on the severity of the fire. If the fire is blowing out of the fire escape windows on the fire floor, you will be in great danger trying to descend to the windows directly above. In this case, the top floor would be a logical decision. You can then move progressively downward. With luck, the fire will have been darkened down by the time you finish the top-floor search or the floor above will have been searched by reinforcements.

A fire in an apartment on the top floor has a different scenario and may call for a change in tactics. Since the interior team will search that apartment, the roof team at times won't be needed for search. The roof team should speed up the search of the top floor and the overall attack by venting the top-floor windows from the roof. This can be done using a pike pole or hook or by tying a heavy tool to a length of rope, then swinging it into the window. (This is often required with double-pane windows.) Conditions on the top floor can be relieved even more by cutting a vent hole in the roof and pushing the top-floor ceilings down. In fact, this will be mandatory at most serious top-floor fires, since fire will rapidly enter the cockloft.

Sometimes, the rapid descent down the rear fire escape can put the members in a position to reach potential victims much sooner than any other method. Particularly when the fire is in the vicinity of the main entrance, entry from the fire escape will mean the best chance of survival for any overcome victims. Otherwise, the interior team would experience delays and difficulty in reaching the rear rooms until the fire is darkened down. In this situation, where people are trapped in rear rooms with fire blocking the front entranceway, it may be possible to go through the adjoining apartment to reach the victims. Adjoining apartments often share a fire escape balcony. If this is the case, ducking out on the fire escape and coming back into the fire apartment can be a quick means of reaching the trapped victims. If not, it may be possible to breach a hole in the common wall between the two apartments to gain access to the rear. This same technique may be used to provide an escape route in an emergency if a firefighter finds himself cut off from the normal exit.

Some of the most dangerous multiple dwellings to operate in are those that have been renovated. Rehabs are often the cause of multiple alarms despite efforts to make them more fire safe. These are generally older buildings that have sound structural bearing members—walls, columns, and such—but whose interior partitions are in such disrepair that it is cheaper to rip them all out and start fresh than it is to repair them. Often the original floor joists are left in place while the ceiling is removed and a new drop ceiling is installed. This allows easier access to heating and cooling ductwork, piping, and the like. It also improves energy efficiency by lowering the height of the ceiling. Fire detection systems are often added, and sometimes sprinklers, but these are usually life-safety systems that don't cover the void.

In effect, what has happened is that a cockloft has been created on each floor. This space interconnects with all the vertical chases in which the piping and ductwork run to create a veritable maze for fire to travel. Fire originating in any of these blind spaces often outflanks firefighting efforts. There are simply too many voids to be able to open all of them at once. In this case, the sprinklers in the public hallways can make life miserable, since they are operated by the heat of the fire above the ceiling, which they are incapable of extinguishing (Figs. 13–26 and 13–27).

Fig. 13–26 When older buildings are rehabbed, the ceilings are often lowered. (Note the black iron hangers.) This void interconnects with plumbing voids to create a combustible maze for fire to spread within.

Fig. 13–27 The collapse of a wood or metal cornice can be a serious danger even at relatively localized fires.

Collapse of multiple dwellings during interior firefighting efforts used to be unusual. Generally, with the older floor and roof systems, the fire conditions had to be so severe that interior forces were chased out before any collapse occurred. This is no longer the case. Two factors have changed the way firefighters must look at multiple dwellings. The first is the vacant building situation. In many northern communities along the Rust Belt, the number of vacant structures of all types is growing. In many cases, these buildings have been left un-maintained and open to the elements for years. Rot from the weather and repeated fires have made many of these buildings unsafe to enter. The second item is lightweight or truss construction (it seems that's all there is today), which is prevalent everywhere, especially in the Sun Belt. By collapsing, these lightweight structures have killed firefighters. Finally, even if the building is a traditional multiple dwelling with wood joists and masonry walls, there is the threat of partial collapse from the metal cornices on many buildings. Wooden members support these cornices in many cases. Fire in the cockloft or cornice could result in the structure falling to the street.

Fireproof Multiple Dwellings

Fireproof multiple dwellings have many of the same problems as non-fireproof versions as follows: high life hazards, large numbers of occupants, difficulties in stretching hoselines, difficulties with stairs, and the like. Fortunately, for the most part, the building is designed to act as our ally, restricting both the vertical and horizontal spread of fire. At the same time, however, the building can cause some unique problems, since it holds tremendous amounts of heat, radiating it back at the firefighters long after the fire has been knocked down. Apartment doors are, in most cases, of sturdier construction than those in older non-fireproof buildings, and often they are better secured. Even recognizing the building as a Class 1 fireproof building may not be so simple. Look for fire escapes and cockloft vents. If you don't find any, chances are it is a Class 1 building (Figs. 13–28 and 13–29).

Scissor stairs pose a common problem in these types of buildings. Building codes often require two means of egress from each floor. This requirement is satisfied (questionably) by putting both staircases back to back with each door on the floor separated from the door to the other stairway by the length of the staircase, usually about 20 ft. This is a space-saving arrangement as opposed to having staircases at opposite ends of the hallway. Scissor stairs create problems for firefighters in several ways. First, if you go to the floor below and attempt to orient yourself to the apartment's location in relation to the stairway door, you are in for a surprise. When you come out of the stairway on the fire floor, your location will be off by the length of the stair run. You must go down two floors to find the same orientation between apartments and a particular staircase because the exit doors continuously alternate positions from floor to floor. In some buildings with two parallel halls, the stairs can alternate from one side of the building to the other on alternate floors.

Fig. 13–28 Non-fireproof buildings can be recognized by their fire escapes and the cockloft vents just below the roof.

Fig. 13–29 Fireproof buildings aren't required to have fire escapes, and their cocklofts are noncombustible.

These changing positions can mean having to travel an extra 20 ft down a heavily charged hallway if you happen to choose the wrong staircase. In addition, where standpipe systems are present in these buildings, there will almost always be one riser that ascends straight up through the staircase. Since it goes straight up while the stairs alternate positions, the result is that the standpipe outlet valves, while present on every floor, will be in one staircase on one floor and in the other staircase on the next floor up or down. Thus, if a company chooses not to attack from the staircase that has the outlet on the fire floor, (which is highly recommended), they will have to go down two floors to find another outlet in that stairway, or else they will have to go down to the floor below and then cross over and hook up in the other staircase. Either option demands additional hose (Fig. 13–30).

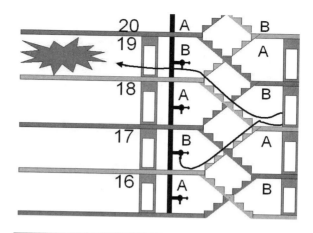

Fig. 13–30 Scissor stairs compound operational problems, especially for the engine company trying to operate off a standpipe. They will usually need more hose to reach the fire.

Fires in these structures are generally confined to the apartment of origin. Most codes require 2-hr fire separations between the apartment and the public hall as well as between apartments. Thus, if the door to the hall can be closed, the threat to most of the rest of the building's occupants is usually at least temporarily eliminated until we can put the fire out. Fire may extend to the floor above, however, most often by auto exposure, and this area, as always, must be checked early. The fire load in these structures is confined to the contents of the apartment, and most fires are relatively routine affairs. They are punishing on the attack team, due to high heat. The interior public hallway is most often windowless, but there is usually no danger of extension, collapse, or severe life threat to anyone outside the fire apartment. In fact, I have seen occupants *inside* the fire apartment survive quite well with a closed door between them and the fire and their windows open.

The most punishing affairs likely to occur in these buildings involve fire that is vented to the outside with the apartment door open and a stiff wind blowing into the fire apartment. In this case, the conditions in the hall will approximate the inside of a blast furnace.

New York City has experienced similarly severe conditions in these type buildings periodically since the 1970s. In the late 1990s, four NYC firefighters and numerous civilians were killed in these extreme wind-driven fires. The department's firefighting procedures manual, "Multiple Dwellings Fires," which was first written in 1979, describes these fires as *blowtorch* fires, and describes tactics to be implemented when encountering this condition. The cause of this condition is high winds blowing flame back into the building through windows that have vented. The blowtorch description is quite accurate. The wind-driven flames blast out of the apartment door at extremely high velocity. The wind striking the side of a building funnels through an involved apartment, increasing the amount of oxygen in contact with the fuel, resulting in a dramatic increase in the burning rate of an object. The resulting flame is then pushed ahead by the wind, venting out of the apartment door and filling the public hall with flame.

The fireball must have an exhaust outlet or the airflow cannot continue, but in a high-rise building, there are numerous possibilities. It is critical to evaluate the effect that any ventilation efforts will have prior to breaking any windows. It is relatively simple to do this from the outside. If any wind is blowing toward the affected windows, don't vent until the fire area is well under control. This will prevent smoldering furniture from being fanned into open flame and driving firefighters out of the area. When venting from inside the building, more information is needed. Make a small experimental opening on the fire floor first, preferably a single pane. When venting from the floor above, be sure to position all doors as they will be found on the fire floor. Otherwise, you won't know the real effect of wind on the fire floor (Figs. 13–31 and 13–32).

Windows in the hallway can be open or may fail to due to heat. In several cases, windows which were open in apartments across the public hall, more than 100 ft from the fire apartment, provided the necessary exhaust outlet when the occupants opened the apartment door to flee. Of course, when the fire department deploys a hoseline onto the fire

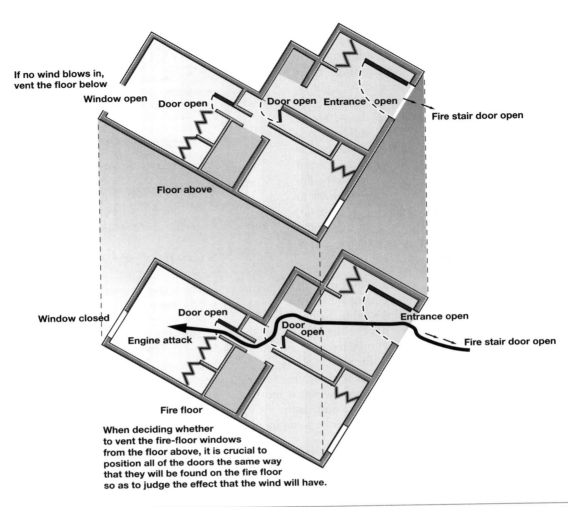

If no wind blows in, vent the floor below

Window open

Door open

Door open

Entrance open

Fire stair door open

Floor above

Window closed

Door open

Door open

Engine attack

Entrance open

Fire stair door open

Fire floor

When deciding whether to vent the fire-floor windows from the floor above, it is crucial to position all of the doors the same way that they will be found on the fire floor so as to judge the effect that the wind will have.

Fig. 13–31 When deciding whether to vent windows from the floor above, in order to know whether the wind will blow into the fire, it is crucial that all the doors on that floor are in the same position as the doors on the fire floor will be.

Fig. 13–32 When a door was opened on the opposite side of this hall from the open door of the fire apartment, a blowtorch of flame filled the entire 145-ft corridor, killing three firefighters.

floor from the stairway, the door to the stairway becomes blocked open and the staircase itself becomes a chimney. This situation is worsened if the door at the top of the staircase is open to the exterior, a practice that is routinely used under less severe conditions in order to lessen the contamination of the staircase and upper floors due to mushrooming. If the wind is not blowing into the windows of the fire apartment, fires in *fireproof* residential buildings are usually extinguished rather *routinely* by a single hoseline, since the fire loading in a typical apartment is relatively light. During a *routine* fire, smoke in the stairs and upper floors is the prime threat to occupants and is easily combated by venting the top of the stair and elevator shafts. During severe conditions, however, the threat is actual flame and high heat, and venting the stair can serve to draw the fire to that stair.

Engine company tactics to be used when faced with such severe conditions have been limited. The problem is that the flame that is blasting down the hall is simply burning gas, similar to a propane or natural gas flame. The water from handlines has virtually no effect on this gas flame. The only way that water can have any effect is to cool the seat of the burning material, but that is out of the reach of the stream, inside the involved apartment. The handlines have no effect on the fire under these circumstances because all the water that they throw is not reaching the fuel, where it can cool it and stop further flame production.

Under these extreme circumstances, an interior handline attack typically involves 8–12 engine companies operating in relays, since firefighters can only spend a minute or two in the hall before they are burned by the high temperature soaking through their protective clothing. No real progress is made in advancing on the fire, despite of the terrible beating the personnel are taking, until the fire has consumed the bulk of the combustibles in the fire area and has began to decay. In most of the past incidents of this type, multiple 2½-in. handlines operating from the staircases have been unable to advance due to the flame blowing at them. The typical result is that once this state is reached, and the lines can advance to the door of the fire apartment, one line inside the apartment readily completes extinguishment. The only objects remaining at this point are those made of steel, since even aluminum objects are found as pools of metal on the floor. The fire has burnt itself out!

Unusual fire situations, such as the blast-furnace hallway, require unusual tactics. Several recent incidents have proved the futility of standard tactics in the face of this wind-driven flame. Numerous engine companies have been beaten up trying to advance two or three 2½-in. handlines down these white-hot hallways, taking more than 30 min to advance 50 ft! If the fire is within reach of an outside stream, it may be best to knock the fire down from outside, something that is nearly taboo in an occupied dwelling. In these cases, there are no longer any live civilians in the fire apartment unless they are in sheltered areas. The fire is blasting down the hall, threatening other civilians as well as firefighters. In low-rise buildings under these circumstances, an aerial stream can be quickly deployed to knock down this heavy fire. Many of the fireproof multiple dwellings are high-use buildings, though, and it is on the upper floors that the winds are most likely to create these blowtorch flames that prevent our advance.

The problem then is, "How do we put an outside stream into operation on the 20th or 30th floor?" The best, most practical answer that I have been able to come up with is to use that antique Navy fog applicator that you are keeping around for tank truck fires. An 8-ft-long applicator pipe is long enough to go from floor to floor of most residential high-rises, yet it is compact enough to fit into most elevators.

To use the applicator pipe on the blowtorch hallway, you need a threaded adapter or a Navy fog nozzle attached to a handline on the floor below the fire. Clear all of the attack personnel out of the halls on the fire floor (they probably haven't gotten out of the stairwell, anyway) then close the stairway doors. Place the Navy applicator out the window, directly below the window on the fire floor into which the wind is blowing. Then hang the fog tip just over the windowsill of the fire apartment and start the water. The impinging jets of the applicator's fog tip have a self-canceling effect, producing very little back pressure, allowing you to keep the applicator in a vertical position with very little effort. The impinging jets also produce a very fine fog mist that is easily carried through the apartment and down the hallway by the force of the wind blowing through the window. To increase the flow of the applicator, always try to use the 2½-in. model (1½-in. pipe) as opposed to the 1½-in. model (1-in. pipe), since it flows about 100 gpm as opposed to about 60 gpm. This flow can be increased further by drilling out the holes on some of the impinging jets, making flows of 150–175 gpm possible. Be sure to drill both sides of a set of impinging jets so that the backpressure remains neutral. Remember, if you put the fire out, everything else gets better. Get that Navy applicator up to the floor below early.

Another option that might be possible involves breaching a hole from the stairway into the adjoining apartment. If this is the fire apartment, operate the nozzle directly into the fire area. If this isn't the fire apartment, continue across it, breaching walls until you gain access to the fire apartment. If the fire apartment is across the hall from the stairway, this method might only get you close enough to direct a hose stream out through a small hole that you punch through the hallway wall and toward the door of the fire apartment.

A third option might be to attempt to reverse the flow of gases, driving them back toward the fire apartment by using the fog nozzle set on a fairly narrow fog as when venting out a window with the line. Team this line with a second line that can operate on straight stream through the fog pattern to push back fire that is well ahead of the hose stream. This effort can be greatly assisted by pressurizing the area behind the lines with PPV fans set up below the fire floor to pump fresh air up the staircase. The top bulkhead and all other doors in this staircase should remain closed. Depending on the physical layout of the stairs and the fire apartment, it may be possible to advance from this attack stairway, pushing fire down the hall ahead of the lines, toward a second ventilation stairway that is past the door to the fire apartment. This requires that this entire ventilation stairway remain clear of occupants and firefighters, and that the bulkhead at the top of this stairway be opened to assist the upward draw. It would also likely require at least one additional stairway to be used as an evacuation route for occupants above the fire, if in fact they must be removed rather than be sheltered in place. In the event of a strong wind (15–20 mph), however, this option is the least likely to succeed. It also requires a tremendous amount of time, coordination, and resources to implement successfully, as well as the proper building layout (Fig. 13–33).

Fig. 13–33 A Navy fog applicator produces a dense fog pattern that will be carried by the wind into a wind-driven blaze. The applicator's length allows it to be inserted into the blazing apartment from the floor below the fire.

Fig. 13–34 The fire window blanket is designed to be lowered into place in front of a window that has wind driving fire into the hallway.

The FDNY has also experimented with a fourth option—stopping the flow of wind into the fire apartment. This is accomplished by draping a fire resistive curtain in front of the offending window from the floor or roof above. This *high-rise window curtain* has shown much promise in test fires, where witnesses described the curtain's effectiveness as "Like somebody turned off the switch to the fire." The curtain the FDNY uses is 10 ft high x 12 ft wide, sufficient to cover the majority of residential windows. It is constructed of a material known as *Hotstop-M,* which will withstand direct exposure to 1,500° for an unlimited time, and 2,000° for up to 15 min. It comes with 20-ft long Kevlar ropes sewn on each corner, allowing it to be lowered into position from two floors above the fire and to be pulled tight by the lower ropes from the floor below the fire. It is weighted with lengths of chain sewn into the lower edge, to help it fall into position even in the face of very high winds (Fig. 13–34).

Buildings with duplex and triplex apartments can severely challenge fire forces. Simply stated, duplex apartments have living space on two floors of a building, while triplexes have space on three floors. This means that there must be a stairway within each apartment from floor to floor. This is an open, unprotected stairway as opposed to the stairway within the fire-rated enclosure that serves the building's exits. The layouts vary drastically from floor to floor in these types of buildings, and it is often impossible without prior knowledge to find the proper floor below the fire to orient yourself. A suggestion: where there are floor layouts that repeat themselves within a structure, have the stair side of the stairway door painted the same color on floors with the similar layout. A unit arriving at a heavily charged floor can look at their side of the door and either drop down to a similarly colored door to get an idea of the layout, or radio to other incoming units to take a quick look at the proper floor on their way up (Figs. 13–35, 13–36, 13–37, and 13–38).

Fig. 13–35 Duplex apartments have an open interior staircase within the apartment. Descending this stair for a fire on the lower level is akin to going down a cellar stair.

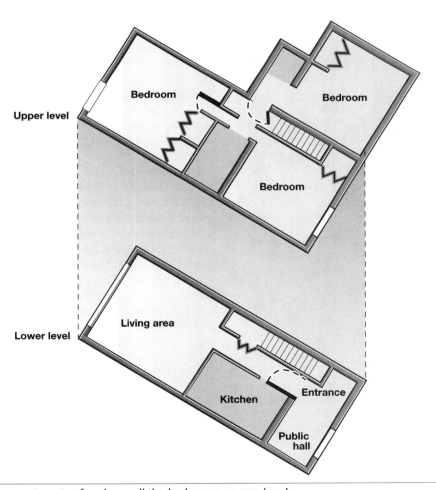

Fig. 13–36 Duplex apartments often have all the bedrooms on one level.

Fig. 13–37 Sandwich apartments only have hallways every three floors. All apartment entrances are located at that level, with one door leading down to an apartment on the floor below, one door leading straight in on that floor, and the third door leading up to an apartment on the floor above.

Fig. 13–38 Sandwich apartments may sometimes be evident from the exterior. Here, the floors with the hallways (3 and 6) are visible due to the continuous row of windows that line the hall.

One of the biggest problems at many of these fires is the large number of tenants who attempt to flee the fire. These people would almost always be far better off if they were to remain in their apartments or hotel rooms rather than flee from the upper stories. Often these people will take elevators past the fire floor, a deadly risk. Otherwise, they clog the stairs, preventing firefighters from using them for attack and ventilation. Many times, the occupants—particularly the elderly, the handicapped, and young children—require firefighter assistance to climb down several flights of stairs, further depleting manpower. The key to success here is pre-incident training, informing residents of the proper action, which is to stay in their apartments or rooms. In a senior citizens' residence in one town, two floor wardens were appointed, themselves residents, to advise the other residents to stay in their rooms whenever the fire alarm sounded. This simple step has alleviated a chaotic scene in the lobby of the complex every time a smoke detector activates. At times, this can be augmented in other premises where a public address system is available in the lobby, announcing directions either to selected floors or to all of them.

A last-resort option for a similar purpose, controlling panic or issuing directives, is to use the public address capability of the apparatus. This has proved to be particularly useful in hotel-type occupancies housing large numbers of transients. There, a single hotel room fire won't be a severe challenge to the attack team, but the resulting attention from the sounding of the fire alarm, to the sirens of the arriving apparatus, to the scurrying of the firefighters, can lead occupants to believe that they are in imminent peril. Using a calm, reassuring tone while addressing occupants at windows can relieve much of their anxiety. Of course, anyone in actual danger must be removed or protected as soon as possible, but this is generally confined to those in the immediate fire area in Class 1 buildings. Search teams will have to account for all rooms or apartments on the fire floor and above, but many times uninvolved floors can be searched by less damaging methods, such as by using a master key. Ventilation of light to moderately charged hallways on these uninvolved floors can often be accomplished by blocking open the hopper door to either a compactor or an incinerator chute on the floor. This avoids contaminating the staircases with smoke, leaving them clear for other purposes, if need be.

Multiple dwellings of all types present firefighters with their greatest life hazards. As such, a tremendous effort is expended in fighting these fires. At each multiple-dwelling fire, take a close, critical look at all phases of your operation. Are these efforts being channeled properly, or are they wasting precious resources of personnel, water, and equipment that could be better used to save lives? That is the bottom line at multiple-dwelling fires.

14 Garden Apartment and Townhouse Fires

In many parts of the country where rapid growth is being experienced, a relatively new type of fire is being encountered, fire in *garden apartment* buildings and/or *townhouses*. Sometimes these buildings are referred to as *condos*, short for condominiums, but that is a very misleading term, referring to the arrangement of ownership and not the layout of the living spaces, which is much more important to the firefighter. These two types of buildings are similar to other residential occupancies, private houses, and older traditional apartments, in terms of fire loading, compartmentation within the apartment, and the need for speed of water application, as well as the many turns likely to be made with the line. There are many unique conditions common to these buildings, however, which may require a shift from typical *house fire* tactics.

At first glance, these two types of buildings may appear to be indistinguishable from each other, but in fact there are some distinct design differences between them that affect our tactics. The term *garden apartment* refers to the fact that unlike older urban apartment buildings, these newer style buildings are usually low-rise multiple dwellings, one to three stories high, although many are being built that range as high as 6 stories of wood frame! They are often surrounded by lawns, trees, etc. Compared to the older inner city tenements that the occupants were leaving, the people moving into these newer complexes felt they were moving into *a garden*. Normally, garden apartments are laid out so that each apartment has its living space on only one floor. There is normally no interior stair found within the individual apartments. In many areas, the entrance stairway to upper floor apartments is exposed to the street or yard. This stair is no less crucial than the interior stair of older multiple dwellings, and its protection must be one of our prime objectives (Figs. 14–1 and 14–2).

Fig.14–1 Townhouses are PDs that are attached to other similar structures on one or both sides. They have all the same problems of other PDs, with additional complications due to lightweight construction and limited or no side ventilation.

Townhouses, on the other hand, are similar to a traditional one-family private home but are attached to other homes on each side. Townhouses typically have two or more floors, (as many as 5 living levels may be found in split-level designs of three-story buildings), connected by an open interior stair. The room size and fire loading are very similar to a more traditional house.

Each of these buildings poses some common problems, which have been shown to repeat themselves over and over across the country. Problems often begin with the layout of the development or complex itself, with poor access for apparatus, which slows hoseline and ladder placement and limits the usefulness of fixed master streams. The closely spaced buildings, often of wood-frame construction and combustible siding, pose a serious threat of building-to-building fire extension. Hydrant spacing and flow are often deficient inside the complex, requiring long stretches to sufficient supplies outside the complex. Gaining access and egress to the complex itself may even be a challenge, with many developments—especially the more exclusive, expensive ones—being built as limited access, high security compounds surrounded by walls or fences. Preplans should be developed for access and water supply (Figs.14–3, 14–4, and 14–5).

Many inner-city firefighters would recognize these buildings as nearly identical to their old nemesis, the *row frame*. One of the key differences between garden apartments, townhouses, and most other dwellings is the attached nature of the combustible exteriors. The interior layouts are similar to homes and *typical* house fire tactics should be employed for the routine room and contents fire, using 1¾-in. or 2-in. hose for speed and maneuverability up stairs and around bends. A major problem develops for some departments that have little experience in garden apartments when the layout of the development precludes the pumper from being close enough to utilize its preconnected handlines. A means of quickly placing a hoseline in operation must be available from both the nearest access point as well as the alternative places where the pumper may have to stop. This requires a bed with several hundred feet of hose in many cases. While 1¾-in. or 2-in. hose is normally the line of choice inside the building, the high friction losses in these lines demand that no more than 300 ft of these lines be used off the engine. If the engines can't always be within 300 ft of the most remote corner of each room on every floor of the development, some 2½-in. or 3-ft hose will be required to cut the friction loss.

The method that I prefer to deal with these long hose stretches is to attach 100 ft or 150 ft of 1¾-in. hose to a break-apart 2½-in. nozzle at the end of a bed of 2½-in. hose.

Fig.14–2 Garden apartments are multiple dwellings that typically have living quarters of each apartment on only one level. Stairs may be open to the exterior. Lightweight construction abounds in these occupancies as well.

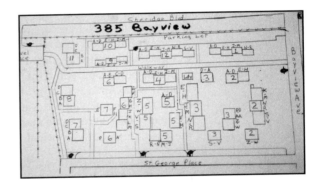

Fig.14–3 Garden apartments are often developed with little thought to fire department access and operations. Preplanning can pay major dividends during an incident.

Fig.14–4 Prepare for long hose stretches to reach remote portions of garden apartment complexes.

Fig.14–5 Closely spaced frame buildings with combustible siding poses potential conflagration hazards. It is not uncommon to find several buildings involved upon FD arrival at these complexes. Note the louver near the peak of each building, which leads directly into the cockloft.

Fig.14–6 The method I prefer for getting water into areas that are out of the reach of preconnected lines is to stretch a 2½-ft line with a break-apart tip. Carry 100 ft of 1¾-in. line to the point of operation and connect to the 2½-in. line for interior use.

The 1¾-in. hose offers the flexibility for making the needed bends and turns inside the structure, and the 2½-in. hose serves to cut the friction loss, allowing hundreds of feet of hose to be stretched. The break-apart 2½-in. nozzle serves as the 2½ × 1½ in. reducer but with added versatility. If heavy fire is encountered, especially fire that is extending to exposures, it is a simple matter to disconnect the 1¾-in. hose and place the larger stream in operation. In the closely spaced, attached rows of combustible apartments present in many developments, once fire vents from a window or door, stopping fire extension becomes critical. Another means of accomplishing a similar task would be to connect the 1¾-in. leader line to a gated wye or *water thief* at the end of a 3-in. supply line. I prefer to use the 2½-in. line and nozzle instead of the water thief for this use. If the 1¾-in. line is not sufficient, the 2½-in. nozzle is immediately available. If a gated wye or water thief is used, and the 1¾-in. line is incapable of knocking down the fire, if may take too long to bring up a 2½-in. hand line and place it in operation (Figs.14–6 and 14–7).

The biggest difference in fire operations in these buildings from other apartment buildings is due to the heavy prevalence of lightweight combustible construction in these occupancies. The great danger to firefighters from lightweight construction is sudden catastrophic collapse. Townhouses and garden apartments built with trusses or other lightweight methods require extremely fast knock-down from a safe area or else collapse is likely to trap fire-fighters. If the fire is confined to the contents of a room, there is little danger of collapse. When fire has begun in or extended to the structural voids however, the danger of sudden collapse should prompt a shift in tactics.

Fig.14–7 If the fire has vented from several windows on arrival and is threatening to spread to exposures, use the 2½-in. line to knock it down. The 1¾-in. line can then be attached to the 2½-in. for overhaul.

A realistic evaluation of the life hazard and the likelihood of collapse should prompt a rapid, overwhelming *blitz* attack from a position outside the collapse zone. Again, the fire killing power of the 2½-in. handline comes into play. A 1¼-in. solid tip at 50-psi nozzle pressure delivers 325 gpm, with sufficient reach and penetration ability to extinguish a fully involved apartment without having to push in under a weakened roof or floor. If available, the elevated master stream of a tower ladder is invaluable. Once the fire has been knocked down, vented, and sufficient lighting is in place to allow firefighters to thoroughly evaluate the structure's collapse potential, it may be possible to enter the structure to complete the search and overhaul. The tower ladder will permit this knockdown from outside the collapse zone of most two- or three-story apartments, since wall failure is not the main danger, but rather floor or roof collapse. Once the main body of fire is darkened down, the 1¾-in. line can be attached to the tip for overhauling (Figs.14–8, 14–9, 14–10, and 14–11).

Fig.14–8 Both garden apartments and townhouse pose a serious threat to firefighters. Can you see what it is?

Fig.14–9 Every IC wishes for the ability to view into a building to examine its construction. Only the ones who have done their homework and preplanned their structures will have this level of knowledge.

Fig.14–10 New hazards require firefighters to remain constantly aware of the construction techniques being used in their area. Document them for future generations.

Fig.14–11 The sheathing on this row of luxury townhouses is a plastic-coated cardboard that is only ⅛-in. thick!

Most garden apartments and, to a lesser extent, townhouses, have a common cockloft, which extends over several apartments within a building, similar to the *row frame*. Standard practice in an old style row frame with fire on the top floor called for 1¾-in. lines to be stretched into the top floor of each exposure to stop extension in the cockloft, while other crews worked on darkening down the original fire. This tactic is not likely to be as successful if the building is built using lightweight methods, since a key weapon is taken away from the fire forces in truss buildings: roof ventilation!

Early cutting of a large hole over the main fire is not possible in lightweight buildings, since firefighters should not be committed to the roof of these structures in the event of fire in the cockloft. Without the needed roof ventilation, firefighters are likely to be driven out of the adjoining building sections. The only place to make a stop under these circumstances is behind a defendable barrier, a *fire wall* or more likely a party wall that extends up through the building from foundation, through the cockloft, at least to the underside of the roof boards and preferably above the roof. From a position behind this barrier, coordinated attack stands a good chance of cutting off a rapidly extending fire, since the roof can be opened on the safe side of the wall, ceilings pulled, and hoselines operated to hit fire that does come through the ever-present holes in such walls (Figs.14–12, 14–13, 14–14, and 14–15).

Fig.14–12 The common cockloft is a threat to the entire building. Once fire gets a head start in this area, it is difficult to stop.

Fig.14–13 Fires that begin on lower floors can also spread to the cockloft and destroy the entire building. Check pipe chases and other voids, just as in older multiple dwellings.

Fig.14–14 A well-built firewall can be a great asset in stopping fire extension. A true firewall will be built of brick and extend at least 3 ft above the roof and any combustible surfaces.

Fig.14–15 Mansard roofs often wrap around division walls outside the building and allow fire to spread to the adjoining cocklofts.

When planning such a defensive strategy, you must be aware of several factors that could defeat your efforts and possibly endanger firefighters. First, beware of mansard roofs, which wrap around such division walls outside the building, and allow fire to sneak around them in large volumes.

Second, be aware that most so-called *fire walls* in these buildings are not true firewalls, in that they are not meant to be totally self-supporting even if the entire fire side of the wall collapses. Townhouses, being PDs, are usually separated from their adjoining neighbors by a common or *party wall*. I am not sure why they are called party walls, whether it is because they are so thin you can hear your neighbors party through them, or more likely, because fire has a party

Fig.14–16 The dividing walls between these townhouses is two layers of gypsum board. Fire spread will be slowed not stopped by such walls.

Fig.14–17 The cinderblock wall being built here is a party wall. The trusses of each building rest on the common wall.

with them. In either case, party walls are problems for firefighters. Party walls have the joists or trusses from both sides of the wall laid on the top of each story either side-by-side or butt end to butt end, and then the space between joists or trusses is supposed to be filled in with masonry. A fire can easily penetrate this wall by burning through the wooden joists or trusses. Thus the need for a hoseline, ceilings pulled, and ventilation. The top floor is an obvious place for this to happen with its large cockloft space, but don't ignore the other floors beneath the top, as they will usually have a *truss loft* that creates the same situation and can occur beneath firefighters working on the top floor (Figs.14–16, 14–17, and 14–18).

Finally, be aware of fire that burns out the joists or trusses on the fire side of a party wall and may result in some structural damage to the party wall itself. If the trusses collapse with sufficient force, it may cause failure of the wall in places. So even though you have lines in place and have opened up the ceilings on the safe side of your defensive fire barrier, do not give up operations on the fire side of the wall. Continue trying to knockdown as much fire as possible from safe positions outside the collapse zone, using large handlines, or better yet, tower ladder streams. Sweep the streams along the ceilings, parallel to and as near as possible to the walls you are trying to protect, without causing further damage to the wall from the stream itself. Realize that in many of the newest-style homes, the party wall consists of two layers of ⅝-in. *firecode* gypsum board on each side of a metal stud. This type of *firewall* is easily damaged by a hosestream operating in close proximity, which could be the reason fire gets by the wall if firefighters aren't careful.

Fig.14–18 After the trusses are set, the gaps are filled in. Any fire that reaches the ends of the trusses will likely extend through this wall and ignite the trusses on the other side. The fire has a party with these walls!

Garden apartments are noted for another common void space that is unusual in most other occupancies: the common cellar, basement or crawl space. The cellar or basement is often the location of the common laundry facilities and is the source of a significant number of fires. The common cellar area has all the fire-spread potential and all the difficulties of limited-access and ventilation of the common cockloft, but is in an even more dangerous location—*under* the entire row of buildings, so the upward spread of fire targets multiple apartments very quickly. Especially in the case of common crawl spaces, stream access is extremely limited. Early cutting of the first floor and use of cellar pipes, distributors or the bent tip may be all that stands between a quick knockdown and a total loss of a large number of apartments.

The largest losses at these complexes normally occur when they are being built. The most dangerous stage of this process, called the open framed stage, occurs when all the wall, floor and roof supports are in place, with the exterior plywood siding and roof decking, but none of the interior gypsum board, doors or windows. This is a stage that allows extremely fast, hot fires to grow and spread rapidly from one surface to another. Time and time again, fires at this stage have devastated blocks of homes to be (Figs. 14–19, 14–20, and 14–21).

My first working fire as a college fire protection student/firefighter occurred in a garden apartment complex in the open-framed stage. At about 7 PM, we were dispatched to a report of a fire in a garden apartment complex under construction. En route we could see a huge column of smoke. This was going to be some job! When we pulled up, there were at least 20 apartments burning. Fire was spreading down the rows of plywood-covered buildings at an alarming rate. The attack began with two, 1½-in. lines flowing a grand total of about 200 gpm of water. The fire laughed at the effort. The ladder company from our station took a position ahead of the fire to try to cut it off but was outflanked as the fire raced past before the ladder pipe could be put in place and supplied. The pumpers on the scene were not equipped with preconnected master streams, or for that matter, even so much as a bed of 2½-in. hose as attack line! I knew that this fire was screaming out for that kind of water.

At this point, three lines were stretched, all 1½-in. lines. The streams of water they were throwing were too short to even reach the fire, and the tremendous radiant heat prevented the nozzleman from approaching any closer. What was needed was a line that could quickly be put in operation and throw a large stream a great distance. A large old brass play pipe, nearly forgotten in the back of

Fig.14–19 The open-framed stage of construction is the most dangerous in terms of conflagration risk. Make sure water supplies are adequate for the hazard and turned on before construction begins.

Fig.14–20 This open-framed development had about 20 apartments involved upon FD arrival.

Fig.14–21 Engine companies responding to fires in buildings under construction should be prepared to place a master stream into operation immediately on arrival.

a compartment, provided the means. With its 2½-in.inlet and a 1⅜-in. discharge behind the main 1⅛-in. tip, it was connected to three lengths of 3-in. hose and quickly placed ahead of the blaze. With the outer tip removed, the 1⅜ in. tip will flow almost 500 gpm. Since we wouldn't be dragging this hose inside the building, a one-person loop was very effective at creating an improvised master stream.

The large line did exactly what it was supposed to do: provide enough volume and reach to hit the buildings ahead of the fire and prevent their ignition, while alternately hitting the head of the approaching fire, knocking it down enough to further reduce the threat of fire spread. I would have preferred a preconnected deluge gun atop the apparatus, but necessity is the mother of invention.

Years later, after several other disastrous garden apartment blazes, the department's apparatus were finally equipped with apparatus-mounted master streams. Why did it take so long? The department was the victim of its own success. They had done so well with 1½-in. hose and fog nozzles when everyone else was using 2½-in. hose that they were lulled into a false sense of security, thinking every fire would go out with the preconnect or else it wouldn't matter and the *surround and drown* ladder pipe would have plenty of time to be set up. That's not always the case.

15 Store Fires——Taxpayers and Strip Malls

ommercial structures pose many of the same problems that non-fireproof multiple dwellings do. Large areas, common cocklofts over the entire building, and Class 3 ordinary construction are all typical features of both. Fire loads are higher in commercial structures than they are in residences, which should prompt an increase in the discharge rate. Fortunately, the civilian life hazard is usually lower. Due to the hours of operation, the occupants are normally awake and will flee if escape is possible. Occupants in cellar areas may find their escape routes blocked if they aren't warned early enough, however, so a thorough search of these areas is mandatory, even though those people may be below the main body of fire. Searches in commercial buildings should not be conducted in the same aggressive manner as required for residential occupancies, since the civilian life hazard is so low.

Conversely, the hazards to firefighters at these occupancies are extremely high. The following chart, adapted from NFPA data, illustrates the dangers in graphic terms: fires in commercial occupancies, with an extremely low civilian life hazard kill four times as many firefighters per incident as do fires in occupied residential buildings, where the high civilian life hazard justifies firefighters taking calculated risks to reach trapped occupants (Fig. 15–1). There is *no* justification for firefighter deaths in commercial buildings where there is no life hazard!

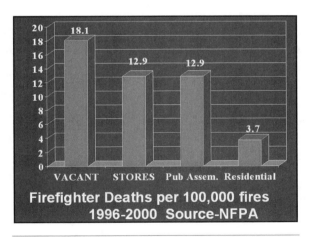

Fig. 15–1 Fires in commercial buildings are nearly four times as deadly as residential fires.

Taxpayer is a term used to describe a row of stores housed within a single building. Although not recognized by any building code, this description has come to be accepted by the fire service as any commercial structure described previously that is built of Class 4 ordinary construction. Newer taxpayers differ from this traditional description in that they use Class 2 noncombustible construction. These newer buildings are often called strip malls. For practical purposes, taxpayers and strip malls can be treated in much the same way, and the terms herein will be used interchangeably.

The term *taxpayer* derives from the practice of landlords who owned a piece of vacant land and constructed a fast, cheaply built structure on it to rent out, thereby generating income to pay the realty taxes while anticipating a future increase in value. These speculative structures were considered expendable temporary buildings, and so weren't particularly designed for fire safety. Although this definition may no longer be true, owing to stricter building codes (often enacted after several catastrophic fires), the name has become a generic description of all one- and two-story multi-tenant commercial buildings (Figs. 15–2, 15–3, and 15–4).

Styles of Taxpayers

Taxpayers may be broken down into two general classes: new style and old style. The older-style taxpayer has a tremendous amount of combustible material in its construction. The floors, roofs, walls, and even ceilings are often made of wood, while the exterior walls are typically brick or cinder block. Usually, the cockloft is open over the entire area of the structure. Partition walls, often of wood lath and plaster on 2×4 studs, usually extends only to the level of the ceiling, leaving the area above open for running utility connections and, of course, vulnerable to extension.

The roof construction on the typical taxpayer is one of four styles: the standard flat roof, the inverted roof, a metal deck on bar joists, or the bowstring truss. Bowstring truss roofs are usually located over larger single-tenant occupancies (supermarkets, for example), which may appear to be part of a row of adjoining stores. In some cases, older single-tenant buildings are subdivided into a number of smaller stores beneath the huge bowstring trusses (Figs. 15–5 and 15–6). In either case, if there is evidence of heavy fire within the bowstring truss, evacuate the entire building and establish suitable collapse zones around the perimeter immediately, for the building will collapse *soon!*

Ventilation operations on either the standard flat roof or the inverted roof should proceed as outlined in chapter 9. An 8×8-ft ventilation hole over the main body of fire will greatly slow horizontal extension in the cockloft. If further relief is necessary, continue cutting additional vent holes. Cutting a trench on a taxpayer roof usually

Fig. 15–2 Older taxpayers often hold as many as 20 stores under one roof, all connected by a common cockloft. They were often built with cellars for storage.

Fig. 15–3 Shopping centers are taxpayers with a parking lot in front. The age of the building determines the type of construction used.

Fig. 15–4 Newer strip malls are usually built without cellars, which means the stores themselves are larger to accommodate storage. The cockloft is still present, though the roof deck may be metal.

Fig. 15–5 When a single bowstring truss failed at this supermarket fire, a 40 x 100-ft opening was created, throwing 12 men into the inferno below. Six survived.

Fig. 15–6 A severe fire in this row of older Class 3 stores was halted by the quick application of more than 600 gpm through handlines. The rest of the row was saved.

Fig. 15–7 This row of Class 2 taxpayers with a metal deck roof had a severe fire in the end store. Roof and interior operations could safely continue in the exposed stores even though part of the roof was collapsing.

isn't practical and shouldn't be undertaken. For a trench to be effective, it must be cut from outside wall to outside wall (or to a firewall) and be subdivided every 4 ft so that it can be pulled. On a taxpayer that is only 75 ft deep, that would require more than 210 ft of cutting. This would be a tremendous drain on resources that could be better used to increase the size of the main vent holes. Remember, the trench is a defensive tactic. You would have to drop back a tremendous distance, giving up everything on the fire's side of the trench to complete such a large task before the fire passes it.

Newer-style taxpayers are often noncombustible buildings, since there is very little to burn in their basic materials. These fall into the category of Class 2 construction. Many of the newer taxpayers are (thankfully) built without cellars or basements. In such buildings, the first floor is poured directly over the earth—slab construction. The benefit to the fire department is obvious—no *chimney* to try to climb down in the event of a fire. The tradeoff for not having a cellar however is increased size of the first floor store itself, resulting in a larger potential fire area. The storage may be separated from the sales area only by a flimsy partition. The roof is usually constructed of corrugated metal decking laid over unprotected steel-bar joists. The interior partitions are most often of plasterboard on metal studs, while the exterior walls are cement block. There is very little fire load when the building is completed, and it would be unusual to get a severe fire going. (You shouldn't overlook the possibility of a metal-deck roof fire, however.) This type of fire can occur in what is really just the shell of the building—the roof, walls, and floor. Once the building is fully occupied, however, the major part of the fire load comes from the materials stored within.

Newer-type taxpayers face the same dangers of fire spread through the cockloft as older types do. In addition, there is the likely problem of early collapse of the unprotected steel roof. Firefighters think of steel as being noncombustible, which it is, but they fail to realize that it is a thermoplastic metal that changes shape fairly easily when it encounters the heat of a structure fire. This is especially true of the relatively thin pieces of steel that make up the typical bar joist and corrugated metal decks. When exposed to fire, these styles of roofs can fail in as little as 5 min. Several sources recommend conducting a strictly exterior operation on bar-joist roofs due to this danger. (In May 2001, the FDNY modified its tactics to order its members not to cut metal deck roofs due to the dangers they pose.) In many towns across America, this would mean giving up every fire in every strip mall they ever went to—a policy with which I don't necessarily agree. I have operated at a number of fires in these noncombustible buildings. I have cut these roofs and operated hoselines beneath them, and I believe that that is the key to success— a coordinated attack involving ventilation and hose streams to cut off the spread of fire (Fig. 15–7).

When steel-bar joist roofs are exposed to fire, the heated steel tends to sag. This provides a warning of impending failure, something that wood trusses don't provide. Wood trusses snap suddenly and drop the roof onto or out from under firefighters. In addition, steel-bar joists are generally spaced relatively close together, from 2 to 6 ft apart as opposed to wooden bowstring trusses, which are typically 20 ft apart. Bar joists come in lengths up to 60 ft however, and from the roof you can not tell which direction they run—front to rear or side to side. The failure of two or three bar joists will cause an area 10 or 15 ft wide by up to 60 ft long to fail, as opposed to the instantaneous 40 ft wide by 100 or more ft long gap created by the failure of a single bowstring truss. Still, the key factor that allows us to operate within a strip mall with a bar-joist roof is that the steel can be protected against failure by applying a hose stream. Firefighters finding heavy fire in a strip mall should begin a coordinated attack from safe positions. A 2½-in. handline should begin the attack on the fire store. The 2½-in. line offers 75–80 ft of reach, allowing you to cool steel that is well ahead of your location. Use the reach and impact of the straight stream to blow through the ceilings and cool the cockloft. If the flow of the handline isn't enough, use a preconnected deck gun.

In the meantime, stretch handlines (generally 1 ¾ in.) into the exposed stores on either side and begin sweeping the cockloft overhead. The objective is to prevent the weakening of the bar joists by heat; therefore, you must cool the steel. To allow these members to remain in these exposed stores, these areas must be ventilated, but this is a difficult task, since the only ventilation likely to be readily available is the removal of show windows at the front. The rear and sides of taxpayers are typically well sealed. That means roof ventilation may be required. On a bar-joist roof, or other unprotected steel roof, you *cannot* cut directly over the fire as we would like to on most ordinary roofs. However, you can back away 60 ft or the width of three average stores from the sagging joists to an area where the hose streams are cooling the steel and cut there.

The key is coordination, since the hoselines must cool the steel to accomplish their task of preventing extension and collapse. If the hose streams are operating and cooling the steel, the roof venting team can do their job providing ventilation so that the hose crew can remain in position. Neither team can function for long without the support of the other. Still, there is no need to give up every strip-mall fire without a fight just because it has a steel-bar joist roof! That's not the case with wood-truss roofs. You should evacuate these completely if heavy fire is in possession of the wood truss or cockloft.

A type of construction being used in the newest taxpayers involves lightweight wood trusses as the roof and/or floor supports. Three firefighters in Orange County, Florida, were trapped when such a roof failed. Two of them died. The stores had been evacuated, and the members were engaged in interior firefighting when the collapse occurred, *9 min* after the firefighters arrived. These lightweight truss-roof systems (and their cousin, the plywood I-beam) are killers, just like bowstring trusses. The tactics developed and used in older-type taxpayers cannot be used in these buildings. The only suitable tactic involves complete evacuation and use of a master stream from outside at ground level to blast the ceiling away and extinguish the fire simultaneously (more on this later). Attempting an interior attack on a serious fire that has involved the trusses is like playing Russian roulette with three bullets in the cylinder.

General Problems with Commercial Fires

Commercial buildings differ from residential buildings in several ways that affect the crews of first-alarm engine companies. Generally, stores consist of larger, undivided areas as compared to residences. A typical store is 20 ft wide and 75–100 ft deep, with a ceiling height of 10–15 ft. Thus, fire has access to a larger area without being obstructed by walls. This creates bigger fires that extend rapidly. Additionally, the fire loading per square foot is

Fig. 15–8 Store fires are very manpower-intensive operations. Roof-venting operations at taxpayers require at least six men and two saws initially. If heavy fire has entered a common cockloft, three or four times as many members and saws may be needed.

Fig. 15–9 First arriving firefighters find heavy fire venting from the middle store in this row, a bad situation due to the need to protect two exposures simultaneously. Handlines are being stretched, and a platform is brought into position.

often much heavier in commercial occupancies. Both of these factors—larger areas and heavier fire loads—require engine companies to use different tactics than they do for house fires (Fig. 15–8).

In residential occupancies, speed is one of the highest priorities in engine company operations, due to the potentially high life hazard. A hoseline must immediately be placed between the fire and anyone who is endangered by it. The smaller fire areas and lighter fire loads of homes permit the use of lighter, medium-size lines, which are easier to stretch. This is *not* the case in commercial buildings. An advanced fire in a commercial building requires large (2½ in.) handlines.

Engine companies arriving under these circumstances must prepare for the long haul, since this is not a simple one-bedroom house fire that will go out with 400–500 gal of water. Each engine should connect to a serviceable hydrant with either their soft suction or via large-diameter (4-, 5-, or 6-in.) supply line. This isn't the place for an in-line stretch of several hundred feet of 2½- or 3-in. supply line. The fact is that multiple large handlines and/or master streams are going to be needed before this is over, so be prepared to supply them. There is little need for speed, either, since there is rarely a civilian life hazard in a store fire.

The use of one or two lengths of LDH by the first-arriving engine has an advantage over a direct hydrant connection in some cases. Although it won't allow quite as many gpm, it allows flexibility in positioning the apparatus. One of the most beneficial places to spot the pumper is in line with and across the street from the fire store. This permits the use of a preconnected master stream or deck gun. The fire that is blowing out the front windows on arrival is an obvious candidate for such a stream. A single person can readily apply upward

Fig. 15–10 Initial attack is begun with a 2½-in. line, as forcible entry is made into the exposed stores on either side to check for extension in the cockloft.

Fig. 15–11 The first attack line is joined by another 2½-in. line, quickly knocking down the fire in the store. Note the lines stretched to the platform as a precaution against fire in the cockloft.

of 500 gpm in this fashion as soon as the apparatus stops. Be certain, however, that the apparatus isn't placed too close, lest the rig become an exposure, and take care not to block out an elevating platform if one is available (Figs. 15–9, 15–10, and 15–11).

If the fire isn't blowing out the windows on arrival, then a handline is needed. For serious fires, this means 2½-in. hose. The 2½-in. line has several advantages over smaller lines at this fire, including the following:

(1) *The volume of the water delivered and the reach of the stream.* A good 2½-in. line delivers between 250 and 325 gpm and has a reach of more than 80 ft. Since the average store is only 75 ft deep, this allows the crew to start putting out fire as soon as they enter the building, cooling structural elements well ahead of their actual location and to continue knocking down fire as they advance.

(2) *Manpower efficiency.* Under these circumstances, a 2½-in. line can be maneuvered with three members. Unlike residences, where bends, partitions, and staircases must be negotiated, the taxpayer or strip-mall store usually has an uncomplicated layout. Flake out the hose in front of the store, be ready for a straight, continuous advance, then call for water. The next stop should be the back door. Two lines of medium-size hose would require at least four members.

(3) *The power of the stream.* Remember that cockloft overhead! It's likely to have fire in it. At times, it may be

necessary to blast a hole through the ceiling to get water on that fire or through a partition that is shielding the fire from the stream. The 1¼-in. solid tip, flowing 325 gpm at only 50-psi nozzle pressure, does a good job of this, improving fire-fighter safety and efficiency.

Fig. 15–12 *The rear of most rows of stores are very heavily secured and require a substantial commitment of personnel and forcible entry tools to vent. This is not the place to send one firefighter with a pike pole to vent.*

Forcible entry difficulties at taxpayer fires are usually greatest at the rear, where there is less light and less traffic, making burglary easier. For this reason, the rear is often well fortified with substantial doors and locks and only small windows, if any. Doors almost always open outward in commercial structures, limiting the usefulness of even HFTs. Rear doors are often secured with fox locks or drop-in bars. These will be evident by the bolt heads penetrating the door. By using a flathead axe and the adze of a halligan tool, you can often shear off the bolt heads and pry open the door. In severe cases, however, and if fire conditions warrant, it may be faster and less damaging to breach a suitable hole right through the cement block by using sledgehammers. Given the difficulty and importance of the task, at least one ladder company should be assigned to the rear of all serious taxpayer fires (Figs. 15–12 and 15–13).

Simply locating the correct store in the rear can be a difficulty in itself. In the front, the officer in command has the benefit of signs and advertising to go by, as well as large show windows to see through. This isn't usually so in the rear. He may order a line taken into the video store with no problem, but a command to vent the rear of this store may result in a great deal of confusion. What is needed is a way of identifying each store at the rear in conjunction with the front. One system that

Fig. 15–13 *The rear of this taxpayer has no signs to let firefighters know which store is which.*

works well is to designate the far-left store (as seen from the front) as store A, and each subsequent store as the succeeding letter. Ask the property owner's permission to stencil that designating letter on the rear wall over the doors and/or windows to each occupancy. Make the letters large and bright enough to stand out at night. This is often better than marking the name of the store, since ownerships are prone to change, but the spaces usually stay the same. In many cases, too, storeowners might object to having their names on their rear doors, since it helps burglars find the high-value occupancies.

The front of many taxpayers is often the easiest area to enter. Often, the door and show windows are of plate glass. For a rapidly extending fire, simply smash the glass. That might be the fastest means of gaining entry. Before doing so, however, look for signs of an impending backdraft—heavy smoke, highly heated windows, no visible fire, smoke issuing under pressure with occasional puffs, or smoke being drawn back into the building. Carefully check these signs at late-night or early-morning fires, where the fire may have been cooking for several hours since closing time.

Fig. 15–14 Force entry in a way that maintains the integrity of the door. If the fire has not vented the door or windows, firefighters should usually not break them until the hoseline has begun applying water on the fire.

Breaking the glass door or window in this case may admit enough oxygen to cause a backdraft explosion, blowing you and your company all the way across the street. A possible backdraft calls for proper coordination of all members. Forcible entry and ventilation at ground level must be delayed until members on the roof have created a good-sized hole and fire is venting from the hole. If a hole is made and no fire shows, you may be venting an area that isn't subject to the severe backdraft conditions that are showing at the front. While the roof ventilation is being done, position and charge 2½-in. hoselines of sufficient length, but keep them clear of the involved store in case a backdraft blows out the plate-glass windows. After the roof venting has been assured, a member with a hook should take out the front windows, standing off to the side of them. The stream may then be directed onto the fire and slowly advanced (Figs. 15–14 and 15–15).

Backdraft conditions are more readily achieved in occupancies that are protected with steel gates, since a security gate can hide a fire in its incipient stage. Security gates also delay forcible entry operations and prevent water from being applied through the windows. When arriving at a row of stores that are all secured for the night, it is a good idea to have the forcible entry team go ahead and force the gates and doors of all of the exposed stores in the row, rather than immediately entering the fire store. In this manner, lines may be readily stretched to cut off extension, if needed. Once one gate has been forced, keep the saw running, or the torch burning, and open the rest of the stores that could be exposed.

Fig. 15–15 This photo illustrates excellent taxpayer tactics. Note: The 2½-in. attack line leading into the fire store, the 1¾-in. line entering exposure 4A, and the member about to vent the front show windows after the lines have been positioned.

These gates should also start you thinking about another recent development that creates the following dangerous firefighting conditions: steel plating on the roofs and walls. In high-value occupancies such as jewelry and camera stores, burglars have been known to go through the walls and roofs, taking the easy way around the gates. In this case, ⅛-in. steel plate is used to make the entire store a veritable vault. If you encounter such plating, extreme temperatures and early collapse are likely results. A change in tactics would be warranted. In this case, and in other potential backdraft situations where roof ventilation isn't possible or will be delayed, create a small triangular opening in the roll-down gate, insert a 2½-in. fog nozzle, and apply water immediately. Direct a narrow to medium power-cone fog stream upward toward the ceiling so as to produce maximum quantities of steam, the indirect method of attack. Be prepared for a violent out-rush of steam under these circumstances. Allow this stream, and several other streams, if possible, to operate for several minutes before opening the gates.

Cellar Fire Operations

Cellar fires in taxpayers are among the most severe challenges that a firefighter can face. The ceiling heights are often low, causing a rapid buildup of heat, even at the floor level. Often, since the area is only used for storage, there is only one cellar stairway. This is usually found in the rear of the store. Occasionally it is reached by means of a trap door in the first floor, and a slide or set of rollers may cover part of the stairs to assist in moving stock. There is often a wide variation between the floor layouts on the first floor and in the cellar. The 100-ft-deep pizza parlor may only require 30 ft in the rear for storage, whereas the adjoining paint store may have an L-shaped cellar, wrapping around the front 70 ft of the pizza parlor. While the stairways to these cellars may be limited, there are often conveyor openings between the floors by which fire can spread upward. The heavy fire loading, the mazelike storage conditions, an absence of quick ventilation, and the lack of a sprinkler system often result in complete burnout once a fire starts in these cellars. The only chance of success is to combine a fast, high-volume attack with immediate cutting of vent holes in the first floor (Fig. 15–16).

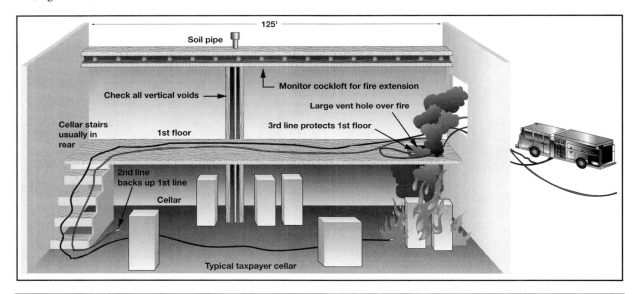

Fig. 15–16 Cellar fires in stores require a coordinated effort and a determined crew to defeat.

Since ventilation will be almost nil initially, using a fog stream should be avoided at all costs, lest it pressurize an area ahead and drive fire back toward the stairway around the sides of the pattern. Use of a fog stream in a highly heated, confined area will also result in large quantities of steam being produced, which is likely to drive the attack team from the cellar. A 2½-in. hose with a 1¼-in. solid tip will usually provide sufficient volume and reach to penetrate and extinguish the seat of the fire, without creating large quantities of steam. When preparing to enter a cellar make sure that you have at least twice the depth of the store plus one length when you stretch the hose, or else you may run short. *Always* charge the line before going down a cellar stairway.

If it is possible to advance the hose deep into the cellar, position a second 2½-in. line at the base of the stairs to protect the escape route of the first line. Also position a member at the top of the cellar stairs to warn the crews in the cellar of any fire lighting up in the stairway or on the first floor. You should begin to cut large ventilation holes, 4×4 ft minimum, as soon as a serious cellar fire is discovered. This means that an additional line will be required on the first floor to protect this opening. The holes should be cut as near to the show windows as possible but out of the routes of ingress/egress so that no one will walk into one. The position just inside the show windows normally puts the vent opening directly opposite the hoseline's entry from the rear stairs, drawing fire and heat away from the line, thereby allowing quicker advance.

A fan blowing fresh air down the cellar stairs behind the attack crew can also greatly aid their advance, but as with any PPV, a sufficiently large ventilation opening must be available opposite the fire. Otherwise, the fire will be blown into the wall and ceiling voids, then up into the first floor and cockloft. By placing the exhaust opening just inside the first-floor show windows, the gases may be vented right out of the windows by a fog stream. This is essential to prevent erupting fire from extending to the first floor. The gases may also be vented through the broken show windows.

You will probably only get one shot at this operation. If you don't darken down the fire in 10 min, you had better plan a new strategy. Cellar pipes and distributors can do for you what the handline can't: darken down heavy fire at the end of the cellar that is opposite the staircase. If the handline attack has been pushed back, don't be afraid to change tactics. Order the handlines withdrawn to a safe area, inside a protected stairway or to the first floor, then place a distributor and/or a cellar pipe in operation over the main body of fire.

Prepare defensive operations at the same time. After the cellar pipe or distributor has operated for 1 or 2 min, shut it down and try to advance the handlines again. If the cellar nozzles did their job, the heavy fire will have been darkened down and the handlines will be able to advance. If not, loss of the structure may be impending. High-expansion foam introduced through as many access points as possible may control the situation. If not, position as many master streams as possible to flood the entire first floor—the cellar is doomed. Be sure to position lines in adjacent cellars to stop extension there, which is likely through the bays between the first-floor joists. The cellars of adjoining stores will also require ventilation.

Masonry Floor Construction

A serious threat to firefighters occurs when masonry floors are placed over wooden floor joists. Terrazzo floors, common in many older drugstores, are typical of this condition. Expect newer versions of this hazard in laundromats, where tons of concrete are often poured on top of wooden floor joists as a base for washing machines. Firefighters have become accustomed to thinking of concrete and masonry as being fire-resistive construction. In a literal sense, it is, since concrete doesn't burn. High-rise buildings built of poured concrete have an excellent record of resistance to fire and collapse, and it is that record that firefighters think of when they encounter concrete. Still, that record is based on one type of masonry—steel-reinforced poured concrete, which behaves very differently under exposure to fire than do other types of concrete and masonry construction. Firefighters need to understand the various types of masonry to predict its behavior under fire conditions (Fig. 15–17).

Fig. 15–17 Cellar fires in buildings that have masonry floors over wooden joist are a recipe for disaster. Such floors collapse without any warning other than the presence of the masonry in a Class 3 or 5 building.

Concrete is a mixture of Portland cement, aggregate (sand, gravel, or stone mixed with cement), and water. When concrete is allowed to harden, it forms a very dense material that is very strong against compressive loads. Concrete has very little strength, however, against either tensile loads or shear loads.

Terrazzo and tile are two other types of masonry floors that firefighters commonly encounter. Terrazzo is a type of decorative concrete in which polished marble chips are the aggregate. Tile floors are set on a base of concrete, known as *mud*. Both types of floors behave the same as other unreinforced masonry floors.

To make concrete and masonry useful for applications like floors, which result in tensile loads in the slab, requires some method of resisting such loads. In modern construction, steel rods, cables, or beams are stretched within the form, then wet concrete is poured around them. When the concrete hardens, the steel acts to resist any tensile loads, holding the concrete together. In turn, the concrete acts as a shield to the steel in the event of fire, since steel rapidly loses its strength when it is exposed to high temperatures. This is known as reinforced concrete.

Older construction methods, still commonly encountered in brick-and-wood-joisted (Class 3, ordinary construction) and wood-frame (Class 5) buildings use wood to support the tensile loads. Concrete, terrazzo, or mud tile floors are poured on top of wooden floor joists for the following variety of reasons: to provide a low-maintenance floor surface, to provide sound or temperature deadening, or sometimes to simplify the leveling of a sagging or an uneven floor! The obvious problem is that, if the floor joists are attacked by fire, the concrete has nothing to help it resist collapse. In fact, the concrete compounds the problem. Its insulating properties may prevent firefighters from realizing that they are operating directly over a raging fire, and the weight of the masonry adds to the potential for early collapse. Such was the case at the 23rd Street Fire, where 12 NYC firefighters, officers, and chiefs were killed when the floor collapsed under them. Similar tragedies have struck all over the country, killing dozens of firefighters in the process.

Concrete, terrazzo, and tile floors each weigh about 150 lbs per cu ft. A 4-in.-thick slab of concrete resting on wood joists adds nearly 50 lbs per sq ft to the load that the beams are carrying. A room 10 ft long and 10 ft wide will have 5,000 lbs of dead load added to its supports before a single piece of furniture is set upon it. This situation is often compounded by a desire to maintain an even floor surface, avoiding a step-up at the entrance to a room with masonry. In one deadly practice that was common in the past, the floor joist was reduced or cut down in the area where the mud floor was to be poured. If the rest of the floor was supported by 2×8-in. joists, some contractors would use 2×4s under the masonry, allowing 4 in. for the concrete and tile to rest while still fitting flush with the adjoining floor surface. These floor joists are dangerously undersized in an area that is now supporting an ever-greater load. You should *expect* catastrophic floor collapse if fire attacks these joists. Unfortunately for firefighters, this condition isn't readily visible from above, and anyone who sets foot on such a floor does so unknowingly.

Experience has shown us that these conditions are likely to be found in specific places in various occupancies. In apartment buildings, for example, the bathrooms often have tile floors. This is one area to observe carefully for signs of collapse. Avoid that area if collapse seems imminent. In PDs, tile is likewise commonly found in bathrooms, as well as in kitchens and rear- and side-entrance foyers. In stores, the front sales areas often feature terrazzo or tile, whereas this relatively expensive feature tends to be omitted from stock areas. One particular occupancy where large concentrations of concrete will *always* be found is in laundromats. The commercial washing machines produce such high torque while operating that they must always be bolted down to at least a 12-in. concrete slab. At times, this slab is properly supported on reinforced concrete pillars, but at other times, it has been found to be entirely dependent on the wood joists for support (Fig. 15–18).

Fig. 15–18 A concrete base for washing machines in a laundromat places tons of weight on wooden floor joists. A serious cellar fire below this area should prompt evacuation of the store and alternate tactics for applying extinguishing agent to the burning cellar.

The single most important action to take is to identify those masonry floors that are supported by wood joists or, just as deadly, by unprotected steel beams. The best time to discover this condition is while performing building inspections or during prefire surveys. Include any such discoveries in a critical information dispatch system (CIDS) or CADS-type program, or enter them into a hazardous building file, clearly stating the location of such conditions. For example, "8-in. terrazzo floor on wood joists throughout the first floor" or "tile/mud floor on wood joists in second-floor bath, NE corner."

Arriving units should carefully evaluate the floor decking directly over the fire. If visibility permits, tile and terrazzo are readily identified. When smoke reduces visibility, however, we must resort to the sense of touch. Take a heavy metal tool, such as a halligan or an axe, and forcefully strike the floor. Wooden floors produce a hollow sound and lend a significant bounce to the tool. Masonry floors produce a pronounced *clank* and give a little bounce to the tool, as well as significant vibration if the tool is all metal.

The discovery of a masonry floor above the suspected fire location must be immediately relayed to the IC for evaluation. If conditions indicate the existence of a serious fire below a masonry floor supported either by wood or unprotected steel, all personnel must be removed from the area directly over the fire until the fire has been knocked down and the structural integrity of the floor has been determined. Lacking prefire information causes the IC to operate blindly.

A discovery of masonry flooring doesn't automatically mean that collapse is imminent. Knowledge of the support system below is required. If the size-up of the building indicates that the structure is of fire-resistive construction, *i.e.*, it is more than 75 ft high, then the masonry floor is likely to be reinforced concrete and therefore safe for operations. If the size-up indicates ordinary construction (Class 3) or wood-frame (Class 5) construction, then unreinforced masonry is most likely. An old two-story taxpayer, for example, was *not* built with reinforced concrete. Operations must proceed with caution based on the assumption that early collapse is likely. An examination opening can be created at a safe area (remote from the seat of the fire or from an entrance doorway) using sledgehammers or jackhammers.

If there is any indication that heavy fire is attacking wood or unprotected steel joists supporting a masonry floor, employ defensive tactics. For cellar fires, high-expansion foam applied through as many areas as possible may control the fire. Cellar pipes or distributors, applied through holes breached from safe areas, may also prove useful. You may also breach adjoining cellar walls to provide access for a stream, although this will create a possible route of extension if the crews are forced to withdraw for any reason.

Once the fire has been knocked down, keep personnel off the suspect surface, as well as out from under the area, until sufficient lighting and ventilation permit a better examination of the floor. Several of the recent collapses, including at least one firefighter death, occurred *after* the fire was knocked down, during secondary searches and overhaul. If it is absolutely essential to operate on or under a slab that has had major damage to its support system, shore up the suspect area.

Collapses are rarely entirely unpredictable. In most cases, the causes of collapse have been known for years. Buildings outlast the careers of even our most senior members. The factors that cause collapse, such as masonry-covered wooden floor joists, or bowstring truss roofs, have existed for decades and even longer in our older cities. There is very little excuse for not knowing about a situation that is likely to kill firefighters if the building has been standing there for any length of time. In the past, one of the highest compliments that could be awarded to a firefighter or an officer was to say, "He really knew his district and its buildings."

Unfortunately, when such a member retired or otherwise left an area, the information that that member had compiled often went with him. With various information retrieval systems available, there is *no* reason for this information not to be available to everyone, from the *probie* (probationary firefighter) to the IC. Go out into your buildings and evaluate them. Be sure that any information about their hazards is forwarded up the chain of command for further action. You may or may not be able to get some of the conditions corrected, but you can at least inform every firefighter who steps into those buildings what the dangers are. The first step—identifying the hazards—is up to you.

Store Fires

Fires on the ground floor are somewhat easier to get at than those in the cellar, but they are just as much of a problem. It is a simple task for fire to reach into the cockloft from the store—then you're in for a race. The tactics for fighting fires in a ground-floor store include using large (2½-in.) lines for moderate fires, venting the cockloft directly over the store and pushing down the ceiling to prevent horizontal extension, and positioning handlines in exposed stores. When manpower permits, one tactic that works well is to have a third medium-size (1½-in., 1¾-in., or 2-in.) line follow the 2½-in. lines, washing down areas that the first hose missed. This is not a back up line, which should be 2½ in. This line is used for overhauling.

For heavy fires in stores, expect the fire already to be in the cockloft. Forget about the involved store and position the handlines several stores down, ahead of the cockloft fire. Be sure to bring in sufficient numbers of long (10-ft) hooks to get the ceiling down rapidly for the entire depth of the store and at least 3 ft wide. You must get ahead of the fire and push it back. Be aware that two or three lines (1 ½-, 1 ¾-, or 2-in.) may be required in each store where you elect to make your stand. A single line can't hope to protect a 75- or a 100-ft-deep store. The main body of fire may be attacked with a preconnected deck pipe or some other master stream. When positioning pumpers for this purpose, don't block out the telescoping platform. Taxpayer fires provide one of the greatest uses for aerial platforms, if used correctly. Remember that the biggest extension problem is going to be in the cockloft. A heavy body of fire in a store can easily require 800–900 gpm to darken it down. By placing the basket of the platform down at the sidewalk, directly in front of the show window, you put in place the most highly mobile 1,000-gpm-plus master stream available—one that can darken down the cocklofts of the adjoining stores and then be rapidly positioned to the front of the involved store with a minimum of time and manpower (Figs. 15–19 and 15–20).

The emphasis should be on getting ahead of the fire and working back. Remember, don't be drawn immediately to the open burning store and neglect the exposures. Nor should you worry about pulling ceilings for the stream. If a 2-in. tip at more than 100 psi can't open the ceiling, you won't be able to do it very readily with hooks, either. Water damage to the exposed stores isn't a concern at this stage. You will put out the fire and save the remainder of the row. If you don't see some improvement fairly soon once you have gotten these streams operating, something is very unusual. It means that the streams may not be hitting the main body of fire, or that there is an exceptionally heavy body of fire, possibly from flammable liquids or gas containers.

Fig. 15–19 The following is an ideal situation for use of the deck gun: heavy fire late at night, no life hazard, no parked cars to hinder apparatus placement. Spot the pumper directly in line with the involved stores 30–40 ft out to leave room for an elevating platform if available.

Fig. 15–20 These firefighters, advancing 1¾-in. hose into a heavily involved store, could end up outgunned. Take the 2½-in. hose for store fires.

Monitor the conditions of the structure, particularly the parapet above the show windows, and be prepared to move the platform basket further out into the street, out of the collapse zone. Don't get stampeded into putting the tower ladder, ladder pipe, or other stream onto fire that is burning above the roof unless it is threatening exposures. That fire is over as far as the building is concerned. It is going away from the danger area. By directing a stream into a hole in the roof, you drive fire back in below that roof. Remember what roofs are built for one thing—to keep water out. Your stream cannot hit any of the fire that is burning up under the roof. All it can hit from the top is a small area directly in line with the opening. By using the water from below, you can hit the fire that is burning up under the intact areas of the roof. If this effort fails, I guarantee you that the roofline attack would have failed as well. All of the water that you would have thrown at it would have been strictly for public relations.

Cockloft Fires

Fires that begin in or extend into the cockloft are probably the major causes of total loss of taxpayer buildings. The rapidity with which fire spreads across this open area can be quite surprising. The fire then drops down into all of the exposed stores or blows the ceiling down on top of firefighters in a backdraft. The following factors compound this problem of fast fire spread: difficulty in exposing the cockloft due to multiple suspended ceilings, potential backdrafts within the ceiling space, and difficulty in locating the seat of the fire when smoke is showing in several stores. In addition, cockloft fires pose the dangers of ceiling and parapet collapse, which may not be obvious to firefighters working in a clear area on the ground floor (Figs. 15–21 and 15–22).

A fire that is confined mainly to the cockloft may give little indication of its severity until it is too late. Firefighters have often entered stores to investigate smoke conditions only to find a small amount of fire visible at the floor level. Advancing toward this area, they have been trapped and killed by ceiling collapses. In at least some of these cases, the falling ceiling didn't kill the members—instead, it pinned them until the fire got to them. This is a particularly dangerous threat in older taxpayers that have multiple hanging ceilings.

Quite often, a new ceiling is added as part of the process when these stores are remodeled. In many cases, these new ceilings are simply suspended from the light wooden framework that was designed to hold only the original ceiling. It won't take a lot of fire in this instance to drop the entire load. To avoid this, make a preliminary ceiling opening at the entrance to each area that has a suspended ceiling. Continue to poke upward with the hook until you are sure that you have reached the roof boards. This requires some experience when encountering layers of wood lath and plaster covered by tin. Be persistent and listen for the hollow sound of the ceiling compared to the solid sound of the roof.

Fig. 15–21 Ceiling collapse at a cockloft fire is always possible. Members trapped below the grid-like construction may be unable to extricate themselves before the area becomes involved in fire.

Fig. 15–22 A great deal of effort is required to pull ceilings in stores. The 2x3 ceiling joists seen here were part of the original gypsum board hung ceiling in this store. It was concealed above a newer drop ceiling that had to be pulled first to expose this traveling cockloft fire.

At times, the smoke in the cockloft can be so heavy that it prevents you from seeing the fire above the layers of ceilings. TICs are a vital asset at fires in void spaces such as cocklofts. Check the ceiling area immediately inside the entrance door and continue checking as you advance. If a TIC is not available, you will have to rely on the pike pole or other hook. This requires examining the head of the pike pole after poking it up to the roof. If it's red hot and shows signs of smoking, or if the handle is charred, then you have a lot of fire overhead. Beware of the possibility of a backdraft occurring in the cockloft. This is a real possibility if you see no fire overhead but feel a great deal of heat. You must vent the roof immediately, and you must get some water onto the fire. This is much easier said than done, of course. Pulling multiple ceilings is a very arduous task under good conditions. When forced to operate wearing SCBA, it can become physically exhausting in very short order.

Traditional tactics at cockloft fires call for getting hoselines in place in the exposed stores, pulling ceilings ahead of the fire, and sweeping the cockloft with hose streams to drive the fire back toward its area of origin. Since this is a very manpower-intensive, time-consuming operation, it is often necessary to skip several exposed stores and surrender them to the fire so as to get far enough ahead to complete pulling ceilings before the fire roars past. Under these circumstances, medium-size lines (two or more per store) are needed to permit rapid repositioning. This requires six or more people per store to pull ceilings, an equal number to man the hoselines, and additional crews to perform roof ventilation and attack the main body of the fire.

The hoseline cannot begin to operate, however, until a fairly sizable opening has been completed. Directing a stream up into a cockloft from the floor through a narrow opening often just worsens conditions. The stream may put out a little fire directly in line with the opening, but there is a venturi effect set up in the process as the stream entrains large quantities of air, blowing it into the cockloft, fanning the fire. Get enough personnel with hooks inside to get a substantial hole open before you begin to apply water. It might be possible for the nozzle man to stand on a table or countertop to direct the stream deeper into the cockloft.

In cases where it is impractical or physically impossible to open the ceiling, you must darken down the fire somehow. Remember those cellar pipes and distributors? They can put water into otherwise inaccessible areas below the roof. This can at least buy you time to find other means of reaching the fire. Meat markets and butcher shops are occupancies where this may be the case. In such stores, it is possible to find ceilings of Formica over plywood to facilitate hosing the rooms down for sanitary reasons. In other areas, large walk-in refrigerators or freezers may take up the entire rear of the store. If you can't get water above these areas by conventional means, it's time to try some unconventional approaches, send the distributors and cellar pipes up to the roof and insert them into the cockloft from above. Prior to starting water to these devices, though, be sure to withdraw all personnel from the areas directly below the involved ceilings to avoid potentially serious injuries

One option that has been successful and that is much more manpower-efficient than pulling ceilings manually is to use a master stream from an elevating platform basket directed from the sidewalk through the front display windows and into the cockloft (Fig. 15–23). A properly supplied tower ladder stream is the most versatile, most maneuverable master stream on the fireground. Two members operating in the tower ladder basket can rapidly apply more than 1,200 gpm to darken down even the heaviest store fire quickly. This same heavy stream can then be rapidly moved from side to side and store to store without shutting down. This is something that no other master stream—and not even large handlines—can do. A stream directed upward from the show windows can be used to blast through ceilings, simultaneously exposing and extinguishing the fire in the cockloft. An elevating platform withstands changes in nozzle reaction much better than a portable deluge set, thus allowing you to whip the stream to get the most effect in punching through the ceiling.

Fig. 15–23 The elevating platform stream used from ground level is the most highly maneuverable master stream available. It is invaluable for fires in the cockloft.

One of the key ingredients to a successful operation in these circumstances is to maintain as high a nozzle pressure as possible through a solid tip. For example, if only 800 gpm were available, the usual tip recommended would be a 1¾-in. tip at 80 psi, a good stream for normal operations. By using a smaller 1½-in. tip at 140-psi nozzle pressure, however, the same 800 gpm achieves far greater penetration and reach. It may be possible in some very old taxpayers to remove advertising signs along the front parapet to gain access to the cockloft. This is another perfect task for the aerial platform. Once the sign has been removed, a stream can be swept across the entire cockloft. On many newer strip malls, there is an all-metal overhanging soffit along the front of the row of stores. This is typically one large, open space, and it may interconnect with the cocklofts of the stores. At times, this area has been found to allow fire to wrap right around a party wall, or even a firewall, and enter the stores on the other side. Opening up the soffit with pike poles is extremely difficult. A far faster method is to use a metal-cutting circular saw from the basket of an elevating platform to cut away large sections of the fascia (Fig. 15–24).

One of the most unrecognized problems with a cockloft fire is the collapse of the parapet wall. This parapet is a continuation of the outer wall above the roof level. It is a freestanding, unsupported wall. Where there are show windows, this wall is carried on a steel I-beam across the opening. Typically, this I-beam connects with columns that are connected in turn to other I-beams that run longitudinally through the cockloft, supporting the roof joists. If a fire in the cockloft heats this front-to-rear beam sufficiently, the beam will expand. Remember, a 100-ft steel beam heated to 1,000° will expand 9½ in. lengthwise. If it is restrained in the back of the building, it pushes that front I-beam out. Obviously, this movement can cause the parapet wall over the show windows to collapse.

What isn't so obvious, however, is that these parapet walls are often given a lot of lateral reinforcement, with the result that, once one part falls, the rest of the parapet may go with it. This can spell danger for anyone standing anywhere along the front of an entire row of stores. I have operated at several such collapses of parapet walls—collapses provoked by wind and old age. Unfortunately, at least one woman who had been walking by on the sidewalk was killed and several others were seriously injured. The warning to firefighters should be clear; however, when serious fire involves the cockloft of a taxpayer or is blowing out the display windows, the entire sidewalk on all frontages should be considered within the collapse zone. Entry into the stores should be made after cooling the steel I-beams, but get off the sidewalk as quickly as possible (Figs. 15–25 and 15–26).

Fig. 15–24 Newer metal soffits (the bottom of the overhang) are extremely difficult to open using conventional techniques. Consider using a metal cutting power saw from an elevating platform to open the fascia (front).

Fig. 15–25 The parapet wall above this show window rests on a steel I-beam. The twist and distortion of the beam, or the expansion of interior beams, can cause the parapet to fall.

Fig. 15–26 The brick parapet wall at this taxpayer fell off without warning and without any fire present, killing a pedestrian. Note the large chunks out at the curb line. An OSHA-approved helmet is no protection against such massive weights.

Although the elevating platform basket can be very beneficial for halting fires in the cockloft, they must be operated early enough into the incident to extinguish the fire. If they have been operating for a great length of time into a cockloft, and if they aren't making any headway against an advanced fire, then they too must be pulled out of the collapse zone.

An additional consideration when dealing with these I-beams is their potential to push through solid walls. I watched at one fire as a hole was pushed through a cement-block wall of an entirely separate building, followed by fire. Fortunately, the ceiling had already been pulled and a line stretched in anticipation of such a possibility. That is one of the secrets of success at taxpayer fires—knowing what is likely to happen and having the foresight to take actions to cut it off at the start. Due to the many dividing walls in our way and the lack of obstructions to the fire, we cannot hope to play catch-up in these buildings. Either we jump ahead or we lose. Taxpayer fires are very manpower-intensive operations because of the many remote areas in which operations must take place. Truck operations at taxpayers are particularly demanding.

Roof operations demand at least six members with two saws, four hooks, two axes, and two halligan tools *immediately,* if you hope to cut the required vent holes in time. Similarly, a store or cockloft fire demands at least three lines almost at once. If the resources aren't available to place and operate these lines quickly, while simultaneously pulling ceilings along the entire depth of a 75- or a 100-ft store, the operation may be doomed before it even begins. Recognize this situation and call all of the needed assistance early. Other units can always be sent home if they're not needed.

At the same time, make the most of what you've got. One 300-gpm handline is often more effective than two 150-gpm lines, due to the added reach and penetration, while using the same number of personnel. The larger open-floor areas permit rapid advance of this large line as opposed to smaller lines that are better suited to residential settings. There is no one single answer to every fire problem. Rather, it is up to the officer in command to devise a strategy that will work for each particular occasion. Good luck!

16 High-rise Office Buildings

This chapter was revised to reflect the lessons learned from nearly a century of high-rise firefighting, including eight high-rise fires that occurred in New York City on September 11, 2001. That's right, eight serious high-rise fires, not just two. In terms of high-rise firefighting, September 11th was a watershed event that taught many new lessons, as well as reconfirmed much of what we already knew in this field.

Most of you know a great deal about the collapses of the twin towers. The worst terrorist attack on American soil received such massive media attention, centering on the gut-wrenching images of the two 110-story towers burning and then collapsing, that it seemed to some that that was all that happened that terrible morning in New York. While the impact of two massive towers collapsing will shake the fire service for many years to come, there are other lessons from that day, and from other serious high-rise fires like the One Meridian Plaza fire in Philadelphia that are just as vital to our knowledge of high-rise buildings and fires in them. These lessons are crucial to all firefighters who respond to any buildings taller than six stories, since it is at that point that most building codes require the construction and fire protection features that are the hallmarks of high-rise buildings. In addition, many smaller office buildings, some only three or four stories high, are built with many of these same features, and we can also benefit from a thorough understanding of the problems, strategies, and tactics that work well in their larger cousins (Fig. 16–1).

High-rise buildings are becoming a fact of life for firefighters all over this country and around the world. The

Fig. 16–1 High-rise buildings are sprouting up every-where, many right in the middle of low-rise communities.

urbanization of many areas, with the resulting increase of property values, has resulted in a new firefighting experience for many departments. As the cost of land rises, builders have found it less costly to build up rather than spread out. This has resulted in high-rises in districts of almost every size.

A department that has previously faced only one- and two-story dwellings may now, or in the near future, be faced with the prospect of having to fight a fire above the reach of its longest ladders. The building may not even be in your first response area, but if your department is a neighbor, you can bet your department will be needed in the case of a serious fire. That fact alone should prompt a major reevaluation of your potential involvement and need for planning for these events, since these fires are far different from any other fire you have fought before, even including fires in high-rise residential buildings. To deal with this problem, fire officers must implement new strategies based on the construction and facilities of the building. Simply bringing in a large number of firefighters whose experience, training, and equipment is limited to room and contents bedroom fires is not likely to produce a good outcome. A high-rise office building fire requires special tactics and equipment, such as 1-hr SCBA cylinders, 2½-in. hose and solid tip nozzles, items that must be prepared in advance of the fire.

Fire safety in high-rise buildings must be the result of concerted efforts by many parties. Architects have attempted to solve high-rise fire problems by using fire-resistive construction, vertical enclosures, and compartmentation. They are defeated, however, by flammable furnishings and finishes, as well as pipe chases and other openings through which fire might travel. Fire and building departments must strive to ensure that the construction of a building complies with the intent of the code. They must assure through inspection and enforcement that fire safety measures aren't circumvented through faulty workmanship or alterations. The building's occupants are the weak link in the chain of fire safety.

High-rises typically serve as hospitals, hotels, apartments, and offices. In any case, a great number of people may be exposed to danger if proper engineering, education, and enforcement are not provided. The occupants of a given building must be made aware of the reasons for, and the necessity of, proper fire prevention procedures, as well as what actions they should take in the event of a fire as follows:

1. Know how to transmit an alarm.

2. Know how to alert other occupants.

3. Know to avoid elevators in the event of fire.

4. Know how to prevent smoke from entering their rooms by sealing cracks and closing vents.

5. Know how to obtain fresh air if they become trapped in a room.

6. If these preventive measures fail, the only thing between the occupants and the fire is an aggressive fire attack based on sound principles.

High-rise strategic plan

A building doesn't have to be 30, 40, or 50 stories tall to be considered a high-rise. Any fire-resistive building exceeding the length of the available ladders—or buildings in which all firefighting must be done from the interior because of a lack of windows—must be accorded the same strategies as are used in tall high-rises. These strategies include the following:

1. Determine the specific fire floor. Often the only information available is very general—for example, smoke on the fifth, sixth, seventh, and eighth floors.

2. Verify the location before committing handlines. This is extremely critical. You could end up having the hoseline stretched to a location only to find the fire originating several floors below or above. Early one morning, I arrived as part of the first-alarm assignment to find heavy smoke on the seventh floor of a 70-story office building in Rockefeller Center. After about 10 min of searching, the fire was located—on the 10th floor! It went to a fifth alarm!

3. Begin controlled evacuation. This step actually has several phases as follows:

    A. Evacuate those people who are in immediate danger. In Class 1 high-rise buildings, this usually means only the fire floor and the floor above, at least initially.

    B. Prevent panicky exit by those not endangered by the fire. Often all of the occupants of a building will begin evacuating at the sound of a fire alarm. In light of the televised tragedy of the thousands trapped above the fire in the WTC and subsequently forced to jump or die in the collapses, total uncontrolled evacuation should be anticipated at future high-rise fires, since many people seem to think that all high-rises will behave the same way. Some of the results of this uncontrolled evacuation are clogged, often smoky stairwells; physically unfit persons walking considerable distances; and a near-riot scene at the ground floor when several hundred people try to enter the lobby, clamoring for news of what is happening. Some means of preventing this include the following:

        • Prefire education of building occupants as to actions they should take.

        • Zoning of the fire alarm system so that alarm bells only sound on the fire floor, one or two floors above, and at the lobby or fire command station.

        • Public address speakers in all stairwells, elevators, and public areas, through which building personnel or fire officers may give directions for evacuation.

    Request extra police presence to assist with crowd control and lobby evacuation at any serious high-rise fire.

    C. Search the fire floor and *all* floors above the fire. The floors may be extremely large, with many occupants. The occupants may include physically disabled persons who need assistance. Be sure to search bathrooms, elevators, stairways, and small offices with closed doors. This requires an extremely large commitment of manpower—a minimum of two firefighters per floor for small-area fires, more for larger areas or heavily charged floors.

4. Gain control of the building's systems.

    A. Place elevators under manual control and immediately return them to the lobby, where they must be searched.

    B. Shut down HVAC systems and place them in a non-recirculating mode (all dampers open to the outside air).

    C. If available, use the building's public address, telephone, and radio systems to get information from the fire area and to assure and direct other occupants.

    D. Ensure that the building pumps are operating to supply proper pressure to standpipe and sprinkler systems.

5. Confine and extinguish the fire. Extra coordination is required in these buildings to ensure that ventilation and hose streams complement each other; otherwise, the fire may drive the firefighters off the fire floor. A large number of personnel are needed to transport equipment to the fire area as well as to relieve firefighters who have been subjected to extreme punishment (Figs. 16–2, 16–3, and 16–4).

Fig. 16–2 Peripheral air supply systems at the perimeter of buildings are a likely area for fire extension to the floor above; they should be inspected for hidden fire.

Fig. 16–3 Flexible ducts feed fresh air in the plenum of the floor below. A fire on one floor will rapidly extend upward through this system.

Fig. 16–4 Standpipe telephones and floor warden stations should be linked to the fire command post (CP) in the lobby. These hardwired systems are often more reliable than fire department (FD) radios.

Types of High-rises

The strategy used to control a high-rise fire depends to a large extent on the building and the effects that its features have on the travel of smoke and fire. Building features vary drastically, depending on the age of the structure. Much like what has happened to Class 5 wood-frame construction over the last 20–30 years, Class 1 construction has evolved into what should properly be divided into the following three distinct categories: Class 1 heavyweight, Class 1 medium weight, and Class 1 lightweight buildings. There is currently no provision in any code that I am aware of for such a distinction, but the differences are certainly present. It is up to the fire service to demand such formal categories (Figs. 16–5, 16–6, and 16–7).

Fig. 16–5 The older high-rise is a very fire-safe structure.

Fig. 16–6 Modern glass towers have the potential to be real towering infernos.

Fig. 16–7 Lightweight high-rise buildings have failed completely in the face of severe fires. Total collapse of a high-rise must now be considered as at least a possibility at every advanced fire in a steel skeleton high-rise. Knowledge of the buildings support system (trusses) and type of fireproofing is required.

In broad terms, these designs fall into the following two eras: pre-World War II and those built afterward. The critical nature of this distinction between styles was made very plainly on September 11, when three lightweight Class 1 buildings suffered total catastrophic collapse as a result of fire. Many people, especially the proponents of continued building of lightweight Class 1 buildings, claim that the collapse of number 1 and 2 World Trade Center was the result of the impact of the jet planes and the ignition of thousands of gallons of jet fuel. While the impact of the planes certainly compounded the likelihood of collapse by removing a good portion of the structural support and probably knocking loose the spray-on *fireproofing* applied to those that remained, that is not what brought the towers down. The inescapable conclusion of this tragic event is that it was a *routine* office building fire that collapsed those towers. Much of the jet fuel was vaporized and burned off during the immediate impact in tremendous fireballs. The rest of it likely burned off in a matter of 10–15 min. The rest of the fires that brought the towers down were from the interior finished and furnishings found in every modern office.

In my studied opinion, the FDNY, including myself, was lucky that a similar collapse of one or the other tower had not occurred at some earlier point in the buildings 30-year life span. The towers were originally built without automatic sprinkler protection for most office floors. Only areas such as shops, storage areas, and public assembly areas like restaurants had sprinklers installed initially. I know—I designed some of them. The buildings were also built with another major fire protection weakness, open access stairs that connected as many as three floors, which would have permitted rapid fire extension from floor to floor from any serious fire that did develop.

Over the course of the 30-year life of these buildings, several serious fires did occur, including several that I operated at, but good fortune and the hard work of the FDNY prevented fire from ever reaching the point where collapse was ever considered a serious threat. In 20/20 hindsight, we probably should have known that collapse was a greater threat than previously thought. In the first and second editions of this book are pictures of the WTC's large open floor spaces clearly showing the lightweight bar joist floors and the access stairs. (In May of 2001, the FDNY issued an amendment to its store fire tactics bulletin declaring that roofs of buildings built with such bar joist construction were too dangerous to cut, at least partially due to the collapse potential.)

A fire at One New York Plaza in 1970 should have served as a warning, since an entire office floor of a *medium-weight* office building (built of steel girder and columns, not trusses, and protected by spray-on fireproofing) had caused the floor over the fire sag as much as 18 in. and required the replacement of more than 150 floor beams that had buckled or torn (Fig. 16–8).

A second warning came in February 1991, when fire engulfed eight floors of the 38-story *medium weight* One Meridian Plaza building in Philadelphia. About 11 hrs after the discovery of the fire, and after three of Philadelphia's bravest had already been killed fighting the blaze, a structural engineer warned the IC of the potential for a pancake collapse of the involved floors. Faced with this evaluation, and having already lost three members, the IC ordered his crews to withdraw. The fire was eventually stopped by a well-supplied sprinkler system on a lightly

Fig. 16–8 The presence of long span lightweight bar joist floor trusses, protected by spray-on fireproofing, seen here in the WTC, should be considered a grave danger in the event of a serious fire.

loaded floor. While the building did not collapse catastrophically as the WTC towers did, it was so severely damaged that it had to be demolished several years later. Unfortunately, because it did not collapse, the lesson of how close we had come was lost.

The validity of the argument for three categories of Class 1 construction was made during two other high-rise fires on September 11th—7 WTC and 90 West St. Completed in 1987, 7 WTC was a 47 story *medium-weight* high-rise. The building was struck by flaming debris from the collapse of 1 WTC (the North Tower). The impact did serious damage to the South facade of 7 WTC and started fires on several floors. The collapse of 1 WTC also sheared a large water main on West Street, resulting in a near total loss of water pressure in the lower Manhattan area. What water that was available from the city's fireboats was being directed at other severe exposure hazards that were not as severely compromised and on the vicinity of the towers where hundreds of rescuers were believed trapped. Since 7 WTC had been evacuated earlier, due to the threat of further aircraft attacking, and since there was no viable water supply immediately available, the commander in charge of 7 WTC ordered firefighters to withdraw. Seven hours after ignition, 7 WTC suddenly collapsed in pancake fashion.

The preliminary federally sponsored probe of the collapses of September 11th, titled "World Trade Center Building Performance Study" (Federal Emergency Management Agency [FEMA] 2002), theorizes that the collapse of 7 WTC was largely due to a fuel oil fire on the 5th through the 7th floors. Emergency power generators were located on the 5th Floor, and the storage tanks were in the cellar, with oil pumped to the generators under pressure at a rate of up to 75 gpm. In this theory, the building might have withstood a normal office building fire, but the investigators feel that the high rate of heat release from the possible fuel oil fire was responsible for the collapse.

I dispute this theory, based on my close personal observation of the building for approximately 45 min prior to the collapse. While I did not have a good view of the south side of the building, I did see the north, east, and west sides. On my arrival at the West Street CP that day, I was ordered to take command of operations inside the Verizon telephone switching center that stood just across Washington Street from the west façade of 7 WTC. Fire in the street (from the burning debris from 1 WTC) was threatening to extend into the telephone building. Using standpipe lines supplied by gravity tanks inside the telephone building, several engine and ladder companies successfully fought to protect that exposure. As I supervised this operation, I had ample opportunity to view 7 WTC close up for about 45 min until I ordered the companies to tie the lines in place as unmanned streams and withdraw to interior positions. This was due to my perception, confirmed by the West Street sector commander that 7 WTC was indeed a potential collapse threat.

Units were ordered withdrawn from surrounding facing exposures, including the telephone building. I was then ordered to proceed out into the debris field where the twin towers once stood to clear the collapse zone around the 550 ft perimeter. As I would approach groups of firefighters and order them back because of the danger of collapse, they would look at me and their usual reply was "Chief, we're a good 150 ft away from that mess, we're OK here aren't we?" What they were looking at was the 9-story high U.S. Customs House (6 WTC) which was heavily involved in fire. What they couldn't see through the smoke from the customs house was the 500-ft high hulk of 7 WTC which loomed over them, an unseen danger. When a momentary shifting of the wind permitted a glimpse of the larger building, many faces instantly registered recognition of the danger.

Still, it's a major challenge trying to clear a 500–600 ft wide collapse zone in an area where hundreds of cops and firefighters are missing and thousands more are searching for and finding their bodies. About an hour after the immediate area was finally cleared, the roar that was unknown to man until that day happened for the third time. In about 15 sec, 7 WTC went down. Fortunately, the evacuation had been complete and the building, like number 1 and 2 WTC had done, fell nearly straight down, causing only a fraction of the additional damage that a lean-over or toppling collapse would have caused (Fig. 16–9).

My observations of 7 WTC prior to the collapse showed only signs of an ordinary Class A combustibles fire, grey to brown smoke. I saw no evidence of large quantities of black smoke that one would expect from a large Class B fire involving fuel oil, especially not on the lower floors. I remain convinced that a fire in ordinary office supplies and furnishings brought down 7 WTC, like 1 and 2 WTC. Admittedly, structural damage played a role in all three structures, as did simultaneous ignition of several floors, but the bottom line is that any fire department that finds itself facing several fully involved floors of a modern office building must anticipate the possibility of total structural collapse as one of the potential outcomes.

Fig. 16–9 Looking up at the West façade of #7 WTC approximately 1 hr before its total collapse, one can see signs of a standard office building fire feeding on Class A combustibles. No evidence is visible of a large fuel oil fire on the lower floors.

In the future, Incident Commanders at fires on the magnitude of the First Interstate Bank, One Meridian Plaza, One New York Plaza, and similar *towering infernos* must give this possibility credence, and the IC must have extensive knowledge of the building at her disposal to facilitate her decision-making. Does that mean that every office building fire is going to lead to collapse? Certainly *not*! Knowledge is the key.

Old style high-rise construction

Prewar buildings were often overbuilt, reflecting the uncertainty of the engineers as to just how much load a support could carry. The general theme was *reinforce everything*. Often a steel I-beam was encased in concrete. Almost forgotten in the events of September 11th is an extremely valuable lesson in this regard: 90 West Street. Built in 1907, 90 West Street is a 25- story high, 125 × 180 ft office building. It sits due south of the WTC site, just across Liberty Street. On the day of the attack, 90 West Street was in the process of being renovated. The entire façade was surrounded by a 25-story high metal pipe and wood plank scaffold that was wrapped in black plastic netting to allow workers to re-point the brick and stone exterior. When the South Tower (2 WTC) collapsed, flaming debris was thrown into numerous floors of 90 West Street, similarly to what happened to 7 WTC from the North Tower. Fires started on numerous floors simultaneously and burned out of control for more than 24 hrs. The building was a severe exposure hazard being located less than 40 ft away from several taller office buildings and hotels.

With the limited water supply available initially, what came to be known as the Liberty Sector command ordered members to concentrate on protecting the exposures, to avoid further extension, lest lower Manhattan become the site of a full blown conflagration. Lines were stretched from fireboats in the Lower NY Bay, supplying handline, master streams, and even standpipes of the nearby exposures (Fig. 16–10). (Normally, pumping salt water into a standpipe should never be permitted, but desperate times call for desperate measures).

The unprotected steel scaffold around the building was a major hazard, being exposed to fire at multiple points, with the wood planks becoming involved at many locations. Chiefs ordered a large collapse zone maintained clear around this building for this reason alone. Everyone expected this scaffold to fall down, but application of hose streams from around the perimeter, operating out of the upper floor windows of the exposures, prevented this from occurring. The hose streams had little effect on interior fire spread however.

All through the night of September 11, into the afternoon of the 12, 90 West Street burned. It was forgotten about by all except those in the immediate sector tasked with preventing further calamity; everyone else was focused on locating survivors and retrieving the bodies of those who had not survived. By the late afternoon of the 12th, the fires inside 90 West Street had subsided to the

Fig. 16–10 For more than 24 hrs, 90 West St. burned after being struck on several floors by flaming debris from the collapsing twin towers. The 90-year-old building remained structurally sound and is about to be reoccupied after being renovated. The taller, more modern building to the left in the photo is being demolished due to mold after exposure to the weather.

point where observers in nearby buildings could begin to see inside the building. What they saw is a valuable lesson in high-rise construction: the building appeared to be structurally intact! There were areas of the façade that had been struck by the collapsing Tower 2 that showed damage, but observation of the floors and columns showed no signs of collapse. After a careful exterior examination using binoculars, a decision was made to enter the structure for further examination, since no searches for victims had yet been conducted therein. Handlines were advanced by carefully controlled groups and fire was extinguished on the several remaining floors. (Several fatalities were located, trapped in the elevators.)

In the FEMA "World Trade Center Building Performance Study," 90 West Street is little more than a footnote, less than 3 pages of a 20-page chapter dealing with "Peripheral Buildings." In my opinion, it is quite possibly the most valuable lesson of the entire day in regard to high-rise fires: high-rise buildings can be built to resist the spread of fire and the threat of collapse! They just can't be built cheaply enough to satisfy the greed of the real-estate industry!

The 90 West Street building is still standing, as is 395 South End Avenue, another heavyweight building that had fire on the second and third floors caused by debris from the South Tower. The South End fire was extinguished by a mere handful of firefighters, aided by building construction that actually served to resist the effects of fire long enough to allow the fuel to consume itself and not fall down.

The prewar era also saw a strong concern for life safety, with numerous means of egress provided, usually remote from each other. This was likely due to the tragic loss of 146 lives in a 1911 high-rise fire, the Triangle Shirt Waist Fire, the lessons of which were lost in the postwar rush for floor space. In prewar high-rises, the stairways were often located

Fig. 16–11 Fire towers (stairways enclosed in a 4-hr firewall and separated from the floor area by a landing that is open to the outside air or a smoke removal shaft) should be used as evacuation stairs and not for fire attack.

in fire towers, that is, a stairway enclosed in a 4-hr firewall, separated from the floor area by a landing that was open to the outside air. Any smoke or fire that blew out from the floor area would vent to the outside or up a smoke shaft rather than enter the fire tower (Fig. 16–11).

The most important prewar building feature regarding fire spread, however, is the absence of central air-conditioning systems that serve more than one floor. Buildings that aren't centrally air-conditioned present far fewer fire problems. First, there are no ducts to pick up smoke at the 3rd floor and spread it to the 1st through the 10th floors. Second, the absence of ducts going from floor to floor means that there are fewer places for hidden fire to travel. Last, there are windows in many of these buildings that can be opened for horizontal ventilation.

The prewar high-rises had the following other features as well that aided fire safety and control: more compartmentation, no hung ceilings or other blind spaces, and fewer electrical and electronic items to act as sources of ignition. (In 90 West Street, the fire on many floors did not spread from the north side of the building to the south side, being stopped by interior walls. This aided in the protection of the Marriott Hotel, located due south of the building.) In short, the older high-rises were originally very safe structures in the event of fire, particularly if the occupants were aware of proper fire behavior, commonly taught in fire drills.

The same cannot be said for the postwar generation of high-rises. As Chief Michael Cronin of the FDNY put it in 1986, while lecturing to a promotion course for NYC fire officers, "The Empire State Building took a direct hit from a B-25 bomber and it was only a third alarm. Today in one of these new buildings, if we have a wastepaper basket burning, it's a third alarm. And if one of these new buildings ever took a hit like the Empire State did, we'd find parts of the building across the East River in Queens." (Chief Cronin knew exactly what he was talking about. Debris from the WTC towers was found across the East River, but in Brooklyn, not Queens, since the towers were located further south than most of the mid-town high-rises.)

New style high-rise construction

September 11 was a watershed event in the fire service in terms of high-rise firefighting. Never before in the 100-year history of high-rise building fires had an occupied high-rise totally collapsed due to fire. In addition to the One Meridian Plaza, One New York Plaza, and the First Interstate Bank fires, several other high-rises had experienced severe fires, including the Andraus Building and the Joelma Building, both located in Sao Paolo, Brazil in the early 1970s.

The Andraus Building was particularly spectacular, as an entire 31-story office building was fully engulfed in flame and burned for many hours until the heavy fuel load virtually consumed itself. Two years later, a fire that began on the 12 floor of the 25-story Joelma building quickly spread upward, igniting all 13 floors above the initial floor in quick succession. This building too had a heavy fire load that was totally consumed, and yet no major structural collapse occurred at either building. Two other buildings in New York City had been struck by large aircraft, the Empire State Building, and 70 Pine Street in lower Manhattan.

That is some of the prologue that led up to September 11: no previous high-rise building fire had ever caused a massive structural collapse. And yet on that tragic Tuesday morning, New York City fire chiefs instinctively recognized the potential for collapse and began to plan for it. In the lobby of Tower 1, a strategy planning session was held between Assistant Chief Joseph Callan, Chief of Safety Albert Turi, and Deputy Chief Peter Hayden. The chiefs discuss the nature

Fig. 16–12 This lightweight floor shows how little mass there is in new buildings. The concrete floor being poured on the metal deck is only 1½ in. thick. Several ¾-in. conduits are buried in the floor, further reducing the slab's overall thickness.

Fig. 16–13 Curtain wall construction consists of panels suspended from the building's skeleton. This creates a gap at the floor levels that fire can travel through from floor to floor.

Fig. 16–14 A view of another curtain wall from the interior. The exterior wall is attached to the concrete floor slab by means of a metal clip every 5 ft.

Fig. 16–15 Looking at the same wall from above reveals the 3-in. gap along the entire perimeter that fire can exploit.

of the event as a terrorist attack and the possibility of collapse. They all agreed on their primary objective, "to get as many people out of the building" as possible and "let it burn up,… we ain't putting this fire out." Within the hour, the first fire-caused high-rise collapse in history showed they were right (Figs. 16–12, 16–13, 16–14, and 16–15).

The major disadvantages of the newer style of high-rises include the use of lightweight building materials including, among other things, spray-on fireproofing in place of poured concrete (the WTC's floors consisted of lightweight concrete on metal decking supported by steel-bar joists). In addition, there is curtain wall construction (where the exterior walls are nonbearing walls attached to the edges of the floors), larger and more wide-open floor spaces, hung ceilings throughout, and core-construction techniques. The impact of these other items will be discussed later in the operations section of this chapter. All firefighters responding to a high-rise structure must be able to evaluate the building for the likely problems it could pose. As a chief officer in the FDNY, I want to know the following key items about the structure in order to determine the strategy I would order, especially if faced with heavy fire in the post-9/11 world:

1. What type construction is involved, heavyweight, medium, or lightweight?

2. Are there any trusses in the fire area?

3. What type of fireproofing is applied to the steel?

Even if faced with another catastrophic fire on multiple floors, I would not be able to tell you now whether I would order an all-out attack or an all-out retreat. That would depend on the type of building, the civilian life hazard, and the nature of the incident including the cause—was it caused by a bomb or impact of a large vehicle, which could have done additional structural damage beyond that caused by the fire. Also, how long is it going to take to accomplish critical tasks such as rescue or fire control? A fire involving the first four floors of a fully occupied office building is going to require a carefully evaluated decision by the IC as to the risks associated with firefighting and rescue efforts. If such a fire would develop in an old style high-rise, such as the Empire State Building, I am sure I would be leading the charge of an offensive attack designed to extinguish the fire and rescue trapped occupants. On the other hand, the same fire in a lightweight building, built with trusses and protected by spray-on fireproofing, might require a more cautious approach, attempting fire knockdown with exterior streams operated with minimal staffing, especially if there are few signs of survivors, and reaching them will take 1½ hrs to walk up 50 floors and another 1 hr to walk back down.

On the morning of September 11, FDNY had no option other than to commit fire forces to the towers to attempt rescue. With dozens and dozens of people jumping 90–100 floors to their deaths, it would have been totally unacceptable to have done anything less than what was done. The public would never have accepted it if a fire chief had ordered his men to stand outside and watch the jumpers without trying to save them. Not only would the public not stand for it, they would have run us out of town on a rail, and rightfully so. That's why they pay for a fire department—to rescue people from fires! On the morning of September 12, that equation changed. We now know that catastrophic collapse is possible. All future ICs at severe high-rise fires (and other terrorist attacks where mass casualties are possible) must weigh their responsibility to the public against their responsibility to their firefighters. This risk/benefit analysis is something we have had to do before at other incidents, it's just that the potential consequences are so high now that the decision carries enormous weight.

At most fires in the modern high-rise, the threat of collapse is rather remote. Instead, the greatest threat to life comes from being unable to perform rapid effective ventilation. Since the late 1960s, architects and building owners have realized that the energy that provides a comfortable environment formerly wasn't being used effectively. It costs a great deal of money to draw air in from the outside, filter out the dust and impurities, heat or cool it, and move it to occupied areas. In effect, this air becomes a valuable commodity. If it can be kept within the building and recycled, the cost of operating the building can be substantially reduced. With this in mind, architects have virtually attempted to seal up these buildings. Windows no longer open and, as an added efficiency feature, the air-conditioning for several floors is often done on one central floor (in the mechanical equipment room).

This presents severe problems in the event of fire. First, if horizontal ventilation is required, you will have to break windows to provide it. Second, and often more critical, is the effect that smoke and heat have traveling through the HVAC system. Even a relatively small fire occurring either within the HVAC system or in proximity to an air intake will result in smoke spreading throughout the building. Smoke will be reported on several floors of the building, requiring lots of manpower and, more importantly, *time* to investigate it and to determine the source. If the smoke is serious enough, it may cause near-panic conditions as several floors suddenly become filled with smoke, and the workers attempt to flee.

Obviously, some means must be found to deal with this spread of smoke. The best method would be to have smoke detectors within the HVAC system that would shut down the system in the event of trouble. Smoke detectors are a requirement of many building codes for high-rises, including New York City's Local Law No. 5. Lacking such required smoke detectors, the IC must achieve the same results manually. To do so, he must have immediate access to the building engineer or maintenance chief. A good policy would be to direct building employees to shut down all HVAC systems immediately upon a report of fire. After the arrival of the fire department, the building personnel can effect measures, under the direction of the IC, to use the HVAC system to direct smoke out of the building. Using an HVAC system for this purpose is subject to many problems and is an extremely complex operation. Using it wrongly can intensify the fire problem and draw it toward safer areas, firefighters, or civilians. Such use should only be undertaken after fire officers, building engineers, and qualified smoke-control design specialists have engaged in serious prefire planning. For an in-depth look at HVAC system management in fire situations, see such sources as the "New York Fire Department's Firefighting Procedures Vol. 1, Book 5, High-rise Office Building Fires," or *WNYF Magazine,* Issue Nos. 2 and 3 of 1983 and Issue 4 of 1984 (*WNYF, With New York Firefighters,* NYFD, 9 Metrotech Center, Brooklyn, NY 11201).

Ventilation at High-rise Fires

You will have to take steps to remove the smoke and heat that build up once a fire starts. Among the options to consider are vertical ventilation up stairways, horizontal ventilation out windows, and ventilation by means of HVAC systems. Any or all of these means may be used, and an option exists to use either natural or mechanical means to move the fire gases. Prior to selecting a ventilation strategy, the fire officer must recall that the following two factors outweigh all others in high-rise ventilation: wind and the stack effect.

Let's look at the types of ventilation available and the advantages and disadvantages of each. As we have discussed, using the HVAC system can provide beneficial results *if* it is properly designed and used. This is especially true in cases of relatively small fires that don't justify breaking the sealed windows. A moderate amount of smoke can be handled in this manner—as long as it isn't so hot as to melt fusible links on fire dampers in the ductwork.

For a large fire, however, it is likely that the volume of smoke produced will overwhelm all but specially designed smoke-removal equipment. It may, in fact, worsen conditions throughout the rest of the area served by the HVAC system. The IC must weigh the benefits to be gained (faster smoke removal) against the dangers posed (risk of spreading or intensifying the fire) before deciding to allow the HVAC system to be used. He must know the following:

1. The exact location of the fire.
2. The floor layout and the location of the stairs.
3. The location of firefighters and civilians.
4. The fire conditions.
5. Whether using the system will endanger firefighters or civilians by drawing fire toward them.

Only after he or she has sufficient information regarding these matters can the IC order an HVAC system to be used (Fig. 16–16).

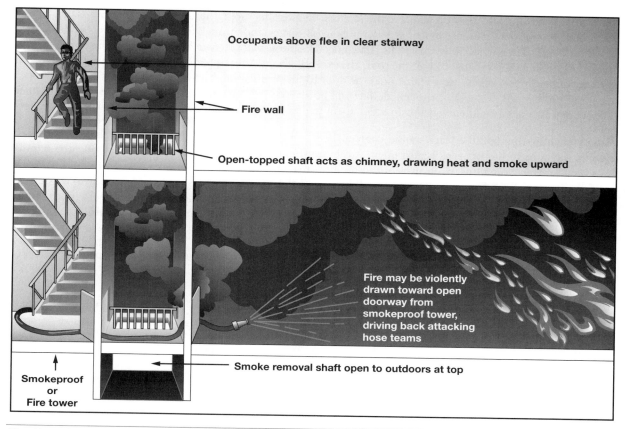

Fig. 16–16 How the stack effect draws fire toward smokeproof or fire towers that are used as attack stairs.

Vertical Ventilation

Opening the top of the building is the easiest type of ventilation to understand. Unfortunately, it is often very difficult to accomplish. In any discussion of high-rise ventilation, the term *stack effect* will soon pop up. An understanding of stack effect and how it relates to both vertical and horizontal venting will simplify the IC's decisions.

Stack effect is the natural movement of air within a building. It becomes noticeable in buildings more than 60 ft high and becomes stronger as the building gets taller. It occurs in these buildings 24 hrs a day, 365 days a year. It is caused by the rising of warm air through stairways, elevator shafts, utility chases, and all other vertical shafts. If it is cold outside, the air at the top of these shafts tends to move out across the upper floors and out of the building. This is normal stack effect, and it can be a very powerful force in very large buildings.

As firefighters, we commonly see very similar behavior at fires in normal, everyday buildings. Smoke from the original fire area rises upstairs to the upper floors, then spreads out through the structure. We call this *mushrooming*. The answer to mushrooming is to provide a vent over the stairway that allows the heated air to escape directly to the outside. The same thing happens in high-rises, but it doesn't require a fire to cause it. If a fire occurs, this stack effect can be responsible for spreading smoke to the upper floors. The usual answer is to provide ventilation at the top of these shafts, which may prevent smoke from entering these floors. To illustrate the importance of top ventilation, consider the MGM Grand Hotel Fire in Las Vegas in 1980. A majority of the 85 victims died on the upper five floors of the building, 15–20 stories above the main fire. If effective top ventilation had been possible at that fire, some of those people may well have survived.

Therein lies the difficulty. How do we get to the roof of a 25-story building? The answer in most cases is that you *walk*. It can be vitally important that stairway and elevator bulkheads that extend through the roof be opened. Send two men to this position with SCBA and forcible entry tools. They may take the elevator to two floors below the fire. From that point on, they must select a stairway that pierces the roof. The exception would be if a blind-shaft elevator was available to take them to the upper floors. When using stairs, hold the door to this ventilation stairway closed on the fire floor until the members are on the roof. If the door is opened, they will be severely exposed and may be overcome. They should clear this stair of any occupants on their way up, directing them to designated stairs for evacuation.

Upon reaching the roof, the team should contact the IC and check the fire's reaction to their briefly opening the door while the door to the fire area is also open. If the reaction is favorable, chock the roof door open or remove it from its hinges. After doing this, open the door to the fire area. Once both of these doors have been opened, fire and heat will be drawn toward this stairway, which will act as a chimney. For this reason, this stairway isn't a good place for the attack line to make its advance. After accomplishing this ventilation, make additional openings at the top of the elevator bulkheads and penthouses. Take care in venting the attack stairs. If fire is drawn toward the attack lines, you will have to close the roof door.

When discussing the ventilation of high-rises, the question of using elevator shafts as vent points often arises. This isn't recommended for the following reasons:

1. The hoistway door on the fire floor must be forced open. This means that the elevator is no longer available for transporting men to the fire area.

2. The open hoistway door is a severe danger to firefighters crawling around in a situation of near-zero visibility. A member could easily fall to the base of the shaft.

3. The hoistway has only a small opening at the top, generally 2–4 sq ft. This isn't enough to vent an entire floor adequately.

4. The hoistway doors on the upper stories will allow heat and smoke out to their floors.

For these reasons, open the elevator room door at the roof to vent any smoke that has entered the shaft. Leave the hoistway door on the fire floor closed.

Two unusual phenomena that sometimes occur in high-rises are the stratification of smoke below the top floors, and inverse (also called reverse) stack effect, which causes smoke to move *down*. The first phenomenon, *stratification*, may occur in sealed buildings when the temperature of the smoke produced isn't sufficient to cause it to rise to the top of the building. As smoke and fire gases travel through a building, they give off heat to the walls and their surroundings. The products of combustion rise until their temperature has been reduced to the temperature of the surroundings. When this stabilization occurs, the smoke and fire gases form layers or clouds within the building. A classic example of this formation reportedly occurred in a 17-story building. The fire was extinguished before sufficient heat had built up to move the stratified smoke to the top floors. Venting this cooled smoke out of the top of the building was accomplished by creating controlled currents of air up the stair shafts and across smoke-filled floors. Similar action may be required when smoke has been cooled by sprinkler discharge (Fig. 16–17).

Inverse or reverse stack effect can occur in air-conditioned high-rise buildings in hot weather. In this case, the temperature inside the building is colder than the outside air. This dense air tends to sink to the bottom of the shafts and, in so doing, may draw smoke from the fire with it. This will result in smoke moving down to floors below the fire. This isn't usually a severe problem, since the difference between the indoor and outdoor temperatures isn't that great. On very hot days, however, it can cause problems if the fire is on the upper floors.

One night Rescue Co. 1 responded to a working fire in a penthouse apartment on the roof of a 30-story apartment house in midtown Manhattan. As we responded North on 11th Avenue, a broad six-lane boulevard, I was able to see fire showing from about four windows. Reporting in to the battalion chief, who had responded on a narrow East–West street, I was somewhat taken back when he ordered me to take my crew "down to the cellar or sub-cellar and find the seat of the fire."

"Down cellar? Chief, the fire's in the penthouse," I replied.

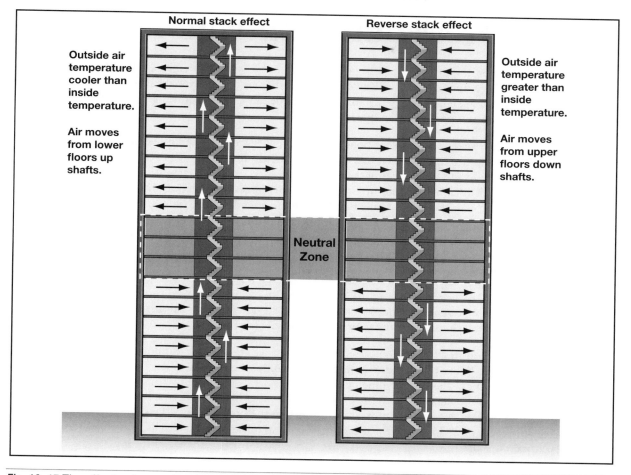

Fig. 16–17 The effects of normal and reverse stack effect on smoke movement and ventilation.

"It can't be" he said, "the smoke is down to the floor in the lobby, it has to be in the cellar," he informed me.

Only after I told him that I had seen the fire out the windows on the roof with my own eyes did he believe me and direct units to make their way up to the penthouse. But even after that he still directed me to have one team "take a look at the cellar anyway, just in case!" Good advice from a very seasoned chief who knew from past experience that a cellar fire could be raging below and have traveled up a shaft such as a rubbish chute or compactor to be seen at roof level. In this case it wasn't. It was, in fact, inverse stack effect at its worst.

The problem that inverse stack creates is almost as bad as landing on the roof of a fire building and trying to descend to a cellar fire to extinguish it. You take a tremendous pounding just trying to get to the fire, you have to wear your mask from the moment you enter the building all the way up the stairs, as opposed to staging just below the fire before finally going on air.

At another fire I operated at in Harlem that had strong inverse stack, two apartments on the 16th and 17th floors were involved. This time everyone knew where the fire was as it was blowing about 20 ft out the windows on the original fire floor, clearly visible to arriving units, and rising up the side of the building, extending via auto-exposure to the floor above. The attack stair became so heavily contaminated that the lobby was untenable without a mask. Fleeing occupants also blocked open the evacuation stair door, rendering it unusable. The situation was deteriorating badly until two electric fans were set up outside the lobby to pressurize it in positive-pressure mode. The two fans quickly reversed the inverse stack effect and cleared both stairways, much to the relief of the crews on the fire floor who had expended their SCBA cylinders on the 16-story hike and ensuing firefight. Note: Even though both of these cases cited occurred in residential buildings, the same thing has happened in commercial buildings.

Horizontal Ventilation

Horizontal ventilation of the fire floor of a high-rise is a complex issue, as well as a difficult task to control. In addition to the effect that breaking windows will have on the fire, there is the additional hazard of falling glass. Broken glass that falls 15 or 20 stories can be extremely injurious—even deadly—to bystanders and fire personnel. This glass can also sever hoselines that are supplying standpipe or sprinkler Siameses.

A serious fire in a high-rise requires a major police commitment to clear the building's lobby and the surrounding streets and sidewalks. Set up fire lines at least one block away in all directions. All fire personnel should stay out of the area around the base of the building unless they are there for a specific task. It is important to ensure that people assigned to tasks such as supplying lines to fire department Siamese connections wear full protective clothing. Operators of pumpers that are in proximity to the building should spend as little time as possible exposed to falling glass. Once the lines are charged to the correct pressure and operating, they should return to the apparatus cab to monitor the engines oil pressure, temperature, etc, and only make brief trips to the pump panel to check pressures or make adjustments as needed. Lines supplying Siamese connections may have to be covered with plywood to protect it from falling glass

The effect of wind and the stack effect are additional considerations. As mentioned earlier, wind at the upper levels of high-rises can be very strong. Venting windows on the windward side can have a disastrous effect. The only way to evaluate this is to duplicate the situation on the floor above or below the fire.

The stack effect is also affected by horizontal ventilation. Positive stack effect will normally allow smoke already entering the stairs on lower floors to continue up and out. If windows are open, this effect may violently blow the fire toward this stairway without the smoke going out the windows. This is especially true of lower-floor fires on cold days, when the outside air is much colder than the inside. Nothing is gained in this case by venting the windows and, in fact, conditions may be worsened if the smoke mushrooms on the upper floors. The same is true on hot days on upper floors. The smoke may be violently drawn into the building toward the stairwells, including the attack stairs as well as the ventilation stairs.

This type of action has resulted in severe punishment for firefighters and has produced injuries. The IC should carefully consider the consequences before issuing an order to vent the windows. A slogan to keep in mind is *High, high; or low, low is a no-no.* In other words, if the fire occurs on the upper floor on a day when the temperature outside is high, or on a lower floor when the temperature outside is low, then venting windows is a no-no, since the smoke will likely go into the building. This isn't a hard-and-fast rule, but it is a guide for the officer in command, who must consider the wind and other factors. The best way to evaluate the overall effect is to make a small, experimental opening. Of course the fire has never heard of the *high, high-low, low is a no-no* rule. It can vent out any window any time it wants until we put a stop to it. Be prepared for the consequences if it does.

When the decision has been made to use horizontal ventilation, the members doing the job should use caution. If possible, make a small initial hole and pull the remaining glass into the building. Using pressure-sensitive tape or contact paper applied to the glass first will hold the larger pieces together and allow you to pull them in. Some buildings have a thin film of Mylar on their windows to act as a sunscreen. This may serve the same purpose as pressure-sensitive tape, and it is already in place. Needless to say, you can't apply tape in the middle of a heavy-fire situation. Members operating near these opened windows should use caution to avoid falling out, pushed by gusts of wind.

Using Elevators

Gaining access to the fire area is the second biggest problem in high-rise fires. Usually it involves using the elevators. Most firefighters know that the public is constantly told not to use elevators in the event of fire, and many firefighters assume that the rule holds true for them as well. In fact, this rule should be slightly modified to reflect the differences between the firefighters' use of the elevator and the public's. The greatest danger in using an elevator during a fire is to have the doors open on the fire floor. Elevators are electrically controlled to stop at each floor in sequence along the direction of travel. If someone on the floor above the fire attempts to use the elevator to descend, and if an electrical signal is received from the fire floor (either by intent or by a short circuit of the call button), the car will stop at the fire floor. There is almost nothing that a civilian can do to prevent this.

The fire department, on the other hand, must recognize the use of elevators as a necessary evil; something to be avoided where possible but used when necessary. Take precautions to lessen the risk in the event that the unforeseen takes place. Some departments ban the use of elevators by fire forces altogether. That is acceptable in shorter high-rises. The highest I have walked is 32 stories in buildings where all of the elevators were OOS. That was after forcing a number of doors and engaging in some serious firefighting on lower floors, when a Mayday call for a downed firefighter was received. By the time we got there another fresh RIT unit was already there. They had found a working elevator. If the downed firefighter had to rely on a unit that had just walked up nearly 40 stories to carry him down stairs, he would have been in trouble. Carrying a mask, forcible entry tools, and a rolled up length of 2½-in. hose, fire extinguisher, or lifesaving rope up 20 stories isn't my idea of fun. And then it's time to go to work!

In buildings higher than 20 stories, using the elevators becomes an absolute necessity if you are going to get anything out of your firefighters. It is also a requirement if you want to keep your reflex time to anything under 45–50 min. Reflex time is the total time elapsed from the receipt of the alarm until an effective stream is flowing on the fire. It includes response and set-up time.

In a single-story home or store, it may be as little as 2–3 min, with response time making up the largest portion of that, especially when using a preconnect. In a high-rise, the response time may be the same, but the following additional elements come into play: removing rolled or folded hose, masks, and tools from the apparatus; walking the equipment into the lobby; recalling the elevators and searching them; verifying the fire floor; ascending to the floor below the fire; and finally, connecting the hoses to the standpipe, flaking it out, and getting water started at the proper pressure.

At the 2001 WTC attack, one elevator remained functioning in the South Tower, carrying firefighters from the lobby to the 44th floor *sky lobby*, its upper terminus. (All 99 elevators on the North Tower were OOS). It took almost 55 min after their arrival for Battalion Chief Orio Palmer, Fire Marshal Ronnie Bucca, (both marathon runners) and the members of Ladder Company 15 to ascend from the 44th floor to the 78th floor (where the plane impacted the South Tower). Radio transcripts show these members had just begun to operate a standpipe line in an attempt to protect a stairway, when the South Tower collapsed, killing them all. Remember that the longer your reflex time, the larger the fire you are likely to face (Figs. 16–18, 16–19, and 16–20).

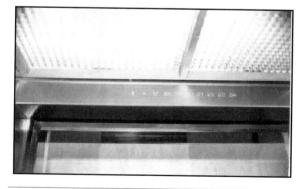

Fig. 16–18 This blind-shaft elevator ascends in a sealed shaft from the first floor to its first stop on 17. If the fire is reported on 17 or 18 you must use the other bank of elevators, which stops at 16.

Fig. 16–19 The elevator-landing key switch recalls elevators to the lobby

Fig. 16–20 Once the car has been recalled, the fireman service switch may then be activated with the key to operate the elevator.

In addition, there are fatigue and logistical problems of mounting an attack 20 or more stories above the ground if you don't use the elevators. Fortunately, at just about the practical limit of ascent by stairway, the following two building features come into play that work to our advantage: the sky lobby (an elevator terminal for separate banks of elevators) and blind-shaft elevators. In the larger high-rise buildings, it becomes very inconvenient and nearly impractical to have a single bank of elevators serve all of the floors. The large number of call buttons in use at any one time would mean lengthy delays as each elevator in turn rose all the way to the highest floor requested before starting down again. As an alternative, builders segregate elevators by the floors they serve.

For instance, elevator A might serve floors 1 through 10, while elevator B serves only floors 11 through 20, riding through a blind shaft between the ground and the 10th floor. Obviously, if there were a fire on the 12th floor, using elevator A to get to the 10th would be safe, since there is no way that that car can open on the fire floor. Not so obvious is the fact that, if the fire is on the eighth floor, elevator B may also be a safe way to get to the upper floors for ventilation and search. The blind shaft at the eighth floor will prevent the car from stopping at the fire floor.

Sky lobbies are found in very large buildings. They are a combination of express blind-shaft and segregated elevators, with an elevator lobby on an upper floor. In the WTC, for example, there were sky lobbies on the 44th and 78th floors. Other elevators provided local service for floors 1–43, segregated as described previously. At a sky lobby, it is necessary to change elevators from the express to a local car for other floors, for example 45–77, or 79–110. Again, selecting an elevator that reaches only to the sky lobby below the fire is a safe use of elevators during a fire. Obviously, firefighters must familiarize themselves with the elevators within their jurisdictions before they hop aboard and start pressing buttons. There are also safety precautions to be taken at all elevator operations, including the following:

1. Attempt to determine the fire floor accurately. Where smoke is reported on several floors, get off two floors below the lowest reported floor. Make sure you know what floors contain access stairs before you get in an elevator. Get off at least two floors below the lowest level of the access stairs if they also serve the reported fire floor

2. If the fire is within eight floors of the elevator lobby or sky lobby, walk up. If there is an elevator available that stops short of the fire floor and brings you within eight floors of the fire floor, use it in preference to an elevator that also serves the fire floor. Walk the remaining seven flights of stairs (connecting eight floors).

3. Ensure that each team that enters an elevator has its car number, unit, and destination recorded by a firefighter at the lobby who is designated as *lobby control*, pending the arrival of an officer dedicated to tracking the location of all units that leave the lobby.

4. Ensure that each team that enters an elevator is properly equipped. Each firefighter should be wearing a mask and have the cylinder turned on. The face pieces shouldn't be in place but should be readily available. Each team should have a set of forcible entry tools as well as a portable radio. In case the elevator loses power, the team can force its way out of the car without assistance or they can radio their position and situation.

5. Immediately upon entering any car, press the *call cancel* button to eliminate any selections that have previously been made.

6. If available, use only firemen's service elevators, which are placed in the proper operating position by a key or other secure means. Firemen's service elevators are common requirements in many codes since the inception of New York City's Local Law No. 5. The specific requirements vary as to operations, however, and it would be pointless to go into detail about a feature that doesn't apply to all elevators nationwide. Suffice it to say that a department needs to get into its elevator buildings and find out precisely how each one works. There may be slight variations installed by different elevator companies, even within the same city. The most important point to make about firemen's service elevators is that they are not fireproof, totally safe devices. They are affected exactly the same way by fire or water in the shaft as any other elevator. All you can rely on the firemen's service switch to do is to remove the elevator from any possible civilian use by canceling any signals received from the elevator call buttons on the upper floors. Freight or service elevators don't make good firemen's service elevators, since they usually serve every floor in the building. A large number of serious fires start in accumulated rubbish in freight elevator lobbies.

7. *Expect* problems when using an elevator. Have the forcible entry tools ready to force open the inside door at the first sign of trouble. This will interrupt the interlock and stop the car's travel right where you are. As soon as the car begins to move from the lobby, press the emergency stop button. The car should stop. If not, pop open the door. If the car does stop, pull out the emergency button and begin the ascent again. Rather than ascending directly to the floor two floors below the fire, the trip should consist of a series of precautionary stops every five floors. These stops serve several purposes, and you should consider them vital to your safety. First, they ensure that the elevator is still responding to your programming. If it fails to do so, pop open the door to stop what might be an express to the fire floor. Second, the stops allow the firefighters to step out and verify the location of the nearest exit stairway in the event that the car later stops at a smoke-filled floor. Third, they allow you to shine a light upward between the car and the wall of the shaft, looking for any smoke, water, or other contaminants entering the shaft. Although it might be possible on some cars to open the roof hatch and shine the light up while the elevator is ascending, the precautionary stops are still necessary for the other two reasons. Just because no smoke may be apparent doesn't mean that you should make the ascent directly. These shafts are often pitch black, anyway, and what appears to be the top of the shaft may be a solid layer of smoke.

8. Once you arrive at the destination, usually two floors below the fire, all of the members should be prepared to instantly don their face pieces and move to the nearest stairway if smoke pours in. When the doors open, be prepared to hit the door close button. If the floor is clear, ensure that you can get into the stairway from the elevator lobby before everyone exits the elevator and returns to the lobby. Don't stand in the car doorway to hold the elevator. Either remain inside the car or lay a length of hose across the doorway to prevent the door from closing fully. Once you are sure that you are at the proper location and have secured access to the stairs, return the elevator immediately to the lobby for use by reinforcements. (Note: A member must remain within the elevator at all times in some firemen's service installations, since the doors will only open in response to the door-open button within the car. If these cars are sent back to the lobby unmanned, they will be unusable by later-arriving units).

Operations

Initial deployment of the first-arriving units is critical. These members have the vital tasks of locating the fire, determining its size and likely paths of travel, and redeploying themselves to confine any extension. The possible life hazards presented by a large-area office building require an extensive search effort. Usually the personnel of at least two ladder companies are required on the fire floor and at least one additional truck company on the floor above to perform search immediately and to check for extension. In addition, extra personnel will be needed to search each and every floor above the fire. Smoke conditions that build up can be lethal many floors above the fire, as evidenced by the MGM Grand Hotel fire in Las Vegas, as well as by the Times Tower fire in New York City in 1961, where two firefighters died more than 20 floors above the fire (Figs. 16–21 and 16–22).

Attack crews on the fire floor will have a tremendously difficult time extinguishing the fire. Floors that are spacious and open at the ceiling level are often a maze of desks and partitions at the floor level. The hose stream may not be able to strike the seat of the fire directly when it is behind such obstructions. Naturally, the heat and smoke are unaffected by these dividers and have free access to the firefighters, baking them as they move in. The preferred hoseline for this situation is the 2½-in. line with a 1¼-in. solid tip. This allows a high volume of water to be discharged at a distance, extinguishing the fire well ahead of the line with only a 50-psi nozzle pressure. This can be critical in high-rise buildings, since NFPA standard 14 states that fire pumps are only required to discharge their capacity at 65 psi at the highest hose outlet, and further limits pressures within a standpipe zone to a maximum of 125 psi.

Fig. 16–21 The large, open areas seen in this photo are available to the fire, whereas firefighters often have to make their way through a maze of desks and cubicles at the floor level.

Fig. 16–22 Early failure of hanging ceilings can obstruct firefighter entry as well as escape. Use of a search/guide rope should be mandatory.

Operating with fog nozzles, particularly on smaller-diameter hoselines, may result in inadequate nozzle pressures and flows, especially at upper stories of a standpipe zone. Although fire department pumpers might be able to increase the pressures on the standpipe, it might not be of much value. It might not even be possible. The standard Class A pumper is rated to deliver one-half of its capacity at 250 psi. That is just barely sufficient to operate a fog nozzle between the 21st and the 30th floors (100 psi nozzle pressure plus 5 psi loss per floor, plus friction loss in the handline, supply lines to the Siamese, and the standpipe itself). As additional lines are stretched from other standpipe outlets, pressures within the system are likely to drop still further, thus demonstrating the value of a nozzle that can put out an effective fire stream (290 gpm) at as low as 40 psi nozzle pressure (the 1¼-in. solid tip).

But what about those fires that won't go out even when you hit it with 300 gpm? The large open floor office is a prime candidate for this problem, where the volume of fire can easily outstrip even the 1¼-in. solid tip. The answer to a large volume fire is no different on the 31st floor than it is on the 3rd floor—large volume hose streams! The problem with the 31st floor fire is that we don't have 31-story tower ladders. (Yes, there are a few 22-story platforms, but they are not the answer to the problem because buildings can always go higher. What do you do for a 60-story building?) The solution that has been tried in the past is the lightweight master stream device. Many of these devices are light enough to be one-man-portable, even when it is necessary to carry it up 30 flights because the elevators are out (Figs. 16–23, 16–24, and. 16–25).

Fig. 16–23 Heavy fire calls for a master stream, whether at ground level or on an upper floor. Here a master stream is operated from a 20th floor standpipe, providing a powerful 700 gpm plus stream with good reach.

Fig. 16–24 Anchoring a master stream against the nozzle reaction can be done using hooks to span a door opening.

Fig. 16–25 Here, the master stream device is secured to anchors shot into the 6-in. thick concrete slab using a powder actuated fastener. See the text for further comments on this technique.

One of the problems with these devices is their tendency to kick back when the stream angle is lowered to only 15–20°, which is necessary to provide reach *and* penetration when the ceiling height is only 8–9 ft high. What is needed is a secure means of anchoring the device against the nozzle reaction. Aside from having six very large engine men throw themselves on the hoselines, there are a couple of possible choices. The first method uses two steel handled halligan hooks to span a door opening. (FDNY ladder companies bring halligan hooks with them anyway, although they won't be used for pulling ceilings until the fire is knocked down.) This allows one of these devices to be employed from a door of the fire stair or the door to the office area and quickly begin application of upward of 700 gpm. (**Note:** Several departments including Boise, Idaho, have experimented with supplying these monitor devices using 1¾-in. hose. Two lines of 1¾-in. hose fitted with 1½-in. to 2½-in. increasers at the nozzle might allow a flow of up to 350–400 gpm, depending upon pumper discharge pressure, pressure reducer valve settings, and length of hose stretched. While this is a dramatic improvement over a 150–200 gpm 1¾-in. handline, it is a long way from what is available from using 2½-in. hose.)

In New York City, tests were conducted on the upper floors of a high-rise building under construction on the west side of Manhattan. Units of the FDNY's 9th Battalion conducted a number of test of various portable monitors, supplied by a pumper in the street more than 20 floors below. The test was aimed at demonstrating the feasibility of supplying such device from a single standpipe riser and to verify the flows and pressures available using a variety of hose lays. All tests began with three lengths of hose, stretched from the standpipe outlet on the floor below the nozzle, following standard department policy. It was very evident that while a single 2½-in. line supplying a portable monitor delivers greater volume and reach than a 2½ in. handline—or a master stream supplied by two 1¾-in. lines could, these devices are best supplied by two 2½-in. or larger lines. That means the second line will likely have to be stretched from the outlet two floors below the fire. In this test, that meant four lengths of hose. The advantage of using 2½-in. hose is quite clear however little more it weighs. It permitted flows of up to 750 gpm to be delivered off a single standpipe riser more than 20 floors above grade without using any specialized high pressure pumper, and utilizing hose and equipment readily available on the first alarm units.

The tests also offered a chance to test another method of anchoring the portable monitor. The devices were set up on the edge of a large open deck, which had no door frames to span with the halligan hooks. Instead, members of Rescue Co. 1 fastened two ¼-in. shank × 2 in. × ⅜ in. diameter threaded powder driven studs into the concrete deck using a .38 caliber powder actuated fastener. We secured the monitor between them using either the device's chain, utility rope, or tubular nylon webbing. This proved successful even in this fresh concrete, thus offering another option for mounting a large volume attack. When I first described this method in a magazine article in 1998, the manufacturer of the master stream used in the test wrote back that his company did not approve of the method. I recommend it as a necessary evil, based on the following data: use the type fasteners described previously: .38 caliber, since the .22 caliber fasteners do not offer consistently solid anchoring in various types of concrete. Use no less than ¼-in. diameter shank by ⅜-in. threaded studs, attaining a minimum 2 in. penetration. This fastening system is designed to provide 4,360 lbs of resistance in the shear mode, as this is being used. That provides more than a 10:1 safety factor against the approximately 385 lbs of nozzle reaction produced by a 1¾-in. tip flowing at 80 psi nozzle pressure. Of course, if conditions permit, other anchoring methods should be used. This idea is intended for use only when you are out of options and the fire is roaring through the building, threatening lives

Unfortunately, not all fires are going to succumb even to this superior stream. In large-floor area offices, particularly those built of core-construction techniques; a complete burnout of the floor may result if a large volume of fire is present on arrival. Core construction is the practice of grouping all of the non-rentable spaces—stairs, elevators, lobbies, electrical, airshafts, and the like—in a central core, thus leaving the more desirable perimeter as open floor space. This practice provides floor spaces in excess of 30,000 sq ft per floor. The result is that, if a fire gains major headway, it can easily spread to the entire floor (Fig. 16–26). (Each floor of the WTC measured 208 × 208 ft, providing more than 40,000 sq ft per floor, often largely open areas.)

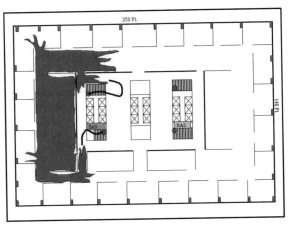

Fig. 16–26 A fire of this magnitude on a center core floor this size will require a substantial commitment of personnel. If the fire cannot be stopped at this stage, and it involves the entire floor, it is likely going to have to be contained, while being allowed to burn down to a reduced intensity.

Even if we assume a light fire load, requiring only .10 gpm per sq ft to extinguish, we can see by multiplying 30,000 by .10 that at least 3,000 gpm is required to extinguish a fully involved floor of this size. Prior to 1993, NFPA 14 only required a standpipe system to be designed to supply a maximum of 2,500 gpm. In the 1993 revision to NFPA 14, that number was lowered to 1,250 gpm! Maybe that's a good thing in light of the potential for collapse of new lightweight buildings, the decision making process as to whether to fight a large fire involving an entire floor is already made for us. We cannot succeed and should know that in advance, and prepare the collapse zone instead of getting killed trying to fight a losing battle. If a fire requires 3,000 gpm to extinguish and the standpipe will only supply 1,250, we cannot hope to extinguish that size fire! A fire flow of 3,000 gpm may well be beyond the scope of manual firefighting, even if the water is available. According to Chief John T. O'Hagan's "High Rise Fire and Life Safety," areas "in excess of 3,000 square ft of a *moderately* fuel loaded area diminish the possibility of success through a direct attack." (Emphasis added.) If a floor of this magnitude is fully involved, the only available alternative may be to get above it and flood the entire floor above with hose stream, to cut off extension while waiting for the fire to consume the bulk of the combustibles, burning itself down to an intensity that can be handled manually.

Even if the floor isn't fully involved, handlines may not be able to penetrate to the seat of the fire. If windows on the upwind side of the building fail due to fire, the attack crews may be unable to advance due to the tremendous heat or flame being blown at them. The same possibility exists where the stack effect draws fire toward the attack stairway. In either case, some way has to be found to get water on the fire and darken it down before full involvement occurs. Several unique options that have been successfully demonstrated are the concrete core drill, concrete cutting chain saw, and the Lorenzo ladder. All are last-resort weapons that can be used from safe areas to get some kind of water into an otherwise untenable fire area. When such conditions exist on an office or commercial floor, the Navy fog applicator with its 100- to 150-gpm flow isn't going to have much of an effect due to the size of the area involved.

The concrete core cutter is a diamond-tipped drill used in high-rise construction to bore holes up to 8-in. diameter through floors. One model requires an electrical power supply and is rather bulky. A smaller, lighter, gasoline-driven model is also available, although the diameter of its cores is limited to 4-in. holes, which are too small to allow a 2½-in. Bresnan or other distributor to fit through. A series of three overlapped holes, made in the form of a shamrock, can produce a sufficiently large opening to pass the distributor.

Both models require a pump tank of water to cool the bits, as well as a means to fasten the drill in place. Normally, this is done by shooting studs into the floor with a powder-activated fastener and then using wing nuts to hold the drill. A 5-in. hole drilled through the floor is large enough for a distributor or a cellar pipe, assuming that the floor above is tenable. A 2½-in. Bresnan distributor operating at 100 psi flows about 485 gpm and covers a radius of 20–25 ft. An alternative would be to drill through the fire floor from below and insert these nozzles on a length of pipe functioning as an exterior stream.

The actual drilling is a relatively fast process: a 6-in. reinforced concrete floor can be pierced in about 5 min. The concrete cutting chain saw is another tool to perform the same task, and may be better suited to cutting through from the floor below the fire. A gasoline-powered model is popular, while a hydraulic unit, powered by a large gasoline-driven pump similar to a Hurst tool power unit also available. Sufficient fresh air to allow the engine to operate smoothly is essential. This may not be practical on the floor above at a fire severe enough to require these type tactics.

The Lorenzo ladder is a readily improvised exterior master stream for use in similar situations. It consists of a standard ladder pipe fitted on a 16-ft extension ladder. The ladder is placed in position out the window of the floor below the fire, using two sets of cables to secure it in place. The stream is a 1¼-in. tip flowing 400 gpm at 80 psi, operated remotely from below by means of a set of halyards. As with any exterior stream, it is crucial to observe its effects. Once the main body of fire has been darkened down, shut down these streams and advance the handlines to continue extinguishment. At high-rise fires, this may require you to post a scout or an observer, either in a helicopter or on a nearby building, in radio contact with the IC and the persons in charge of the exterior streams (Figs. 16–27, 16–28, and 16–29).

Fig. 16–27 This core drill is set up to drill through from the floor below the fire. A stream may then be directed upward through the hole to cool the fire area.

Fig. 16–28 A gasoline-driven core drill as seen here may be suitable for drilling from the floor above to apply a distributor.

The primary danger from using exterior streams is that they will drive the fire toward interior forces. That danger shouldn't exist in the case of the high-rise. The interior attack lines have tried and failed to make the fire floor and have been withdrawn to the stairwells. The danger of opposing streams does exist, however, from *interior* lines. It is even possible in core-construction techniques for a single hoseline to drive fire around the core and back on itself. In this design, there is often a common corridor that wraps completely around the core, connected by the elevator lobbies. A hoseline advancing out of a stairwell in one direction can push fire around the corridor and right back toward the open attack stair door, almost like a dog chasing its own tail. The resulting heat coming from the rear can force the attack line to retreat rapidly to the stairway. The only solution may be to fight your way out of the stairway with two lines: one to advance and one initially to remain at the stairs to protect the rear.

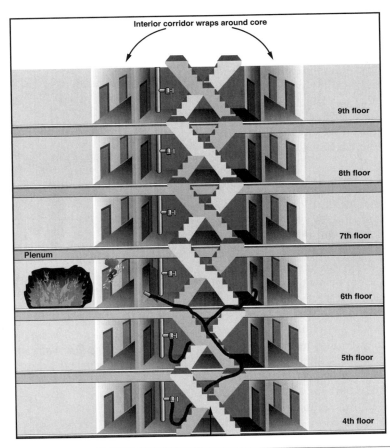

Fig. 16–29 Care and great coordination must be used to avoid opposing hose streams if lines are stretched from scissor stairs on two sides of a core.

Another common problem with core construction is the presence of scissor stairs, which alternately exit the core on opposite sides. It will be necessary in this case, when staging for operations on the floor below the fire, to set up and connect to the standpipe on the side of the building that is opposite the fire. An additional type of stairway commonly found in high-rise office buildings is called an access stair. This is usually an open, unprotected stairway leading from floor to floor within a single-tenant occupancy.

It is common for businesses to rent more than one floor. The public exit stairs are usually designed not to allow reentry on any floors aside from the ground-floor exit or other specific floors. In this case, if an employee wanted to take a form to another floor, it would be necessary to go out to the public lobby and take the elevator up or down, then reenter. This is a very common arrangement for security reasons. To avoid waiting for elevators and having to travel outside the office, an extra convenience stairway is often added. Since it isn't a required exit, it is often unprotected. From a fire protection point of view, it is a nightmare. The slightest amount of smoke will almost immediately trigger alarms on all of the connected floors. If a serious fire develops, the available floor area doubles or triples, and with it, the threat to life.

Buildings that have access stairs between floors should have a mandatory, fire-rated enclosure installed, or else have all of the floors fully sprinklered, with a separate deluge system designed to protect the openings between floors. This information should be conspicuously noted at the ground-floor lobby, and members entering the elevators should be sure that the floor they're selecting as their elevator exit point isn't connected to the fire floor by an access stair (Fig. 16–30).

Fig. 16–30 An access stairway is an open, unprotected stairway between two or more floors in a high-rise. Note the plenum space shown above the ceiling where piping, ducts, and wires are run.

Control of High-rise Operations

High-rise fires pose unique challenges to fire departments because of the high hazards to life and property, the logistical difficulties, and various features of construction. The complexity of the fireground demands a strong command presence—someone who is well versed in the tactics that may be required. In addition, all firefighters responding to a serious high-rise fire should also be trained in the unique problems and operations that they may be likely to face. This factor, training all levels of firefighters, may be the most crucial item when preparing for a high-rise fire. Chief Daniel Murphy, onetime commander of New York City's Third Division, in charge of many Midtown Manhattan high-rise fires, once told several visiting chief officers that, like almost all other fires, the actions of the first-alarm companies—and particularly their company officers—will determine the outcome. New York City has a 56-page tactics bulletin on high-rise operations.

Chief Murphy pointed out that such a plan is great on paper and that things usually go well when it is followed completely to the letter, but if just one of the first-alarm units commits its line to the wrong place or uses a wrong stairway as an evacuation route or if any of a number of other items go wrong, then the best plan in the world isn't going to rectify the situation. That is why a large number of chief officers are required at this type of operation to provide experienced leadership in each crucial area. For example, in New York City, the signal 10–76 is used to indicate a *possible* working fire in a high-rise. The command response to this signal is a deputy chief and four battalion chiefs. If a second alarm is subsequently required at that fire, an additional battalion chief, another deputy chief, and a senior staff chief will also respond (Figs. 16–31 and 16–32).

Fig. 16–31 The lobby CP at a serious high-rise fire will be a complex command and control operation. Proper staff support is critical. Record the identity of all units that leave the lobby, along with their destination and route taken.

Fig. 16–32 The availability of floor diagrams and hardwire communications systems are major requirements for the high-rise CP.

The deployment of all of this "brass" should be made with specific objectives in mind. The IC should be located in the building's lobby, where the CP should be established. Establishing the CP in the lobby of a high-rise office building has many advantages over setting up in the street. Modern high-rises usually have a highly sophisticated building control system. This should be operable from a location in or just off the lobby that allows incoming units to report in to command and stage safely away from the dangers of falling plate glass. The building system controls are often built into a desk referred to as the fire command station. From this location, the IC has immediate access to such vital information as floor plans, status of the HVAC system, fire alarm and extinguishing systems, elevator return panels, public address or two-way voice communication with upper floors and elevators, standpipe telephones, regular telephone lines and a myriad of other sensors and intelligence that is not available in the chief's car or command van. It is essential to dedicate a knowledgeable chief officer just to the duty of monitoring and controlling the building systems to match the IC's strategic goals. He or she should be assisted by the senior building engineer and if available a firefighter or officer to act as a communicator/coordinator. This group's ICS designation is the systems unit.

An operations officer, usually one of the first-arriving chiefs other than the IC, should be sent to the floor below the fire to assume direct command over operations on the fire floor and the floor above. Building blueprints should be available at the lobby CP and, if available, a copy of the floor plan of the fire floor (and plans of the floor above) should be sent to the operations post. In this manner, communications between the CP and the operations post will have a common reference point. Early establishment of several channels of communication between the lobby CP and the operations post should be a priority. Radio communications in high-rise buildings are often inferior without special devices, such as cross band repeaters or special antennas in the building. As an alternative, use building telephones, security radios (often designed especially for use in these buildings) or standpipe telephones. Due to the likely amount of radio traffic, use a separate channel to communicate between the operations and CPs, one that isn't in use by companies on the fire floor. The purpose of the operations post is to take the strategy that is being developed by the CP and implement it on the fire floor while directing the specific tactics to be used.

For instance, if heavy fire is present on the south end of the building, the strategy may be to prevent it from spreading northward. The operations officer will decide whether that means bringing lines all the way from the north stairwell or whether additional lines operating from the south stairwell can do the job. Obviously, the operations post must keep the CP constantly informed of the tactics being used and the results of them. One of the major functions of the operations officer may be to push the advance of the attack teams. For this reason, and actually to observe conditions, the operations officer will have to operate at least part of the time on the fire floor. The operations post must then be assigned at least two members, usually a chief and an aide, to handle communications. This arrangement allows the aide to monitor the command channel on his or her radio, while the operations officer uses the fireground frequency to control the units.

As mentioned previously, keeping the lines moving forward is one of the operations officer's primary duties. The heat conditions at these fires can limit the time that a member can remain on the fire floor to only a few minutes. Arranging for the orderly rotation of fresh personnel to act as relief on the lines will greatly improve the ability of the lines to advance. The operations officer must then have a readily available pool of personnel from which to draw. If an attack is being made from more than one location or stairway, a separate attack director may have to be designated at each location, each reporting back and responsible to the operations officer (Figs. 16–33 and 16–34).

Maintaining the resources needed to keep the attack moving is the duty of another sector commander, the staging area officer. A forward staging area should be set up approximately three floors below the fire to maintain the needed supplies and personnel on standby to feed up to the operations officer. The forward staging area should be located in an area that provides ample storage space to segregate full and empty air cylinders, resuscitators, hose, forcible entry and overhaul tools, as well as personnel. It should be in an area that allows good communication with both the operations post and the CP, using as many means as possible (similar to the CP). If practical, the staging area should be in close proximity to the attack stairway. Again, a member should monitor both the fireground and the command channel to anticipate and acknowledge requests for supplies.

Having a reliable method of transporting equipment to upper floors in case the elevators become unreliable is the role of yet another sector, dubbed stairwell support. In the event that the elevators are lost, items like spare air cylinders, hose, portable deluge sets, and additional personnel will have to be transported manually to the staging area. At times, rather than have personnel who will soon become attack personnel carry this extra equipment, it is better to assign additional personnel to the task. Generally, one member is assigned for every two floors above grade to the fire.

If the fire is on the 24th floor, 12 members are assigned for stairway support. These personnel, operating in a safe area, can remove most of their protective clothing (always keeping it close by). Their task is simply to shuttle material upward to staging. Each member is responsible for moving the equipment only two floors, then coming down for another load. Since they aren't burdened by bunker gear and SCBA, and since 50% of the time they are coming *downstairs* (usually only lightly burdened, if at all), these personnel can move more equipment faster than having each group carry their own.

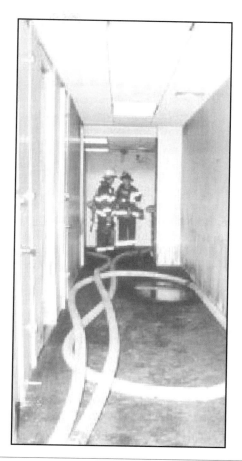

Fig. 16–33 The attack chief must coordinate the movements of multiple units operating from his assigned stairway. The operations chief coordinates the actions of different stairways.

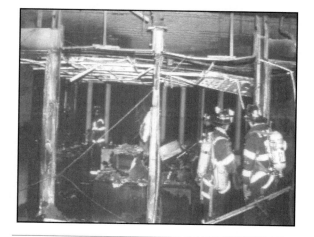

Fig. 16–34 Successfully extinguishing a serious high-rise fire depends greatly on a coordinated operation, supported by many personnel playing vital roles beyond the fire floor.

This is more effective than walking 24 floors wearing protective clothing and then being expected to go to work at the top. Those who are to be the attack personnel can make their way up the stairway without having to drag along spare air cylinders, extra hose, and the like. Thus, they arrive at the staging area less fatigued than they would have otherwise.

In areas where the on-duty strength of a department doesn't permit assigning 10 or 15 firefighters to the task, the logistical effort becomes even more critical. When you are shorthanded, you need to get as much out of your suppression personnel as possible. In this case, it may be possible to use non-suppression personnel for the stairwell support function—for example, police officers, as long as they operate in nonthreatening environments.

As a rule of thumb, the staging area should maintain a reserve force of at least two attack teams and two support teams of three members each, equipped for forcible entry and overhaul. If necessary, these can consist of members who have been rotated off of the fire floor, once they've had their SCBAs serviced and have been given sufficient time to recuperate. This isn't the best idea, although it may be necessary in crucial circumstances. A high-rise fire is likely to be the most severe challenge a department has to face and, as a result, drastic actions may be necessary. Early establishment of an EMS sector within the staging area is a must. Realize that using the elevator may be curtailed at any moment by water in the shafts. Be prepared to treat life-threatening injuries (primarily cardiac-related) on the upper floors until the victim can be carried downstairs, if necessary. That means getting an Advanced Life Support Unit to staging as early as possible.

You should also establish a third major sector at serious high-rise fires. Dubbed the search and evacuation post (SAE), it should be established in a reasonably safe area above the fire. Also staffed by two members, similar to the operations post, its duties are to direct and control the activities of all fire forces operating over the fire floor and the floor immediately above. These members should establish the same communications network with the CP as the operations post and the staging area, and they should also have communications with the units under their command. These members must verify and record the results of searches made on each floor. They should be informed of any reports of the location of victims still unaccounted for so as to direct search teams to each location. Being above the fire, they must be prepared to operate in smoky conditions. They should consider bringing a spare SCBA cylinder with them when setting up. An excellent choice of floor for the location of the SAE is a sky lobby or other floor not served by an elevator that also serves the fire floor, thus allowing passage past the fire floor by elevator with reasonable security. Constant communication must be maintained between the SAE and its operating forces, as well as between the SAE and the operations post, to allow all members to retreat in the event that the situation gets out of control.

Despite all of the advances in construction technology and fire protection in recent years, the possibility of an actual towering inferno remains real, even without the specter of terrorism. At the First Interstate Bank Fire in Los Angeles, five floors of the building were totally destroyed. Only a Herculean effort by a large, well-equipped fire department stopped the fire from totally gutting the rest of the upper floors. Many, many departments can find themselves faced with a serious high-rise fire. Although most won't be able to muster the response of Los Angeles or other big cities, the tactics they use will be essentially the same.

Buildings under Construction, Renovation, and Demolition

17

Fires in buildings under construction, renovation, or demolition pose unique challenges to firefighters in terms of structure, fire load, and operations. Very often, fires in these types of buildings result in substantial loss and account for a large percentage of casualties among firefighters. Considering the fire situation, this shouldn't come as a surprise. During any phase of work in progress, conditions are ripe for a fast-spreading fire. Large quantities of combustibles are exposed, from piles of lumber to floor, roof, and wall assemblies, as well as piles of boxes, sawdust, refuse, and scraps. Often, such flammable materials as adhesives, sealants, and finishes are stored for temporary heating. Ample sources of ignition are also present, in the form of welding and cutting tools, portable heating units, and temporary wiring. Often, so-called controlled fires of scrap lumber are present as well. When the source of ignition reaches a supply of fuel, the travel of fire is often rapid due to the unlimited air supply inherent to an open, possibly windowless building. In addition, the fire protection components required of an occupied building, such as sprinkler, standpipes, and fire doors, may not be present. Once ignition occurs, the fire forces are in for a spectacular workout.

Buildings under Construction

Fortunately, major fires in buildings under construction aren't a common occurrence. Most of the sources of ignition at the sites are only present when construction crews are on the scene working, hence fires are spotted while they are still small, and prompt action can prevent extension. At most sites, power to the entire area is turned off at the end of the workday and remains off until the next workday (Figs. 17–1 and 17–2). Thus, many of the potential ignition sources have been removed. Past experience has also shown the value of posting a fire watch in the vicinity of hazardous operations, such as cutting or welding. The fire watch shall remain in the area looking for signs of hidden fire for 30 min after the work has been completed. Most fires on these sites, then, occur because of accidents, such as a broken fuel line to a space heater, or carelessness, such as placing a heater too close to combustibles. Arriving firefighters should be aware of the major hazards likely to be present at these sites and how to handle them. This is best accomplished by frequent visits to the site, since work can progress rapidly and quite drastically alter the situation you encounter. Watch for the following items, and plan your strategy around them.

Fig. 17–1 Difficulties present at this construction site include a standpipe Siamese totally obstructed by the cinder blocks on the left, only one working elevator (which is shut down after hours), and a large quantity of combustible materials scattered about.

Fig. 17–2 An elevator shaft with a wooden guard. There may well be no guard on the fire floor. This one happens to be on a floor 20 stories to the next stop!

The entire volume of a structure within the outside perimeter is often exposed to any fire that begins. Often there are no firewalls in place until quite late in the construction. The floors often have large penetrations to permit access and passage of construction material. Extremely rapid vertical spread of fire is likely. Fortunately, these same conditions also ease our fire control operations. Even as the openings permit the spread of fire, they also allow for rapid, nearly effortless ventilation, and the lack of partitions allows a hose stream to sweep an area for its full range. The stream of choice, where the fire is within reach, is the elevating platform. From outside the structure, it can deliver a high-volume, long-reach stream that should knock down an entire floor area. Properly applied, a 2-in. solid tip will discharge more than 1,100 gpm over a 100-ft-deep floor. Although this is an excellent stream for serious fire conditions, using it involves some precautions, particularly in buildings under construction.

The first activity should be to clear all personnel from areas in the streets below on the sides opposite the nozzle. With no substantial side walls in place to restrain them, the large hose streams can send tons of debris showering down on surrounding areas. Items weighing more than 100 lbs can be propelled in this manner. Large-caliber streams used at close range can also dislodge scaffolding or shoring, precipitating collapse.

When ordering elevating platform streams into operation at buildings under construction, renovation, or demolition, it may be necessary to vary the usual sequence of stream operation to cut off fire extension. Normally, when a fire has hold of more than one floor and a platform stream is to be used, it should begin on the lower floor and progress upward. This works well in a finished structure. The steam produced on the lower floor will rise and may aid extinguishment above. In addition, upward extension is relatively slow due to the firestopping of vertical openings. Neither statement is true of a wide-open building. The vertical spread of fire may be so rapid that, if the lower floor is attacked first, the fire will race right past the reach of the stream before it can be brought to bear on the upper floors. In the wide-open condition found in these structures, the extinguishing effect of steam is nonexistent. Cut the fire off. The stream runoff will pour down the myriad vertical openings, checking extension. You may then bring the stream down into position to attack the lower floors.

Watch for the presence of these vertical openings. In addition to being a means for the rapid upward spread of fire, they are also hazard to firefighters working near them. Be aware that OSHA regulations specify the way that such vertical openings must be protected. Often the guard around such openings is made of light wood or even rope. These work fine if you're walking around in broad daylight, but crawling around in smoke and darkness after knocking down a fire is another story. Any fire will readily remove the guard, allowing a member to crawl right into the shaft. These vertical openings pose still another problem: Nearly all of them are vertically aligned, so that burning debris from the fire floor will rain down on firefighters attempting to make their way up through them.

Fig. 17–3 Nearly vertical wooden construction ladders may be the only access from floor to floor at many sites. These ladders are sometimes pulled up to the second floor at night to prevent unauthorized access.

Determine access routes to the upper floors. When conducting your site inspections, try to determine how you can reach the upper floors after the workday is through. Whereas a completed, occupied building might have three or more staircases, access during the time when work is in progress may consist of only a single set of construction ladders made up of 2×4s (Fig. 17–3). It is a common practice to remove the lower ladder at night to prevent entry to upper areas by vandals and children playing in the area. Usually, if a fire department ladder is put up to the second floor, the remaining ladders are in place. But it is still something of a treacherous operation to ascend to an upper floor with tools and hoseline on one of these ladders, which may project through the future elevator shaft. Since these ladders may go from the basement to the roof, a misstep could mean a severe fall. If the fire is within reach of the department's ladders, aerial or portable, it is usually quicker and safer to make the initial access off the department's ladders.

Buildings higher than the reach of an aerial ladder are usually provided with hoists, similar to elevators. Two concerns are involved in using these hoists. The first is that they, too, are often shut down at night and on weekends, making for a long walk up. The most important item, however, is to recognize that there may be two types of hoists on the job site: one for people and one for materials and equipment only. The people hoist will always have a control position in the car—the operator will ride up with you. The equipment hoist won't have a control position in the car. Since it isn't equipped with many of the safety devices found on an elevator, it isn't meant to be ridden (Figs. 17–4, 17–5, 17–6, and 17–7). Be sure to select the proper hoist!

Fig. 17–4 Only passenger elevators should be used for access during an emergency. Material hoists should not be used. Learn the difference.

Fig. 17–5 Operations at a high rise under construction should be conducted with caution. There is rarely a civilian life hazard at these fires.

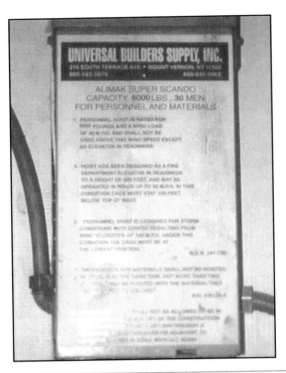

Fig. 17–6 Look for the passenger label and capacity when stepping into any construction elevator. Do not overload.

Fig. 17–7 A material hoist is operated remotely, and has few of the safeties of a passenger elevator. They should not be used for transporting personnel.

Watch out for large quantities of flammable liquids or gases. Particularly in colder seasons, large quantities of liquefied petroleum (LP) gas may be present on the site. LP gas is used to fuel salamanders, tar kettles, torches, and various other devices. Often, dozens of 100-lb cylinders may be present on a large job. When pouring concrete in cold weather, the entire building may be wrapped in canvas tarpaulins or plastic sheeting to keep the heat in and the elements out, thus speeding the curing process of the concrete. These same tarps can keep leaking LP gas in, allowing an explosive mixture to build up. Fires in areas where LP cylinders are present should be fought with large streams from shielded locations (Figs. 17–8 and 17–9).

Fig. 17–8 Buildings are sometimes wrapped in tarpaulins to allow cement to cure in cold weather. LP gas is often used to fire the heaters. A leak here would no longer be an outdoors leak, as defined by many codes.

Fig. 17–9 Make sure the standpipe riser keeps pace with construction. This one (at floor level behind the wood fence) is three floors below the work, and the top of the 6-in. pipe is wide open!

Ensure that fire protection systems are keeping pace with construction. As a general rule, sprinkler systems will be among the last items completed since the heads aren't installed in hung ceiling areas until the ceiling itself is installed. The same may be true for fire doors and partitions in areas that aren't yet rented. As an absolute minimum, however, the standpipe riser and outlets should progress to within two floors of the highest level under construction. This is a required part of many construction codes, but it is often overlooked.

At the same time, many codes don't specify that the standpipe be kept wet, especially if the primary water supply is to be from a gravity or pressure tank on a roof or an upper floor. In this event, all that may be provided is a fire department connection (FDC) unlike any with which you are familiar: just two black pipes with swivels sticking out through a fence or a tarpaulin. The FDC may be the *only* source for putting water into the system until the building is nearly complete. Be sure that it is clearly marked and visible, that it is kept unobstructed, and that all fire personnel are familiar with its location. Be prepared to pump into the first floor standpipe outlets if need be. Since the standpipe is most often dry during construction, it isn't unusual to find the floor outlet valves left in the open position, especially the ones in the cellar, which are used as drains for the plumber. In the event of fire, when the hoses to the fire department connection FDC are charged, the water will pour out of these valve outlets until they are shut off manually.

First-arriving units at a fire in a building under construction should make it a habit to test the first-floor outlet for water before proceeding upward. If no water is found, it would be wise to assume that the riser is dry and that numerous valves will be open. Have a member check each floor outlet while ascending toward the fire. Two other conditions are found involving the standpipe systems of buildings where work is in progress: The top of the riser is left open without a shutoff of any kind, which will render the system useless, and a sectional valve may be installed to divide the system into two or more zones. These valves may be closed to facilitate pressure testing of a segment while leaving the other segments intact or for other reasons. If they are left in the closed position, however, they may place the system out of use. These valves are all of an indicating type, usually Outside Stem and Yoke (OS&Y) style. They should be examined if you have a problem obtaining water from a system that is known to be charged. Their location may vary. A sectional valve is usually on a mechanical equipment floor, but it may be in a staircase as part of the riser. The only way to be sure of the location is to trace the pipe, checking for water flow as you go. The shutoff valve will be somewhere between the outlet that has water and the first one with no water.

Structural Problems of Buildings under Construction and Renovation

A building in which work is in progress is often more prone to structural failure and collapse than it would be in its completed, occupied state. This is due to a number of factors, including the lack of floors or roofs. A completed structure relies on each of its major structural components to brace and support the others. A wall that is carrying a load (a bearing wall) is more stable than a wall with no load (a free-standing wall). This is because of the downward compression on the wall as well as the horizontal bracing provided by the floor and the roof. Structures that are missing all or parts of their floors or roofs may be unstable even without being exposed to fire.

When fireproof coatings or membranes are missing, steel may be exposed to rapid rises in temperature, which can lead to failure. Steel I-beams and columns often receive a spray-on fibrous coating that protects them from this heat. If this coating is missing, the steel will lose strength at approximately 1,500°F. Often many so-called noncombustible buildings rely on a hung ceiling to provide a membrane-type fire resistance between the occupancy and the otherwise unprotected steel. The hung ceiling tiles often aren't installed until much of the rest of the construction has been completed and the occupants have begun moving in (Figs. 17–10, 17–11, and 17–12).

Fig. 17–10 Sprayed-on fireproofing is often scraped off by workmen as they make modifications to a structure, leaving the exposed steel prone to failure under fire conditions. This major girder in an unsprinklered hotel could endanger the entire building.

Fig. 17–11 These steel cables, secured in position with wooden 2x4s, provide temporary support for a large structural steel section. Fire attacking either the cables or the 2x4s could precipitate failure. Just 800° is all that is required to cause most cable to fail under tension.

Fig. 17–12 This post-tensioned concrete structure under construction is at a vulnerable stage. The steel cables that apply tension to the floors have been placed under load but are left exposed to the heat of any fire.

Steel bracing cables, used to provide temporary support until additional bracing or framing is in place, are particularly of interest in buildings under construction. Exposure of these cables to fire is quite likely to cause extensive collapse, since the cables fail readily, and they are often the primary supports or bracing. A heavy fire in an area in which numerous steel cables act as supports demands either immediate cooling of the cables or prompt evacuation of the surrounding area.

Perhaps the most serious collapse threat in buildings under construction occurs in structures that, when finished, would provide the greatest resistance to collapse: poured-in-place reinforced concrete structures. These buildings are created by erecting a wooden mold in the shape that the concrete is to assume, then filling this wooden form with wet concrete. The concrete has considerable weight, yet almost no strength at this stage. The strength develops as the concrete dries or cures. It takes about 48 hrs, depending on temperature, for the concrete to harden enough even to be able to support is own weight. It is during this 48-hr period that the structure is most vulnerable to collapse due to fire. The floor below the most recently poured concrete is a veritable lumberyard of plywood and 4x4 posts. Since the forms are built as needed, there is often a good deal of scrap wood and sawdust lying around from the cutting and fitting of the formwork. The 4x4 posts are placed every 4 ft in each direction to support the weight of the concrete. The entire ceiling is made of plywood, which has often been treated with a combustible coating so that the concrete doesn't stick to the wood (Figs. 17–13, 17–14, 17–15, and 17–16).

Fig. 17–13 The poured-concrete building under construction can be a firefighter's death trap. The upper floors are literally a wood-frame building holding up tons of cement. Note the concentrated loads formed by the bundles of 4x4s.

Fig. 17–14 In poured-concrete buildings, 4x4 posts are used to hold up the plywood mold.

Fig. 17–15 This floor has cured for at least 48 hrs, but the 4x4s should warn firefighters that collapse is still possible.

Fig. 17–16 This floor has been totally stripped of shoring, and firefighters shouldn't have to worry about collapse from a fire here.

Any fire originating or extending to this area will spread rapidly. Failure of the 4x4 support posts is likely to result in collapse of the entire mass of concrete above. If the fire doesn't cause this failure, there is still the danger of the posts being dislodged by firefighters operating hose streams or during overhaul. The weight of this mass of concrete, reinforcing, and formwork collapsing onto the floor below could overload it, causing the floor below to collapse in turn onto the one below it, setting up a chain reaction that could go all the way to the lowest floor.

A progressive collapse of this nature took down all 27 floors of a building under construction in Bailey's Crossroad, Virginia, killing 14 construction workers. They could just as easily have been firefighters had a fire precipitated the collapse; instead it appears that workers manually removed the critical shores. It takes concrete approximately 28 days to reach its greatest load-carrying capacity. After the first 48 hrs, the plywood formwork is stripped off to be reused on a higher floor. For the next 16 to 20 days, the floors will still be braced with 4x4s, spaced every 8 ft. This should indicate to fire forces that any fire on or above these braced floors could cause a collapse. After the concrete has cured sufficiently to allow the 4x4 posts to be removed, the threat of collapse from fire exposure is greatly reduced, but it hasn't been eliminated.

Concrete of any age exposed to heavy fire is subject to *spalling*—the separation of chunks with explosive force, caused by the expansion of water trapped in the concrete as it is turned to steam. Very substantial chunks, weighing 100 lbs or more, can collapse as a result of exposure to fire. Fortunately, it usually only occurs where there is direct exposure to flame, thus reducing the danger to firefighters, since it is unlikely that a firefighter would be in an area with large quantities of open flame without an operating hose stream. Cooling the heated concrete reduces the chances of spalling, but this can take tremendous quantities of water even after the flame has been darkened down.

Danger of Partially Occupied Structures

The trend of allowing part of the building to be occupied while work is in progress poses great challenges to firefighters. Such a practice may be acceptable in a one- or two-family home, with several windows or doors leading from each occupancy, but it creates disproportionate life hazards in larger structures, particularly when the work is taking place on a floor below the occupied levels. Buildings under construction don't begin to generate a positive cash flow until the floors have been rented. Consequently, there is great pressure to get the tenants in and paying rent as soon as possible. Construction and occupancy patterns, however, are sometimes different.

Whereas the upper floors of a structure may be the last to be built, they are often the first to be rented, due to the desirability of a view and the prestige associated with having offices on the top floor. In the meantime, the lower floors often remain unoccupied until a later date, with the result that these lower floors become storage areas for construction materials being used to finish the other floors, creating a heavy fire load. To compound the problem, most construction codes allow these floors to remain unsprinklered, even though the finished areas, often with lighter fire loads, require sprinklers (Fig. 17–17). The feeling is that it would be unreasonable to ask the builder to have sprinklers in an area where no tenant is yet paying for them. I say this is utter nonsense.

Fig. 17–17 This all-wood construction shanty would provide quite a bit of fuel to expose the occupied apartments on the floor directly above.

It is a very simple matter to install a basic, exposed-pipe sprinkler system with 1-in. outlets leading to each head, instead of having the head mounted in the fitting directly with its ½-in. thread. This is what is going to be installed later, anyway! All that is necessary is a slightly more conservative estimate of spacing when designing the initial layout. At one high-rise building in which I was responsible for the design, a basic loop was installed with the sprinkler lines spaced only 10 ft apart (100 sq ft per head). This seemingly excessive spacing was soon accounted for as the occupancies began constructing partition walls for offices, closets, and the like. The more liberal spacing of the initial requirement soon proved that it would have been unsatisfactory. As a result, considerable savings were realized in design and construction, since the larger-diameter supply main could be laid out and installed while there were still very few obstacles in the way. Later, once the wall layouts had been decided, it was a simple matter to run a smaller 1-in. pipe over, (meaning sideways) several feet to provide a required head instead of having to redesign or relocate larger piping.

On the other hand, if builders object to installing so-called temporary sprinkler systems, there is an alternative: Forbid the use of any floor above the levels where all fire protection systems are not operational. This may sound like a severe restriction, especially to the real estate industry, but it is the only way to ensure the safety of the occupants above. Remember that the spread of fire is unusually rapid in buildings where work is in progress. This is so even in partially occupied buildings where all of the doors and windows are in place. Stairwell doors and doors between occupancies are often blocked open to facilitate the passage of material from one area to another. The result may be that a fire contaminates all of the stairways in the building, trapping any and all occupants that happen to be caught above it.

Although the movie *The Towering Inferno* may have overdramatized the fire, it accurately depicted the foolishness of allowing the occupancy of areas above the location where all fire protection and building features have been installed *and tested*. The only way to prevent the actual occurrence of such a catastrophe is through a combination of proper engineering of the structure and its protection systems, enforcement of the applicable codes with an eye toward fire prevention and extinguishment, and educating the owners and occupants of the buildings about the proper actions to take in the event of fire.

Buildings Undergoing Renovation and Demolition

The most dangerous building in the United States to fight a fire in is the vacant building. The firefighter death rate at vacant building fires is more than 5 times that of fires in occupied residential buildings. This is due to many factors, including old age, damage done by weather, vandals and previous fires, removal of structural supports during renovation or demolition (Figs. 17–18, 17–19, and 17–20) and the rapid fire spread caused by removal of doors and other fire barriers such as ceilings. All in all, these are firefighter traps. I know. I speak from experience. I was nearly killed several times in these structures. When a chief warned me to "be careful" after one particularly close call, I replied, "Chief, nobody is more careful than me in a vacant building, but we still almost got killed." I meant it. Nobody hates operating in a building with all these hazards more than I do, and I really, really don't want to get anybody hurt when there is only property at stake, especially property that may be in the process of demolition anyway. The fact is however, that no matter how careful you try to be, if you operate in these dangerous buildings, you are taking extreme risks that probably can't be justified if all the civilians are out.

Fig. 17–18 This building has been completely gutted for renovation. The hazards to firefighters are many, including open floors, extremely fast vertical fire spread, and weakened walls that can collapse.

Fig. 17–19 After renovation or rehab, the same building retains is classical old style look, indicative of standard Class 3 construction. It contains a big surprise for unsuspecting firefighters, however...

Fig. 17–20 ... during the rehab, the original 3x12 wooden joists were all removed and replaced with C-joists, lightweight steel floor supports that fail readily during a fire.

How do you stay safe at these structures? It takes a major attitude adjustment. Just because it looks like a building, doesn't mean it is a building. Particularly in the case of buildings undergoing demolition, the *building* is nothing more than a pile of rubbish that is going to be carted away the next day or week.

Two other firefighters and I were nearly killed during a fire in a building under demolition many years ago. Fire suddenly extended toward a group of us who were waiting for water while attempting to stop extension. In the scramble for a safe haven, three of us were knocked into the staircase. Unfortunately, the *staircase* consisted only of a few iron stair risers, the treads and everything else had already been removed by the demolition crew to speed their dropping debris to the cellar for removal. We were lucky. The few remaining risers held us and prevented us from falling seven stories to the cellar. We should never have been in that building in the first place. Fortunately, the FDNY has long since revised its tactics for fires in vacant buildings and other structures where there is no civilian life hazard and no property of major value to save (Fig. 17–21). All it took was an attitude adjustment brought about by a number of tragedies. Don't you pay for the same lessons.

After a series of tragedies at vacant buildings in the late 1970s and early 1980s, the FDNY began a program to inspect and mark every vacant building in the city with a system that indicates the hazards present and serves to indicate to firefighters what type of attack should be conducted. The building is marked with a box near the main entrance and any other remote entrances. An empty square indicates the building is of normal structural stability, and normal interior operations may be conducted, depending on the extent of fire found on arrival (Figs. 17–22 and 17–23).

Of course, the fact that the building is marked indicates it is vacant, and operations should be conducted more slowly and with greater caution than at a building with a life hazard. A box that has one line drawn diagonally through it indicates a building with some serious hazards, such as holes in the floor, and signals that all operations should be conducted from the exterior if possible. If the fire is small and interior operations are necessary, they should be conducted only after a careful examination for hazards. A building marked with the X as shown in Figure 17–23, indicates a building with severe hazards, and all firefighting should be done from the exterior. If a small fire requires a crew to enter, permission must be sought from the IC. A careful examination must be conducted prior to entry, and operations shall be conducted with as few people as possible, minimizing exposure time to dangers. Use of exterior streams is preferred at all times, something not ordinarily done at occupied buildings. What you must keep in mind though is that *we are the only life hazard at a vacant building!*

Fig. 17–21 *This is the view up the remains of a staircase from the cellar through the roof seven stories above. The building was undergoing demolition when it caught fire. I nearly fell the entire height when sudden fire extension caused an unplanned retreat.*

Fig. 17–22 *Vacant buildings pose severe dangers to firefighters. The FDNY marks every vacant building in the city to indicate the dangers to members and what type tactics might be allowed. A box with one slash through it indicates hazards that should prompt extra caution in the event it is necessary to enter the structure. Note the cinder block sealed windows.*

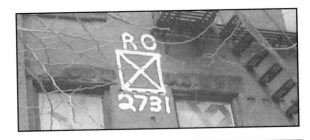

Fig. 17–23 *The X in the box on this building indicates only an exterior attack should be conducted. The RO indicates the roof is already open (vented) so no one should have to go on the roof.*

18

Fire-Related Emergencies: Incinerators, Oil Burners, and Gas Leaks

Natural Gas

A large percentage of the total number of fire department responses each year involve incidents in which, while there is a need for emergency assistance, there is either no fire present or the fire is within its normal container. These may be termed emergency responses. A working knowledge of how to handle these emergencies is vital to the efficient performance of duty and may, in fact, save lives, both firefighter and civilian. Gas leaks are among the most common of these emergencies.

Natural gas is formed as a result of nature's action on organic matter over millions of years. Texas, Oklahoma, California, Louisiana, and West Virginia are the largest domestic producers of natural gas. Once it has been tapped from the ground, it is then distributed to consumers nationwide via pipeline.

Natural gas is primarily methane (more than 90%), and it behaves much like pure methane. It contains other gases as well, including ethane (up to 5%), carbon dioxide, and nitrogen. The presence of ethane in natural gas can sometimes be used by fire departments and utility companies to pinpoint the source of persistent gas odors that have no apparent source. A sample of the area is collected and analyzed in the laboratory. If the sample contains natural gas, the utility must look further for the source. If no ethane is present, the source isn't natural gas, but possibly swamp gas or other hydrocarbon vapors.

Natural gas as it comes from a well is colorless and odorless, yet everyone has "smelled gas." That is because an odorant is added at a precise rate, so that as little as 1% of gas in the air can be detected. This odorant is generally mercaptan compounded with sulfides. As little as ¼ lb of odorant can treat up to 1 mil cu ft of natural gas. Obviously, this is a powerful compound.

This process of dispersing a liquid in the natural gas has resulted in some unusual circumstances of which the fire service—and in some cases, the public—should be aware. First, the odorant tends to be lost as it travels long distances with the gas. Some of it breaks down chemically, while some of it condenses on the inside of the pipeline. At various points along its route, usually where the local utility hooks up to a transcontinental pipeline, odorant must be added. A leak at the odorant area is a flammable liquid spill, and a very pungent one at that.

An unusual case occurred some years ago involving a large relief valve on a 12-in. gas line that malfunctioned and was venting large quantities of gas to the atmosphere. As the apparatus left its quarters, about half a mile from the scene, its crew encountered a very heavy odor of gas. This was frightening, but something didn't seem right. The members pondered the dilemma, knowing how gas is supposed to behave. Its vapor density is .60, lighter than that of air. The gas should have been rising rapidly, not settling down a half-mile downwind. In fact, it *was* rising, but as it drifted by, the tiny mist like droplets of odorant were settling to earth—a gas odor that couldn't be detected by a combustible gas indicator.

Common Natural Gas Emergencies

Natural gas emergencies can be divided into the following three basic categories: inside leaks, outside leaks, and leaks resulting from fires. Although each kind has its own dangers and unusual concerns, the greatest danger results from interior gas leaks because of the potential for explosion.

Outside leaks are the next most dangerous. Gas may collect in buildings, manholes, or other spaces and possibly cause an explosion. Gas that has ignited is probably the least of firefighters' problems. Although it may ignite nearby exposures, the potential for explosion is extremely small.

Distribution systems and device related problems

Gas is shipped from wells across the United States via large-diameter, high-pressure transmission pipelines. These transmission lines operate at pressures of 350–850 psi and are remotely monitored and controlled from central locations. A rupture of one of these lines is a major emergency, often resulting in fire. By virtue of remote sensors that monitor the pressure, pipeline operators will usually be able to detect a leak immediately and isolate the section that has been damaged by remote-control valves.

Because of the large diameter of the lines, the length between valves, and the high pressure at which they operate, however, considerable *line pack* will result in leakage for an extended period even after the valves have been shut off. Fire department operations should be directed toward evacuating the endangered area, protecting exposed buildings, and examining these buildings for evidence of gas seeping into them. No attempts should be made to close any valves without the specific direction of knowledgeable pipeline personnel (Fig. 18–1).

Gas is received by the local utilities at various points along the pipeline, called *city gate stations*. These stations reduce the pressure from the pipelines down to the utility company's transmission system pressure of 60–150 psi. These lines continue to district regulator stations, usually in an underground vault, where the pressure is further

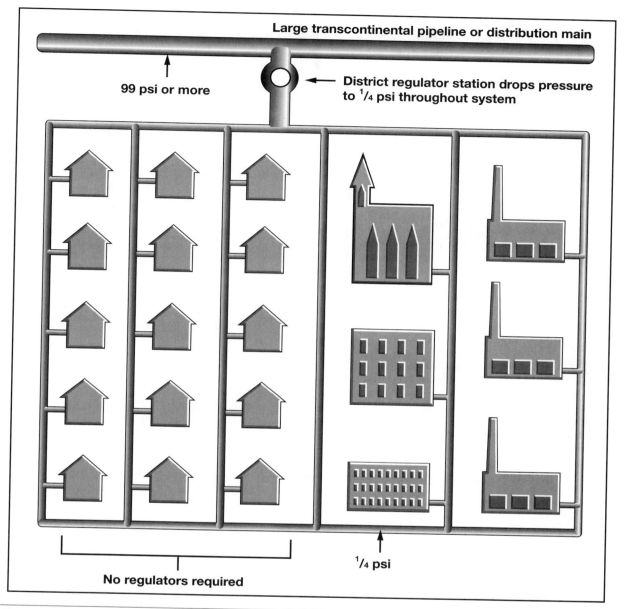

Fig. 18–1 Illustration of a low-pressure gas system.

reduced for the distribution system. The pressures in the distribution systems vary according to type and age. The older systems are known as low-pressure systems, operating at ¼ psi. The gas mains are all at the ¼-lb pressure, and no further reduction of the pressure occurs. The gas flows through the distribution mains, shutoffs, meters, and appliances at this same low pressure (Fig. 18–2).

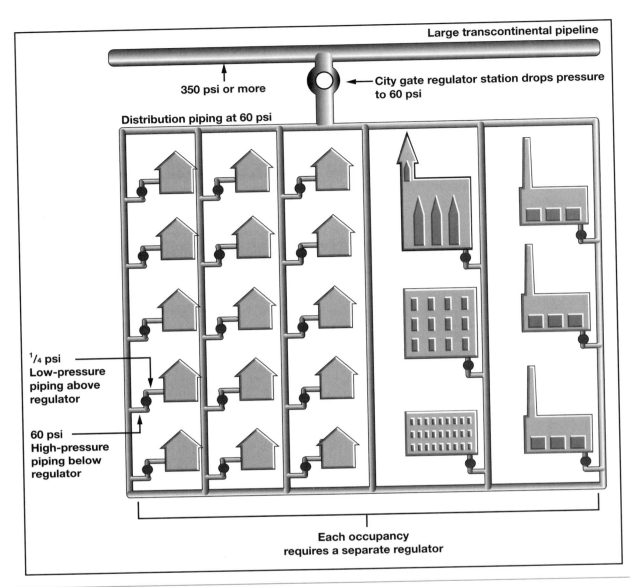

Fig. 18–2 Illustration of a high-pressure gas system.

In newer systems, the pressure in the street mains is higher, between 2½ and 60 psi, and is termed *high-pressure gas*. Gas appliances cannot use this high pressure, so a reduction must also take place at each building. Thus, each gas line going into the premises from the street has its own regulator to reduce the pressure to ¼ psi.

There are two kinds of regulators. The newer ones are spring-loaded to maintain a steady ¼ psi against a diaphragm. The older kind is called the dead-weight regulator. It uses a column of mercury to maintain the pressure on the diaphragm. Dead-weight regulators are rapidly being replaced. As mechanical devices, both kinds of regulators are subject to failure. Should this occur, excess pressure (above ¼ psi) flows through the regulator, with most of it flowing out through a vent in the regulator body. The regulator should hold the pressure to the appliances in the 2- to 4-psi range, but that is still much higher than ¼ psi. This excess pressure can make existing burner flames much, much larger, igniting nearby combustibles. The sudden surge may otherwise blow out the pilot and burners, allowing for a buildup of un-ignited gas (Fig. 18–3).

Regulators located inside a building have the vent line piped out through the wall to a *peck vent* or a *fisher vent* to make sure that gas from a regulator failure doesn't build up indoors. Fire department units can recognize a regulator failure by the gas odor, as well as the hissing sound coming from the vent. The following actions should be undertaken in such an instance:

1. Notify the utility.

2. Shut off the gas.

3. Search the involved premises for fire extension, a buildup of gas, and overcome victims.

4. Vent the area as needed.

Natural gas isn't toxic, but it will displace oxygen and, if the quantity is large enough, it can cause asphyxiation. In years past, some utilities used manufactured gas, which contained carbon monoxide. In those days, it was possible to commit suicide by turning on oven jets with the pilots blown out and inhaling the carbon monoxide laden manufactured gas. Today, natural gas doesn't pose as great a health hazard.

Fig. 18–3 A typical high-pressure gas service consists of (from bottom left) a service cock or valve, a regulator (note the vent pipe extending to the left), and a meter.

One problem with natural gas utilities that truly brings with it the potential for mass disaster and casualties is the failure of a district regulator, which is found on all types of gas distribution systems. When a district regulator fails in the open position, the pressure on the downstream side is increased, resulting in overpressure to every gas service in the area. As a result, the fire department may be faced with a deluge of gas-related fires, leaks, and explosions. During one such failure in Chicago, building after building in an 11-sq-block area experienced gas problems resulting in several explosions, numerous multiple-alarm fires, and countless gas-leak calls. Fire personnel, particularly dispatchers, may be the first to recognize what has happened from the sudden pattern of alarms being received. The fire department cannot do anything directly to solve the problem other than notify the utility company of the locations where the incidents are occurring. Through remote pressure-monitoring devices that sense the pressure changes downstream of the regulator, they will most likely know about the failure before we do. They may not, however, know how extensive the problem is.

Fire department actions in such a case will involve massive operations for a large number of incidents in a small geographic area. Upon recognizing this type of incident, get large numbers of additional resources rolling toward the vicinity of the incident. Set up a staging area outside the affected area. Units will be required for firefighting, structural collapse from explosions, and for shutting down gas services to every building within the area. Use fire and police PA systems to notify occupants of the affected buildings to turn off all gas appliances and building supplies immediately. This may prevent some incidents. These are rare events that will require close cooperation with the local utility.

Fire officers must recognize the pattern of these alarm responses and take decisive action to prevent the incidents from escalating. Units may have to be assigned to a block with the task of entering and shutting off the gas services to every building on it. These personnel will require forcible entry capability and SCBA, and they should split up to cover as many buildings as possible. The utility company will have to be consulted to determine the boundaries of the zone that is affected by the particular regulator, and utility crews will have to shut down the malfunctioning device.

Eventually, personnel will have to enter every building within the affected area to turn off the gas service valve before gas can be restored to the area. The fire department should ensure that each structure has been searched for overcome victims, particularly if the event occurs during the sleeping hours.

Tactics at inside leaks

What actions must be taken in the event of a natural gas leak in a structure? In what order and by whom? Any plan of action should include the following:

1. Notify the utility and request their estimated time of arrival.

2. Determine the intensity of the leak and when it was first noticed.

3. Determine the extent of evacuation required.

4. Eliminate the sources of ignition.

5. Locate the source and stop the flow.

6. Search and ventilate.

The first priority on report of a gas leak should be to notify the utility. Their expertise and equipment will often prove invaluable. Request their assistance early. This should be a predetermined action by the dispatcher, without waiting for fire units to arrive at the scene.

A thoroughly trained dispatcher can also often begin steps 2 and 3. While questioning the caller, he should ask, "How strong is the smell of gas?" and "When did you first notice the smell?" A strong smell, or one that has just appeared, indicates a potentially serious situation. In this event, the dispatcher should advise the caller to leave the premises and await the fire department. This won't be so obvious to the caller at other times, and the determination of intensity and required evacuation will have to be decided by the officer in command (Figs. 18–4, 18–5, and 18–6).

Fig. 18–4 The only reliable way to determine the dangers of a gas leak is with a combustible gas detector.

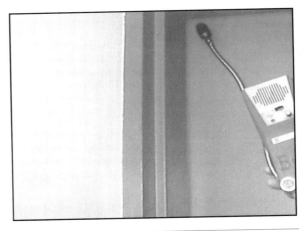

Fig. 18–5 This detector can be used to trace the source of leaks. Check for natural gas at the upper levels-it's lighter than air.

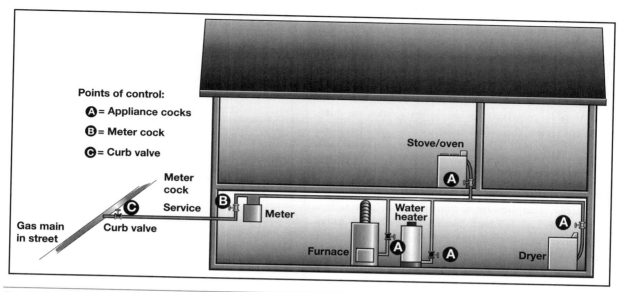

Fig. 18–6 An illustration of typical residential gas piping showing several potential points of control.

Expect the worst if you are met by a heavy odor of gas. Expose as few people as possible, meaning evacuate promptly when necessary. A faint whiff of gas, or a leak that has come and gone for the past few days, may allow more leeway for performing a leak examination. A combustible gas detector is an essential tool at any incident involving odors of gas, since the odorant may have been filtered out of some underground leaks, and even the best nose can't tell the difference between 3 and 4% gas in the air. Most of these devices will indicate the concentration of gas in the air, expressed as a percentage of the lower explosive limit (LEL). One type of gas detector the FDNY uses does not directly display the concentration of gas at all but has proven to be exceptionally useful in pinpointing the origin of even small leaks. The FIT 8800 detector sounds an audible alarm that rises in frequency as the concentration of combustible gas increases, thus allowing the user to follow the sound to the source of the leak. Many times an odor of gas is reported in the hallway of an apartment house, with no known source. This detector, held against the edge of closed doors can locate the correct apartment for entry.

For faint odors, always check the condition of pilot lights as a first action. If you suspect a leak, pour soapy water over the suspect area to try to confirm its presence. When you find a leak, always try to isolate the area as close to the leak as possible. The problem can be stopped by turning the quarter-turn appliance valve nearby, leaving the rest of the premises unaffected. If that isn't possible, move back along the supply piping to the next point of control—the meter wing cock, generally another quarter-turn valve just past the meter.

Common gas service is often present in multiple-tenant occupancies such as apartment houses and shopping centers. All of the tenants have their own meters and meter cocks. Additionally, where the service enters the building, there will be a master valve called the service entrance cock, which stops the flow to all tenants. Use this valve with discretion, since a minor leak at an appliance doesn't warrant shutting off 20 or 30 apartments. Conversely, if there is a major leak and you experience difficulty in determining which meter controls which apartment, the service cock provides the fastest means of control (Figs. 18–7 and 18–8).

Fig. 18–7 A large leak in a building with numerous meters may require shutting down the entire building. For most leaks though, isolate the individual apartment by closing its meter cock.

Since natural gas has a flammable (explosive) range of 4–14% in air, the mixture must be kept out of this range. Remember, the odorant permits us to smell as little as 1% gas in the air. That 1% is still 25% of the LEL. Only 3% more gas is needed to create the danger of explosion.

All that is required to ignite an explosive gas/air mixture is the tiniest of sparks or an open flame. Something as small as the spark that occurs inside a light switch when it is thrown on or off or the static spark created after walking across a carpet, can ignite this gas/air mixture. Don't throw any switches or ring doorbells in gaseous areas. Also, beware of your own portable radio. Don't use it to call for help from within the gas area, because most radios aren't rated for use in flammable atmospheres. Most fire department pagers, on the other hand, are safe. Also beware of cell phones, turn them off or leave them on the rig when investigating gas leaks. You never know when someone else is going to call you! (**Note:** Intrinsically safe—nonexplosive—phones are available and should be used by fire personnel. I was personally thankful for this device when I was operating at a major gasoline barge explosion on Staten Island. As several members and I approached a leaking gasoline pipeline to assess the damage to it, one of my office staff called to see if we would be needing additional supplies or personnel. Whew!)

Anything electrical is a potential source of ignition. Attempt to disconnect power to the building if it can be done safely. This may involve tripping the main breaker (if it's remote from the gas area) or cutting the service entrance wires outside the building. Most utilities advise fire departments, as well as their own people, not to pull the meter, since it is connected by a pipe to the inside of the building where the gas is located. Gas has sometimes drifted to the meter pan and been ignited when the meter is pulled. Remember, don't throw a switch inside the area with gas around. When outside at the meter, always assume that there is gas in the pan (Fig. 18–9).

Fig. 18–8 In an emergency, the entire gas supply to all of the occupancies in this building can be shut off by turning the service entrance cock. It is located just above ground level on the right, below the tee where the piping splits off.

Fig. 18–9 This pile of debris is all that remains of two buildings, one of them a four-story brick structure that exploded as the result of a natural gas leak. Six people were killed, including passersby in the street.

Actions 5 and 6, locating the source and stopping the flow and searching and venting, require a great deal of input at the scene. Some items to consider are the following: Where should venting be done? Who should vent? Should the building be shut down at the curb valve? (A curb valve controls the gas service into the building. As its name implies, it is often located near the sidewalk or curb. Not all buildings have curb valves. As a guide, low-pressure services and services with outside meters often don't have curb valves, whereas high-pressure, inside services often do.)

Determining where to vent is simple. Gas is lighter than air, so vent at the upper areas. Don't forget to vent the blind spaces at the top, such as attics and cocklofts. When venting, don't cause unnecessary damage, but be thorough. Opening windows is generally sufficient. It is rarely necessary to break them. Don't use smoke ejectors to suck gas out of the area, even if they are labeled explosion proof. The motor may be explosion proof, but the blades may strike the housing, or some other mishap may occur. If additional flow is needed in venting, use the fans outside to blow fresh air in.

Those who are searching the building for occupants, the source of the leak, and the location of the shutoffs can usually vent at the same time. Don't over-commit manpower. Send in only enough persons to do the job. Everyone else should stand by in a safe location, such as on the opposite side of a pumper, where they will be less exposed if the building does blow.

Where to vent depends on several factors that must be determined in the field. Generally, begin venting as soon as possible, but weigh how large a leak is present, the sources of ignition and whether you can control them, and the status of the gas/air mixture—is it below, within, or above the flammable range?

Some years ago, a 2-in. high-pressure (55 psi) gas main broke inside a 200- by 600-ft, fully occupied department store at 3:30 PM. Firefighters were met by the store manager and plumber who described the conditions, location, and evacuation, and then promptly fled! After ensuring that the emergency response of the utility companies (both gas and electric) had been launched, the initial actions included evacuation and closing off all of the gas entering the building, thus removing pilot lights as a source of ignition. A four-man crew equipped with SCBA and a leak-stopping kit entered the building and discovered that there was no possible way to stop the leak except by the street shutoffs, since the curb valves had been covered over.

Additionally, the combustible gas indicator showed 100% of the lower flammable limits at the front door and over the explosive range in the interior. With the evacuation complete, it was time to recognize a hopeless case. All of the units were withdrawn approximately 300 ft, to an area behind the apparatus, to await the shutdown of gas and electricity. The crew from the electric company was ready to do their job first, and as they did, something occurred that no one had anticipated. When the electricity was disconnected, it automatically kicked in the relays to the emergency lights, and instantly, dozens of potential sources of ignition occurred throughout the building as the relays closed in each lighting unit. About 45 min. later, the gas supply was finally shut down, and two 4x8 skylights on the roof were vented.

Luck was with the firefighters that afternoon. The gas had gone through the explosive range without hitting a single source of ignition. It was above the explosive limits when, by cutting the power, potential sources of ignition were introduced all over the store in the emergency lights. Venting was delayed until all sources of ignition were removed, then the gas was brought back below the explosive range by venting.

Of all the decisions to be made, where to stop the leak—inside or outside—has the most variables. The practicality of either means must be determined, and a curb cock isn't always present. The decision to send personnel inside should be made after considering the risks. Whether it's natural gas, liquefied petroleum gas (LPG), gasoline vapors, or any other flammable vapors, unless there is a known life hazard present, treat the situation as the potential time bomb it is and expect the worst. Again, use the minimum number of people needed to do the job, and make sure that they are properly trained and equipped with SCBA, forcible entry tools, explosion-proof lights, and an 18- or 24-in. pipe wrench.

Gas emergencies usually involve at least one engine company. Although its personnel might be used to perform the previous duties, enough personnel must be available to stretch a line, if needed. The line should be long enough to cover the entire building. It shouldn't, however, be placed where it might be exposed to blast damage. With this in mind, try to position the apparatus to provide the most shielding effect for the pump operators and nonessential personnel. The water supply should be consistent with the expected involvement. Remember, the building is being filled with a flammable gas, and heavy fire should be expected, although it may be localized near the source of the gas after any blast.

Tactics at outside leaks

Leaks outside of structures, although not as common as inside leaks, can be just as dangerous. That's because the gas takes the path of least resistance as it tries to escape to the atmosphere. Often that path is along the gas service pipe or other underground lines into buildings or manholes. This migration is insidious because, as the gas travels through the ground, it tends to be deodorized as the soil filters out the odorant. Underground leaks tend to migrate great distances before they are discovered. This is especially true in areas that are largely paved over.

I recall firefighters being unable to recognize a musty smell as gas leaking into the cellar of their own firehouse. This highlights the unreliability of the human nose as a gas detector. The nose can easily detect 1% of gas in the air if you walk in from clean air, but after a short length of time, the nose becomes desensitized even to dangerous concentrations. A well-maintained combustible gas indicator (explosimeter) is the only way to check suspected areas safely. It is preferable to use two separate meters to check a given area, since any single meter could be unreliable (Fig. 18–10).

Winter is particularly dangerous in many areas. Once the frost line penetrates more than a few inches, it effectively puts a lid on the soil, making migration more likely. Gas lines are subject to greater stress during periods of repeated freeze and thaw and may fail more often. In the event of an underground leak, check the basements of all of the buildings in the surrounding area. Use a reliable gas indicator, and pay particular attention to areas where service lines penetrate the foundation. When positioning apparatus, make sure neither the apparatus nor the operator is over a manhole.

Fig. 18–10 At underground gas leaks, be sure to check the surrounding exposure cellars, wherever any utility pipes enter the foundation. Gas can migrate a great distance underground, especially during freezing weather.

Many outside leaks are caused by contractors excavating in the area. Generally, the safest course of action is to:

1. Notify the utility.
2. Approach from upwind.
3. Stop the sources of ignition.
4. Await the utility.

Do not touch any valves located in the street. If absolutely necessary, you can stop leaks in smaller, low-pressure steel lines by using various plugging and patching devices, but you should do this only as a last resort. Under no circumstances, however, should fire department personnel ever attempt to stop a leak in a plastic pipe. Whenever any fluid travels through piping, it creates a static electrical charge in that pipe. With steel pipe, that current is drawn off and safely dissipated to ground by the pipe itself. Plastic pipe is an insulator, however, and it is probable that the pipe will have a static charge of up to 30,000 volts. A person grabbing the pipe to apply a plug will likely discharge this current, creating a spark, igniting the gas.

Tactics at fires

Operations at fires should be similar at indoor and outdoor operations. As always, the first steps are to call the utility company, control evacuation, protect exposures, and let the fire burn until the supply of gas is shut off. Small fires may be extinguished with dry chemical, halon, or CO_2, if necessary, to get to a valve or to save a life. For larger fires, fog streams can be used to approach the valves. Use care when placing hose streams where excavations have ruptured the gas line. Try to keep unnecessary water out of the pit, since utility crews may have to get in there to stop the leak, and the water could compound their problem (Fig. 18–11).

Fig. 18–11 The solder holding this gas meter melted when exposed to a relatively minor fire in this cellar, rapidly creating a very large fireball. That fire further melted the adjoining aluminum meter. Take 2½-in. hose for cellar fires.

Unusual Problems Involving Natural Gas

Try to solve the following problem based on information from the preceding sections.

You are in command of a fire department rescue company assigned to a report of a major gas leak. As you approach the scene, you notice several gas utility crews excavating in the area and are met with a heavy odor of gas for several blocks. The first companies on the scene have investigated, but no leak has been discovered. Many occupants of several large apartment buildings have begun to self-evacuate, fearing an explosion due to the strong gas odor.

You begin sampling with combustible gas indicators and contact the utility crews to see whether they have a leak. The answer is negative on both accounts. No readings on the meters and no leaks on the mains, and the gas company's meters and office monitors show no leak, either. But the odor is still very heavy. Now what do you do?

"Start investigating upwind!" you say, but the direction upwind is also uphill. Still no leaks, and now you are up to the edge of a college campus. Could the answer be there?

Yes, it could. A short ride to the chemistry lab and a little detective work reveals that a small chemical spill has occurred. A pint bottle of methyl mercaptan was emptied into the storm sewers, and its vapors have permeated the neighborhood. "Case solved!"

Another unusual situation to be aware of involves the vapor density of gas. As mentioned previously, gas is much lighter than air, and usually it rises rapidly. During extremely cold weather, however, and especially when it is cold for extended periods, the demand for gas is at its peak. In some cases, pipelines and gas reserves are insufficient to meet this demand. During such a period, some utilities mix in LPG gas to make up the difference between supply and demand. The vapor density of propane is 1.52; for butane, it is 2.01. Depending on how much of these gases are added to the natural gas, the vapor of this *peak shaving gas* will get heavier. At times, it may approach 1.0, which is the weight of air. This may mean that leaking gas won't rise as quickly as expected and it may require mechanical ventilation. Contact your local gas utility to find out if they ever have occasion to use *peak shaving gas* or any mixture of other gases with natural gas.

Natural gas is a safe, clean, efficient fuel when it is properly contained and used. When it is allowed to accumulate where it doesn't belong—where heat and oxygen are available—the potential for tragedy exists. Such catastrophes don't occur often, considering the tremendous volume of gas used every day and the astounding network of pipes and storage facilities that cover this continent. When something does go wrong, however, cooperation between the fire department and the utility companies is essential.

Liquefied Petroleum Gas

A homeowner decides that today is the day that he will drag the barbecue out from the garage and treat the family to the first flame-broiled steak of the season. Without any thought of inspecting the device for safety, he lights up the match and fires this baby up. What this hapless individual fails to realize is that there is a propane leak somewhere near the neck of the cylinder. It could be due to a faulty regulator, loose connections, a ruptured hose, a split fitting, or any one of several other causes. At the moment, the exact cause doesn't matter, but what happens next does. When the gas reaches the burner head (or when burning grease drops down to the gas), the ignited leak immediately exposes the metal components at the top of the cylinder, the steel shell of the tank, the brass valve, and the aluminum valve handle and regulator. At this point, the fire is usually a small one, particularly if the leak is on the low-pressure side of the regulator. The homeowner may successfully extinguish it at this stage simply by turning off the valve. Many times this isn't done, however, either because of the person's lack of knowledge or for fear of injury. It is usually at this point that the homeowner calls the fire department. Simply applying a stream from a garden hose could prevent these conditions from getting out of control; however, most citizens don't realize how drastically things are beginning to change.

One of the first things that happen when flame exposes the upper part of the cylinder is that the pressure begins to rise inside. About the same time, the aluminum or pot-metal valve handle melts away, to be followed shortly by failure of the aluminum pressure regulator. The fire may have been burning for 2 min at this point. The alarm has just been transmitted by the dispatching center. With the melting of the valve handle and regulator, the problems escalate swiftly, with the fire increasing in size as the full cylinder pressure pushes against the leak. This further raises the pressure inside the cylinder, weakening the steel tank shell. The first apparatus is just leaving the firehouse.

After a 4-min response, the unit arrives at the block, and based on the jet-like roaring noise, decides to lay a hydrant supply line before pulling up to the house. It is now 9 min after ignition. The officer orders a 1¾-in. handline to be stretched while she investigates the situation. The roaring noise is caused by fire venting from the cylinder's relief valve, which is shooting flames 6–8 ft across the yard, igniting the side of the attached garage and a nearby fence. The hoseline is flaked out and charged. It is now 10 min after ignition. The officer knows that, in spite of the extending exposure fire, the first priority is to cool the shell of the cylinder. She orders the nozzleman to apply the stream in a fog pattern to the blazing cylinder. Just as the nozzle is opened, there is a tremendous blast that severely burns three firefighters standing in the driveway and ignites the home. The cylinder has undergone a BLEVE. To avoid these casualties, firefighters need a working knowledge of the properties of LPG, the cylinders in which it is shipped, and the tactics for dealing with the various scenarios (Fig. 18–12).

Fig. 18–12 LPG is stored in single walled containers with the product inside at room temperature. The thin steel shell will fail if heated, creating a BLEVE.

LPG is a mixture of several of the hydrocarbon gases, mostly propane, but with quantities of butane and possibly ethane mixed in as well. Depending on the region of the country you are in and the time of year, the percentages of the various gases in LPG may vary, but the greatest portion of it will consist of propane, and thus it will behave very similarly to propane. Pure propane is colorless and odorless. To make leaks detectable by humans, an odorant is added, as it is to natural gas. Unlike natural gas, however, propane is heavier than air and it will collect in low points, possibly traveling great distances, hugging the ground as it travels.

Propane has a wider explosive range than gasoline: 2.1% LEL and a 9.5% upper explosive limit (UEL). Since it turns into a gas at minus 44°F, it is always ready to ignite if it finds the conditions to be correct. Although propane tends to sink, air movement in a given area may result in explosive concentrations at nearly any point in an enclosed room, under certain conditions. The gas/air mixture may be too lean at the ceiling to ignite, and at the floor it may be too rich to burn, but at some point in between there definitely exists an explosive mixture. All that is needed is a source of ignition, which could be a light switch, a spark of static electricity, or a pilot light on a gas burner, among others.

When this happens, it is called a vapor-air explosion, and it can level very substantial buildings. A vapor-air explosion is far more devastating than a leak that ignites immediately, or even a cylinder that BLEVEs. This is because the gas has had the opportunity to spread out farther and mix with more air, thus enlarging the area that ignites at one time and greatly increases the pressure that results.

One such propane-air explosion in Buffalo, New York, involving a 500-lb tank, leveled a four-story brick factory and several nearby structures, killing six people, five of them firefighters sent there to control the leak. In another similar incident in Brooklyn, a single 20-lb barbecue tank fell down a flight of stairs into the cellar of a plumbing supply store. Its valve was knocked off during the fall. The cylinder emptied, and the escaping LPG filled the entire cellar before it reached a source of ignition. The resulting explosion leveled three attached brick buildings and heavily damaged two others. Three people, including a passerby, were killed in that blast (Fig. 18–13).

It is because of incidents like these that many jurisdictions ban the indoor storage of LPG. In New York City, for example, anything larger than a handheld torch cylinder requires special permission and will only be allowed under severe restrictions of occupancy, location, ventilation, and other criteria. Even the small amount in a hand torch can do considerable damage, however, because of the nature of the gas and the way it is shipped

Fig. 18–13 A vapor-air explosion resulting from the leak of a single 20-lb barbecue propane tank destroyed these three brick and wood joist buildings in Brooklyn. The explosion killed several civilians.

Liquefied propane cylinders, as their name implies, contain propane in the liquid and gaseous state. Propane gas can easily be compressed by putting it under pressure. If it is compressed enough, the gas turns into a liquid. In doing so, it shrinks to 1/270th of its original volume as a gas, making it very convenient to ship large quantities. When the pressure is released, such as by opening the valve, the liquid begins to boil off into a vapor. As it does, it now expands back into a gas at a rate of 270 parts for each part of liquid that evaporates. It is this rapid expansion of liquid into gas that can wreak havoc in a confined space. Consider a leak in a confined space of only one single quart of liquid propane, which is just slightly more than the amount in a hand torch. That single quart will evaporate into 270 quarts of pure propane vapor. This vapor mixes with air in amounts up to 47 to 1 (2.1% LEL) to produce 12,857 quarts of explosive gas-air mixture that requires only the slightest spark to detonate!

The first thoughts of many firefighters regarding LPG, however, are often of a different kind of explosion than the vapor-air explosion described previously. A BLEVE is a particular threat to firefighters, since it often occurs *after* the arrival of the fire department, unlike most vapor-air explosions. Most propane leaks don't become violent vapor-air explosions as late in the incident as did the two described previously. Usually, either the gas ignites almost immediately, in which case there is a routine fire and no explosion, or else the occupants take steps to stop the leak and prevent ignition. Once the gas ignites, however, a chain of events that has been repeated many times begins to unfold. The outcome of the event depends on the timing and on the training of the responding firefighters. Almost always, the first thing that happens after ignition of the gas is that fire exposes either the gas cylinder involved or nearby cylinders. This starts the clock ticking.

As mentioned earlier, LPG cylinders contain propane in both its liquid and vapor states. Like all liquids, the LPG sits at the bottom of the container while the vapor occupies the top. Liquids, even propane, are good coolants. As the shell of the cylinder is heated, the area in contact with the liquid is kept cool by the propane inside it. The heat being absorbed by the liquid propane raises its temperature, making it boil faster, thus increasing the pressure on the gas at the top of the cylinder. The pressure inside a propane cylinder is directly related to the temperature of the liquid inside of it. At 70°F, the pressure in the cylinder is about 100 psi. If the temperature goes up to 100°F, as on a very hot day, or if the cylinder is heated by sunlight or fire, the pressure in the cylinder skyrockets to 190 psi. If the temperature of the liquid were to continue to rise, so too would the pressure, until the cylinder reached its failure point. At this point, a BLEVE would occur.

A BLEVE occurs when a liquid above its boiling point is released suddenly from its container and the superheated liquid expands rapidly to become a vapor. The resulting energy release is tremendous and may launch parts of the container in any direction. BLEVEs can occur with *any* liquid in a closed container, even water. When flammable or combustible liquids are involved, there is the additional damage caused by the fireball. Even without this, however, the container shrapnel and scalding liquid can kill everyone in its path. This is exactly what happens when a steam boiler explodes. Boiler explosions are BLEVEs, and they have killed dozens of people without any fire. Cylinders that BLEVE can travel great distances. In one instance, a fire captain was killed when an exploding LPG cylinder rocketed through the side of the van in which it was being transported, traveling more than 50 ft before striking him and continuing along its path of destruction.

A BLEVE is the result of too much heat being applied to the cylinder in the wrong location. Although this heat raises the pressure inside the cylinder, it isn't usually the pressure rise that directly causes the BLEVE. Most often, the cause is the loss of strength of the metal cylinder when it is heated. In this case, the pressure pushes through the softened metal just as if it were a much thinner piece of material. Therefore, installing a pressure-relief valve on the cylinder won't always prevent a BLEVE from occurring, even though it operates and keeps the pressure inside at around 250 psi. If the metal is heated enough, it will fail at even lower pressures than that. The only way to prevent a BLEVE is to keep the metal shell from overheating. The propane liquid inside the cylinder does this for us on the bottom liquid area, but the upper vapor spaces cannot absorb heat from the shell. Unfortunately, this upper vapor space is where the cylinder control valve, relief valve, and discharge outlet are all located, thus creating potential leak areas (Fig. 18–14).

Fig. 18–14 The only way to prevent a BLEVE is to apply a sufficient quantity of cooling water to the exposed steel shell to reduce the temperature of the steel and the pressure of the gas inside. For large tanks, this requires at least 500 gpm at the point of flame impingement.

A typical barbecue fire scenario

Tactics at LPG incidents should focus on the unpredictability of the situation. Cylinders exposed to direct contact with flames, particularly when it occurs in the upper vapor spaces, are subject to BLEVE in as little as 10 min. In addition to the BLEVE hazard, there is the probability of sudden operation of the relief valve, which can shoot a blast of fire at approaching firefighters. As in all firefighting operations, the strategy must be the same as follows: protect life, protect exposures, *then* worry about extinguishment. The life hazard here may be less apparent than normal. Due to the spectacle that a well-involved LPG cylinder creates, civilians in close proximity usually back away on their own. The noise level can become uncomfortable in the immediate area. Still, backing up 50–60 ft isn't sufficient. Clear the area for at least 150 ft in all directions.

The evacuation of directly exposed structures should be done routinely and should include the homes on either side, as well as those that adjoin the backyard. Evacuation beyond this point generally isn't necessary, since the cylinder fragments are unlikely to penetrate three frame walls. Take care when approaching the area. Seek some information from the homeowner regarding the state of the cylinder: How long has it been burning, how full was it, and, if it isn't plainly visible from the street, where is it in relation to the house or garage? Particularly at leaking-cylinder incidents, pay attention to open cellar windows that can let gas enter, possibly to be ignited by a pilot light. The first engine should ensure a continuous supply of water, preferably a hydrant capable of supplying three or four 1½-in. or 1¾-in. handlines at minimum. In the event of a serious escalation of the incident, at least one line will be required within the exposed building, one outside to protect the structure, and one to control the cylinder fire.

Rather than approaching the cylinder like marshals of the old west at high noon, firefighters should approach it as though they were members of a SWAT team flushing out a sniper. Use all available cover to shield your advance. Hide behind a parked car, the corner of the house or garage, or any other substantial object until you have been applying cooling water for at least 2 min. Chances are then, the fire will have greatly decreased in intensity. If it is working properly, the spring-loaded relief valve will have shut itself off as the streams cool the cylinder.

Selecting stream patterns requires a degree of care. Use the reach of the stream to place some distance between yourself and the cylinder, at least initially. For this purpose, a straight stream or a narrow fog may be useful. As the distance between the nozzle and the cylinder decreases, however, the angle of the fog pattern *must* be widened. In serious cases, where the nozzle can't hit the cylinder until you are only 5–10 ft away because the cylinder is hidden behind the house or garage, you may have to begin the initial attack on a full-wide fog. If the pattern you use is too tight, there is a very real risk that you will knock over the cylinder. This may cause the situation to worsen, since liquid propane instead of propane vapor will then be discharged. In addition, you may wind up chasing the burning cylinder all around the yard with the hose stream, causing it to ignite other exposures.

Assuming that all goes well, however, and that the initial application of water succeeds in controlling the fire, resist the temptation to rush the affair by closing the valve manually. Chances are that you won't be successful anyway! Remember one of the first things that happened after ignition—the valve handle that melted off? Even with a pair of vise grips or a wrench, success isn't assured. There is a rubber O-ring inside that valve that almost ensures a leak once it has been burned away, no matter how you tighten the valve. There is the possibility that the valve itself may blow out of the cylinder if it is handled improperly after being highly heated. The best course of action at this point is simply to slow things down. Take a moment to evaluate your options. What happens if you just put out the fire and the leak continues? Where will the gas accumulate? Are there any sources for reignition? How is the fire behaving? Is it burning steadily or subsiding slowly? Depending on the circumstances, you may elect to continue to allow a controlled burn, or you may elect to extinguish the fire. Extinguishing the fire isn't recommended in built-up areas where gas can accumulate and find sources of ignition.

On the other hand, if there is plenty of space and a good breeze blowing away from the exposures, it may be perfectly acceptable. All members should be aware of the consequences of an uncontrolled leak and know what to do about it, since quite often the fire is inadvertently extinguished in the initial attack. In this event, continue applying fog streams to divert the gases away from danger spots such as cellar windows, as well as to dilute the gas concentration.

Remember the one advantage that you have on your side: the size of the container. Although these 20-lb cylinders can produce a sizable fire, they aren't the eternal flame. They have a limited supply of fuel. Once it's gone, the incident is over. In tests held at the Nassau County Fire Academy on Long Island, full 20-lb cylinders exposed to direct-flame impingement on the vapor space burned themselves out in less than 20 min.

In these three fire exposures, the gas escaped through both the cylinder valve (when the rubber O-ring burned away) and at the relief valve. It should be pointed out that, although no BLEVE occurred at these three tests, there is no guarantee that the next cylinder to be exposed wouldn't experience a BLEVE under similar circumstances. The lesson to be learned is that, by showing some patience, the department will be much more likely to return to quarters with all of its members intact.

Oil Burners: Their Operation and Hazards

Gases aren't the only problems that firefighters encounter. Heating systems account for 21% of all fires. Oil burners are also subject to failure.

Firefighters encounter numerous emergencies involving oil-fired heating systems during the heating season. To understand and perform our duties efficiently at these incidents, we should acquire a working knowledge about the oil itself and the equipment that it feeds.

Grades of fuel oil

Fuel oils come in several different grades, from No. 1 to No. 6. Among the factors that determine the grade is the flash point of the oil. It wouldn't be very desirable to have a 275-gal tank of highly flammable liquid inside a person's home. Fuel oils, therefore, have flash points that are relatively safe.

No. 1 oil, the lightest of the fuel oils, has a minimum flash point of 100°F and is similar to kerosene. No. 2 oil is the most common oil for one- and two-family homes. It has a minimum flash point of 100°F, but more impurities. No. 3 fuel oil, a former intermediate grade, is no longer standard. No. 4 oil may be used in midsize applications, apartment buildings, and factories and has a somewhat higher minimum flash point of 130°F. Even with this relatively high flash point, No. 4 oil usually requires no preheating. No. 5 oil also has a flash point of 130°F, but it has more impurities than No. 4. It may be preheated to get the oil to flow easily. The heaviest heating oil is No. 6, with a minimum flash point of more than 150°F. It requires heating to get it to burn, as well as to flow smoothly. Both Nos. 5 and 6 are primarily industrial oils used in large plants and apartment complexes.

When the oils are stored, unused steam or hot water from the boiler is piped to heating coils within the oil storage tank. These heating pipes serve to preheat the oil, at times to more than 212°F. Firefighters should be very cautious when operating around these high-temperature tanks and piping, since they can cause burns. Also, these oils are often above their flash point, and any release will result in the travel of flammable vapors looking for a source of ignition.

The oil burner

For the oil to burn, there must be heat, fuel, and oxygen present in sufficient quantity. For example, one way to get a gallon of No. 2 oil to burn is to pour it into a container, heat it until the oil is at or above its flash point, and then pass a spark into the vapor space within the flammable range. The oil will ignite and burn with the surrounding air, producing heat and great quantities of smoke. This is a totally unacceptable and impractical way of heating a home or business, however, for it would be expensive to heat the oil to 100°F, and the large amounts of smoke generated by the incomplete combustion would back up into the premises.

The modern oil burner, on the other hand, burns oil at a much more efficient rate. Its high-pressure oil gun atomizes the oil, eliminating the need to preheat the oil and eradicating the heavy black smoke so common to burning oil in the air.

Oil burners consist of a firebox and an oil gun. The burner receives its supply of oil from the tank, burns the oil, and transfers heat to a useful place, generally to make hot water or steam. Any products of combustion are then vented up a smoke pipe to the chimney and out to the atmosphere. This is done automatically, and the system shuts itself off as directed by a set of built-in controls. The components of an oil burner are divided into the following two systems: oil handling and controls.

The oil-handling system. The most common burner is the high-pressure gun type, which combines several functions. Oil is drawn from the tank through a pipeline to the oil pump in the gun. A shutoff valve on the supply line is found at the tank, and often at the burner as well. This pump supplies oil to the nozzle at pressures ranging from 100 psi in homes to 300 psi in large installations. The high-pressure discharge acts much like a high-pressure booster line, breaking up or atomizing the oil into very fine particles and exposing more surface area to combine with oxygen. This gun also contains an air fan to blow fresh air in through the oil spray at the same time. At the end of the gun is a source of ignition, either a gas pilot light or, more often, two high-voltage electrodes (5,000 or more volts) that create an electric arc. With fuel and oxygen present, a signal is now needed to light the pilot and begin the chain reaction, producing fire. This fire, when it occurs, is meant to take place within the firebox, a chamber lined with firebrick to control the fire within. Inside the firebox are coils, around which the heated air flows. The heat is transferred to the water within these coils. This water is then used to warm the premises or perform a work function. The unused heat and gases formed in combustion rise out of the burner into the smoke pipe, a metal duct that connects the burner to the chimney.

The control system. The main or primary control features a starting circuit and a running circuit that control the operation of the fuel pump, fan, and igniter. It also has a safety circuit, which shuts down the pump and fan if the igniter malfunctions or fails to light after 90 sec.

The thermostat is the most common of the limit controls. Its function is to send a signal to the burner that fire is required. When the thermostat senses a demand for heat, it goes to the *on* position. When the heat demand has been satisfied, it goes to the *off* position. There are two other limit switches that you may encounter, the Pressuretrol and the Aquastat. The *Pressuretrol*, used to sense pressure within a steam system, may have a minimum as well as a maximum setting. The maximum setting, usually the only setting used in residential occupancies, shuts down the burner to avoid dangerously high pressures that could rupture the piping. In some large commercial applications, however, there may be some processes that require a steady supply of steam at a useful pressure. In this case, the low setting may keep the burner operating even through there is no need for heating the premises.

Aquastats are found on steam and hot-water systems. They perform roughly the same function as the thermostat and Pressuretrol: to keep the temperature of the water within set limits by turning the burner on and off. All three controls send their messages through the primary control.

The remote control, or emergency switch, is extremely important to the fire department, for it is the preferred means of interrupting the operation of the burner. The emergency switch cuts off all power to the pump, fan, and igniter, usually from a relatively safe area such as the top of the cellar stairs or outside the burner room.

A stack switch is a bimetallic strip that is designed to sense the presence of heat in the smoke pipe. During the initial call for heat, the primary control operates on the starting circuit; that is, oil pressure and airflow are built up and the arc or pilot flame is started. If the oil ignites and heat flows up the stack, the stack switch senses this and switches the primary control to the running circuit, stopping the pilot flame or arc. The main purpose of the stack switch, however, is to act as a safeguard, keeping the oil and air mixture from building up in the box if no fire is present. It does this by stopping power to the pump and fan. On newer burners, the stack switch may be replaced by an infrared sensor at the burner head, serving the same purpose.

Potential burner problems

Since the oil burner is a relatively complex machine that requires maintenance, there are several possible problems that the fire department may be called to handle.

The first, and probably the easiest, problem to handle is the smoking burner. To burn cleanly, oil requires the proper dispersion (atomization) and the proper ratio of air to fuel. Each gallon of No. 2 oil must be mixed with almost 2,000 cu ft of air to burn efficiently. If anything interferes with this ratio, the result is likely to be *soot*. This can mean clogged nozzles, clogged air supplies, worn-out pump and fan motors, and improper or contaminated oil supplies. The smoke may back up out of the firebox, prompting the owner to contact the fire department.

The procedures for firefighters at such an incident are relatively simple. On approach, observe the chimney for thick smoke. Remember, burners that are operating properly burn fairly cleanly. On arrival, verify from the occupants the cause for alarm, make a quick size up, and locate and turn off the emergency switch. Turn off the oil tank valve, make an examination to verify that the cause was minor, and advise the occupant of the need for adjustments by a qualified serviceman.

The second condition to which firefighters are often called is delayed ignition, or *kickback*. In this case, the thermostat calls for heat and begins the start cycle. Oil and air are discharged into the firebox, but for some reason, ignition doesn't occur immediately. Oil vapors fill the chamber and travel up the smoke pipe, while the primary control is trying at the same time to recycle the start-up. If it is successful and does produce a source of ignition, the entire cloud of oil vapor will ignite almost at once. This will be accompanied by a loud thud, similar to an explosion, which may knock the smoke pipe loose from the chimney or blow open the burner door. Either one will allow smoke to enter the burner room. Burning oil may have pooled in the bottom of the firebox or run out onto the floor. If fire has extended to nearby combustibles, handle the incident as at any cellar fire, keeping in mind that the burner must be controlled by shutting down the power and fuel.

More often, though, the fire is confined to the burner and the immediate area. Tenants frequently report an *explosion in the cellar*. Arriving units are often met by occupants reporting a loud bang and thick black smoke. At such an incident, firefighters should take the following steps:

1. Shut off the remote control.

2. Enter the basement to examine the area.

3. Stretch a handline as a backup.

4. Use a portable extinguisher (AFFF, dry chemical, CO_2) only if oil is burning *outside* the burner.

5. Ventilate the area.

6. Shut off oil at the tank.

7. Examine the area for extension of fire.

8. Advise the homeowner to call a serviceman.

If the fire is burning inside the firebox, allow it to burn itself out. Using water or dry chemical inside the firebox may do unnecessary damage. Remember, the firebox is *supposed* to have fire in it, so this is not abnormal. Water in particular is dangerous. Cooling hot cast iron with water could cause the box to crack, allowing steam, hot water, or both to spray firefighters.

The white ghost

The least common oil burner response is also the most dangerous. In the case of the white ghost, a truly life-threatening emergency exists for both occupants and firefighters. A *white ghost* is a cloud of vaporized oil and air heated above its flashpoint and out of its container, looking for a source of ignition. Such a cloud is usually produced when a burner that has been running at peak for a long period shuts down and is shortly thereafter called on for more heat. This is fairly common during severe cold spells. If there is a delay in ignition, however, the oil/air mixture in the firebox is vaporized by the highly heated walls of the burner, creating a fog-like mist with the smell of fuel oil. The vaporization causes the mixture to expand, often filling the surroundings with a highly flammable combination of heated oil vapor and air. If this finds an ignition source, there will be an explosion that can blow down walls and floors.

Fire units encountering this kind of situation should immediately take the following steps to prevent ignition and to protect life:

1. Evacuate the entire building immediately.

2. Do not enter the cloud for any reason.

3. Shut down the remote control, using SCBA and a fog nozzle open in a wide pattern as protection.

4. Use the fog nozzle to saturate and cool the cloud.

5. Vent the area.

6. Secure any other sources of ignition.

7. Shut off the fuel.

As with all oil burner incidents, you should direct the responsible person in charge of the building, preferably in writing, to have the burner inspected by a qualified repairman.

Oil burners may bring about numerous responses for many departments during the winter season. By and large, they are routine alarms that result in only minor damage when they are handled properly. If the basic, necessary procedures aren't followed, however, excessive property damage, injury, and even death may ensue. Wear SCBA and call for a handline if you have any doubt about the severity of the emergency. Don't take these seemingly routine alarms lightly.

Determining the Origin of Odors or Smoke

When I joined the fire service in 1970, there was a saying that went "This would be a great job if it wasn't for the smoke." Two technological developments have made smoke much less of a problem than it was 35 years ago: masks that allow us to breath in even the heaviest of smoke and TICs that allow us to see through it. While smoke can still be a major problem for firefighters, knowledgeable firefighters, can use smoke to their advantage if they can read its message, for smoke can help firefighters determine the location and intensity of a fire and suggest possible strategies.

Often the first indication that firefighters have of a working fire is a column of rising smoke, sometimes visible for several blocks. Black smoke suggests the presence of burning petroleum-based products. Large volumes of dense, black smoke at the level of the roof could signal the involvement of roofing materials. Lighter volumes of smoke at the same level could be the result of a defective oil burner. Light to moderate quantities of black smoke in basements often indicate an oil burner malfunction and should prompt firefighters to bring a Class B extinguisher with them. Historically, black smoke from the interior of a residential building often meant the presence of accelerant, but this clue is no longer reliable because of the proliferation of plastics.

Common Class A materials produce gray to light brown smoke when oxygen is present. When less than sufficient oxygen is available, large amounts of dark gray or yellow gray smoke are produced. This is an indication of potential backdraft, especially if the smoke is issuing under pressure and being drawn back into the building. Many materials produce smoke of other colors, such as compounds containing oxides of nitrogen, which give off a reddish brown hue, but these are unlikely to be found at residential or routine commercial fires (Figs. 18–15 and 18–16).

Fig. 18–15 The color of the smoke is an indicator of the fuel burning. This grey-brown smoke suggests Class A material.

Fig. 18–16 This large column of black smoke suggests a Class B fire, but it has a lot of grey mixed in. In fact it is a car inside a garage, a mixed fuel load.

The movement of the smoke can tell a firefighter about the intensity of the fire. Heavy, rolling clouds violently twisting skyward indicate extremely hot smoke from an intense fire deep in the building. This is frequently followed by fire igniting through openings out of which the smoke is pouring. Firefighters should use extreme caution when entering to avoid being caught in a flashover.

Wispy smoke, usually light in color, indicates a fire in the incipient state. Quick use of an extinguisher should solve the problem before it poses a danger. However, be prepared to stretch a line to back up the extinguisher. Smoke settling or hanging in low spots is known as cold smoke. It is found at fires in sprinklered areas or where a fire has been fully or partially extinguished. Fires that are partially extinguished give off great amounts of carbon monoxide, often an unrecognized danger to firefighters. The necessity of wearing SCBA isn't always obvious at these kinds of fires, but it is essential to protect yourself against the odorless carbon monoxide.

Smoke often masks the seat of the fire, making it difficult to determine the fire area. Since smoke normally rises, the lowest floor of visible smoke will usually have some fire on it. However, firefighters should make a point of checking at least one floor below this level. A small fire often originates on the lower floor, and smoke and fire spread upward before breaking out. A heavy smoke condition present throughout the building with no fire visible frequently indicates a cellar fire. This is even more apparent if smoke is issuing from the chimney, especially during non-heating seasons. Although these observations can start first units looking in the right direction, they should be verified internally.

The human nose, while not as sophisticated as that of other animals, is still a marvelous sensory organ. Firefighters quickly learn to identify commonly encountered aromas. Often, the fire is detected in such an early state that flame and smoke aren't visible. Odor is the only clue that a problem exists. This is especially true with the increasing presence of smoke detectors.

How, then, is the source located and the solution devised? Consider the following scenario and develop a course of action.

It is 11:45 PM on Thanksgiving night. Your company responds to a report of smoke on the third floor of a four-story brick and wood-joist apartment building. Nothing unusual is apparent on arrival. A few lights are on, no smoke is showing, no people are waving, and there is no indication of fire. On the third floor, there are six apartments with the doors closed, and there is a faint odor of smoke. How would you proceed?

The most logical step might be to request the dispatcher to contact the calling party and pinpoint where the smell was first noticed. The caller can determine whether it is getting stronger or dissipating, and possibly aid in locating the source. At the same time, the respondents can investigate such sources as light fixtures and the incinerator or compactor chutes in the public hall.

Sometimes the distinct odor of certain fires will point firefighters in the right direction. Recognizing an odor makes determining the origin simpler. Once firefighters have smelled certain burning odors, they will always recognize them. For example, burning food, garbage, mattresses, and oil burner backups all exhibit unmistakable smells.

It is important to exercise discretion when alerting sleeping families. In the Thanksgiving scenario, the odor was reported on the third floor and wasn't noticed while passing the first and second floors, so the investigation began on the third floor. Pressing a nose to each door might identify the location. If this is unsuccessful, then knock on each door. A quick look in each apartment is sufficient. The offending pot burning on a stove is usually located in an apartment where no one is home or the resident has fallen asleep.

A fire escape offers another view of the structure, and it permits forcible entry through windows once the source has been located. This minimizes damage and generally doesn't affect the security of the apartment. Remember, often no one is home. If forcible entry is required, arrange with the police or a responsible tenant for security of the apartment before leaving the scene. When entering from a fire escape, it is wise to announce yourself as "the fire department" before you actually set foot inside, lest you be mistaken for a burglar and face violent resistance.

Another fairly common smoke odor comes from overheated fluorescent light ballasts, which often don't give off visible smoke until later in the emergency. Light ballasts contain a cooling oil that gives off an oily smell when it is overheated. On older types, this was often PCB oil, so you should wear SCBA when handling these seemingly minor incidents. Although locating the defective unit can be complicated by the height of the fixture, TICs or other types of heat scanner devices can be used that allow a firefighter, standing on the floor, to search rapidly along the rows of light fixtures. Without a heat scanner, firefighters must inspect each fixture.

The search can be narrowed by following a few visual hints. Begin by looking for fixtures with intact bulbs that are glowing faintly. Fixtures without bulbs won't have a flow of current and usually aren't the problem. Fixtures that are operating correctly can be the source of the problem, but this isn't as common as dim fixtures. Look at the fixtures for defects. Dark-colored oil on the outside of a fixture, or on the light diffuser of a recessed type, indicates a problem. On surface-mounted fixtures, a dark smudge around the sides, particularly near the vent holes, indicates soot deposits. As a last resort, feel each fixture. A normal ballast is warm to the touch. A defective unit will be hot.

Once the defective unit has been identified, take steps to isolate it. First, ensure that there hasn't been any ignition of nearby combustibles and prevent any further flow of current. Generally, this entails gaining access to the ballast, feeling around it for abnormal heating, and opening the electric circuit. Shut off power to the fixture, remove all of the bulbs, open the cover, and disconnect all of the wires leading to the ballast. Cover the exposed ends of the black and white wires separately and examine nearby combustibles for hidden fire.

Electrical fires are common occurrences that can usually be handled easily. To locate the source of the problem, consider where the smell is first detected, then examine all of the devices in the area that are on or had been on recently.

Confusion occurs in larger buildings when odors of smoke are reported in several areas. This can happen if elevator motors overheat, allowing smoke to drift through the shaft as the cars move. Another common occurrence is an overheated motor of an HVAC system that dumps smoke out through the ductwork. In both cases, send someone—with a radio, if possible—to locate the machinery spaces as quickly as possible. Fires usually occur where equipment is found. They don't normally begin in the middle of shafts or ducts. If the electrical equipment is on fire, as opposed to being merely overheated, use nonconductive extinguishing agents, preferably a clean agent that won't damage the equipment or compound cleanup problems. Halon or CO_2 are both well suited for this, although halon shouldn't be used for routine incidents because of the effect that they have on the earth's ozone layer. Disconnect the power as soon as possible.

Other fires that often result in reports of smoke on several floors involve two other kinds of shafts. These are incidents involving incinerator shafts, in which the shaft becomes blocked by rubbish, and fires in trash compactor chutes. It is important to differentiate between the two kinds of incidents. One is a fire that can spread easily to nearby combustibles; the other is less likely to do so. Still, both are abnormal, life-threatening conditions.

Incinerator shafts are common in many older, multistory buildings. They allow rubbish to be dropped through a fire-resistant door into a *chimney* to be burned at the base of the shaft. The doors to the shaft are located on each floor, usually in or just off of the common hallway.

A fire in an incinerator isn't necessarily an emergency. In fact, many incinerators have auxiliary burners, usually natural gas or propane, to make the rubbish burn hotter and, therefore, cleaner. Incinerator problems develop when an oversized piece of rubbish is forced into the shaft, blocking the escape of smoke and gases up the chimney. The gases start to bank down and push their way out of the chute through the paths of least resistance, usually the hopper doors on each floor.

The solution is to locate the blockage and remove it. Examine each floor where smoke is reported. The halls in these areas must be checked to make sure that the smoke isn't coming from an apartment fire, as well as to search for life hazards and fire extension. The amount and toxicity of rubbish generated in even a small- to medium-size apartment building or hospital is often staggering. Large quantities of plastics are present, as well as aerosol cans and other containers of flammable liquids.

The use of SCBA in incinerator fires should be mandatory. Why take any smoke from somebody else's garbage fire? While searching these areas, a quick examination of the chute door will reveal whether firefighters are above or below the blockage. If smoke comes out when the door is opened, you are below the blockage. Proceed up until you reach a floor where no smoke comes out when you open the door. Then you are either above the blockage or the blockage has freed itself. To free the rubbish, use a hook or several sash weights tied on a retrievable chain to try to push it down.

Prevent any part of your body from getting into the shaft, because residents above may unknowingly drop more debris. SCBA also protects the face and eyes in case glass shatters or containers explode when you open the door. Such occurrences are common. If attempts to push down the blockage fail, it might be possible to burn it out. Go to the floor above the blockage, ignite a newspaper, and drop it onto the pile. Should all else fail, put out the fire and have maintenance personnel clear the blockage. Remember that most incinerators have an auxiliary burner (often gas fired) that must be shut down before extinguishing the fire. Otherwise, gas will fill the shaft and could explode.

Compactors resemble incinerators, consisting of shafts with hopper doors on each floor to convey rubbish to the basement for removal. Compactors differ from incinerators, however, because they aren't designed to hold fire at any time. Smoke coming from a compactor indicates that immediate action is needed, since the fire could spread throughout the building. Stretch a charged line to the first floor above the fire and operate it into the shaft, if needed, to prevent extension there. At the same time, members should go to the basement and work on extinguishment.

Compactor rooms pose the following hazards to firefighters: high-voltage electrical equipment, high-pressure hydraulic hoses, moving rams that can shear off an arm if it is caught in the compactor, falling debris, and exploding bottles and cans. Use extreme caution when operating in these rooms. The quick identification of a compactor fire can be the difference between a routine fire and a rapidly spreading multiple alarm.

Many jurisdictions require automatic sprinklers in compactor chutes, but they are often turned off due to the frequency of such fires and the difficulty of replacing a fused head in the shaft.

The chimney is another shaft that often travels the height of the building and frequently causes reports of smoke. Oil burner problems, gas furnace malfunctions, and difficulties with wood and coal stoves are all common occurrences. Gas-fired units are clean burning and usually have no problem of fire extension. They are, however, subject to gas leaks and backup of fumes from a clogged or defective chimney. The fumes in this case are less odorous than those from an oil burner and may be difficult to recognize.

Responding to Carbon Monoxide Alarms

In recent years, another non-fire emergency has emerged causing numerous fire department responses: the carbon monoxide (CO) detector. This device has been available in some form or other for more than 50 years, beginning with Colorimetric detector tubes. These required that the ends of a glass tube be broken off and a sample of suspected atmosphere be drawn in through a mechanical pump. In the early 1990s, self-contained battery-powered detectors, similar to smoke detectors, became commercially available at a reasonable cost. Following the well-publicized death of tennis star Vitas Gerulitis from carbon monoxide poisoning, the demand for CO detectors skyrocketed. Unfortunately, the technology used in the first generation of these detectors was far from foolproof. Their design led to many false alarms and low-battery signals. A jittery public, frightened by the death of a celebrity, and instructed to do so by the manufacturer, has often contacted the fire department to check for carbon monoxide when these devices have activated. Fire officers responding to such an alarm must have a plan of action as well as an understanding of how these devices operate and the symptoms of CO poisoning.

Most of the responses to CO detectors will be minor, non life-threatening incidents, but I have carried out several overcome victims and several more bodies. These incidents must be treated as being potentially lethal. The most common CO detectors are the self-contained battery-operated units that look just like smoke detectors. The next time you respond to this sort of alarm, make sure that you aren't telling the occupants that their smoke detector must be broken because there is no smoke present. Carbon monoxide is colorless, odorless, and tasteless. The only way to detect it is with a calibrated carbon monoxide meter.

If you respond and find an activated CO detector, you should first ask the occupants whether they feel ill. Actually, this step should have been taken by the dispatcher on receipt of the alarm and if anyone reported *flu-like* symptoms, an EMS unit should also have been dispatched. CO exposure often produces the following flu-like symptoms: fatigue, headache, nausea, dizziness, and mental confusion. If anyone is reporting these symptoms, the building should be evacuated until it can be checked for CO with the proper meters. If you don't have the proper meters, call for them, either from additional fire department units, mutual-aid companies, or the local gas utility. There is no alternative to testing with meters. If you let the occupants back in without having done a thorough examination with proper detection equipment, and if some of them subsequently die, the IC may well be held criminally liable.

Most of the battery-powered units will activate either for low levels over long times (15 parts CO per 1,000,000 parts of air, over either an 8-hr time frame or a 30-day time frame, depending on when the detectors were made) as well as high doses (100 ppm in 90 min to 400 ppm in 15 min). Unlike the 120-volt detectors, however, they won't reset quickly. Firefighters are used to smoke detectors that will rapidly reset if the smoke is cleared from the area. Battery-powered CO detectors are designed to function like the human body, which absorbs and holds on to carbon monoxide for long periods of time, releasing it slowly. The sensor in these battery-powered units is a gel that absorbs CO. It takes 24–48 hrs for this gel to release the accumulated CO before it will reset. In some cases of high-dose accumulation, the sensor won't reset and must be replaced.

Once you have determined that there is a potential life hazard and have removed it, you need to take steps to locate the source of the CO. Ask the occupants what fuel-burning appliances were in use before and during the time that the detector was activated, and whether they ventilated any areas. Start your meter readings at the front door (with a minimum of two mask-equipped members, preferably two CO meters, and at least one portable radio), then proceed to the location of the CO detector. Record all readings by ppm, location, and time. Check around each and every fuel-burning appliance as well as any chimneys.

If you find a defective or leaking appliance or flue, shut it down and continue your meter readings, since there could easily be other sources of CO. *After* the readings have been completed, you should begin venting the area. Inform the owner (preferably in writing, with your unit keeping one copy) of the location and description of the source of the CO and the need to keep it shut down until it has been repaired. After the CO readings have dropped to an acceptable level (9 ppm for residences, 35 ppm for commercial occupancies where no one sleeps), the building may be turned back over to the occupants.

Wood-burning Stoves

Wood-burning stoves involved in creosote fires are also possible sources of fire extension. A severe creosote fire rapidly reaches blowtorch proportions and can damage a chimney to the point of allowing fire to the outside. Shut off the air intake to the stove and check for extension along the length of the chimney. If the air intake cannot be shut off, or if a severe fire exists in the chimney, the fire department will have to extinguish the fire to prevent extension to the combustible surroundings. Commercially available chimney fire extinguishers can be placed in the firebox and ignited, suffocating the chimney. ABC-type dry chemical extinguishers can be used through the firebox opening, directed up the chimney, but they can create an excessive mess in the occupied area.

Another alternative, if the top of the chimney is accessible, is to drop plastic sandwich bags full of ABC dry chemical powder into the top of the chimney. As the bag falls, the plastic melts and releases the powder, extinguishing the fire. At other times, a special low-volume, fire mist-producing chimney nozzle may be required. Avoid introducing streams of water from standard hose nozzles into a chimney fire, however. The large drop in temperature produced by standard nozzles will crack the red-hot chimney flue. The next time the flue is used, the fire will have an easy path out of the chimney.

Even when they are operating properly, wood-burning stoves can be a minor nuisance to responding firefighters. A report of *smoke in the area* must be investigated. In the past, the well-known odor of burning wood was a signal to arriving firefighters that they "had a job." Take a second to investigate the odor before radioing in with a *working fire* signal. That column of smoke may be from the chimney or the backyard.

19 Structural Collapse

Structural collapse is one of the most feared occurrences on the fireground. It is often swift, often deadly, and can occur with little apparent warning. Although collapses that injure and kill firefighters aren't a common occurrence, they account for a large percentage of multiple-casualty incidents. The following collapses have accounted for some of the largest firefighter death tolls in U.S. history:

- Chicago Stockyard collapse in 1910—23 dead
- New York City's 23rd Street collapse in 1966—12 dead
- Boston's Vendome Hotel in 1972—nine dead
- Brooklyn's Waldbaum's Supermarket in 1978—six dead
- Hackensack Ford Dealers in 1988—five dead
- WTC in 2001, where 343 New York firefighters and medics made the supreme sacrifice

The toll is likely to continue, if not increase, due to changes in the methods of construction that place a greater value on lightweight materials and lower cost than on the lives of firefighters and occupants when those materials are exposed to fire. Firefighters and officers have a responsibility to themselves, their fellow members, and their families to learn to recognize the warning signs of impending collapse and the causes of it, as well as to prepare themselves for the time when a collapse does occur. They must also learn the steps that can be taken to rescue those unfortunate enough to be caught in the debris (Fig. 19–1).

Fig. 19–1 Lt Vinnie Ungaro of Squad 288 FDNY works atop the smoldering North Tower, seeking survivors of the worst fire-related collapse in history.

Collapses occur for different reasons and in different ways in the various classes of buildings. Learning to recognize the building types can enable you to predict how they will behave with respect to collapse under fire conditions. The five classes of buildings can be arranged in an order of increasing or decreasing resistance to collapse. This can give you a general idea of how stable the structure is, as well as how much time you can devote to interior firefighting before impending failure of the structure dictates the withdrawal of interior forces.

Buildings that show the greatest resistance to collapse are those of Class 1 fireproof construction. These buildings have fire-resistance ratings of up to 4 hrs. Generally, they are built using some type of skeletal framework, either of poured concrete or steel I-beams. Collapse in these buildings is usually rather localized. Either the concrete on the ceiling spalls or several of the I-beams sag. The load that is being supported, however, generally remains in place. When these structures are exposed to heavy fire, they are supposed to provide sufficient resistance to collapse to allow time for evacuation and fire control. As the collapses of # 1, 2 and 7 WTC have shown, the type of lightweight construction that is so prevalent in modern buildings may or may not provide the time needed for a long stair climb, evacuation and fire control, as well as a long descent before collapse occurs.

There have been other cases in high-rise office building fires where the fire totally consumes all of the combustibles on the fire floor but hasn't caused total collapse. As described in chapter 16, this is only true of heavyweight buildings. In addition to the three WTC building collapses, One Meridian Plaza in Philadelphia suffered a near collapse and had to be razed, after a malfunctioning standpipe system prevented the fire department from mounting an offensive attack. In this case, with eight floors totally gutted, the building's structural engineer warned that there was a potential for total collapse of the upper floors. All of the firefighters were withdrawn, and the fire burned itself out without collapse, but the structural damage was so severe that the building couldn't be reoccupied. The floor-support beams had only 2-hr-rated, sprayed-on fireproofing, and still they endured nearly 11 hrs of fire exposure. In February of 2005, a 32-story high-rise in Madrid, Spain was totally gutted in a severe fire, and only localized collapses occurred. This building was built with a reinforced concrete skeleton for most areas, as opposed to the lightweight steel skeletons of many modern buildings. The most severe collapse danger in Class 1 buildings occurs in poured-in-place concrete buildings under construction when fire involves the wooden formwork of the most recently poured floor. Such a collapse could quite possibly cause a pancake collapse of the entire structure (Fig. 19–2).

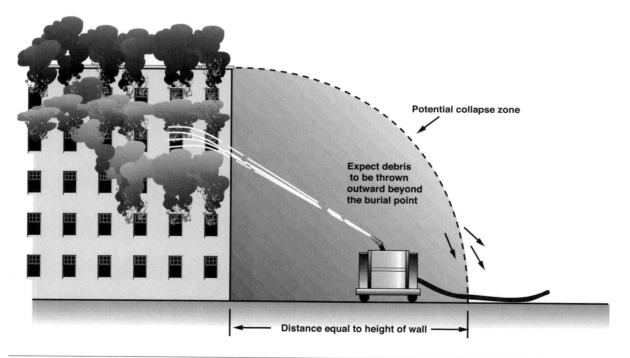

Potential collapse zone

Expect debris to be thrown outward beyond the burial point

← Distance equal to height of wall →

Fig. 19–2 When determining the size of a collapse safety zone you must estimate the height of the facing walls, remember that large debris will project even greater distances.

The second-best construction type, as far as resistance to collapse is concerned, is heavy-timber construction, Class 4. These buildings are generally quite stable due to the size of their load-bearing members, which are generally 12 x 12 wooden columns and brick walls. In most cases, manual firefighting will have long since shifted to exterior operations before there is a danger of collapse—the volume of fire required to burn through these large structural members would simply drive firefighters out of the building. An increasingly common exception, however, is the building that has been the scene of numerous repeated fires in the same area, usually in old industrial buildings that have been abandoned by their former owners. In this case, what may seem to be a rather minor fire for such a large, well-built structure could be the one that results in a relatively large collapse. Collapses in these buildings are often large-scale ones, with both walls and floors being affected. The impact of upper floors collapsing on lower floors can cause each successive floor to collapse. At advanced fires in these structures, apparatus should be removed from the collapse zone. The means the *entire* area for the full length of the frontage, for a distance out from the wall at least equal to the full height of the wall (Fig. 19–3).

Fig. 19–3 The belief that brick walls will crumble as they fall is a myth.

The third most collapse-resistant building is of Class 3 ordinary construction, or the standard brick and wood joist. These buildings are generally more prone to burn-through than collapse, especially under normal floor-load limits. That is, the floor or roof sheathing above burns through long before the floor joists fail. It is unlikely in this case for firefighters to be working on the falling surface, although they may be working below, unaware of the conditions overhead, which can be shielded by hung ceilings or obscured by smoke or fire. A distinct danger in this type of construction is the imposition of unusually high floor loads, such as may be found in a plumbing supply warehouse, and any concentrated loads, such as from roof-mounted air conditioners and signs. These concentrated loads can result in early failure of localized areas without advance warning. Any member discovering such loads should immediately report them to the officer in charge, who should order interior forces out from the area below until the stability of the floor or roof has been assured.

The fourth category in order of resistance to collapse is Class 5, wood-frame buildings. This surprises many people, who suspect that Class 2, metal or noncombustible buildings, are more resistant to collapse than either wood frame or even heavy timber. That is incorrect. The standard wood-frame home, built with dimensional lumber, is more likely to burn through, chasing personnel out, before it collapses. This statement applies only to standard construction methods, however, and not to those types of buildings that, even though meeting the technical definition of wood frame, use lightweight truss construction methods instead. For the purposes of collapse resistance, a special category containing all lightweight and truss-construction methods should be considered (Figs. 19–4 and 19–5).

Fig. 19–4 Certain occupancies, such as plumbing supply stores, have very heavy, concentrated floor loads. Expect early collapse and withdraw members in the event of a serious fire.

Fig. 19–5 When it is highly heated, steel twists and sags, dropping its load. When cooled, it freezes in its new shape as it contracts.

This leaves Class 2 construction, metal or noncombustible, as the least resistant to collapse when exposed to fire. This is principally because large quantities of unprotected steel are used in its construction. If the steel were protected with some sort of fireproofing and/or encasement, it would no longer be a Class 2 structure, but a Class 1. Steel's behavior under fire conditions creates several problems from the point of view of collapse. The first is that steel expands when it is heated. A 100-ft-long I-beam heated uniformly to 1,000°F will expand 9½ in. lengthwise. This can push down columns or walls, as well as punch holes through walls through which fire can spread. When steel is further heated to about 1,500°F, it will lose its strength, either dropping its load or twisting and sagging. Any beams that depend on it will fall. Moreover, when the steel is cooled back to its original temperature, it will contract back to its 100-ft length while retaining its distorted shape. If an I-beam were severely twisted and then cooled by a hose stream, the steel would shrink, possibly enough so that the ends of the beams would no longer be resting on their original supports.

Collapses tend to occur in similar fashion in similar types of buildings. Those with a protected steel or concrete supporting system resist total collapse rather well. These are called *framed structures*, since most of their weight is carried on the frame or skeleton. Collapses that occur in framed structures are usually very localized—only the area between the two supporting members. Of course, even this small an area can kill you if you happen to be below the section that falls (Fig. 19–6).

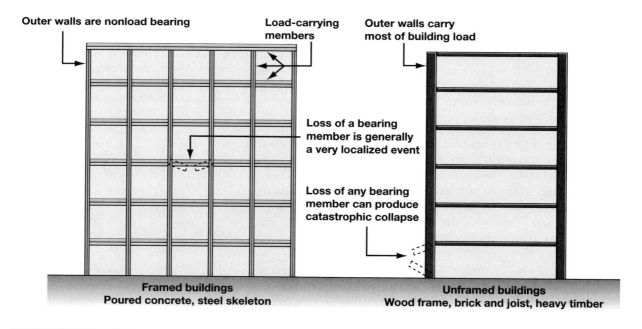

Fig. 19–6 Framed and unframed structures are affected differently during collapses.

Structures that are built so that most of the weight rests on bearing walls are known as unframed structures. The collapse of a bearing wall tends to cause extensive collapses as the floor's supports are taken out. In all types of structures, the failure of a vertical member, such as a column or a wall, is usually more serious than the failure of a horizontal member, such as a beam, since the column is more likely to support other structural members. Structural elements can be arranged in a priority order when it comes to evaluating the effect their failure will have on the structural stability of the rest of the building. This Hierarchy of Structural Components is as follows, ranging from most important to least important:

1. Bearing walls

2. Columns

3. Girders (which support beams)

4. Beams and joists

5. Floor or roof decking

Causes of Collapse

Structural collapse can result from a wide variety of causes, and it can occur before the fire department arrives, as in the case of a gas explosion. Collapses can occur during the initial fire attack or even after knockdown, while a unit is overhauling. Many times, various accounts would indicate that a collapse occurred without any warning, but that is usually not the case. It is more likely that the IC and the firefighters failed to understand the causes of collapse and to recognize that those factors were present. The fire itself may at times be the only indication that a collapse is likely. If you are aware of the causes and warning signs, however, you will know when it is time to take action to avoid having people caught in what is almost certain to ensue.

The following items are among the most common causes of structural collapse on the fireground.

1. Structural weakness due to faults in design, shoddy workmanship, and illegal or improper renovations

The removal and failure to adequately replace a bearing wall, coupled with further alterations to another section of a bearing wall, was the main cause of the Vendome Hotel disaster in Boston. As one noted fire service author puts it, "One of the greatest danger signs to a firefighter is a 40-yd dumpster parked in front of a building." Go in and find out what renovations are being done. Try to evaluate what kind of effect they will have during a fire, particularly on the structural integrity of the building. Get the code enforcement people involved if the work is being done without a permit. A building permit usually ensures that an architect or structural engineer has at least reviewed the design for any obvious weaknesses.

2. Fire damage to wooden structural members

The failure of wooden bowstring trusses alone has been responsible for dozens of firefighter deaths, including Hackensack Ford 1988, Waldbaum's 1978, and Richfield Park, N.J., 1968. Add to this the number of roof and floor failures due to burn-through, and the danger is tremendous. Remember, wooden structural elements burn through at a rate of about 1 in. for every 45 min of open-burning time. If fire can attack two sides of a joist, for example, the standard 1½-in.-wide (2×8) joist will burn *completely through* in less than 45 min, and will have lost its load-carrying ability long before that.

3. The heating of unprotected steel

Like wood, steel is affected by heat. Lightweight steel-bar joists can lose their strength in as little as 5–10 min of fire exposure. Fortunately, steel sags prior to total failure. This may provide sufficient warning to firefighters working on a steel deck, but it may not be visible to people working below due to smoke obscuring their view. Steel structural elements are serious risks unless provided with substantial protection in the form of *fireproof* coatings or encasing.

4. The cooling of highly heated cast-iron columns or facades

Cast iron is an older construction material that poses unique difficulties for firefighters. Cast iron is a very brittle material that has great compressive strength but almost no shear strength. Thus, it is unsuitable for use as beams, but it was once a favorite material for columns. Its ability to be readily cast in a mold and then bolted together in sections made it the first material to be used for prefabricated buildings, such as the loft buildings built throughout the Northeast in the late 1800s. Unfortunately, when cast iron is highly heated and then suddenly cooled, as by a hose stream, it can shatter, causing failure of either the column or the cast-iron bearing walls. Either case could be catastrophic.

Note: In his book, *Building Construction for the Fire Service*, NFPA, 1982, Professor Francis Brannigan disputes this explanation of cast-iron building collapses. Instead, he theorizes that flaws in the castings have resulted in many columns being much thinner on one side than on the other. Unequal rates of expansion due to the rise in temperature, therefore, could result in cracks in the column. An additional factor could be the force of a high-pressure stream striking what is known to be a brittle material (Fig. 19–7).

Fig. 19–7 Old cast-iron buildings have been the scene of several catastrophic collapses. The exterior walls, as well as the interior columns, are made of cast iron, a very brittle material.

5. Explosions of fuels, explosives, or from backdraft

Even relatively low-order explosions can cause vast increases in the pressure inside an enclosure. As little as one psi exerted laterally on a brick wall can topple it like a house of cards. Any location where flammable gases or explosives are stored demands shifting tactics away from the offensive to taking up cautious, defensive positions. If a potential backdraft is suspected, all firefighters must take the proper actions to prevent it, venting at the top prior to entering below, or take the precautions described in chapter 3 to avoid being caught by these destructive events (Fig. 19–8).

6. Overloading of the floor and the expansion of absorbent materials

Many manufacturing processes require large quantities of materials that readily absorb water. Such materials as rolled newsprint, baled rags, and baled cotton are a few examples. When hose streams are applied to these items, or when runoff water drips down from operations above them, there are two effects that threaten to cause collapse.

Fig. 19–8 A backdraft explosion in the basement of this hardware store caused this collapse that killed three firefighters.

The first is the weight that can be absorbed by these materials. Many bales weigh more than 1,000 lbs dry and can absorb their full weight, or more, in water. This added weight alone can cause collapse if the floor load is even close to the allowed maximum. The second effect can be just as much of a problem: When these rolls or bales absorb water, they tend to swell up. This swelling can push out walls and knock columns out of plumb.

7. Overloading of the floors and roof

This can occur in buildings other than those housing absorbent materials. In fact, this has happened at structures even without fire being present. The previously mentioned plumbing supply store, with boxes and boxes of cast-iron fittings, as well as pipe and fixtures, presents an extremely heavy load. Accumulated snow is also problematical in many areas. Many times, however, the biggest offender at fire operations is the runoff from hose streams. A 1,000-gpm master stream adds approximately 8,500 lbs to the building *each minute*. If you don't see a tremendous amount of water pouring out of the doors, chances are that it is remaining in the building. An accumulation of water 1 ft deep over an area 20 ft wide by 20 ft long represents nearly 25,000 lbs added to the load (Fig. 19–9).

8. Cutting or removing structural members during overhaul

One factor over which firefighters have direct control is the cutting of structural members during overhaul. We must avoid removing any structural components unless they are obviously not supporting any weight other than their own. There is rarely any reason to do so, anyway. Careless overhauling could result in at least a partial collapse around the members.

Fig. 19–9 Many large-caliber streams with little runoff is a classic indicator of potential collapse.

9. Vibration and impact load

Firefighters are often called on to operate in areas that are only marginally stable. Although you may be able to walk gently across an area that is damaged, you must avoid jumping onto it or causing any other type of impact load. When a force is applied almost instantaneously, the result is a much greater shock to the structure than if the same weight were applied gradually. If forced to operate in an area where structural weakness is suspected, use methods and tools that produce little or no impact load or vibration. For instance, if it is necessary to cut a piece of flooring, use a circular saw rather than an axe or a reciprocating saw. At times, both impact loads and vibrations can come from external sources. Beware of how close to a building the elevating platform operates. On occasion, they have knocked down parapets and roofs by bumping into them. Vibrations from nearby rail lines and highways can also cause the failure of an already shaky structure.

10. Miscellaneous causes

High winds, flooding due to storms, and damage to water mains can cause collapse under some circumstances. Other times, the impact of high-velocity hose streams can knock structural supports out of place entirely or simply displace them enough to let some other factor come into play. Then, of course, there is just plain old age. Few buildings are made to last more than 100 years. Many have already passed that milestone and are simply tired of standing. Keep in mind that gravity is always seeking to flatten vertical objects. The job of firefighters is to ensure that no one is in the way when nature finally takes its course and demolishes a structure in what we call collapse.

Collapse Indicators

The way we avoid being caught in a collapse is to examine the structure constantly, as well as its surrounding circumstances (most notably the fire), for indications that the building is losing the battle against gravity. These indications may be physically visible items such as cracks in walls, or they could be noises indicating movement, or they might only be conditions such as a heavily loaded floor. Each indicator taken separately might not raise sufficient alarm to provoke the evacuation of personnel; however, when viewed as a whole, they could be warning signs that the building is in danger. For this reason, it is critical that the officer in command promptly receive notification of any of the following indicators of potential collapse.

1. Occupancy by problem businesses

We know from past experience that certain occupancies present unusually heavy floor loads. Plumbing supply stores, appliance dealerships, and printing shops all have large concentrations of heavy items. Other occupancies, such as supermarkets, car dealerships, bowling alleys, and churches, tend to have truss construction. Although the type of occupancy alone isn't reason enough to withdraw all forces, it should trigger a careful examination for additional signs of possible collapse.

2. Construction

Of all the other possible warning signs of collapse, only one is more serious than the presence of truss construction—a structure that contains explosives. Fortunately, fires in such occupancies are rare. Moreover, they are usually over before we arrive. Fires in truss-construction buildings, however, aren't so rare and, in fact, are probably going to become *the* fire problem of the 21st century as the buildings of today begin to deteriorate. A new style of firefighting will have to emerge to combat fires in these structures. (Or maybe it's not so new. It involves standing back and lobbing water from outside the collapse zone at any fire that seriously involves the trusses.) Other new styles of lightweight construction, such as particle board I-beams, will deserve the same technique. Always bear in mind that the designers of these products, the builders who put them in, the code enforcement agency,

Fig. 19–10 This townhouse under construction reveals many of the following lightweight elements: metal wall studs, floor and roof trusses, and gypsum wall sheathing that will later be invisible to responding firefighters.

and the insurance companies that tolerate their use don't care a bit about the effect that their collapse will have on a firefighter. Their concerns are with the bottom line (Fig. 19–10).

3. Overloaded floors

Constantly observe the floor loading to see whether it poses a risk of collapsing to the floor below. Heavy equipment, water-absorbent stock, and built-up water are all prime indicators of potential collapse.

4. Heavy fire burning for more than 20 minutes

A quick glance at the structure may not indicate the extent of the fire or how long it has been burning, but if heavy fire is obvious and you haven't put it out in 20 min, you should consider withdrawing your forces. When the first members come out of the building with their low-air pressure alarms ringing, it is time to make a decision. Their reports on what is happening can give you some guidance. If they say that most of the fire has been knocked down, you can probably let them continue. On the other hand, if they say that they can't find the seat of the fire, or that they haven't been able to put a big dent in it, start to pull them out. The fire has either outflanked you or it is bigger than you, or both. This is a guide rather than a hard-and-fast rule, but it tends to be on the conservative side. Remember that the 20-min time frame doesn't begin on your arrival, but rather, when the fire reaches the open, flaming stage or flashover. Even the 20-min time limit is too long for fires involving lightweight construction. In this event, a 5-min time limit might be appropriate, which may mean no entry at all if your response time approaches 3 or 4 min. All of these time limits depend as well on actual observations of conditions. If the conditions are severe enough, you may have to pull out the troops much sooner. If the situation is very clear, you may elect to hang in there a while longer.

5. No appreciable runoff

If you have been using numerous outside streams and there is only a trickle of water coming out the front door, something is definitely wrong. All of that water is collecting somewhere. You'd better find out where before it brings the floors crashing down on your head.

6. Cracks or bulges in walls

Cracks in walls are obvious signs of weakness. Unfortunately, by the time they appear, collapse may be imminent. On the other hand, the crack may have been present for the last 40 years. Pay special attention, though, to cracks that are expanding or lengthening. These are definite indications of movement and usually should prompt evacuation if they are extensive (Fig. 19–11).

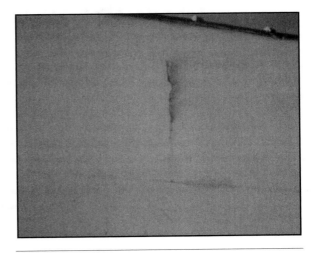

Fig. 19–11 Cracks or bulges in walls are definitely problem signs; they must be monitored during operations to determine whether they are expanding.

7. Water or smoke seeping through a solid brick wall

This indicates a buildup of pressure on the interior as well as a weakness in the wall construction. This combination can result in failure of the wall.

8. The roof pulling away from the wall

This may be evidenced by the relatively clean wood on the ends of beams that were set into the wall or by gaps in the roofing material at the wall joint.

9. The roof sagging or feeling abnormally soft or spongy

Obvious sagging of floors or roofs is one of the last warning signs before collapse. It should prompt evacuation of the endangered area and the area immediately below. Sponginess, however, can be a natural condition on some types of roofs, particularly inverted roofs. This condition calls for further evaluation. What type of roof is it? How much fire is present under the roof? How long has the fire been burning in the roof space?

10. Any obvious movement of floors, walls, or roofs

Again, it may be too late to reach safety once these signs are noted. Pay attention to the earlier warning signs, such as the type of construction and the duration and extent of the fire.

11. Noises

Creaking or groaning sounds are often heard in wood buildings as movement occurs. Cracking timbers may also be audible. Deep rumbling noises may sometimes be heard accompanying partial collapses in remote areas of the structure.

12. Plaster sliding off the walls or plaster dust hanging in the air

Plaster is applied over a lath that is held rigidly to the structure—the plaster *itself* is not. When shifting occurs on the wall, the plaster may be wiggled off in fairly large chunks. If plaster dust is evident in an area that has otherwise been untouched, suspect that subtle movement of the structure is taking place. This may be further indicated by windows that suddenly crack on their own or doors that swing by themselves. The distortion of the door and window frames is the result of being pushed over. This is prevalent in Class 4 construction, wood-frame buildings.

Establishing Collapse Zones

Once the chance of collapse has been recognized, the IC must take the necessary steps to remove everyone from its path. This should be done well enough in advance to allow for a neat, orderly withdrawal from the structure as opposed to a helter-skelter rush. An orderly withdrawal allows time to locate all of the members and ensure that everyone brings out his or her equipment. In an emergency evacuation, it's drop your tools and run! The portable radio is the usual means of initiating the withdrawal. After issuing the order to evacuate, the IC should contact each unit individually and confirm that they have received the order. If a unit fails to acknowledge, its members must be sought out and advised. Emergency evacuation requires some additional means of communicating for those who aren't equipped with radios, as well as for those who may not have heard the call because of background noise. One method is to instruct all apparatus on the fireground to turn on all audible warning devices—sirens, air horns, electric sirens—for a full minute, or until so ordered by the IC. Members within the structure should be able to hear that kind of racket and recognize it as the emergency evacuation signal. Members who have withdrawn from the structure should rejoin their companies and prepare for a head count.

Using audible warning devices this way isn't without problems. The noise can prevent members from hearing verbal or radio messages that may be critical. In one instance, members evacuated a building because they heard sirens only to find that another responding unit had arrived making lots of noise.

A couple of new devices on the market may offer some help in sounding the emergency evacuation signal. Some radio makers are offering an alert tone that can be activated and transmitted over all of the radios on the fireground. The second aid is the Target Exit Device™ (TED). As mentioned in an earlier section, firefighters take the TED into the building with them and leave it in proximity to their exit point. When armed, the unit flashes a strobe light and emits a 95-decibel chirp 4 times in 2 sec, then remains silent for 8 sec. This cycle repeats, allowing operating members to gauge their distance and direction to the exit point. Some models have been equipped with a wireless remote control activating device. By this device, the IC can switch all of the units on the fireground from the standby mode to the alarm mode, thus signaling the evacuation. Whatever system your department uses, the signal must be known by all of the members. Moreover, it must be heeded.

Simply getting the members out of the building, however, doesn't ensure their safety. Collapsing walls can fall great distances and can project debris even further. One firefighter was killed when a collapsing wall, too short to reach him, toppled a tree that *was* long enough. What is required is the establishment of safe exterior collapse zones. Once you have made the decision to shift to an exterior attack, based on the threat of collapse, there is no reason to allow firefighters to be in close proximity to the building. Back off and use the reach of the stream. At one incident, a lieutenant was killed and two firefighters were seriously injured by a wall that fell *after* they had withdrawn from the building. They had been ordered closer again to direct an exterior hose stream through a doorway—this on a vacant building that had been the scene of many fires and wasn't much more than a standing rubbish pile anyway.

The size of the collapse zone must be at least equal to the size of the facing wall. If a wall is 30 ft high, the area in line with that wall must be cleared of all personnel for the full length of the endangered area, and for a distance out from the wall of *at least* 30 ft. Many types of walls will hinge at the base and fall outward in a single, massive piece. Old rules of thumb in estimating collapse zones counted on brick walls crumbling like a curtain. They do in some cases, but more often, they lean over full-height. If a collapse zone is based on less than a full-height estimate, it is likely that any members inside that zone will be killed. Masonry walls weigh approximately 80–130 lbs per cu ft. A very small piece falling from a relatively low height can easily kill, despite your state-of-the-art helmet (Fig. 19–12).

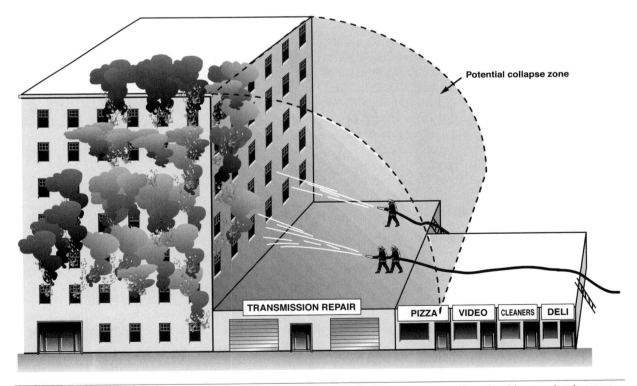

Fig. 19–12 Exposures within the collapse zone must be evacuated. Here, the evacuation should extend at least up to the video store.

Don't believe any of the tales you've heard about being able to remain in line with windows or stepping into doorways to avoid the falling debris. If any such incidents have occurred, the survivors owe their continued existence to something bordering on the miraculous. Pay attention to members operating on aerial devices. Although the apparatus turntable may be positioned outside the collapse zone, members operating on ladders or from platform baskets occasionally tend to get themselves too close to buildings. They forget that the top of a wall often falls outward. They must avoid positioning themselves any closer to the building horizontally than they are vertically, from the top of the building to the lowest part of their device. Otherwise, a falling wall could strike the device, injuring them and damaging the apparatus (Figs. 19–13 and 19–14).

At times, even a collapse zone that is the full height of the wall is too short. Collapse initiated by an explosion is the obvious instance where this is true. There is one particular type of building, however, that has proved deadly beyond what were once thought of as the safe collapse zones. These are buildings with bowstring truss roofs in which the hip rafters span from the end trusses down and out to a bearing wall. (See chapter 9.)

Early one summer evening, I was monitoring the Brooklyn fire radio when I heard a battalion chief request the box to be transmitted for a *fire in a vacant garage*. Since my unit, Rescue Co. 2, would be assigned if this incident escalated just one step, I turned up the radio slightly, not really expecting to go because of the description of the vacant garage. I was thinking of a one-car private garage. Within a minute, an engine company officer came on the radio, excitedly giving the signal for a working fire, "10–75!" My unit was assigned, but I was still thinking that we weren't going to do anything. "It must be this guy's first night as a lieutenant," I recall thinking.

Fig. 19–13 *Two firefighters operating in the basket of an elevating platform were seriously injured when a falling wall caught their bucket, shearing it off the boom and carrying it and the men to the ground. The boom of the tower is visible at the extreme upper left.*

Fig. 19–14 *This elevating platform is obviously operating in a dangerous position. Order it moved to safety.*

That thought was dispelled when the battalion chief transmitted a second alarm. "This is no one-car garage," I said to myself. That particular box was a pretty good ride away from our firehouse, and before we arrived, I was shocked to hear a ladder company officer come on the radio with an urgent message to the dispatcher to transmit a third alarm. "Get us some ambulances and two more rescue companies! We've had a collapse and we've got two firemen buried!"

The building turned out to be a vacant auto-repair garage, about 60 x 80 ft, with brick outer walls and a bowstring truss roof. This fire occurred in the same FDNY division as the Waldbaum's fire that had killed six Brooklyn firemen. The members were keenly aware of the dangers of bowstring truss roofs, and yet here we were, digging out two badly injured firefighters. What wasn't recognized here was the effect that the hip rafters resting on the brick wall would have when the truss that was supporting the other end failed. What was proved in the investigation of this incident, as well as the subsequent evaluation of several other similar collapses, was that the hip rafters act as levers on the top of the wall they rest on. In this case, the wall was measured to be only about 17 ft high. The members had thought themselves relatively safe as they set up a portable deluge gun in the middle of the street, approximately 24 ft from the wall. When the truss failed and brought down the back of the hip rafters, the rafters were instantly propelled outward and upward, sending the wall it rested on outward with explosive force. The two members were totally buried by the wall at a distance 1½ times the height of it. They suffered numerous broken bones, burns, and lacerations. Fortunately, both survived.

The primary lesson here is that you can never be far enough away from a wall as described previously—one with rafters resting on a truss that has the potential for sudden collapse. Responders must take defensive positions. Ground-level forces must withdraw to a distance at least 2½ times the height of the wall—a distance quite unnecessary for most collapse zones. This position may not be possible at many buildings due to narrow streets or side-yard setbacks. In that case, there are only two other options in most instances: the flanking positions off to the sides of the threatening wall, or a position above the wall, as in the basket of an elevating platform. The flanking position must ensure that firefighters aren't endangered by other, less obvious threats. Certainly it wouldn't do to move out of the collapse zone of one wall into the collapse zone of the intersecting sidewall. Beware of trees, telephone poles, and overhead power lines that could be brought down by the primary collapse. The elevating platform, as well as positions in higher exposed buildings, isn't totally without risk, either. When roofs such as these fall in, there is a sudden release of a huge fireball, in some cases 50–60 ft high. Don't be perched directly over this roof, particularly on the downwind side. Otherwise, you could be severely burned (Fig. 19–15).

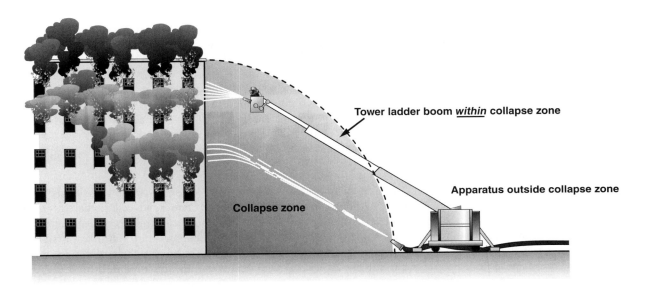

Tower ladder boom _within_ collapse zone

Apparatus outside collapse zone

Collapse zone

Fig. 19–15 Elevating platform booms are susceptible to collapses even though the rig is in a safe area.

One additional lesson came out of this near-tragedy. Some sources, quoting information from the Waldbaum's fire and the Hackensack fire, have stated that bowstring truss roofs are relatively safe to operate on, offering at least 20 min in which to conduct offensive operations before collapse becomes a concern. This particular incident proves such an assertion to be untrue.

This collapse was reliably witnessed and timed. The first report of this alarm came from a battalion chief who was passing by on department business. He saw smoke emanating from the building and radioed the *transmit the box* request for three engines and two ladders. He then exited the car, retrieved his fire gear, and went to investigate. Later that night, he told me:

> *When I went in the door, all I had was four or five separate rubbish fires spread around the building. If I had had just one line, even a booster line, it would have been an 18 (only one engine and one ladder company required). As I proceeded toward the far end of the building, the flames from one of the piles that had a few tires in it licked up and reached the roof joists. From there it ran right down to the far end and was blowing out the bay door there in under a minute. That's when the first-due engine gave its 10–75.*

At that point, I sent the second alarm, seeing how fast it spread and knowing we had an occupied exposure attached on the right. Everybody knew their jobs and the engine didn't fool around at all. They went right for the big guns. They had a hydrant right there and put the deck pipe on the exposure. We had a tower on the way in, so they started setting up the New Yorker (portable deluge set).

Then the whole thing came down.

The total elapsed time from flashover to collapse was *8 min,* as verified by tapes of the radio transmissions. Although this building was vacant and most likely had weather damage to the trusses that hastened their collapse, the lesson is clear. These are dangerous, dangerous buildings that are totally unpredictable as to *when* they will collapse. But they *will* collapse. Don't be in their way when they do!

Collapse Rescue Operations

At times, despite all our vigilance, our precautions aren't enough, and firefighters become trapped by falling debris. When this happens, the events that take place during the ensuing few minutes will likely determine the survival or demise of one of your peers. The actions that you take must be thought out in advance and practiced at training sessions. They may require some additional policies or procedures. The time to take these steps is *now.* No department, large or small, expects to lose a firefighter, yet each year many do. By taking some positive actions, you may be prepared for the day that tragedy strikes your department.

In the event that firefighters are trapped or disabled in the vicinity of fire, one of the first actions to take is to bring as many streams to bear in that area as possible. Although there might be a risk of flooding a cellar where firefighters are located, the first priority is to keep them from burning to death. For this purpose, and several others, there is nothing that can compare to a tower ladder. Set it up as directly in line with the location of the victim as possible. The tower ladder stream is the most versatile, maneuverable heavy stream available. Its boom may be placed right through show windows or overhead door openings, driving fire ahead of it with its 1,000 gpm-plus stream. If there is fear of a secondary collapse, the stream can be locked in position and the basket left unmanned while operations are conducted from the turntable. In this case, a follow-up handline may enter below the boom, shielded somewhat by the boom overhead or alongside. Obviously, this would only be done under the most extreme circumstances, but if having to save trapped firefighters isn't extreme, what is?

Another advantage of having the tower ladder in place is that the boom can possibly be used as a tie-off point for hoisting objects that are pinning members. Again, this should only be done as a lifesaving measure when nothing else is readily available. All apparatus manufacturers will advise against using their unit as a crane, but if a member is pinned, you must do what you can to get him or her out alive. To make this operation safer, however, it is wise to take out all of the removable weight from the boom, including the tools and personnel in the basket. By jettisoning everything that isn't bolted down, you make the full working load of the basket available for hoisting. This can be up to 1,000 lbs or more. This rating can be increased somewhat if it is possible to brace the boom with cribbing or planking. In this case, use the boom only as a fixed anchoring point. Do not attempt to use the hydraulic system to raise the load. Instead, use a chain fall, come-along, block and tackle, or some other hoisting device to do the lifting. Remember, under these circumstances, *everything* is expendable.

Search teams sent in after downed members must be specially selected and equipped. This is the place for the best you've got available. Realize that this is likely to require extensive effort, and that you should summon help immediately in nearly every case. At the very least, get EMS units, prepared for advanced life support, on the scene. If a specially trained heavy rescue company is available, even if it requires mutual aid, get them on the way. This should be staffed by paramedic-level firefighters (Fig. 19–16).

Fig. 19–16 Structural collapse rescue is an extremely dangerous task. Close supervision of rescuers and coordination between units is critical to safety.

As a minimum, ensure that the searchers are all equipped with radios and masks, and that they carry with them a set of forcible entry tools, lights, TIC, and a search rope. The search rope must be deployed so that, once the victim has been located, reinforcements can quickly follow it if the first team experiences a mishap while making the rescue. In addition, the search team should consider taking along a spare SCBA for each victim in the event that they have run out of air or that extrication will be delayed. Obviously, at least one member of the search team should be an EMT or at least proficient in first aid.

Avoid sending every available member rushing in to join the search. This can result in delays in bringing needed equipment to the scene once the victim has been located and may expose an unnecessary number of people to undue risk without materially improving the chances of removing the victim. Remember, the best means of locating the victim is to concentrate on the location where he was last seen or assigned, rather than simply starting at one end and moving inward. For this same reason, firefighters should be taught to avoid freelancing on the fireground, lest they be missing with no one having any knowledge of their location.

A backup team should remain in a safe area, fully equipped with extrication and patient-transport equipment, to wait for the first team to locate the victim and advise what items are needed. Once this has been communicated, the backup team can then rapidly follow the guide rope to the proper area. The backups would also serve as a rescue team for the first team. Once all known victims have been accounted for, it is time again to change the pace of the operation, shifting back to a more cautious, controlled attitude.

When conditions permit and manpower is sufficient, it would be wise to continue fire suppression efforts simultaneously with rescue efforts. At times, the performance of various attack support functions can be very beneficial to the rescue effort. For example, suppose that a member has fallen through a hole in the first floor into a well-involved cellar. Two 2½-in. handlines are in place at the top of the hole, and a ladder is about to be placed into the opening. In this case, it might be very beneficial, if sufficient personnel are available, to cut a large vent hole in the first floor directly over the fire. In this situation, by venting directly over the fire and drawing heat and fire away from it, you will relieve the tendency for the first hole to act as a chimney.

Consider using high-expansion foam, especially in below-grade areas, to protect firefighters who cannot be removed immediately and to extinguish fire in their vicinity. Although foam requires additional tactics when it is used for routine extinguishment, it might be useful in this case simply to place a shield between the trapped members and the fire while efforts are made to free the victims. If the rescuers are unable to enter the area immediately, the foam blanket flowing along the low areas may shield fallen members from fire spreading around them, as well as provide them with a source of breathable air. As soon as conditions permit rescuers to enter the area, the foam can be broken down, first by sweeping the ceiling and then progressing down the foam blanket with a narrow fog pattern or a straight stream. Tie a search rope to the trapped members so others can find him in the foam.

The sheer variety of ways that firefighters find themselves trapped prevents us from developing a specific action plan for each instance. However, it is practical and highly desirable to develop a general plan of action. More importantly, such a plan will address the required psychological preparation that is necessary to make an immediate push when an organized rescue effort is required. Often the result of this preparation is the establishment of a separate rescue company, in which highly motivated personnel are selected, trained in the necessary search and rescue techniques, and provided with the proper equipment. Where having a separate unit isn't practical, at least hold a number of training sessions where you interrupt the usual smooth flow of the drill with a rescue scenario. The time it takes to react to this news and organize a search and rescue effort may startle you the first time you try it. After a number of such sessions, the suddenness of the event will no longer faze the members, and they will begin to react more efficiently.

Collapse Rescue Plan

Actual rescue operations should be based on the following proven effective plan of action. This sequence of events has been used for more than 35 years with excellent results by FDNY in numerous collapses, both fire- and non-fire-related. It is designed to provide the greatest chance of survival to the greatest number of victims while using the most efficient deployment of manpower. It consists of five separate stages of operations, which should nearly always be carried out in order. To be thorough, and to be certain that no victims are overlooked, all five stages should be carried out under fire department direction and control, even though much of the work in the last stage will likely be carried out by private contractors. The five stages of the collapse rescue plan are as follows:

1. Reconnaissance

2. Accounting and removal of the surface victims

3. Searching voids

4. Selected debris removal and tunneling

5. General debris removal

Reconnaissance

As mentioned previously, the best chances of locating a victim occur when you know in which area to start looking. This is part of reconnaissance, or *intelligence gathering*. It is actually a size-up of the collapse and rescue problem. Before anyone is allowed to go scrambling into the building, the officer in command must get some answers to some very important questions. What happened? Where? Who's missing? Where were they last seen? Can they possibly be alive? What help do you require? Where can you get it? What is the situation with the fire, secondary collapse, explosions, or other dangers? All of these items should be clearly defined before you send more people in to what may be a similar fate. Although it is very possible for people to survive even the worst collapses, you shouldn't risk additional lives if you are only going to recover a body. For instance, if a 12-in. brick wall fell outward and witnesses saw members struck directly, it would be improper to send a rescue team after the bodies while there was still a threat of further collapse in the same area.

Other considerations in the reconnaissance phase include the following:

- The construction of the building and the likelihood of a secondary collapse

- Any problem occupancies that may pose additional dangers, such as pesticides or flammables stored in the area

- Any problems with utilities

Note all of these at this time, and take action to secure them. The collapsed debris can have live wires entangled in it, threatening trapped members as well as rescuers. Broken gas lines can allow explosive mixtures to build up or they may simply intensify an existing fire. Broken water mains can flood basements and cellars. One difficulty in controlling the utilities at collapse scenes is that the controls may be buried under the debris. Immediately request utility repair crews in the event that they have to shut off the services in the street (Fig. 19–17).

Fig. 19–17 Any easily removed surface victims should be assisted immediately without detracting from other search efforts.

Accounting and removal of the surface victims

At serious incidents that truly occur without warning or time to evacuate, there may be many victims in various degrees of entanglement. After completing the reconnaissance (or simultaneously, if conditions and manpower permit), the next item to attend to is the removal of those victims who are lightly pinned. In some cases, they won't require much assistance, as when people on the roof *ride the building down*. They will most likely be in shock, however, and usually should be examined before being returned to duty. Many times, a member in this situation won't realize that he is injured until after the situation has calmed down.

Immediately designate an officer to keep track of all members who leave or are removed from the debris. This officer should at least get the member's name and where she was working when the collapse occurred. If time permits, ask the victim to point out her escape route. If members who were working near it are missing, such information may give you the most direct route to them. The member designated to obtain this information is the victim tracking coordinator. He should make note of any injuries and, if transportation was involved, to what hospital and by whom. Having an accurate list of who was on the fireground and accounting for each member by way of a roll call can tell you who is missing. However, if in the confusion after the collapse a member is transported out of the area without anyone's knowledge, other members may be jeopardized looking for a nonexistent victim. Be sure to account for all members and call off the dangerous operations once everyone has been accounted for.

Searching voids

When buildings collapse, there are almost always void spaces created in the debris where victims may be sheltered. These voids result from very strong items holding up the debris above them. The item can be part of the building, such as a foundation wall or a heavy piece of machinery. The result is that a clear area exists alongside that object.

The following are four common types of voids: the lean-to void, the V-shaped void, the pancake void, and the individual void. The lean-to void results from the failure of the support at one end of a floor or roof. If one bearing wall blows out while the other remains intact, the floor will drop at that end. Everything that was on the floor will be thrown into a heap at the low end. There is an excellent chance of survival for members who were operating on the floor below if they were near the remaining wall. Members who were on the falling floor may also survive if they aren't crushed by heavy objects. The worst chance of survival is for anyone operating on the lower floor in the vicinity of the failed wall.

The V-shaped void results when a floor fails in the middle, usually due to overloading or being burned away under load. In this case, the walls and floors above usually remain intact, although any floors below may collapse because of the impact of the floors above. The shape of the collapse produces two voids on the sides of the debris, both of which should be searched. Victims on the collapsing floor not in the vicinity of the collapse will be thrown toward the center along with all the loose debris. Victims directly below the collapse will have the least chance of survival, while those along the perimeters will have the greatest (Figs. 19–18 and 19–19).

Fig. 19–18 A lean-to floor collapse can have very large voids along the intact wall.

Fig. 19–19 V-shaped collapses may have smaller voids along each wall.

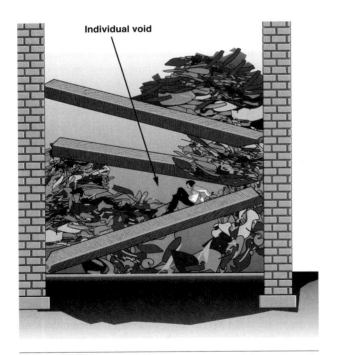

Fig. 19–20 Pancake collapses can have voids scattered throughout the debris, depending on the nature of the materials in the debris.

Pancake collapses are so named because the floors of the structure fall in layers, resembling a stack of pancakes. Although this may appear to present an unstable situation, that is not the case. Numerous rescues have been made from pancake collapses. Survival depends on the presence of strong objects nearby to keep the weight of the entire load from landing on the victim. Items such as refrigerators, washing machines, and display cases can hold up the floor above, which often remains intact. It may be possible to crawl in or out of the space between floors (Fig. 19–20).

Locating the victims of pancake collapses is often somewhat simpler than in other cases, since the floors usually maintain their physical aspect and the victims aren't as likely to be thrown as they would if the floor were to incline. It may be possible to crawl in or out of the space between floors.

Individual voids may be found in any type of collapse. They result from spaces formed by a series of strong objects that prevent collapse into that particular section. Their presence is more difficult to locate than the other types, since they occur at random, depending on the layout of partitions and furniture.

The search of void spaces should be made immediately after the surface victims have been removed, or simultaneously, if sufficient manpower is available. Remember, the fire will be trying to fill those same void spaces, so the survival of the victims depends on your reaching them first, preferably with a hoseline. Persons who are pinned should be provided with sufficient spare SCBA or, better yet, a hoseline-supplied air source in the event that conditions worsen. Attach a search line to lead others directly to them. If the conditions are severe and the entanglement is extensive, a surgeon may have to consider amputation to remove the victim. This should be a last resort (Fig. 19–21).

Fig. 19–21 At times, substantial voids exist within debris piles. These will have to be explored for pinned survivors. Void exploration must be closely monitored and should not involve tunneling or cutting structural elements without specialized training and support.

Selected debris removal and tunneling

The search of voids is done to take advantage of natural openings remaining in the debris. This is the least manpower- and time-consuming method used to reach specific locations where victims are located or suspected to be. It may involve breaching a hole through a wall, cutting through a roof or a floor, or tunneling through mixed debris. The objective must be clear so as to know in which direction to operate. This wouldn't be a useful maneuver if you had no idea where the victims were. This is an even more dangerous task than void search, since in debris removal, you may be moving an item that is supporting a load, possibly causing a secondary collapse. Use extreme care, and only permit specially trained members to perform these tasks. Expose only the absolute minimum number of personnel necessary to perform each task, and rotate these members out for a rest at frequent intervals.

Tunneling and removing debris involve arduous labor, requiring fresh personnel with clear minds. It is very easy for a member to overextend himself while trying to reach a trapped victim. It is the officer's responsibility to prevent that, since fatigued firefighters make fatal mistakes. In this case, the officer should assume a strictly supervisory role, monitoring the progress and the condition of the members, as well as arranging for the necessary support operations. If she gets involved in manual labor, these other tasks will suffer, as will the rescuers and the victims. When possible, multiple avenues of approach are useful. Tunneling is a very uncertain task. You may invest hours in one avenue of approach only to find that the tunnel is blocked by an impassable object. Manpower, equipment, and the stability of the debris are all-important factors. You shouldn't allow an action that could have a negative effect on either the rescue team or the victim. However, if it is practical, two or more approaches could increase the chances of success. Just be sure to have each team keep the other informed of any problems that they are creating.

When conducting tunneling or trenching operations, you will probably end up cutting through layers of various types of debris, depending on what type of building has collapsed. Certain tools lend themselves to different types of debris. Jackhammers, circular saws, and torches are required for Class 1 buildings; chain saws are appropriate at most buildings with wooden floors and roofs. Still, no matter what type of building is involved, large power demolition equipment such as backhoes and cranes must be forbidden at this stage. Although a manual operation is very time-consuming, it is necessary as long as you suspect that there may be survivors. Whenever you are using various tools, you must exercise extreme care to avoid injuring the trapped victims while penetrating the material that is adjacent to them.

Once you begin to get close to the victim, all work should proceed with only hand-powered tools, unless the object and the victim are visibly clear and no injury is likely. Along this line, tools that produce no exhaust fumes, such as electrically powered tools, are preferable to gasoline-driven units. Besides emitting no fumes, electric tools are easier to put down and pick up again without the hassle of having to shut them off and restart them; moreover, they lack the noise of a gas engine.

Fig. 19–22 This free-standing wall remains hanging four floors above the collapsed building. It could easily fall on rescuers.

Fig. 19–23 Firefighters search remains of a collapsed garage after installing emergency shoring.

Fig. 19–24 Rescue 1 firefighters use a Searchcam remote video probe to examine voids at this gas explosion/collapse that killed four people.

Tools that perform a task without producing showers of sparks are usually preferable to those that do (such as using a sawzall rather than a torch to cut steel bar). Of course, the right tool for each task will require the judgment of trained personnel based on these criteria, plus speed, working room, and other factors—hence the importance of a well-trained rescue team. Provide sufficient lighting to ensure that the operation proceeds safely. If there is a threat of secondary collapse, shore up the area, then tie off the debris to a substantial object, or if possible, pull it over to where it will do no further harm. Once all of the live victims have been removed, the selected debris removal operation should cease. This includes situations where you reach a victim who is obviously lifeless or who has been pronounced dead by a physician. To risk the lives of members to retrieve a body is unjustified. Equally unjustifiable is the removal of tools and equipment that were used to shore up the area. Jacks and other supports may be very unstable, and trying to remove them could result in additional injuries (Figs. 19–22, 19–23, 19–24, and 19–25).

Fig. 19–25 A firefighter uses a seismic/acoustic sensor to search for possible trapped victims at a collapse.

General debris removal

You should only begin to remove the general debris when you are certain that there are no other survivors. This operation should be completed under the direction of the fire department, even though it may require the use of heavy equipment. At times, there may still be persons missing when this operation is begun. This assumes that all of the voids have been searched and that selected debris removal and tunneling shows that there is no chance of survival.

Specially trained search dogs can be useful in making this assessment, as well as in locating victims, if fire and smoke permit. A variety of acoustic and seismic sensors are available to detect faint noises or vibrations that trapped survivors may make. This requires absolute silence in the area to succeed. In this case, cordon off a designated area and keep the crowds well away. Also shut down non-essential apparatus and remove all personnel other that the technical search team from the debris pile, in order to eliminate extraneous vibrations and noise.

General debris removal is the most gruesome task yet, removing sections of debris and depositing it in the designated area, where firefighters must search through it before it is scooped up again and hauled away. This depressing task is obviously necessary when there are still persons unaccounted for, but it should also be performed at any fast-moving fire that results in early collapse, since there may be persons present who haven't yet been missed. It will be necessary in this case to examine every inch of the structure, right down to the foundation, to avoid overlooking a victim who turns up a week later when the demolition crew finally shows up for the non-emergency demolition that you didn't order (Fig. 19–26).

Fig. 19–26 Firefighters rake through piles of debris to search for missing persons after heavy equipment has deposited a load of debris during the general debris removal phase.

Street Management at Collapses

A major building collapse is an overwhelming event for the first-alarm units. They are faced with dozens of concerns, the possibility of numerous casualties, and massive destruction. Annually, fire departments and other emergency responders across the country are called upon to respond to the scene of building collapses caused by fires, gas explosions, terrorist bombings, and other causes. The initial actions that these units take will have a great effect on how well a given incident is handled. The critical actions that these units must take are street management and fire control. They are related and must be kept in mind by all responding units and agencies.

Simplified, street management is the art of putting essential units in the right place and keeping unnecessary units out of the way, to allow continuing access for later-arriving equipment. Street management will be hampered by the scope of the destruction, by a crowd of curious spectators, and by the sheer volume of responding rescue personnel. It is these rescue personnel over whom we have direct control and who can be trained in advance. Older, more densely populated areas can be subject to gridlock conditions where nothing moves. That is where coordination and cooperation of all responding agencies is critical to success. It is up to the officer and the driver of each and every vehicle approaching the scene to take the proper action to ensure continued smooth access to the site. Units should take positions similar to those described as follows, according to the function and specific conditions at the scene.

First-alarm engine companies must position themselves to protect the occupants and rescuers against the threat of fire. This means that they must ensure a continuous supply of water and stretch protective handlines. Fire actively burning within the rubble of the structure is the highest priority, since it threatens anyone who is trapped. Exposure protection is the next priority, followed by the extinguishment of any burning vehicles in the vicinity, since these add greatly to the confusion and pose the risk of further injury. At scenes where there is no active fire, stretch precautionary lines in sufficient number and length to cover the entire operation, since there is always a chance of fire from leaking gas, arcing wires, and ongoing operations. In the event of a suspected bombing, the protective hoselines may be positioned behind apparatus or other substantial shielding to protect them from secondary blasts set to injure personnel.

The pumpers themselves should be placed outside the fire block for several reasons, the first being that the truck itself isn't needed. Pumpers have thousands of feet of hose—use it! The front of the building must be kept clear for the needed aerial platforms, heavy-rescue trucks, and later, heavy equipment.

Second, a pumper stands a better chance of maintaining its water supply if it isn't spotted on the block where the collapse occurred. A serious collapse or explosion will rupture water mains in the building. With the sidewalk covered with tons of debris, the only way to prevent flooding of the cellar and drowning people who are trapped there may be to shut off the water mains on the block. Both first- and second-arriving engines should take separate hydrants outside the block, each on separate mains, if possible. They should make preparations for supplying a large-caliber stream (elevating platform).

The third and, if needed, fourth engines should take similar positions at the rear of the structure. Later-arriving pumpers won't normally be required for their pumping capacity but for their manpower. Position them well away from the immediate operations. Set up staging areas with secure parking spaces for vehicles that aren't in use. Don't block street access with the pumpers.

Similarly, aerial ladders should also be kept away from the front of the building unless they are needed for specific rescue purposes. Aerial ladders do bring equipment that will be essential, such as power saws, generators, lights, and portable ladders; but for the most part, this equipment can be carried in by hand. Unneeded aerials should be parked with the pumpers and out of the way and in secured areas (Figs. 19–27 and 19–28).

Fig. 19–27 The area in front of the collapse scene should be kept free for an elevating platform if available. They are tremendously versatile units that can reach over debris piles to lift patients or transport heavy tools.

Fig. 19–28 Heavy stream application is one of the vital tasks for platforms at explosions and collapses.

The positions in front of the collapsed structure should be taken by elevating platforms—the telescoping type, if possible. These tower ladders are invaluable. They provide highly maneuverable master streams that can be used to protect victims and rescuers from fire. They are essential for removing severely injured survivors from either heights or from deep within the destroyed area. They can be used to travel above an unstable debris pile for survey, access, and transportation of heavy rescue tools. Rescuers independently supported on the basket can begin cutting and removing unstable pieces of debris without placing themselves in danger below objects that may fall. In short, they are priceless pieces of equipment. Unfortunately, they are of relatively limited length—usually 100 ft maximum. Therefore, they must be placed close enough to be able to do their job, yet out of the range of secondary collapses. If the building is large, use several units.

Also, in close proximity, but not directly in front of the building, should be one or more heavy-rescue apparatus and/or specialized collapse shoring vehicles. The nature of their equipment, which is much heavier than that of a ladder company, as well as the inventory of tools that they will supply, justifies positioning them in close proximity to the operations area. As with all other vehicles, however, they should be placed outside the collapse zone in case of secondary collapse (Fig. 19–29).

Ambulances have several concerns, which will likely require that they be split up according to function. The first two or so ambulances will likely begin the triage of victims and initial treatment. Position them in an area that is in proximity to the structure, visible from the scene, if possible but out of the way of rescue and firefighting operations. They should establish medical command and initiate triage. At first, these units should provide treatment supplies for other units as needed. Additional units should stage away from the operations site, in an area where egress can be made once patients have been loaded for transport. If available, bring in a mobile disaster unit for additional supplies. If not, personnel from later-arriving ambulances should walk in with all needed supplies on their stretchers. Keys and drivers must remain with the ambulances at a secure staging area for the transport of patients.

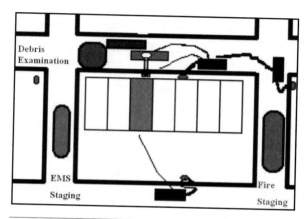

Fig. 19–29 Street management at a collapse scene should ensure access to the debris pile by elevating platforms, heavy rescues and later, heavy equipment. Pumpers should take hydrants away from the building. EMS units should have an access route into and out of the scene.

Heavy equipment, such as cranes and front-end loaders, and eventually dump trucks, must also be assured access to the site. All units must be aware that access paths that seem wide enough for firefighting apparatus may not be sufficient for heavy equipment. Do not block access routes with parked vehicles, especially if they are left unattended. Tow trucks should be staged several blocks away so that they may promptly remove abandoned vehicles near the incident site. They may also have to move unattended emergency vehicles that are impeding operations.

Obviously, police assistance is critical at these incidents. It must be stressed to the police that their assistance in ensuring the smooth flow of emergency traffic will far outweigh any volunteerism at the debris pile. Crowd control must be established early, as must traffic control, for several blocks around the site, and particularly those streets designed as access routes. On major incidents, a frozen zone may have to be established, where no non-emergency vehicles are permitted to move. This can be essential for reducing vibration, which could trigger a secondary collapse. In the event of a bombing, it will help preserve evidence.

These actions, coordinated through a joint CP and with the cooperation of all personnel, will improve the odds of survival for anyone still alive in the debris pile. In the event of a serious collapse or explosion, all of the community's resources will be brought into play. Each responder must understand the importance of his actions as part of a team effort, working in unison to ensure the safety of all.

Safety Precautions during Collapse Operations

1. Shut down all utilities.

2. Monitor the atmosphere for flammable or toxic gases, as well as sufficient oxygen.

3. Prohibit smoking.

4. Remove all non-essential personnel.

5. Control the spread of fire, if present. If there is no fire, be prepared for it.

6. Maintain constant surveillance of weakened walls, floors, and other building components. Use a surveyor's transit to monitor suspect areas for movement.

7. Eliminate all vibrations from nearby highways, rail lines, unnecessary apparatus, and the like.

8. Do not cut or remove major supports. Work around them if possible.

9. If you absolutely must cut supports, brace and shore around them and prepare for secondary collapse by removing everyone but the personnel performing the cut.

10. Rotate personnel frequently, every half hour or less.

11. Maintain communications between rescue teams as well as between rescuers and the victims.

12. Seek expert assistance.

It has been said that the responsibility for members being caught in a collapse rests with the officer in command. Although this may be partly correct, all members have an obligation to themselves, their families, and their peers to avoid placing themselves in areas where collapse is likely. Additionally, all members must act as the eyes and ears of the officer in command, reporting any conditions warning of impending collapse. Some of the reasons members become victims include the macho attitude, as well as ignorance of the causes and warning signs of collapse. Hopefully, the reader will by now be more aware of these. The notion that firefighters can put out *any* fire fails to take into account that this isn't a fire but an unstable structure that can kill you. There are buildings that we as firefighters just don't belong in. The sooner we accept that fact, the sooner we will stop dying in numbers needlessly.

Table 19–1 Structural collapse rescue plan checklist.

Establish CP.

Designate staging area, assign company officer to command.

Transmit preliminary report, including the location of the CP and the staging area, to the dispatcher.

Request EMS and PD supervisors to respond to CP.

Ensure ALS ambulance is assigned.

Perform initial size-up.

- What caused the collapse? (circle) explosion, fire, construction mishap, other _____

- Check for radiation/chemical hazards at explosions.

- What type construction has collapsed? (Cl.3, etc) _____

- What floor is the collapse located on? _____

- How large is collapse area? _____

- What is the Occupancy? _____

- What Occupancy Hazards are present? (Hazmats, etc) _____

- Are there reports of people trapped? How many? _____

- Where are people trapped? _____

- Is it an interior or exterior collapse (Circle)

- Is there fire? Yes / No

- Is there a gas leak (smell, sound, or visible signs) Yes / No

- Are there electrical shorts (wires arcing)? Yes / No

- What is the status of the building stability? _____

- Manpower and equipment requirements. _____

Second Notification to Dispatcher

Notify dispatcher of nature of incident and need for additional FD units or other agencies resources.

Transmit 2nd alarm/request additional units if conditions warrant.

Manage the Collapse Rescue Plan

1) Ensure necessary hand lines are stretched. Units must be able to protect the entire collapse area with hand lines or TL streams.

2) Assign units to immediate rescue / removal of surface victim

   ID of Units Assigned_____

3) Assign units (2nd due ladder to begin) to shut off all utilities

4) Request utility companies to respond via dispatcher.

5) Special Call a tower ladder if one is not assigned. Depending on size of building, more than one TL may be needed.

6) Have a 3½-in. supply line stretched for use by the TL.

7) Survey site for indications of possible secondary collapse and other hazards. (Identify hazards here) _____ _____

8) Identify and establish collapse danger zone.

9) Begin Street Management,

   a) Make sure units do not block the street.

   b) Seal off ends of street to all but units IC requests

   c) Only vehicles that the IC authorizes should enter the street.

   d) Eliminate vibrations. Shut down:

      1. Trains

      2. Nearby traffic (buses, trucks, etc.)

      3. Construction site equipment.

      4. Apparatus not in use.

10) Assign CFR-D engine companies to first aid for victims.

11) Start witness/victim interviews using the Collapse Survivor Interview Form.

12) When surface search is completed, start void searches.

13) Evaluate need for additional help (3rd alarm, etc.)

14) Transmit progress reports.

Assignment of Additional Chiefs

Identify Chiefs assigned to each position.

A) Victim Removal Officer _____ Batt. _____

B) Communication Coordinator_____ Batt. _____

C) Safety Coordinator _____ Batt. _____

D) Victim Accounting Coordinator_____ Batt. _____

E) Staging Area Chief _____ Batt. _____

F) Logistics Chief _____ Batt. _____

Table 19–2 Collapse survivor interview form.

Was anyone else in Building/Area?

How many?_____ Who?_____

If Time permits, get a description of each missing person

Person 1 _____

       Age _____ Sex _____ Clothing _____

Person 2 _____

       Age _____ Sex _____ Clothing _____

Person 3 _____

       Age _____ Sex _____ Clothing _____

Person 4 _____

       Age _____ Sex _____ Clothing _____

Where were the missing people before the collapse? Room?

Did you see/hear them after the collapse?

Was there any warning of the collapse?

Describe the area others were in, color/type of floors, walls, furniture, etc.

How did you get out? Ask them to point out their escape route if possible, it may be a way in to other victims.

20 Fire Department Roles in Terrorism and Homeland Security

Terrorism is an ugly, evil word. Striking at innocent civilians to further some twisted political or criminal agenda is anathema to what firefighters do and stand for—helping those in need regardless of race, religion, political views, social status, or any other artificial criteria. If the victim is a human being, they are our highest priority. Period. We in the fire service have the task of dealing with the results of terrorist attacks; digging victims out of the rubble of collapsed structures, extinguishing the fires that result from bomb blasts, and providing patient treatment and transport to medical facilities.

In the process of performing our missions, we are often exposed to danger; secondary collapse, fire, and blood-borne pathogens, to name a few. But now, the specter of far more deadly attacks lurks dimly in the shadows, threatening not only our cities, but also the emergency responders that seek to restore order. All members of the fire service must recognize their role in responding to these events and take the necessary steps to maximize the safety of the public as well as our own personnel. We have seen first hand the devastation within the ranks of first responders that the terrorists created at the WTC (Fig. 20–1).

Fig. 20–1 America's first responders are truly engaged in a war against terrorism.

Fortunately, we do not have to just sit helplessly waiting for the next attack to come; there is something we can do to fight back. We can pay attention to what is going on around us and notify law enforcement of our observations. All Americans are part of our Homeland Security effort, each and every one of us. The fire service may not play as large or as visible a role as the FBI, the CIA, or the Transportation Security Administration's screeners at airports, but in some ways we have a role that these agencies are unable to fulfill: We go right into people's homes and businesses all the time, responding to fires, medical emergencies and other emergencies such as water leaks, meaning the people have generally not had a chance to prepare for our arrival.

It is reported that what would have been potentially one of the most devastating plots against the United States to date was averted through the alert actions of a Philippine firefighter who responded to a minor incident at the location where Ramsey Yousef, the mastermind behind the 1993 bombing at the WTC, was engaged in preparations to blow up 12 U.S. airliners over the Pacific Ocean. The firefighter is said to have sensed that something was not right about the situation and reported it to the police, who swooped down and arrested the plotters before they could carry out their deadly plan. It is that kind of involvement that we need from every resident of this country if we are going to win this "Third World War," as some noted analysts have called this conflict.

Fire personnel have a distinct advantage over the general public in this area. In the event of an emergency, we have immediate access to any building we respond to. Under these circumstances, terrorists are likely to flee the scene to avoid having their cover blown before they are ready to act. On the other hand, personnel conducting routine fire prevention inspections of buildings could also discover a plot in progress.

The Fire Department is the only government agency that routinely conducts inspections of existing structures, and we should maximize our opportunities to examine buildings where such illicit activities as mixing explosives or chemical agents are potentially occurring. Personnel that suspect something is amiss should take no overt action under these circumstances. Continue your inspection as though everything was progressing normally; perhaps even give some tips about housekeeping. Wait until you are well away from the premises and out of sight before contacting authorities. (Every emergency vehicle should have the number of the local police detective squad and the nearest FBI office displayed in the cab. I do not recommend calling 911, as the call takers there may not rate your information as a priority requiring response of patrol cars, given the volume of calls they receive from the public.) Try to use a hard-wire telephone, as opposed to a cell phone or the radio if possible, to avoid being overheard by any bad guys monitoring those frequencies.

Potential Indicators of Ongoing Terrorist Activities

Protective clothing

- Masks
- Chemical suits
- Rubber gloves

Laboratory equipment

- Beakers and other glassware
- Ice baths (glassware sitting in ice)
- Petri dishes
- Distillation equipment

Improvised ventilation equipment

- Enclosed fume hoods

- Glove boxes

- Exhaust fans temporarily located near sinks, stoves, etc.

Precursor chemicals

- Fertilizer containing ammonium nitrate, acetone, diesel fuel, peroxides (*i.e.* hydrogen peroxide)

- Cyanides (sodium cyanide, potassium cyanide, etc.)

- Acids

Many of the previous items are found in legitimate businesses all around the country. For example, metal plating shops use cyanides and acids and would appropriately have the proper protective clothing, ventilation equipment and laboratory facilities to deal with these potentially toxic materials. The same equipment found in the cellar of an apartment house or in a rented locker at a storage facility should prompt an immediate alert to law enforcement.

A Historical Perspective

Terrorism has been around for many years. In fact, the attacks on the WTC, the Pentagon, and the failed attack that came to rest in the fields of western Pennsylvania were only the latest in a string of attacks, primarily bombings, that go back at least until the 1920s, including attacks by such groups as the Weathermen or the Puerto Rican nationalist group Fuerzas Armadas de Liberación Nacional, or Armed Forces of National Liberation (FALN), stretching nearly uninterrupted all the way back to a bombing on Wall Street in the 1920s. The New York Stock exchange was the scene of a chemical attack in the 1930s, when tear gas was dispersed into the ventilation system, a precursor of what we are likely to face in the future.

The causes that prompt a terrorist to act are too numerous to consider here. They can range from religious, as the current assault on the West by Muslim terrorists illustrates, to abortion foes to environmentalists, as a recent wave of arson and explosions linked to the Earth Liberation Front (ELF) has shown.

The 1920 Wall Street bomb was approximately 100 lbs of TNT wrapped with window sash weights. It killed 40 people and wounded 300, but there was a distinct lack of serious structural damage. The attack was never solved, but was attributed to anarchists. FALN, a Puerto Rican nationalist group set off 49 bombs in NYC in 1974–1977, including a bomb at Fraunce's Tavern (the scene of George Washington's farewell speech to his officers after the American Revolution) in 1972 that killed 4 people. The Weather Underground or *Weathermen* set dozens of bombs around the country, culminating in the explosion of their West Village bomb factory in 1970. The Weathermen moved away from targeting people and aimed their bombs more at property, particularly government-owned buildings. Similarly, the FALN did not cause many deaths in their campaign. Fewer than 100 people died in the more than 100 bombs set off by the Weathermen and the FALN combined. This was intended to avoid alienating the public to their causes. Al Quaeda has no such compunction.

The United States has been the target of extreme nationalist groups for some time, even though we had no direct involvement in the region at the time. Croatian nationalists detonated a bomb in a locker at LaGuardia airport in December of 1975 that killed 11 people. The use of bombs by terrorists is increasing and changing, however. Recent world events point to the possibility that we could encounter another wave of bombings. These would likely be different than bombing campaigns we've experienced in the past.

One of the *trademarks* of the attacks attributed to al Quaeda is the use of a *swarming* technique involving multiple simultaneous attacks designed to overwhelm any defensive measures and outstrip the capacity of emergency responders to deal with the consequences. The attacks of September 11th, using four separate airliners were the most graphic example of this method. The Madrid train bombings in March of 2004, which saw 10 bombs detonate nearly simultaneously on four different trains, killing nearly 200 and wounding 1400, is another example. The 1993 bombing at the WTC was the first demonstration that radical Muslim terrorists are a different breed than any we've experienced in the past. The fact that the 1993 WTC bomb only killed 6 people was due primarily to their mistakes—not their intentions. Their real goal was to topple the North Tower into the South, bringing both buildings down. It took them 8 more years, but they accomplished their goal. This is a new breed of terrorists, with money and support.

Factors that put our cities in the terrorist's bulls-eye are as follows:

- Large concentration of people (targets)

- Immediate, worldwide media coverage likely

- A large community for the terrorist to blend into with anonymity, both before and after an attack

The new terrorism

While terrorism may not be new, it has taken several deadly new twists in recent years. The 1995 bombing of the Murrah Federal Office Building in Oklahoma City confirmed how easy it is to mix and deliver a massive bomb and avoid detection until it is too late. That devastating blast also put to rest the notion that it was only places like New York City and Washington D.C. that had to prepare for the possibility of confronting a terrorist attack. That same year, federal authorities broke up a radical Muslim plot to wreak massive destruction on New York City from a series of simultaneous bombings of bridges, tunnels, and office buildings. In 1996, another group of Muslim terrorists was captured in Park Slope, Brooklyn only hours before they planned to detonate a bomb in the Flatbush Avenue terminal, a main subway and commuter railroad hub.

Also in 1996, America received another lesson that our counterparts in locations such as Israel and Northern Ireland have known for years—that responders, including rescuers, are targets of the terrorist. At a bombing of an abortion clinic in Atlanta, Georgia, a secondary device exploded some time after the original blast. Its intent was clearly to kill or injure those who came to help at the first blast. Firefighters must recognize the potential for a secondary device at any obvious bombing and report any suspicious devices. Normally police will handle the search for secondary bombs, but we could discover them while we conduct fire control, search for victims, or first aid operations. Be aware that such devices are often placed around the perimeter of the blast site, aimed at CP personnel. Do not disturb any suspicious devices, report them to the incident command structure for relay to police, and try to position your personnel away from the projected damage zone.

Perhaps more disturbing than this targeting of rescuers is the threat of use of *weapons of mass destruction* (WMD) or chemical, biological, radiological, nuclear, or explosive (CBRNE) weapons by terrorists. As horrific as the WTC and Oklahoma City attacks were, the fact is that there are people who would like to kill even more Americans, in ways that attract even more media attention, groups who have sought, acquired and used nonconventional weapons designed to kill thousands of people. Unlike conventional explosives, these weapons may not leave any visible evidence of their presence, yet they will kill anyone who comes in contact with them. Fortunately, firefighters have some means of protecting themselves against *some* of these weapons. They key is to *recognize their presence* and take the *proper precautions*. All firefighters must be aware of the potential for a terrorist event, either conventional or utilizing WMD or CBRNE at *any* incident. Suspicion saves lives.

Certain locations are more likely to be a terrorist target than others. In the past, terrorist's favorite targets[1] have focused on are the following:

World landmarks. WTC, United Nations, Golden Gate Bridge, Statue of Liberty, Fraunces Tavern, Seattle's Space Needle, etc.

Financial institutions. The New York Stock Exchange, American Stock Exchange, commodities exchanges, bank headquarters, etc.

Media and corporate headquarters. ABC, NBC, CBS, Mobil Oil, Exxon, etc.

Ideological targets. Government offices especially IRS, military bases and recruiting offices, abortion clinics.

Educational research labs. Especially those conducting animal research, Columbia University, Stanford.

Al Quaeda's training manual has expanded on this target list, to include a wider array of targets, many of which were *off-limits* to other terrorist groups who often sought to limit civilian casualties. Among al Quaeda's potential targets are the following:

Energy facilities. Nuclear power plants and conventional electrical generating sites, refineries, pipelines, and tank farms.

Transportation facilities. Airports and aircraft, rail hubs and trains, bridges, shipping and port facilities.

Soft targets. Recent attacks overseas have struck apartment houses, hotels, and other residential buildings. Places of worship, particularly synagogues, have also been targeted.

Al Quaeda and similar Muslim extremist groups have declared religious war on the United States, Israel, and the West in general. This distinguishes them from previous terrorist groups in the United States, which typically sought to change the factors they resented by forcing the rest of the country to recognize their cause and discuss it in an open, democratic fashion. Largely, they used terrorism more as a way of attracting attention and followers than severely crippling their opponents.

That is not what the new crop of Muslim terrorists seeks. According to Col. Russ Howard, US Army, coauthor of *Terrorism and Counterterrorism, Understanding the New Security Environment*, the old terrorists "did not want large body counts because they wanted converts. They wanted a seat at the table. Today's terrorists are not particularly concerned about converts and don't want a seat at the table." Howard quotes former CIA Director James Woolsey: "They want to destroy the table and everyone sitting at it." For this reason *every* incident at *any* location that meets the following criteria should prompt *suspicion* and extra awareness by fire personnel to the potential that a terrorist event is in progress.

1. Any recognizable landmark or building as described previously.

2. Any response to a crowded public location, *i.e.* sporting events such as the Super Bowl or Rose Bowl, rush-hour subway platforms, Times Square on New Years Eve, concerts in theaters or other large gatherings.

3. Any reports of "persons overcome by fumes" or "chemical odor in the building/area."

4. Any bldg/site where an explosion has occurred (including reported *transformer* explosions).

5. Any response where a threat or other intelligence indicates a motive.

Of course fire departments respond to alarms at public locations thousands of times each year, most often without incident. Yet we have seen how first responders have died at these attacks. It only takes one incident to kill you (and in some cases your families) with some of the weapons that are likely in the hands of some terrorist today. Intelligence estimates are that an attack using WMD or CBRNE is an increasingly likely event. The only way to avoid becoming a fatality at these incidents is to be *alert*, be *properly trained*, and to *react instinctively and correctly* to the first indication that *this* response is *not* a routine response!

Types of terrorist WMD agents and CBRNE weapons

Chemical agents are generally fast acting and will likely produce a very visible display of their effects. People are likely to know very quickly that an attack has occurred, and there is a real possibility that fire department units could be called to the scene of a chemical attack, either to investigate an odor or to treat civilians complaining of symptoms or because the event involves an explosion, smoke, or fire. These agents can kill with extreme rapidity. The agent used could be a dispersal of a military agent like the nerve agents or it could be a highly toxic industrial chemical that is shipped through your city, such as chlorine or hydrogen cyanide. The doomsday cult Aum Shinrikyo has demonstrated that chemical attacks on civilian targets are possible.

Here are some potential chemical agents and their associated identification code letters.

- *Nerve agents*—Sarin (GB), soman (GD), tabun (GA), VX, GF

- *Blister agents*—Mustard gas (H, HD, HN), lewisite (L), phosgene oxime (CX), HL, HT, Other variations of blister agents, chemical names not commonly used (symptoms often delayed 4–48 hrs).

- *Choking agents*—Phosgene (CG), chlorine (CL)

- *Blood agents*—Hydrogen cyanide, (AC), cyanogen chloride (CK)

- *Tear gases, crowd control*—CN, CS

Biological weapons

Biological agents are living organisms that cause disease. Most likely, other than for overt releases that are accompanied by a threat (such as the anthrax letters that were sent to several media outlets and government facilities in late fall of 2001 and the Ricin letters of 2004) fire department units will not be aware that a biological attack has occurred. This is because they are virtually undetectable in the field with current technology, and most persons exposed will not show any symptoms for a period of several days after exposure. Paramedic and ambulance personnel encountering a number of patients exhibiting similar unusual symptoms may be the first to detect an attack. The death rate after exposure to some of these agents is nearly *100%*. Fortunately, the prior attacks and the numerous *suspicious white powder* runs they spawned, some hoaxes and some from genuinely concerned individuals, have heightened many departments' awareness of the potential for such incidents and increased our preparedness for such attacks in the future (Fig. 20–2).

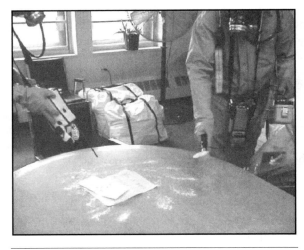

Fig. 20–2 The anthrax attacks of 2001 have highlighted the need for first responders to properly prepare for the possibility of WMD attacks.

Some potential Biological agents.[2]

Viruses—Hemorrhagic fevers, Marburg, Ebola, machupo, dengue fever, smallpox

Bacteria—Anthrax*, bubonic plague, tularemia, and brucellosis

Toxins—Botulism, Ricin*, staphylococcus enterotoxin B (SEB)

(* indicates agents that have been used in recent terrorist attacks)

Radiological agents

Radioactive elements such as plutonium, uranium, cesium etc., and including spent radioactive waste, as well as weapons-grade material, is increasingly available, whether by finding its way out from behind the old Iron Curtain or being found in everyday equipment in American industrial sites. While it is possible that an atomic bomb could be detonated here, it is not likely. However, that is not the only type of threat involving radiological material. A somewhat more likely scenario involves the explosion of a so-called *dirty bomb*, or other dissemination of radioactive materials. A dirty bomb is not an atomic bomb but consists of highly radioactive material wrapped around an ordinary explosive. When the ordinary explosive detonates, it causes a *normal* explosion, not an atomic bomb blast. This fills the air and the surrounding structure with tiny particles of highly radioactive debris. While it leaves no visible sign of contamination, contact with the debris, particularly through inhalation, is deadly. Radiation detection equipment is essential to first responders in order to even recognize what has taken place!

Nuclear

Members of the intelligence community describe the threat that many people envision as the real weapon of mass destruction—the explosion of a multimegaton nuclear bomb in the heart of an American city—as an unlikely event. Formulation of such devices is extremely difficult to accomplish. Theft of such devices or purchasing them on the *black market* in usable condition is also described as very unlikely. That does not mean *impossible*, of course, but there are far more likely scenarios to prepare for. The detonation of a small, improvised nuclear device, in the low kiloton range of yield, something that would do major damage, but not wipe out an entire city, is a somewhat more likely, although still remote, possibility.

One of the things that fire department planners and emergency managers have to consider is the possibility of a detonation of a crude nuclear device. Most people that I know think that if such a thing happens they "may as well bend over and kiss themselves goodbye," because there will be nothing left of them or their city. That is not entirely true, especially in the case of smaller devices. There will be a very large job for the fire department and other emergency responders. There will likely be large areas of the city that are not obliterated, where large numbers of potential survivors may be found, and large conflagrations that may be have to be fought to defend other areas that have survived the blast from incineration. All is not lost immediately.

Explosives

In spite of the horror that the exotic WMDs inspire, the fact is that high-order explosives are still the terrorist's main weapon of choice. As pointed out, this is nothing new, and terrorists tend to deal with methods they are reasonably certain will succeed. While Aum Shinrikyo attempted dispersals of anthrax as well as Sarin, and gained worldwide media attention, the fact is these WMD attacks were a dismal failure, killing only seven people, but causing the destruction of the group in the ensuing criminal investigation. Terrorists don't like to risk failure or detection, never mind both. Explosives are a much more reliable weapon, and they are readily available, as opposed to the requirements needed to develop almost any of the other WMDs.

Fire Department Response to Bombings and Explosions

The current worldwide political climate has seen much unrest, which has resulted in the use of car bombs and other explosive devices aimed at civilian targets. Israel, Northern Ireland, Spain, and even New York City have experienced a large number of explosions over the years that have taught many lessons in how to respond and operate at these events. The effect of a terrorist blast has the potential to cause far greater harm to rescuers than the same size explosion caused by natural gas or other accidental causes. Given the current threat levels around the world, all fire personnel must approach each and every explosion as though it is the result of a bombing. If you operate as though it is a bombing initially, you will do all the things that would need to be done for an accidental explosion, as well. And you and your fellow responders will be safer.

Incident management

At any explosion, the immediate priorities are lifesaving and control of imminent hazards such as fire or secondary collapse that threatens further loss of life. In conjunction with these rescue-related activities, a rapid preliminary examination for the likely cause of the event is vital, since a bombing creates the following additional problems for responders that are not present at accidental explosions: the threat of secondary devices, potential for WMD involvement, crime scene considerations, to name a few.

One of the most important tasks that can be undertaken at a bombing, given the nature of the new terrorism, is to disperse the crowds of onlookers, as well as avoid concentrations of responders that might be a target of a secondary device or a suicide bomber. Request law enforcement assistance immediately to establish a secure perimeter around the incident scene. The first arriving fire officer has prime responsibility for rescue activities at the incident site and should assume the role of initial IC for all explosions. All other agencies responding to an incident should assist as requested by the IC.

If the incident appears to be the result of a criminal or terrorist act, as soon as lifesaving efforts have concluded, the role of IC should transition to law enforcement and/or federal authorities, such as the FBI and Bureau of Alcohol, Tobacco, and Firearms (BATF). It is the mission of these agencies to determine the cause of the incident and to investigate any criminal acts. Therefore, it is essential that the scene not be indiscriminately damaged or disturbed after rescue operations are completed, other than to make the area safe. When the rescue and removal of all persons involved in the incident has been accomplished and all immediate life-threatening hazards are mitigated, the scene should be secured and all fire personnel withdrawn, pending direction from law enforcement and/or federal authorities. Investigators should be accommodated as soon as their presence does not interfere with ongoing rescue efforts. The IC will determine when an area is safe enough to permit personnel of other agencies such as utilities to enter an area to begin the process of restoration of services and recovery. Request law enforcement to provide site security.

Objectives of the fire department at explosions

The objective of the fire department is to protect the safety of the public and emergency responders by extinguishing any resulting fires, removing any persons entrapped in the affected structure or vehicles, triaging, treating, decontaminating, and transporting any injured persons. The preservation, gathering, and documentation of evidence are a prime consideration of law enforcement, and so must be one of ours as well. These two objectives must coexist in the same space.

In the event an explosion has occurred, the following actions must be taken:

- Rescue or evacuation of endangered persons
- Monitoring of the scene to detect the possible presence of radioactive or chemical agents
- Application of first aid to the injured, then triage, treatment, decontamination (decon) if needed, and transport
- Safeguard operating forces from further explosions or collapse
- Extinguishment of fire.
- Mitigation of other life threatening hazards, gas leaks, hazmat releases, live power lines, etc.

- Primary and secondary searches of all affected structures
- Stabilization of collapsed/damaged structures to prevent further death or injury

Potential hazards to be dealt with

- Secondary explosions
- Fire
- Structural collapse
- Hazmat release/exposure
- Biological hazards/blood-borne pathogens
- Ruptured gas, steam, water, or sewer lines
- Falling glass from surrounding buildings

Preparation for Handling a Bombing Campaign

Fire departments respond to numerous explosions each year, most often without incident. Yet a bombing is very different from an accidental gas explosion. The only way to avoid becoming a fatality at these incidents is to be *alert*, be *properly trained*, and to *react instinctively* and *correctly* to the first indication that *this* response is *not* a routine response! We must learn how to avoid the dangers that may be present. One way to do that is to study the way others with experience in this area deal with the problem. A response to an explosion at such landmarks as the United Nations Building, Sears Tower, Golden Gate Bridge, Seattle's Space Needle, etc, should automatically prompt fire personnel to expect that a bombing caused the explosion. We must be just as cautious at an explosion in a PD or storage warehouse however, since that is where the devices are built in secrecy then transported to their target. Proper steps must be taken to protect the safety of civilians and fire personnel when it is suspected that an attack has occurred.

Facts Related to Recent Bombing Campaigns

Most bombings are antipersonnel bombs

They consist of a relatively small amount of explosives wrapped with large amounts of shrapnel. These produce relatively minor structural damage, primarily glass and light metal. Shrapnel and flying glass cause most injuries. Large numbers of casualties require response of a large number of responders capable of administering the ABCs of first aid: open *airways*, check *breathing*, and stop life-threatening bleeding (*circulation* problem). Mass casualty kits, which stock large quantities of bandages and oxygen administration equipment should be available to supply the large numbers of first responders. Triage of critical patients is vital since resources will most likely not permit personnel to provide intensive advanced life support methods for large numbers of patients. Vehicle fuel tanks are vulnerable, and bombs can be augmented to cause fire with propane cylinders etc, which would require large numbers of burn treatment kits. Massive vehicle-borne explosive devices, such as those used in the Oklahoma City or WTC bombings, are more difficult to make and carry out. But they produce even greater casualties and severe structural damage that require advanced collapse rescue techniques.

Bus and other mass transit facility

This type of bombings is a serious threat for loss of life. Most of these events are intended for maximum casualties with minimum material. If a bomber can penetrate into a crowded, enclosed area, casualties can be high, but even in severe bombings there are often survivors. On arrival the scene is likely to be chaotic. Do not assume that there is nothing you can do to save lives. But you must not get tunnel vision. Perform a cautious size-up, watching for additional threats until law enforcement secures the scene.

Extrication is not normally required at bus bombings, or is not difficult. These are primarily multi-casualty incidents (MCI), large-scale medical responses for trauma and burns. Engine companies, EMS and ladder companies can normally perform all the required first aid and any required extrication. The same device detonated inside the lead car of a speeding train inside a tunnel will have a much greater effect. In that event, advanced extrication skills and equipment would likely be required if the blast results in a derailment. Don't get suckered in however. Remember the other possibilities, *i.e.* secondary device, dirty bomb, chemical attack, etc.

The tendency toward secondary attacks aimed at the rescuers has been increasing. Request the bomb squad or emergency ordnance disposal (EOD) team to establish a safe corridor for the entry of rescuers. Situational awareness is important. Are you being drawn into a killing zone in an ambush? Get law enforcement to secure a safe corridor. The fire department IC *must* request adequate site security from law enforcement. Make this request over your radio to your dispatcher for relay to law enforcement superiors, since on-scene personnel can be distracted by other events around them.

Response Consideration

Clues in debris

While responding, watch for clues in debris for causes and nature of an event. Note things that indicate the nature of the explosion, like deformation of a car or bus body, total blow out of glass, or damage that is not seen at a routine accident or fire. Look for clues such as debris strewn on branches of trees or overhead wires that indicate something blowing up. A car gas tank that ruptures and *explodes* does not produce that effect.

Note people in area

Watch particularly for *armed people*, or people wearing protective clothing, masks, etc, as they flee the scene. At a recent drill simulating a chemical attack in a subway, two *terrorists* walked up out of the subway exit right past entering responders without anyone giving so much as a glance toward them. They were wearing gas masks coming out of the subway!

Beware of ambushes

Initial units should avoid getting into a bottle neck where an ambush could be likely. If suddenly you find yourself being blocked in by civilian vehicles, seek shelter! Try to position apparatus so that they can be driven forward out of any danger area that develops. Especially if it has been confirmed that an event is in fact a bombing, this precaution should be a primary consideration. Given the propensity for multiple attacks demonstrated in recent years, the first unit to identify an explosion as likely being a bomb attack must notify the department-wide command structure—for relay to all units—so they can anticipate further attacks and take proper protective measures (Fig. 20–3).

Position apparatus

Position apparatus with specific objectives in mind as per the collapse rescue plan. Position engines on hydrants but not blocking streets, or intersections. Aerial ladders should be positioned for rescue as needed. If not needed, position them out of the way to allow access for tower ladders, a collapse unit, heavy equipment, etc. Elevating platforms should be specially called as needed to cover collapse areas but must be located out of the secondary collapse zone.

Check for hazards

Check for hazards as you approach. Special units such as hazmat teams must realize that they are the only ones capable of detecting many chemical and radiological agents during the initial stages. This identification is critical to the safety of all responders. The effects of some agents (mustard, etc.) do not appear until hours after exposure. By then, some of your people could be dead if not warned immediately! Get off the rig with all meters ready to go! Every time! There may be people in need of assistance on your arrival. If you do not have enough people to cover the meter assignments, the injured will have to wait until you cover these crucial metering assignments. By taking the readings you may save the victims' lives as well as your own, since you might be able to identify what is affecting them and identify the need for an antidote or treatment (Fig. 20–4).

Fig. 20–3 Firefighters responding to explosions should have a very high **situational awareness**. Know who comes in with you and who goes out. Be aware of the actions of bystanders to avoid secondary attacks.

Fig. 20–4 The potential use of chemical weapons has prompted some departments to purchase special chemical agent detectors. They require proper training as well as operating protocols.

Secure the area

Request police to secure the area and provide a path that is clear of secondary devices. Firefighters must recognize the potential for a secondary device at any obvious bombing and report any suspicious devices. In New York City, the FDNY does not search for bombs. But that doesn't mean we couldn't find one! Normally police will handle the search for secondary bombs, but we could discover them while we conduct fire control, search for victims, or perform first aid operations. Be aware that such devices are often placed around the perimeter of the blast site, aimed at CP personnel. Do not disturb any suspicious devices; report them to the incident command structure for relay to police and try to position your personnel away from the projected damage zone. Be alert! All members must know how to react to a secondary device. You must instantly take steps to protect yourself and all those around you.

If you discover a bomb or explosive device

- Do not disturb anything!

- Do not use the radio or handie talkie to report the device. You may set off an explosive!

- Immediately evacuate the area for at least 300 ft in all directions.

- Notify law enforcement (Fig. 20–5).

Fig. 20–5 Fire departments could discover a secondary device while performing patient care or fire control operations at the scene of an explosion. Know how to react to the presence of a bomb. Do not use the portable radio within 300 ft.

Immediately warn everyone around you of the device, and inform them of the need to *not* transmit on the handie talkie. (Electric blasting caps may be sensitive to the radio energy.) Immediately evacuate the area and report the device to the IC and the police. Do not go back in to show anyone where it is. As part of routine operations at bombing scenes, the police must be requested to secure a perimeter against additional suicide bombers and to sweep the area around our operations for secondary devices. In large structures, they should also sweep a path to the blast area through any corridors that operations will go through.

Maintain situational awareness

- Know who came in with you and who goes out.

- Watch the actions of bystanders.

In the 1980s, members of a black separatist group stole firefighter uniforms and tried to buy a used fire truck. Their plot intended to follow FDNY units in to a false alarm at Kings County Hospital to free a prisoner who was being held there. Police broke it up before the plot could be carried out. In Israel, ambulances are known to be used by terror groups to transport explosives, arms, etc, since they're rarely stopped and searched. Recently, suspicious people tried to buy a used FDNY ambulance in New Jersey.

Maintain clear access routes in and out of the scene

These incidents attract huge crowds of spectators and media, as well as large numbers of emergency vehicles. There is a pressing need for organization of responders to ensure that proper access is maintained to allow later-arriving specialized units like collapse units to be able to reach their correct operating location. Ambulances will also have to have clear egress from the scene to transport injured persons to hospitals. Request that a law enforcement superior officer report to the CP to establish crowd control and traffic routes.

Apparatus staging

The early establishment of a secure remote staging area that multiple alarm units report into is essential to ensuring open access to the scene. Many units will not need their apparatus, but their personnel will be used for treating and assisting in the transport of non-ambulatory patients. Apparatus should be properly parked to permit traffic flow while personnel await assignments. The personnel may be directed to proceed on foot to their operational assignment if traffic congestion is severe. Request law enforcement to send resources to secure the staging area from potential attack.

Interagency coordination

Early establishment and clear identification of the CP is vital to organizing the proper coordination of outside resources. The first arriving IC must notify the dispatcher of the location of the CP and request that all responding agencies be notified to send their ranking representative to that location.

Injured to be triaged

At large-scale incidents, you will not have sufficient resources immediately on scene to treat every one. Triage is crucial to minimize duplication of effort. At chemical, radiological, or nuclear events, this can be critical to rescuer safety. Use *kick triage* to limit the rescuers exposure. You want to spend as little time in any contaminated area as possible, and limit the number of rescuers exposed to the absolute minimum.

Persons caught in explosions are likely to suffer a number of serious, potentially life-threatening injuries, notably lacerations, broken bones, other trauma-related injuries, and burns. Many of these injuries only require administration of basic first aid skills such as bandaging to save a life, at least initially. Patients that require advanced techniques such as intravenous administration, defibrillation or even CPR are likely not to survive, given the shortage of such trained personnel and transportation resources at large scale events. The purpose of triage is to identify those injured people that can survive only if given the level of treatment that is *available*, and get them that treatment, while those who will survive anyway wait for further resources. Those that will not survive given the resources that are available should not prevent someone else who could survive from receiving the treatment that is needed to save their life. If sufficient resources are available to treat all casualties simultaneously, that sort of hard decision making is not as crucial, but triage is still required to ensure those in the greatest need are treated first.

At an MCI, use triage tags to rapidly identify those that need treatment and transport first while avoiding duplication of efforts.

Command Procedures

Incident commander

Establish fire department CP. Designate a staging area and assign a company officer to command the staging area until the arrival of a staging chief. Notify the dispatcher of the nature of the incident and the need for additional fire units or other agencies' resources. Evaluate the situation to determine the extent of the fire department commitment at the incident. The commitment of resources should be based, in part, on the following:

- Amount of fire that remains to be extinguished.

- The number of civilians injured.

- Any remaining hazards that must be resolved after initial operations have stabilized the scene and *after* the criminal investigation has been conducted. (**Example:** Broken window glass in the surrounding area that must be secured. This should not interfere with the crime scene investigation. It may have to be conducted after the investigation.)

- Other ongoing demand for resources, such as other attacks, fires, etc.

- Requests from law enforcement for assistance.

Assignment of additional positions

An officer should be designated as the triage unit leader, monitoring the status of triage efforts, providing estimates on the number and condition of patients to be treated, and coordinating movement to medical care at the appropriate treatment area.

Another officer should be designated as the treatment unit leader, coordinating activities within the treatment area, requesting sufficient personnel and supplies to handle the number of casualties in the following three treatment categories:

- immediate treatment required (red tags)

- delayed treatment (yellow tags)

- minor treatment (green tags)

This person must coordinate with the patient transportation group supervisor to ensure patients do not leave treatment areas until there is sufficient transportation available (ambulances, helicopters, buses, etc.) to move them to further treatment as needed.

A patient transportation group supervisor should be designated to ensure that sufficient transportation resources are available and avoid overloading the nearest hospitals with patients. This person should also ensure that proper tracking of patients is recorded.

An officer may have to be designated as the extrication unit leader. That position should oversee any required extrication and request sufficient resources to remove patients from immediate danger to triage.

Company Operation

Engine companies must secure a hydrant supply

Be prepared to protect exposures from a safe distance. If a structure is affected, all handlines should be 2½ in., for their added reach and volume, allowing faster knockdown from a safe position. Attack vehicle fires with 1¾-in. handlines from a distance. (Use foam if possible) Use the reach of the hose stream for protection, and minimize overhauling to that which is absolutely necessary. Remember, this is a crime scene, and you must not wash evidence down the street. Be guided by instructions from the bomb squad or EOD if present. Have the law enforcement personnel secure the area before attacking any strictly outside fires. Avoid over-commitment of personnel until size-up is completed

Treat the entire area as a crime scene

Do not disturb evidence. If they're dead, do not move victims except to help survivors. Rapid triage of a large number of injured is vital to the survival of critical patients. You cannot afford to waste resources dealing with people who are already dead. In addition, the crime scene team will need the scene as undisturbed as possible for their reconstruction. If you must move a body or other evidence, try to memorize as many details about location, description, position, etc. as possible.

Assume every explosion is a *dirty bomb* until you prove otherwise with meters

While the prime targets discussed earlier should immediately catch our attention and prompt us to follow this bomb protocol, don't ignore the other explosions. The residential building in the suburbs could be the bomb factory. Remember that the 1993 WTC bombing, the Weatherman's Greenwich Village townhouse, and the Muslim terrorists in Park Slope in 1996 all were plotted and built in residential areas.

Secondary devices

When further explosions are expected or deemed possible, the following procedures should apply:

- Incipient fire may be extinguished and personnel withdrawn until the area is surveyed by bomb squad personnel.

- Where large area or room fires are involved, heavy streams may be used from a distance or from behind substantial shielding, using the reach of the stream to increase the safety factor. Consider using unmanned monitors.

When standing by or in staging, personnel and equipment should be located in a safe area, strategically positioned for quick deployment but out of the blast zone and the path of any shrapnel or flying glass. "Apparatus and personnel should be positioned after consultation with bomb squad personnel, keeping in mind that a blast is likely to cause a power failure, resulting in elevator and fire pump failure. Building engineers should be consulted for keys and to determine shut-off valve locations for sprinkler systems, standpipes, domestic water, gas and electric services. Water damage from broken piping has proved heavy in past explosions."[3]

Hazmat

Use hazmat units for mitigation of leaks and spills, monitoring the area for the presence of radioactive or chemical threats. If a bomb has caused no fire and little blast damage, suspect that it was used to disperse a biological agent! Have a hazmat unit respond and take samples in conjunction with the police bomb squad for laboratory analysis. Most biological attacks are likely to be covert in nature. No one will know of the release until symptomatic patients start to appear in numbers. It is possible though, that a group could purposely explode a bio-bomb in order to get mass media coverage and to demonstrate its ability to deploy such weapons or to disperse a biological in order to complicate rescue efforts at a standard bombing. There is no guarantee that an explosion will destroy a biological. Again, first responder actions will determine whether they fall victim or not. Wear your personal protective equipment (PPE), take samples, and follow all the proper steps outlined previously.

Blast Damage, Size-up, and Expected Effects

Arriving units should look for signs of the destruction of concrete floor slabs, columns or girders, which is usually not seen in gas explosions. Look at the occupancy. Is this a structure that normally uses materials that could cause an explosion, such as flammable liquids and gases? Or is this an office building, which has far fewer reasons for an accidental explosion to occur. Listen to what bystanders and survivors are saying, but don't accept everything you hear as the truth. No one is going to come out of a bomb factory blast and tell you they were making an explosive and it went off by accident! They will always say it was a gas explosion! Proceed as though every explosion were the result of a bombing until other causes are determined.

Table 20–1 is a BATF chart intended to illustrate the potential quantities of explosives that could be involved and highlight the danger zones. The lethal air blast range is the minimum distance to evacuate. Just the blast pressure can kill anyone inside the lethal air blast range, without being struck by shrapnel. Even at the specified evacuation distances, members must be behind a very substantial shield to protect against shrapnel injuries. *No one should have a clear line of sight to the suspected device or vehicle!* Evacuation should follow the yellow guides, but that may be impractical. Some are a mile radius. In many cities, the blast effect may be compounded by the fact that there is likely another large building across the street from a targeted structure. This will reflect and amplify the blast effect, as well as create another building to be dealt with.

Secondary threats

Until such time an area has been designated as being *safe* from secondary devices, avoid touching, moving, or striking suspicious items that may be in the area in which one is operating. Limit the number of personnel operating in this area until deemed safe by bomb squad or EOD personnel. All persons leaving the incident site should be treated as potential perpetrators. They should all be searched for weapons/explosives before being loaded into ambulances or being brought to casualty collection points. Never touch suspicious devices or items; request police assistance.

Evidence Preservation

Recognize potential evidence. Secure evidence found with the victims or in the ambulance or at a hospital, until it can be turned over to law enforcement. Consider embedded objects as possible evidence. Victims' clothing of may also contain evidence such as explosives residue.

Table 20–1 BATF-recommended evacuation distances for explosive devices.

ATF	VEHICLE DESCRIPTION	MAXIMUM EXPLOSIVES CAPACITY	LETHAL AIR BLAST RANGE	MINIMUM EVACUATION DISTANCE	FALLING GLASS HAZARD
	COMPACT SEDAN	500 Pounds 227 Kilos (In Trunk)	100 Feet 30 Meters	1,500 Feet 457 Meters	1,250 Feet 381 Meters
	FULL SIZE SEDAN	1,000 Pounds 455 Kilos (In Trunk)	125 Feet 38 Meters	1,750 Feet 534 Meters	1,750 Feet 534 Meters
	PASSENGER VAN OR CARGO VAN	4,000 Pounds 1,818 Kilos	200 Feet 61 Meters	2,750 Feet 838 Meters	2,750 Feet 838 Meters
	SMALL BOX VAN (14 FT BOX)	10,000 Pounds 4,545 Kilos	300 Feet 91 Meters	3,750 Feet 1,143 Meters	3,750 Feet 1,143 Meters
	BOX VAN OR WATER/FUEL TRUCK	30,000 Pounds 13,636 Kilos	450 Feet 137 Meters	6,500 Feet 1,982 Meters	6,500 Feet 1,982 Meters
	SEMI-TRAILER	60,000 Pounds 27,273 Kilos	600 Feet 183 Meters	7,000 Feet 2,134 Meters	7,000 Feet 2,134 Meters

ATF I 5400 1 (01-99)

Explosions are very hazardous incidents, regardless of their cause. Until the incident scene has been investigated by trained personnel, equipped with the proper devices to detect the potential hazards that may be present, all members should take all necessary precautions to protect themselves, their fellow firefighters, and the public from possible escalation of the incident. Take cautious, deliberate action, based on intelligent decisions to minimize the impact of such events on your city, its residents, and your department.

Dirty Bombs and Radiological Dispersal Devices

Radiological dispersal devices (RDDs) known as *dirty bombs* use conventional explosives to disperse dangerous, but not fissionable, radioactive material. In the event of a bomb blast containing radioactive material, the main threat is physical damage (trauma/burns) caused by the explosive, *not* the radioactivity. The dirty bomb will generally cause more psychological injury to the population than physical injury from the radiation.

Limitations on the size of the explosives employed and the amount of radioactive material used would tend to limit the size of an area that would be affected. Radiation levels could be high in the immediate vicinity of the blast but would be vastly lower than levels produced by a nuclear weapon. The levels would drop as you move further away from the dispersal site, but areas downwind could be contaminated. Acute radiological symptoms would not be as evident, and patient contamination levels would also be much lower than after a nuclear explosion. Treatment of life-threatening injuries should not be delayed simply because the patient is contaminated with radioactive material. Unlike contamination with biological or chemical agents, there is little danger to rescuers who do not inhale radioactive materials. Treat the patient's life-threatening injuries first then wash decon them (Fig. 20–6).

Fig. 20–6 Radiation detectors should be available at every emergency scene, whether it is terrorist-related or a simple industrial or medical accident involving increasingly common radioisotope use.

Dirty bombs are overt disseminations of radioactive materials. By attracting emergency personnel to the scene with the bomb blast, they tend to limit their actual effectiveness. Since equipment to detect the presence of radioactive materials is readily available in most cities, we can take steps to protect ourselves and the public. Steps can be taken to reduce the spread of contamination, including *one of the most important—extinguishing any fire that sends plumes of contaminated smoke skyward.* If possible, a fog spray from a master stream or elevating platform should be directed at the smoke plume as soon as possible, to knock down as many radioactive particles as possible, thereby limiting downwind spread. Keep everyone out of the runoff water. Radiological terrorism could also be accomplished by an attack on a nuclear reactor. In the event of an attack on an operating reactor, the potential for release of far greater levels of radioactive materials than have been discussed here are possible. However, maximizing the protection factors described later in this chapter applies to all incidents.

Covert dissemination of radioactive materials

Covert dispersal could be accomplished by dispersing powdered or liquid radioactive materials by a variety of mechanisms, *e.g.* through HVAC systems of buildings, spreading radioactive materials in public spaces such as subways or by leaving sources of radioactive materials hidden in areas where members of the public congregate. Symptoms of radiation sickness may not appear for hours or days after exposure, depending upon the dose received. In such cases, only those directly exposed to the materials are in danger. They are not contagious, although they could be contaminated. Detection of a covert dissemination could be delayed until symptoms started to appear in groups of persons that could be traced to a particular site. In New York City, fire department units are instructed to help detect such a covert attack by utilizing their radiation detection meters *every time* they are out of quarters, such as during building inspection activities, securing the company meal, etc. This provides good practice with the instrument and familiarizes the firefighters with routine *background* radiation readings in their response area. The drain on the batteries is minimal and is well worth the expense.

Health effects of radiological materials

Radioactive materials of the type most readily available are not suitable for building a nuclear bomb without extremely difficult, highly advanced lab facilities, if at all. Their main hazard is contamination. The contamination hazard, if not guarded against and decontaminated, could lead to the following radiation hazards at varying degrees, depending on the absorbed dose:

- Psychological impact
- Chemical poisoning
- Short-term effects, radiation burns, and radiation poisoning
- Leukemia, cancer, bone destruction, birth defects

Other effects of radiological materials

- Environmental and property damage
- Economic costs of cleanup

Operations at Suspected Radiological Incidents

The basic tasks at a dirty bomb explosion are the same as at any other explosion. While we should be aware of the presence of radioactive contamination as soon as the radiation meter detects it, the mere presence of radiation should not prevent units from carrying out assigned duties. Fire personnel can operate in a radioactive field as long as they take the proper precautions. We are exposed to radiation all the time, from man-made sources such as medical X-rays to naturally occurring *background* radiation. The key to safe firefighting operations is to avoid breathing contaminated material and monitor your exposure with a dosimeter, limiting exposure to safe levels.

Force protection factors

To protect yourself against the effects of radioactive materials, maximize the following protection factors:

- Time — Spend as little time in the hot zone as possible. Monitor dosimeters to keep track of your exposure, and leave the area for decon when your dose approaches 25R (maximum under emergency conditions).

- Distance — Stay as far from the source as possible to accomplish your task.

- Shielding — Use your mask to shield your lungs, your bunker gear to shield your skin, and buildings to shield your body in general.

Establish the perimeter of the cold zone at the 1.0 m/R (milli-roentgen or 1/1,000 of a roentgen) line. This is a very low level, but it is well above most normal or *background* levels, so it indicates the presence of an additional source. This level is also the point at which most radiation detectors and dosimeters sold to first responders begin to sound an alarm. Be sure to search in a circular pattern, moving slowly toward the hot zone. Walking directly toward a source could be disastrous, since you may be getting *low* readings where you are, only because the source is shielded from your line of sight by heavy obstacles. Turning a corner could expose you to a high dose. Wear dosimeters and check them periodically. The maximum dose a firefighter should absorb under emergency situations is 25 R! (This is 25,000 times the level when a 1 m/R alarm sounds.)

Protection against contamination

Firefighters in bunker gear and SCBA are very well protected against radioactive fallout or contamination and should be safe, as long as they limit their exposure to gamma radiation. Most of the likely radioactive material that could be used in a dirty bomb scenario is low-energy material that does not produce high levels of gamma radiation. The main hazard from this material is from inhaling or ingesting (swallowing) it, where the alpha and /or beta particles can become trapped in the body. The SCBA protects against both inhalation and ingestion. The cartridges for air-purifying respirators (APR) are intended to protect the wearer against inhaling and swallowing radioactive particles if they are P-100 rated cartridges. The APRs should only be used after the area is tested for the presence of other dangerous materials. Persons remote from the immediate blast site will be adequately protected from inhaling contaminated dust by wearing an APR. External contamination can be removed simply by showering and shampooing with soap and water.

Radiological warning signs

- Observations by operating personnel

- Placards, labels

- Specialized packaging

- Radiation detection devices—dosimeters

Look for heavily shielded or extremely heavy packages, especially in relation to their size. At fires, watch for puddles of molten metal, which could be melted lead shielding. Use your meters; they're for your protection. Remember to approach in a circular pattern to avoid being caught by a radioactive source that was shielded from your meters by structural elements until you walked into its line of sight. Every effort should be made to ensure that all victims and witnesses are interviewed by police officers as to what they saw before and during the event. Emergency responders must be aware that the perpetrators may be among the victims. After treating and or decontaminating a person, fire and medical responders should direct them to law enforcement for follow-up questioning.

Chemical Agent Incidents

Operations at suspected chemical attacks

The threat of chemical attack, either using a military agent such as Sarin, or an intentional release of a toxic industrial chemical such as chlorine or hydrogen cyanide (recall that an *accidental* release of methyl isocyanate killed thousands in Bhopal, India in 1985) is perhaps the most likely WMD attack that Americans might face. A number of potential scenarios can be envisioned that must be planned for in order to avoid unnecessary loss of life among responders as well as civilians.

Incident objectives

The objective of fire personnel at a chemical attack is to control the situation by removing any live persons entrapped in the area, then triaging, administering the appropriate antidote if applicable, decontaminating, treating, and transporting any injured persons. Operations may include mitigating any chemical dispersal devices that are still dispersing agent, assisting law enforcement and/or FBI as requested, and after all necessary evidence collection procedures have been concluded, ensuring proper decon of the affected area to ensure the safety of the public.

Potential hazards to be dealt with

- Hazmat release/exposure

- Biological hazards/blood-borne pathogens

- Secondary explosive devices

- Armed assault

Clues to detecting a chemical attack

As with any response, all members must conduct a size-up of the operation. This process begins with the receipt of the alarm. Dispatchers as well as responders must listen carefully for clues that might indicate an attack has begun! Be alert for dispatch information that gives reports of more than one person unconscious (*people overcome*), reports of chemical odors or fumes (*unknown odors*), or other indicators that a chemical attack might be underway. All personnel should be aware of the location of the response. At a fire, the location and time *usually* indicate the potential life hazard. In the case of a terrorist attack, this is *usually* true as well but not always! At times, an attack may purposely be made when few civilians are present. This could be done for many reasons; to convince authorities that the terror group is serious and that their demands must be met in order to prevent future, more deadly attacks, or to put a hated institution out of business without turning the public's wrath against the responsible group by killing large numbers of civilians. For this reason, the lack of a large number of civilian victims should not be taken as a guarantee that all is safe. Keep your guard up!

Initial response

Approach the area and stage the apparatus and equipment upwind and away from any potential sources of contamination such as ventilation gratings or exhaust fan discharges. Generally, only the first due engine and ladder should approach the immediate location. Additional units should stop farther back from the incident location; a shift of the wind could suddenly force large quantities of vapors or other product directly at responders.

Upon arrival at the scene of an incident, continue your size-up. An unknown odor in the suburbs is *probably* a gas leak or other minor incident, but if the same condition occurs at any of the targets previously described, you should expect a terrorist attack! Remember, though, that much of the preparation for an attack is likely to be done remote from the actual target. Obvious explosion damage should automatically prompt a protective response. Find out who called. What caused the first problem? Take several moments on arrival to observe conditions, especially the occupants. Examine civilians leaving the area for the early indicators of a chemical attack, found in the table at the end of this chapter. The signs and symptoms they display can often identify victims of nerve agent attacks.

One reality of a chemical attack on a crowded location is that the first sign of the attack will usually be *biological indicators*—people or animals that were affected by the release. You do not need air-monitoring equipment to see the effects that most of these chemicals have on people, and if it can do that to others, it can do it to you as well! Finding 30 people lying on their backs frothing at the mouth and convulsing is *not* what we usually see at an *unknown odor call*. That would be an easy case to identify however.

A well-planned attack would not be that obvious, initially. Probably, it would look much like the typical *unknown odor*. Be aware that an atomized mist, such as delivered by a typical *roach bomb* is a nearly perfect means of spreading most chemical and biological agents. Be particularly wary when going below grade, since most *chemical* weapons are heavier than air, making subways and cellars of buildings more inviting targets. *Wear your bunker gear, gloves, mask face piece on, hood up, and earflaps down to cover all skin!* Many of the nerve agents are so deadly that a single drop of liquid on bare skin can cause a horrible death! The only way to avoid becoming a fatality at these incidents is to be *alert*, be *properly trained*, and to *react instinctively and correctly* to the first indication that *this* response is *not* a routine response! Be sure to avoid unprotected skin contact with any victim; they may have liquid on their clothes or skin.

Signs of a chemical attack

- Visible vapor clouds

- Biological indicators (dead/dying birds, pigeons, animals, people)

- Multiple victims down for unknown reasons

- Victims displaying symptoms listed at the end of the chapter

- Positive indication on military M-8 paper or M-9 tape, or a chemical agent monitor (CAM) (confirm CAM readings with M-8 or M-9)

Operational Procedures and Strategies

Injured to be triaged

Injuries at a chemical agent attack will vary with the type agent used. Military warfare agents or toxic industrial chemicals produce many symptoms that are in some cases similar initially to less harmful agents such as tear gas or pepper spray. The key to handling all mass casualty chemical incidents is to remove the patients from the contaminated environment as rapidly as possible and begin removing the contaminant from their bodies, *without contaminating any rescuers* (Fig. 20–7)!

In many fire departments as well as military units, rescuers wearing special chemical protective clothing (CPC) have been trained to use noxious stimulus or *kick triage* to determine the victim's responsiveness and thus estimate the victim's potential for survival in a chemical agent environment.

- If the victim responds to voice or gentle touch, the victim is probably still viable and should be removed from the area immediately. With medical treatment, this patient will likely survive.

- If the victim only responds to painful stimulus, the victim requires immediate advanced treatment, and the availability of such advanced treatment will determine whether the patient survives. In a mass casualty setting, where there are more patients than there are advanced medical personnel, this patient is not likely to receive the advanced care they require. Efforts are better spent saving more patients in the previous category.

Fig. 20–7 Units arriving at the scene of a chemical attack need specialized protective clothing for their own defense. The monitors are for the safety of other responders, outlining the contaminated area, and identifying the agent used so the proper treatment can begin.

- If the victim does not respond to painful stimulus, the CPC-equipped members should leave the victim in place and proceed to the next patient. This patient is likely to die, even if she is still breathing, even if she received the most advanced treatment available.

- Deceased victims should be left in place until the criminal investigation is completed.

- The *walking well* are those persons who are exhibiting no signs or symptoms of exposure, but who are requesting treatment simply because of fear. They were somewhere near the event and feel they might possibly have been exposed. They should be directed to a designated area nearby where they can be interviewed by law enforcement and health care personnel and be given instructions on how to care for themselves. The number of walking well is likely to be many times the number of persons actually affected.

Command Procedures

Command options

Depending upon conditions found upon arrival and resources available, the IC can initiate one of the following two modes of operation:

1. Rescue operation

2. Defensive operation

If living victims exhibiting nerve agent exposure symptoms can be seen or heard, then the IC can consider starting the rescue mission.

If no live victims can be seen or heard, conditions are such that rescuer entry may result in uniformed personnel becoming casualties with no positive result in reducing civilian casualties. The IC should initiate defensive operations.

If there are large crowds, persons in panic, and obvious injured, do the following:

- Establish ICS and IC Post

- If by observation, you suspect a chemical agent involved, order units responding to approach accordingly and have units report on the situation (situation report or *sitrep* for short) at their location in an attempt to define the perimeter of the incident area.

- Order all units other than the first and/or second arriving units not to enter the hazard area, until initial reports are received from the first arriving units estimating whether entry with SCBA and bunker gear is safe or not.

- Define the operational area.

- Isolate the area, control site access, vehicular, pedestrian, and building entrances/exits of surrounding properties.

- Control/limit ventilation of surrounding properties; shut down air intakes.

- Coordinate with law enforcement; request joint entry to search for secondary threats (devices).

- Request security for locations used for emergency gross decon, casualty collection points, safe area/area of refuge/shelter from weather, CP, staging area, set-up/dressing area for hazmat units.

- Develop a preliminary operational plan, and communicate it to all agencies.

- Announce locations of CP, decon, casualty collection points, safe areas/areas of refuge/refuge from weather, to on-scene personnel.

- Order engine companies to set up for emergency gross decon in secure location, away from affected buildings and exhaust fan discharges, upwind/uphill from incident area, and have them announce location selected on Handie Talkie and department radios.

- Order ladder companies to facilitate evacuation, attempt to enlarge exit access by forcing/removing gates, fences; provide a sitrep on conditions if possible, and perform forcible entry for follow-on forces equipped with CPC.

- Direct evacuees to decon area; use loudspeakers, bullhorn, signs, and barrier tape.

- Monitor time on air of all units, effect relief in sufficient time for decon.

- Monitor radiation levels in addition to chemical meters.

- Call for additional resources as necessary.

IC Considerations

Establish communications with units on scene prior to arrival of the IC and remain aware of their location and status through continuous updates. The IC should consider the potential for secondary attacks aimed at responders. He or she should immediately request a police supervisor to report to the CP and request that the police establish a secure perimeter around the CP. The CP should be located upwind and away from any potential sources of contamination. It should also be located away from any items that could be used to conceal a secondary explosive device, such as mailboxes, trash containers, dumpsters, etc., if possible.

Avoid creating large congregations of responders, which could become targets of secondary attackers. Direct uncommitted companies to remain with their apparatus, as opposed to standing near the CP. Locate staging area several blocks away, in a secure area. (Request PD to respond to this area and conduct search and perimeter security.) Staging should be upwind and remote from the incident site. Transmit necessary multiple alarms and/or special calls to provide sufficient resources to cover the number of patients that may require decon and patient triage, treatment, and transport. Request EMS superior officer to the CP to coordinate medical care.

SCBA cylinder demand will be high. Designate one or more SCBA cylinder air supply depots in safe, upwind areas. All responding units should be directed to bring all their spare cylinders to these locations for use as needed.

Designate a decon officer to ensure that decon facilities are established as needed around the perimeter of the incident, and ensure that all personnel operating in the warm and hot zones are aware of the location of these facilities.

First-arriving chief officer

Establish fire department CP.

Notify dispatcher of the nature of the incident and any need for additional units or resources.

Begin the chemical attack action plan outlined as follows:

1. Isolate the area.

2. Establish zones. Reduce the spread of contamination/cross contamination.

3. Facilitate the evacuation of people who are ambulatory.

4. Conduct rescue of injured victims as personnel/PPE equipment/conditions permits.

5. Perform rapid triage and removal of injured patients.

6. Provide emergency gross decon to all contaminated people.

7. Obtain a rapid identification/classification of the chemical agent involved.

8. Treat symptomatic patients with appropriate antidote as ordered by medical control authority.

9. Utilize appropriate decon solution for final decon.

10. Coordinate and cooperate with law enforcement on crime scene issues.

11. Initiate recovery operations/consequence management when appropriate.

Assignment of additional chiefs

- A chief officer should assume the role of forward operations officer. This chief must be trained and equipped to operate in chemical protective clothing (CPC) and will control and coordinate operations inside the hot zone upon her arrival.

- A chief officer should oversee establishment of the decon sector.

First Alarm Tactical Operations

Protective measures and procedures

The best protection is to avoid contact with any suspected agents. If possible, stay out of any spills, clouds, or areas where fumes are located. Officers should directly supervise all members. Companies should operate as whole units; no member shall operate alone.

Tests performed by the U.S. Army's Soldiers Biological and Chemical Command (SBCCOM) with military-grade chemical weapons on bunker gear have shown that full bunker gear and SCBA should protect a rescuer for a limited time—if the rescuer enters an area where there are *still live victims*. The presence of living people indicates that the concentration of the chemical agent in that area is not likely to penetrate an SCBA and protective clothing in the limited time this team should be in the area. According to SBCCOM, a fully protected rescuer, operating only in areas where there are still live victims should be able to perform rescues for up to 30 min without suffering serious effects, as long as the rescuer is decontaminated promptly after exiting.[4]

If a dispersal device is still dispersing agent when fire department units arrive, however, they shall make no attempt to stop it. They shall immediately retreat and begin decon. Bunker gear is not satisfactory protection against the concentrations of agent around a still-operating dispersal device. The concentrations around an operating dispersal device can penetrate bunker gear and SCBA and kill the wearer.

If you find that you have inadvertently entered an area that is likely an attack site, perform the following actions immediately.

- If you are fully protected, *i.e.* mask on, all skin covered (you should be!) alert other members, officers, and the IC to the situation. If you are not protected, first get your mask on and cover up, then advise others. Officers who observe patients displaying the signs and symptoms outlined should transmit "mayday—gas attack" via handie talkie.

- Leave the area immediately, preferably via a route that does not expose you to further contamination. Do not walk through any puddles of liquid or any visible vapor cloud.

- Call for a handline to flush you off, while you are still breathing air from your SCBA. If no hoseline is immediately available, use either water or AFFF extinguishers.

- Strip off and isolate all contaminated clothing, again while you are still breathing air from your SCBA.

- Report to medical personnel after decon for examination, treatment, and monitoring of vital signs.

An important consideration is that the severe effects of some chemical agents (*i.e.* mustard) may not appear for as long as 48 hrs after exposure. An attack could conceivably combine several agents, such as nerve, mustard, radioactive etc. Use all PPE and limit exposures. Officers must monitor their member's on-air time. Officers must ensure members are deconned before their air supply is depleted.

Initial survey team

A two-person survey team may have to perform an initial reconnaissance of the reported area. They should don their SCBA, including their face pieces; hoods should be in place, with earflaps down and coat collars up. All exposed skin should be covered. (Latex gloves worn under the leather firefighting gloves add an extra layer of protection to a vulnerable area.) This initial survey team should *not* commit to entering an area where all they see are unresponsive persons!

If live, moving victims can be seen or heard, the survey team should make a rapid reconnaissance of the area. Mark your route with search ropes or fire line tape to allow backup rescue teams to follow you. If possible, identify the source and nature of the problem. Get an idea of the number and severity of casualties. Relay this size-up to the IC. Avoid physical contact with any patients until after they have been decontaminated. Have the *walking wounded* help remove those in worse shape to an outside area where they can be *deconned* and then treatment can begin.

Do not attempt to physically remove any unconscious persons; the contact with them and the activity could dislodge your protection, exposing you to a lethal dose. Facilitate the evacuation by forcible removal of obstructions to egress. Force open any padlocked gates etc. (take bolt cutters) to speed evacuation and allow later CPC-equipped members to gain access.

If no live victims can be seen or heard, this is an indication that either everyone is out and there is no need for rescue or the concentration of agent in the area is potentially lethal. The SBCCOM tests show that bunker gear will protect the wearer only for a maximum of *3 min* in a lethal concentration. This limited time can be critical to members who find they have entered an area that is full of agent and only dead victims. If that is what you encounter, immediately retreat and proceed directly to decon. If no indications of live victims can be detected or a unit comes across an area with several dead victims and no survivors, the members in bunker gear must cease their entry and retreat immediately to a decontamination area. Units in specialized CPC-carrying chemical agent detection meters will have to conduct any further entry for reconnaissance in that area.

Once the initial size-up is completed, the survey team should withdraw to a safe area. They should undergo emergency gross decon. After decon, report to the IC for further debriefing and be prepared to brief other responding units about conditions observed, access routes, etc, and medical monitoring.

Backup team

Before the initial survey team enters any potentially contaminated area, a backup team and a decon team must be in place. This backup team should first help dress the initial entry team, ensuring all exposed skin is covered. Then working together, they will don their own SCBA and protective clothing (do not go on-air immediately) and await orders from the entry team. They should take a Stokes or SKED stretcher and a length of utility rope with them so they can drag the stretcher instead of carrying it. Their first priority should be the safety of the initial survey team, including ensuring the initial team is properly deconned before they take off their SCBA and bunker gear. They should be instantly available to conduct rescue operations if a member of the survey team encounters a problem.

Decon team

The primary duty of the decon team will be to ensure that a decon area has been established so that at the end of their air supply, emergency responders can safely be taken off air! They are responsible for ensuring that this area is set up to avoid future contamination as follows: away from exhaust fans, out of the path that victims are being taken, and preferably remote from and uphill of any decon areas set up for civilians.

To set up this area, use following:

- A charged 1¾-in. or booster line with a fog nozzle

- Water and/or AFFF extinguishers if no hoseline available

- Resuscitator for use on emergency responders

- A supply of antidote if available (*i.e.* mark 1 kits for nerve agents)

After the decon team has ensured that decon capability is ready, they should assist the backup team in dressing in. They shall also act as the air and communications monitor for their members, *writing down* what time the members go on air and making sure that they return to begin the decon process before their vibralert operates. With 45-min bottles you can allow a 20-min time frame for operations and allow 5 min for decon. With 30-min bottles you can allow a 10-min time frame for operations and allow 5 min for decon. The members *must* be completely deconned before they remove their SCBA and bunker gear! In extreme circumstances (very large numbers of victims) this decon team may have to assist in victim decon. However, it is critical to the safety of all responders that they set up the decon area for fire personnel before performing any other tasks. All members must be fully dressed, all skin covered, with SCBA on, when deconning anyone. Do not touch anyone until they have been deconned!

First arriving engine company

The first due engine company should stage the apparatus and equipment upwind and away from any potential sources of contamination such as exhaust fan discharges. If observations of signs and symptoms indicate such, alert other responders with a radio report, declaring a hazardous material terrorist attack. They should spot at a working hydrant and connect to it, establishing a positive water supply. The members should don full protective clothing including SCBA. They should not enter the contaminated area unless ordered by the IC. They should provide immediate gross decon for ambulatory victims with a large caliber fog stream employed from the apparatus master stream. Another fog nozzle can also be placed directly on an apparatus discharge to reduce drains on manpower. A handline with a fog tip should also be stretched to provide additional coverage.

Direct persons who have been deconned to casualty collection points staffed by EMS and later-arriving units for triage. Take an approximate head count of ambulatory victims. Provide estimate of number of ambulatory victims to the IC (Fig. 20–8).

Fig. 20–8 Washing with hose streams is an important step in removing contamination from victims of a chemical attack. Firefighters in bunker gear **may** be able to assist in this task—if they are fully protected—and assist people who are still moving on their own, then undergo decontamination themselves.

Third and later arriving units

These units should stage the apparatus and equipment upwind and away from any potential sources of contamination. They also should spot at working hydrants and connect to them. The members shall don full protective clothing including SCBA. They should also initiate a mass decon operation as outlined previously if needed. Given the number of remote exits many theatres, stadiums, and other large venues have, the second, third and even fourth engines may have to operate independently, each at its own entrance, to ensure that decon facilities are available quickly at these locations. These units will have to be augmented by later-arriving units in setting up more complete decon corridors.

Direct persons who have been deconned to casualty collection points for triage. Take an approximate head count of ambulatory victims. Provide an estimate of the number of ambulatory victims to the IC.

Incident decision making

When deciding whether to initiate a rescue or defensive operation, the IC must determine the specific hazards to firefighters and civilian victims. ICs should review the guidelines recommended by SBCCOM in their improved response program (IRP), in order to understand the implications and consciously decide what type of operation the units under their command shall be permitted to perform. The report by SBCCOM provides estimations of residual vapor hazard potentially faced by first responders 30 min after chemical release. It does not address aerosol or liquid hazards nor compound effects through multiple simultaneous exposures.

If a chemical agent detector capable of accurate near real-time vapor qualification is not available, the IC must operate based on other indicators. These include as follows:

- Signs and symptoms of victims

- Reports from escaping victims

- Knowledge of the room, area, or occupancy size

- Airflow characteristics of the room or occupancy

Factors that indicate a rescue operation may be practical

- A period of time has elapsed since agent release (10 min or more).

- Self-evacuation continues.

- People who self-evacuated are in panic but not symptomatic.

- You observe *living* victims with nerve agent exposure symptoms.

- Victims are located in line of sight near the entrance/exit.

- Victims can be heard calling for help.

- Survivors indicate other viable victims are nearby.

- Victims are not trapped/entangled and do not have traumatic injuries.

- The area attacked is well ventilated.

- There is access to the area that will not impede evacuation.

- Decon site for rescuers is established.

- The room or area where the victims are located is directly accessible without having to go through other rooms or stairwells, where agent concentrations could be much higher.

- There are units available on the scene trained and equipped to operate in a terrorist-induced hazmat environment.

- CPC fast unit is immediately available on the scene.

- No indication of secondary device/booby traps.

- HD/mustard agent is not present as indicated on CAMs or M-8/M-9 paper. HD is a type of mustard blister agent that will register as such on a CAM

Factors that indicate a rescue operation may *not* be practical

- Self-evacuation has ended.

- No *living* victims can be seen or heard.

- Survivors indicate there are no more viable victims in the attacked area.

Factors that indicate a defensive operation should be considered

- This is a second attack or a pattern of attacks is currently occurring.

- Uniformed personnel of initial alarm units have become casualties.

- Secondary attacks have occurred or are ongoing.

Defensive Operations

The strategy that guides a defensive operation for a chemical release in a large venue can be difficult to accept at first glance, but it is grounded in solid military tactics. In the military, the concept is known as *force protection*. It means protecting the troops from situations where they face insurmountable odds and risk being wiped out, so that they can fight on in another situation where they can win. It requires all members to recognize that they are more valuable alive and doing good, than dead and doing no good. To paraphrase General George S. Patton, commander of the famed Third Army in World War II, "The job of a soldier is not to die for his country, it's to make the other poor dumb bastard die for *his* country." The threat of terrorism requires that we steel ourselves to that purpose.

This strategy requires the IC to acknowledge that non-ambulatory victims injured in a chemical attack cannot be saved. In reality, in a mass casualty chemical attack, this decision has already been made by the chemical itself. The natures of the chemical agents are such that if they have affected the victim sufficiently to render them unconscious, there is virtually no chance of survival in a mass casualty setting at this stage in our response. The amount and degree of advanced medical treatment required for each victim and the amount of time it takes to reach, remove, and decontaminate these victims before they can receive this level of treatment means that there is nothing that we can do to save them. They are already too far gone to save.

Defensive operations should be initiated when the following occurs:

- Large loss of life is evident on arrival.

- First responders have become casualties on arrival.

- This is a second attack or part of a series of attacks.

- Secondary devices or attacks are in progress or likely.

- Limited resources exist on the scene, *i.e.* no units are available with the training and equipment needed to operate in a terrorist induced chemical agent environment.

We do not normally begin operations in a defensive manner. In all likelihood, first responders will begin rescue or evacuation operations. All responders must be alert to retreat to defensive operations when they recognize the factors listed previously.

Defensive Tactics

Initial rescue actions should concentrate on removing able-bodied persons from immediate danger. When the probability is high that a victim cannot be saved or is already dead, rescue should not be attempted. The danger of exposure to potentially deadly unknown chemicals makes the risk unacceptable. Units should report in with all backboards, stokes baskets, trauma, and oxygen supplies. These supplies may be deposited in a medical staging area as directed. They should also bring a spare SCBA cylinder with them for each member, which will be deposited in a designated air supply depot.

Provide information to other incoming units, initial reports to dispatcher, including the location and number of patients, and all signs and symptoms observed. Attempt to approach the area from upwind/uphill if possible. Do not remain downwind/downhill. Transmit the best approach into the incident over the apparatus radio and handie talkie.

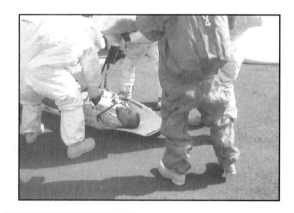

If able-bodied persons are encountered, facilitate evacuation by forcible removal of obstructions to egress, gates, fences, etc. (take bolt cutters and rebar cutter in addition to other forcible entry tools). Attempt to advance until the last able-bodied person is encountered. If unconscious or non-ambulatory victims are discovered, or if unidentified pools of liquid or vapor are encountered, do not approach. Define the area as the exclusion zone, evacuate the area, and report this location to the IC. Attempt to mark the location as hazardous to warn other responders, then report to the street for immediate gross decon and medical evaluation (Fig. 20–9).

Fig. 20–9 In a scene that would have been unimaginable to most firefighters I knew 35 years ago, firefighters must now train for such threats as chemical, biological, and radiological attacks. As the nation's first responders, we all must stay ahead of the learning curve that these new threats pose.

A terrorist attack is only one potential problem that fire departments might be called to deal with. Given the pace that our society is evolving, we will undoubtedly face other serious challenges in the future. Continue to prepare for your responsibilities as though your life depends on it. It does! Stay safe.

Table 20–2 Chemical agent signs and symptom quick comparison chart.

Military Nerve Agents (Sarin, Soman, Tabun, VX)
If vaporized or aerosolized, their onset can be observed in seconds. If a liquid, onset can be observed within minutes. Even small amounts (several droplets) of liquid nerve agent contacting unprotected skin can be severely incapacitating or lethal if the victim or responder is not decontaminated rapidly (minutes) and treated medically.

Nerve Agent Signs and Symptoms—S-L-U-D-G-E-M, Agitation, & Respiratory Distress
S—Salivation (drooling)   L—Lacrimation (watery, teary eyes)   U—Urination (involuntary, uncontrollable)   D—Defecation (involuntary, uncontrollable)   G—Gastrointestinal distress (cramps)   E—Emesis (vomiting)   M—Muscle twitching (convulsions)   Dimmed vision (ocular miosis)   Sweating   Respiratory arrest, rapid, fatal effect   Toxic Industrial Materials Organophosphates (pesticides, insecticides i.e. parathion)   • Same chemical properties as nerve agents but concentrations and quality are lower. Toxicity depends upon concentration.   • Similar signs and symptoms as seen in nerve agent exposure. Likelihood of victims surviving is greater than for nerve agent.   • Often fatal injury, long and painful recovery, but can be medically treated, detoxified.

Vesicants (Blister) Agents (Mustard gas, Lewisite) Signs and Symptoms:
Immediate Pain followed by blisters or onset may be delayed 4–48 hrs.   • Burning, itching, or red skin   • Prominent tearing and burning and redness of eyes   • Shortness of breath   • Nausea and vomiting

Asphyxiants (pulmonary agents) (cyanide) Signs and Symptoms:
Patients may appear to be gasping for air   • Altered mental status   • Shortness of breath   • Extremely difficult to breathe   • Seizures   • Cardiac arrest   • Death in 1–3 min

Incapacitating Agents (Tear Gas)
Self-limiting, self-correcting, symptoms usually clear up in 15–20 min from onset   Signs and Symptoms: • Burning pain of skin and eyes.   • Irregular respiration

Notes

[1]   Indicates site of previous terrorist attacks in the United States.

[2]   Indicates an agent that has been used recently in terrorist attacks, although none widespread.

[3]   From FDNY's All Units Circular 190, which was written at the height of the 1970s wave of bombings by the Weathermen and FALN.

[4]   Bunker gear and SCBA, even when fully worn, is no guarantee of protection against chemical agents, just as it does not protect fully against flame. It may provide protection against some agents for limited periods of time and is certainly better than breathing contaminated air or allowing bare skin exposure, but it is not an acceptable alternative to CPC. Even in the criteria tested in the SBCCOM scenario, rescuers can suffer injury; but with proper treatment it should not result in life-threatening injury.

Glossary

2 in/2 out – The U.S. OSHA has mandated that at least a two-person team be on standby to rescue anyone wearing respiratory protection to enter any atmosphere that is immediately dangerous to life and health (IDLH). This rule applies to firefighters. It is known as the 2 in/2 out rule. It does not mean that for every two firefighters in a building, two must also be outside, but that at least two rescuers must be available at every scene where members are working in dangerous atmospheres.

A

ABC extinguisher – Dry chemical extinguisher suitable for use on the three most common classes of fire, ordinary combustibles, flammable liquids and gases, and live electrical fires.

ABCs of first aid – Airway, breathing, and circulation.

Access stair – Stairs within a high rise building that connect one floor to another within a tenant's space. They do not serve as required exits. They are often unenclosed, allowing rapid fire travel from one floor to another.

AFFF – Aqueous film-forming foam.

APR – Air purifying respirators.

ALS – Advanced life support. Prehospital medical care capable of such advanced techniques as intravenous drug administration, endotracheal intubation, and cardiac defibrillation.

Atmospheric pressure – The pressure exerted on the surface by the weight of the atmosphere above. At sea level, it is generally 14.7 psi.

Atrium – An open space that connects three or more floors vertically, with each floor open to the common area below. Any fire, smoke, or gas immediately threatens all exposed floors.

B

Backdraft – An explosion of accumulated fire gases (smoke) generally caused by admitting air into a superheated environment.

Balloon framing – A construction method previously used in some parts of the country that involved exterior vertical wall studs that ran continuously from the foundation sill up to the attic eave. This created an uninterrupted path for fire spread in the cellar to quickly reach the attic.

BLEVE – Boiling-liquid expanding vapor explosion. A BLEVE is a result of heating liquids to above their boiling point in a closed container. This increases the pressure within the container. The exposing fire weakens the container walls, and if sufficiently weakened, the internal pressure blows the container apart in a violent explosion. A liquid does not have to be flammable to BLEVE. Containers of water will BLEVE if sufficiently heated. If a flammable liquid BLEVEs, the escaping liquid/gas will ignite, creating a fireball.

BTU (British Thermal Unit) – A measurement of the amount of heat required to raise the temperature of 1 lb of water by 1°F.

C

CADS – Computer-aided dispatching system.

CAFS – Compressed air foam systems.

CAM – Chemical agent monitor.

CBRNE – Chemical, biological, radiological, nuclear, or explosive.

CFR-D – Certified first responder-defibrillator.

Channel Rail – A vertical steel column in a Class 3 building that runs from the foundation to the cockloft. Channel rails support upper floors. They are usually enclosed with wood, creating a ready path for fire to travel vertically.

CIDS – Critical information dispatch system, an input system for computer- aided dispatch.

Class A foam – A foam/water/air solution used for combating fires in ordinary combustibles.

Class B foam – A foam/water/air solution used for combating fires in flammable and combustible liquids.

COAL WAS WEALTH – Construction, occupancy, apparatus and manpower, life hazard, water supply, auxiliary appliances, street conditions, weather, exposures, area, location and extent of fire, time, and height.

Cockloft – A horizontal concealed void space located between a building's top floor ceiling and the roof. Fire entering the cockloft will spread out horizontally over the entire area and burn the roof off, often destroying the top floor in the process.

CP - Command post.

CPC – Chemical protective clothing.

D

decon – Decontamination.

E

ELF – Earth Liberation Front.

EMS – Emergency medical services.

EOD – Emergency ordinance disposal.

EPA – Environmental Protection Agency.

F

FALN – Fuerzas Armadas de Liberación Nacional or Armed Forces of National Liberation.

FAR – Feedback-assisted rescue.

FAST – Firefighter assist and search team.

FBI – Federal Bureau of Investigation.

FD – Fire department.

FDC – Fire department connection.

FDNY – City of New York Fire Department.

FEMA – Federal Emergency Management Agency.

Flame spread rating – A measurement of the relative speed that flame will spread across the surface of a given material. Higher flame spread ratings mean fire will travel faster across that surface than a lower rated surface.

Flowing pressure – A measurement of the pressure in a water main or hose as water move through the pipe or hose.

Friction loss – Pressure loss resulting from turbulence within the water stream as it moves through a pipe or hose. Usually expressed in psi loss per 100 ft of hose for firefighting.

G

Garden Apartment – A low-rise, multi-family residential structure that often has a common cockloft over the entire structure.

gpm – Gallons per minute.

H

hazmat – Hazardous materials.

Head pressure – Pressure in a column of water resulting from the weight of the water above the measuring point.

HFT – Hydraulic forcible-entry tool.

Hook – A family of tools used to extend a firefighter's reach, such as a pike pole, halligan hook, Boston rake, or others used to pull ceilings open, break windows, and pull open roof boards among other uses.

H-T – Handie-talkie, a portable two way radio.

HVAC – Heating, ventilation, and air conditioning systems. Typically found in high-rise office buildings, atriums, malls, etc. HVAC systems move large quantities of air, which can be contaminated with smoke. They can wreak havoc if they move smoke to areas away from the immediate fire area.

Hydraulic overhauling – The practice of using the hose-stream to apply large quantities of water to soak a structure to prevent rekindle, rather than using fire-fighter to open up the structure for closer examination. Hydraulic overhauling should be maximized in vacant buildings, and minimized in occupied buildings.

I

IC – Incident commander.

ICS – Incident command system.

IDLH – Any atmosphere that is *immediately dangerous to life and health*. See 2 in/ 2 out rule by OSHA.

ISO – Insurance Service Office.

Insulspan™ panels – A new type of roof construction, which combines the elements of roof support, insulation, and decking in one element. They are a concern for firefighters, since the extreme thickness of the panel prevent cutting with circular saws, and the huge unsupported spans pose potential collapse dangers under fire conditions.

L

LAFD – Los Angeles City Fire Department.

Laminar flow – Smooth flow of a fluid in a hose, conduit, or pipe. Laminar flow generally occurs only at low flow rates, and produces low friction losses.

LDH – Large-diameter hose.

LOVERS U – Laddering, overhaul, ventilation, forcible entry, rescue and search, salvage, and control of utilities, all of which are generally considered ladder company functions.

LP – Liquefied petroleum.

M

MAPP – Methylacetylene propadiene, a type of fuel gas used in cutting torches instead of acetylene.

MCI – Multi-casualty incidents.

MD – Multiple dwelling.

MOVE – A black separatist organization in Philadelphia in the 1970s. A police siege of their compound set off a raging fire that destroyed more than 50 homes.

Mushrooming – The fire behavior that occurs when heated gases rise vertically from their source until they encounter an obstruction such as the roof or ceiling. Once stopped from rising, the gases spread out horizontally until they encounter obstructions such as walls, then they begin to descend from the top down, filling the upper levels first.

N

NFPA – National Fire Protection Agency.

O

OIC – Officer in command, same as incident commander.

OOS – Out of Service, unusable.

OSB – Oriented strand board.

OSHA – The U.S. Occupational Safety and Health Administration.

OS&Y Outside stem and yoke or outside screw and yoke, a style of valve used on fire protection piping to visually indicate its position. When the stem projects from the valve it is open.

OV – Outside ventilation.

P

PASS – Personal alert system.

PCB – Polychlorinated biphenyls.

PIV – Post indicator valve.

Piston effect – The movement of air in front of a moving object inside a closed shaft, such as an elevator, or a train in a subway tube.

PPE – Personal protective equipment.

PPV – Positive-pressure ventilation.

Pressure sensitive tape – Specially reinforced tape, in approximately 24 in. wide roles that can be applied to the inside of plate glass windows in high rise buildings in order to prevent shards of glass from falling to the street below if it is necessary to break a window for ventilation.

PRV – Pressure reducing valve.

Private dwelling or PD – A dwelling housing only one or two families.

psi – Pounds per square inch.

PTO – Power takeoff.

R

Rain roof – A new roof built over an existing leaky roof below. It creates a new void space between the roofs which is inaccessible to firefighters below pulling ceilings.

RDD – Radiological dispersal device.

Reflex time – The amount of time elapsed from the moment the alarm is received by the fire department, until the first effective operations are being conducted on the fire ground, generally measured by the beginning of hose stream operations. Reflex time includes alarm processing time, response time, the time to position at the fire location, and arrange hose line for attack. It is not the time of arrival in front of a high rise building, since we may have to walk up 20 floors before beginning operations.

Rehab – A building which has been substantially renovated or *rehabilitated*. Rehabs have many more void spaces than the building originally contained and may be fitted with other features such as HVAC systems and energy efficient windows, which compound firefighting problems.

Relay pumping – Using one pumper, located at a water source, to relay water to another pumper closer to the fire.

Residual pressure – Pressure available in the water supply mains when water is flowing.

RF – Radio frequency.

RIT – Rapid intervention team.

S

SAE – Search and evacuation post.

SBCCOM – Soldiers, Biological, and Chemical Command. A research unit of the U. S. Army that has examined the impact of chemical warfare agents on firefighters protective clothing.

SCBA – Self-contained breathing apparatus.

Scrub area – The surface area of a building that can be accessed from the basket of an elevating platform.

sitrep – Situation report.

Smoke generation rating – A measurement of the relative amount of smoke generated by the burning of a given material. Higher smoke-generating ratings mean greater quantities of smoke will be produced by that material than by a lower rated material. The measurement does not calculate which of the two materials produces the more toxic smoke, just which produces more of it.

SOP – Standard operating procedures. Planned actions intended to be taken in a given situation. The key word is standard. Do not use SOPs for non-standard events.

SRO – Single-room occupancy.

Static pressure – Pressure available within a water main when water is not flowing.

T

TED – Target Exit Device™.

TIC – Thermal imaging camera A device that registers variations in temperature as a visual image, most often resembling a black-and-white film negative, where light-colored objects are hotter than their darker colored surroundings. TICs can allow firefighters to see through smoke with certain limitations.

Townhouses – Single family homes that are built in a row and are attached to similar units on one or both sides.

U

UEL – Upper explosive limit.

V

VES – Venting, entering, and searching.

W

Water hammer – A shock wave within a water main or hose, created by suddenly starting or stopping the flow of a large quantity of water. Water hammer can burst hose, mains, and damage pumps, causing injuries to firefighters.

Wetting agent – A water additive that reduces waters surface tension, allowing greater penetration of Class A combustibles.

WIV – Wall indicator valve.

WMD – Weapons of mass destruction.

WTC – World Trade Center.

Index

B

C

D

I

N

Q

R

S

T

W–Z